D1179641

Digital Signal Processing with Examples in MATLAB®

SECOND EDITION

THE ELECTRICAL ENGINEERING AND APPLIED SIGNAL PROCESSING SERIES

Edited by Alexander D. Poularikas

The Transform and Data Compression Handbook
K.R. Rao and P.C. Yip

Handbook of Antennas in Wireless Communications
Lal Chand Godara

Handbook of Neural Network Signal Processing
Yu Hen Hu and Jenq-Neng Hwang

Optical and Wireless Communications: Next Generation Networks
Matthew N.O. Sadiku

Noise Reduction in Speech Applications
Gillian M. Davis

Signal Processing Noise
Vyacheslav P. Tuzlukov

Digital Signal Processing with Examples in MATLAB®
Samuel Stearns

Applications in Time-Frequency Signal Processing
Antonia Papandreou-Suppappola

The Digital Color Imaging Handbook
Gaurav Sharma

Pattern Recognition in Speech and Language Processing
Wu Chou and Biing-Hwang Juang

Propagation Handbook for Wireless Communication System Design
Robert K. Crane

Nonlinear Signal and Image Processing: Theory, Methods, and Applications
Kenneth E. Barner and Gonzalo R. Arce

Smart Antennas
Lal Chand Godara

Mobile Internet: Enabling Technologies and Services
Apostolis K. Salkintzis and Alexander D. Poularikas

Soft Computing with MATLAB®
Ali Zilouchian

Signal and Image Processing in Navigational Systems
Vyacheslav P. Tuzlukov

Medical Image Analysis Methods
Lena Costaridou

MIMO System Technology for Wireless Communications
George Tsoulos

Signals and Systems Primer with MATLAB®
Alexander D. Poularikas

Adaptation in Wireless Communications - 2 volume set
Mohamed Ibnkahla

Handbook of Multisensor Data Fusion: Theory and Practice, Second Edition
Martin E. Liggins, David L. Hall, and James Llinas

Discrete Random Signal Processing and Filtering Primer with MATLAB®
Alexander D. Poularikas

Advanced Signal Processing: Theory and Implementation for Sonar, Radar, and Non-Invasive Medical Diagnostic Systems, Second Edition
Stergios Stergiopoulos

Digital Signal Processing with Examples in MATLAB®, Second Edition
Samuel D. Stearns and Don R. Hush

THE ELECTRICAL ENGINEERING
AND APPLIED SIGNAL PROCESSING SERIES

Digital Signal Processing with Examples in MATLAB®

SECOND EDITION

Samuel D. Stearns and Don R. Hush

CRC Press is an imprint of the
Taylor & Francis Group, an informa business

CRC Press
Taylor & Francis Group
6000 Broken Sound Parkway NW, Suite 300
Boca Raton, FL 33487-2742

Printed in the United States of America on acid-free paper
10 9 8 7 6 5 4 3 2 1

International Standard Book Number: 978-1-4398-3782-5 (Hardback)

Library of Congress Cataloging-in-Publication Data

Stearns, Samuel D.
Digital signal processing with examples in MATLAB® / Samuel D. Stearns, Donald R. Hush. -- 2nd ed.
p. cm. -- (Electrical engineering & applied signal processing series)
Includes bibliographical references and index.
ISBN 978-1-4398-3782-5 (hardback)
1. Signal processing--Digital techniques--Computer simulation. 2. MATLAB. I. Hush, Donald R. II. Title.

TK5102.9.S719 2011
621.382'2--dc22
2010042257

Visit the Taylor & Francis Web site at
http://www.taylorandfrancis.com

and the CRC Press Web site at
http://www.crcpress.com

This book is dedicated to the cause of promoting peace on earth.

We hope its contents are used to further this cause.

Contents

Foreword to the Second Edition xv
Foreword to the First Edition in Memory of Richard W. Hamming (1915–1998) xvii
Preface to the Second Edition xix
Preface to the First Edition xxi
Authors xxv

1. Introduction 1
1.1 Digital Signal Processing 1
1.2 How to Read This Text 3
1.3 Introduction to MATLAB® 3
1.4 Signals, Vectors, and Arrays 4
1.5 Review of Vector and Matrix Algebra Using MATLAB® Notation 6
1.6 Geometric Series and Other Formulas 13
1.7 MATLAB® Functions in DSP 16
1.8 The Chapters Ahead 17
References 17
Further Reading 18

2. Least Squares, Orthogonality, and the Fourier Series 19
2.1 Introduction 19
2.2 Least Squares 19
2.3 Orthogonality 24
2.4 The Discrete Fourier Series 26
Exercises 33
References 37

3. Correlation, Fourier Spectra, and the Sampling Theorem 39
3.1 Introduction 39
3.2 Correlation 40
3.3 The Discrete Fourier Transform (DFT) 42
3.4 Redundancy in the DFT 43
3.5 The FFT Algorithm 45
3.6 Amplitude and Phase Spectra 47
3.7 The Inverse DFT 51
3.8 Properties of the DFT 52
3.9 Continuous Transforms, Linear Systems, and Convolution 57
3.10 The Sampling Theorem 62

3.11 Waveform Reconstruction and Aliasing ... 64
3.12 Resampling ... 72
3.13 Nonuniform and Log-Spaced Sampling ... 76
Exercises ... 84
References ... 87
Further Reading ... 88

4. Linear Systems and Transfer Functions ... 89
4.1 Continuous and Discrete Linear Systems ... 89
4.2 Properties of Discrete Linear Systems ... 89
4.3 Discrete Convolution ... 92
4.4 The z-Transform and Linear Transfer Functions ... 93
4.5 The Complex z-Plane and the Chirp z-Transform ... 96
4.6 Poles and Zeros ... 101
4.7 Transient Response and Stability ... 105
4.8 System Response via the Inverse z-Transform ... 107
4.9 Cascade, Parallel, and Feedback Structures ... 109
4.10 Direct Algorithms ... 111
4.11 State-Space Algorithms ... 114
4.12 Lattice Algorithms and Structures ... 116
4.13 FFT Algorithms ... 124
4.14 Discrete Linear Systems and Digital Filters ... 129
4.15 Functions Used in This Chapter ... 130
Exercises ... 131
References ... 135
Further Reading ... 136

5. FIR Filter Design ... 137
5.1 Introduction ... 137
5.2 An Ideal Lowpass Filter ... 138
5.3 The Realizable Version ... 139
5.4 Improving an FIR Filter with Window Functions ... 142
5.5 Highpass, Bandpass, and Bandstop Filters ... 148
5.6 A Complete FIR Filtering Example ... 150
5.7 Other Types of FIR Filters ... 152
5.8 Digital Differentiation ... 152
5.9 A Hilbert Transformer ... 154
Exercises ... 155
References ... 159
Further Reading ... 160

6. IIR Filter Design ... 161
6.1 Introduction ... 161
6.2 Linear Phase ... 162
6.3 Butterworth Filters ... 163

6.4 Chebyshev Filters ... 167
6.5 Frequency Translations ... 173
6.6 The Bilinear Transformation ... 177
6.7 IIR Digital Filters ... 180
6.8 Digital Resonators and the Spectrogram ... 185
6.9 The All-Pass Filter ... 189
6.10 Digital Integration and Averaging ... 189
Exercises ... 193
References ... 196
Further Reading ... 197

7. Random Signals and Spectral Estimation ... 199
7.1 Introduction ... 199
7.2 Amplitude Distributions ... 200
7.3 Uniform, Gaussian, and Other Distributions ... 204
7.4 Power and Power Density Spectra ... 209
7.5 Properties of the Power Spectrum ... 213
7.6 Power Spectral Estimation ... 216
7.7 Data Windows in Spectral Estimation ... 221
7.8 The Cross-Power Spectrum ... 223
7.9 Algorithms ... 226
Exercises ... 226
References ... 229
Further Reading ... 230

8. Least-Squares System Design ... 231
8.1 Introduction ... 231
8.2 Applications of Least-Squares Design ... 232
8.3 System Design via the Mean-Squared Error ... 235
8.4 A Design Example ... 239
8.5 Least-Squares Design with Finite Signal Vectors ... 242
8.6 Correlation and Covariance Computation ... 244
8.7 Channel Equalization ... 247
8.8 System Identification ... 250
8.9 Interference Canceling ... 253
8.10 Linear Prediction and Recovery ... 257
8.11 Effects of Independent Broadband Noise ... 261
Exercises ... 263
References ... 270
Further Reading ... 271

9. Adaptive Signal Processing ... 273
9.1 Introduction ... 273
9.2 The Mean-Squared Error Performance Surface ... 275
9.3 Searching the Performance Surface ... 276

9.4 Steepest Descent and the LMS Algorithm 281
9.5 LMS Examples 288
9.6 Direct Descent and the RLS Algorithm 291
9.7 Measures of Adaptive System Performance 296
9.8 Other Adaptive Structures and Algorithms 300
Exercises 301
References 306
Further Reading 307

10. Signal Information, Coding, and Compression 309
10.1 Introduction 309
10.2 Measuring Information 310
10.3 Two Ways to Compress Signals 312
10.4 Adaptive Predictive Coding 314
10.5 Entropy Coding 319
10.6 Transform Coding and the Discrete Cosine Transform 328
10.7 The Discrete Sine Transform 335
10.8 Multirate Signal Decomposition and Subband Coding 342
10.9 Time–Frequency Analysis and Wavelet Transforms 352
Exercises 356
References 361

11. Models of Analog Systems 363
11.1 Introduction 363
11.2 Impulse-Invariant Approximation 364
11.3 Final Value Theorem 368
11.4 Pole–Zero Comparisons 370
11.5 Approaches to Modeling 372
11.6 Input-Invariant Models 374
11.7 Other Linear Models 382
11.8 Comparison of Linear Models 386
11.9 Models of Multiple and Nonlinear Systems 389
11.10 Concluding Remarks 397
Exercises 397
References 401
Further Reading 402

12. Pattern Recognition with Support Vector Machines 403
12.1 Introduction 403
12.2 Pattern Recognition Principles 406
12.3 Learning 411
12.3.1 The Independent and Identically Distributed Sample Plan 412
12.3.2 Learning Methods 413

12.4 Support Vector Machines ... 417
12.4.1 The Support Vector Machine Function Class ... 417
12.4.2 The Support Vector Machine Learning Strategy ... 420
12.4.3 The Core Support Vector Machine Algorithm ... 423
12.4.3.1 Constructing the Primal, Dual, and Dual-to-Primal Map ... 424
12.4.3.2 Margin, Support Vectors, and the Sparsity of Exact Solutions ... 430
12.4.3.3 Decomposition Algorithms for the Dual Quadratic Programming Problem ... 433
12.4.3.4 Rate Certifying Decomposition Algorithms ... 435
12.5 Multi-Class Classification ... 436
12.6 MATLAB® Examples ... 436
Exercises ... 445
References ... 450

Appendix: Table of Laplace and z Transforms ... 453

Index ... 461

Foreword to the Second Edition

The combination of digital signal processing (DSP) techniques with today's computing capabilities allows us to address many difficult problems. For example, computers can "understand" human speech, translate it to another language, and then synthesize the translated speech in a new voice. Satellites can transmit information from one part of our world to another in the blink of an eye. Medical doctors can diagnose, and surgically correct, heart valve problems in a baby before it is born. Biometric signals, such as fingerprints and iris scans, can be stored in a "smart card" to allow positive identification of the person using the card.

The capabilities that allow solutions such as the ones mentioned above all use DSP algorithms to extract information from signals collected from the environment around us. These algorithms are based on fundamental principles from mathematics, linear systems, and signal analysis. This book will guide you through the mathematics and electrical engineering theory using real-world applications. It will also use MATLAB®, a software tool that allows you to easily implement signal-processing techniques using the computer and to view the signals graphically. Digital signal processing and MATLAB together open up a fascinating new world of possibilities.

The authors of this new edition, Sam Stearns and Don Hush, detail unique perspectives of signal processing and its applications. Sam Stearns spent much of his career solving difficult problems at one of the finest laboratories in the United States, Sandia National Laboratory in Albuquerque, New Mexico. He also brings a lifelong passion for education. Sam has been a teacher, guide, mentor, and friend to me for over 20 years. I shall always be in his debt for introducing me to the wonders of DSP.

Don Hush and I were colleagues for a number of years in the Department of Electrical/Computer Engineering at the University of New Mexico (UNM), in Albuquerque, New Mexico. We published a few papers on signal processing together, and enjoyed encouraging each other's graduate students. Don also brings unique experiences from another of our country's great laboratories, Los Alamos National Laboratory, in New Mexico. The Los Alamos National Laboratory continues in the tradition of its Manhattan Project and works on challenging problems in national security.

The reader of this text is fortunate to be guided by two wonderful teachers who translate the issues and understanding of using signal processing in the real world to examples and applications that open the door to this fascinating subject.

Dr. Dolores M. Etter*

* Dr. Delores M. Etter, a former Assistant Secretary of the Navy for Research, Development, and Acquisition, is now the Texas Instruments Distinguished Chair in Engineering Education and the Director of the Caruth Institute for Engineering Education at Southern Methodist University in Dallas, Texas. She is the author of a number of engineering textbooks, including several on MATLAB.

Foreword to the First Edition in Memory of Richard W. Hamming (1915–1998)

The information age in which we are living has underlined the importance of signal processing, while the development of solid-state integrated circuits, especially in the form of general-purpose minicomputers, has made practical the many theoretical advantages of digital signal processing. It is for these reasons that a good book on digital signal processing is welcomed by a wide range of people, including engineers, scientists, computer experts, and applied mathematicians.

Although most signal processing is now done digitally, much of the data originates as continuous analog signals, and often, the result of the digital signal processing is converted back to analog form before it is finally used. Thus, the complex and often difficult to understand relationships between the digital and analog forms of signals need to be examined carefully. These relationships form a recurring theme throughout this book.

The theory of digital signal processing involves a considerable amount of mathematics. The non-mathematically inclined reader should not be put off by the number of formulas and equations in the book, because the author has been careful to motivate and explain the physical basis of what is going on and, at the same time, avoid unnecessarily fancy mathematics and artificial abstractions. It is a pleasure, therefore, to recommend this book to the serious student of digital signal processing. It is carefully written and illustrated by many useful examples and exercises, and the material is selected to cover the relevant topics in this rapidly developing field of knowledge.

R. W. Hamming
Bell Laboratories
Murray Hill, NJ

This foreword is included in memory of Professor Richard W. Hamming, one of the world's great mathematicians and a pioneer in the development of digital signal processing. The fact that his comments on signal processing, written in 1975 for the progenitor of this text, are just as relevant now as they were when they were written is a testimony to Dr. Hamming's foresight and genius.

Samuel D. Stearns

Preface to the Second Edition

The preface to the first edition, which has been altered slightly to update its contents, follows after this section. It is the main preface to this book. The purpose of this second edition, its intended audience, and our reasons for writing it are the same as those described in the preface to the first edition. In this second edition, we have attempted to correct and improve the original text in response to comments from colleagues, students, and other friends who have suggested ways to clarify and improve the book. We have also added topics that have become the basis for current DSP applications of which we are aware, as well as several topics that, as we have been informed repeatedly, "should have been included" in the first edition. These include a chapter on modeling analog systems and a chapter on pattern recognition (as used in discrimination, detection, and decision making) using support vector machines, as well as sections on the chirp z-transform; resampling; waveform reconstruction; the discrete sine transform; logarithmic and non-uniform sampling; and other items, including an appendix containing a table of transforms that is more comprehensive than the table provided in the first edition.

Most but not all of the material in this second edition is appropriate for an upper division or first-year graduate student in DSP. The book has been written primarily to introduce the subject to readers who know some mathematics and science, but have little or no experience with DSP. Some parts of the book, especially Chapters 8 through 12, are more advanced than the rest and require a more solid background in mathematics and statistics. Chapters 8 and 9 are better understood if the reader has some understanding of linear algebra. Chapter 12 will be difficult for most beginning students in DSP unless they know a lot of statistical analysis; on the other hand, it is a unique and valuable resource for advanced students who want to work in discrimination and pattern recognition.

In addition to the help we have acknowledged in the preface to the first edition, we are grateful to all the teachers, engineers, and students who, after using the first edition, offered valuable comments as well as suggestions for improvements and new material. Several colleagues were kind enough to read and comment on parts of the manuscript, including Professor Ramiro Jordan of UNM, as well as Dr. Chris Hogg, David Heine, and Dr. Mark Smith of Sandia National Laboratories. We also wish to thank Frank Alexander, James Howse, Reid Porter, Clint Scovel, and James Theiler of the Los Alamos National Laboratory. Finally, we could not possibly have completed this project without support from our homes. We are grateful beyond words for the care and assistance given by our wives, Mary and Sylvia, in our work. We also thank Jennifer Civerolo for reading and editing the page proofs.

MATLAB® is a registered trademark of The MathWorks, Inc. For product information, please contact:

The MathWorks, Inc.
3 Apple Hill Drive
Natick, MA 01760-2098 USA
Tel: 508 647 7000
Fax: 508-647-7001
E-mail: info@mathworks.com
Web: www.mathworks.com

MATLAB® functions and examples are available for download at http://www.crcpress.com/product/isbn/9781439837825.

Preface to the First Edition

This book is intended to be a text for courses in digital signal processing. It is written primarily for senior or first-year graduate engineering students, and it is designed to teach digital signal processing and how it is applied in engineering.

A progenitor of this text, *Digital Signal Analysis*, was published in 1975. At that time, there were only a handful of texts in the signal processing area, but now there are many texts, and we must give reasons why still another is needed.

One reason is that there is always room for another text in an area like digital signal processing, which is still expanding and finding new applications, provided the text is up to date and covers the fundamental areas, yet does not leave gaps in development that must somehow be bridged by the reader. We endeavored in this text to provide a complete development of the main subjects in modern digital signal processing. These subjects, we believe, have changed over time in nature and relative importance, and this book is meant to respond to these changes and present the main subjects, or "basics," applicable in today's technology.

It is more challenging now than it used to be to produce a text on DSP, because the field has grown in so many directions. It is more difficult now in a single book to cover the relevant theory and applications, and yet present the whole subject in a constructive manner, that is, in a manner conducive to teaching that proceeds logically from one topic to the next without omitting derivations and proofs.

A second and perhaps more important reason for this text, in particular, lies in our understanding of current applications of digital signal processing to engineering problems and systems. In most applications, digital signals, which are simply vectors or arrays with finite numbers of discrete elements, are worth processing and analyzing only because they represent discrete samples of continuous phenomena. That is, the engineer who applies the techniques described in a text like this is normally working with at least two, and usually all three, of the following operations:

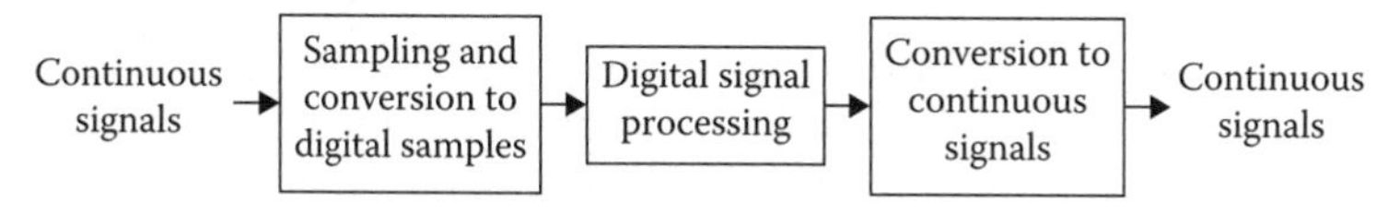

Our premise is that the reader will find the text more useful because we do not treat the central operation above, that is, digital signal processing, as a subject in isolation. We try always to relate digital signal processing to continuous signal processing and to treat digital signals as samples of physical phenomena, just as the engineer must do when he or she applies digital signal processing to solve real problems.

Other important features of this text include the use of MATLAB® to provide examples of signal processing, digital system design, solutions to exercises, coding and compression algorithms, etc. The MATLAB language has become a standard in signal processing, because it is so easy to understand and use. Its plotting and graphics functions make it ideally suited to signal processing applications. We provide a brief MATLAB tutorial in Chapter 1, so that even if the reader does not use MATLAB, he or she can easily read and understand the algorithms and examples in the text. All of the examples and functions presented in the text, as well as answers to exercises, are included with the software provided for the reader on the CRC Press website (www.crcpress.com).

As indicated in the table of contents, other features include areas that are now considered basic but are not always covered in signal processing texts, including an introduction to statistical signal processing, the discrete cosine transform, an introduction to time-frequency and wavelet transforms, coding and compression of signals and other digital data, least-squares system design, and an introduction to adaptive signal processing.

The author of this text is greatly indebted to a number of students, colleagues, and friends, who have been patient and interested enough to comment on large portions of the manuscript. Several have reviewed the entire manuscript and suggested many changes and improvements, in particular, Professors Chaouki Abdallah, Scott Acton, Majid Ahmadi, Victor DeBrunner, Dr. Robert Ives, Jeff Kern, Chip Stearns, and Dr. Li Zhe Tan. The students are those who have attended classes where the manuscript was first used, at the University of New Mexico, and in short courses taught at Sandia National Laboratories as well as elsewhere in industry. Other friends and colleagues who have encouraged the author and commented on the text's contents include Dr. Nasir Ahmed, Professors James Matthews, Wasfy Mikhael, Marios Pattichis, Balu Santhanam, Michael Soderstrand, and Dr. Stanley White.

The author is also thankful for his children, who kept asking, "How's the book coming?" (and were really interested in the answer), and to his wonderful wife, Mary, who has patiently read and helped correct the entire manuscript.

Finally, in writing this text, the author has tried to adhere to the descriptions and ideals implied in the two Forewords: First, we hope this project will honor the memory of Richard W. Hamming, one of the world's great mathematicians, who made so many contributions, who inspired so many through his teaching, and who laid so much of the foundation on which signal processing is built today. Second, in this book, we have strived to meet the high standards of clarity and logic in teaching and research set by Professor Delores M. Etter, formerly Deputy Director of Defense Research and Engineering and now Director of the Caruth Institute for Engineering Education at Southern Methodist University in Dallas, Texas.

MATLAB® is a registered trademark of The MathWorks, Inc. For product information, please contact:

Authors

Samuel D. Stearns is Professor Emeritus at the University of New Mexico. He has been involved in adjunct teaching and research at UNM since 1960.

His principal technical areas are DSP and adaptive signal processing. His most recent occupation is teaching these subjects at UNM and in industrial short courses, as well as consulting.

Previously, Dr. Stearns was a distinguished member of the technical staff at Sandia National Laboratories. He retired in 1996 after 27 years at Sandia.

Dr. Stearns has taught and advised dissertation research at several universities in the United States, including Kansas State and Stanford Universities; and the Universities of Central Florida, Colorado, and New Mexico. He has guided the dissertation research of over 25 doctoral students at UNM and elsewhere.

Dr. Stearns is a Fellow of the IEEE. His fellowship citation reads, "For contributions to education in digital and adaptive signal-processing systems and algorithms." He has served in various IEEE activities and has published a number of papers in signal processing, adaptive signal processing, and related areas. He is the author or coauthor of the following texts:

Digital Signal Processing with Examples in MATLAB® (2003)
Signal Processing Algorithms in MATLAB® (1996)
Signal Processing Algorithms in Fortran and C (1993)
Digital Signal Analysis, 2nd Ed. (1990)
Signal Processing Algorithms (1987)
Adaptive Signal Processing (1985)
Digital Signal Analysis, 1st Ed. (1975)

Don R. Hush received his BSEE and MSEE degrees from Kansas State University, Manhattan, Kansas, in 1980 and 1982, respectively, and his PhD in engineering from UNM in 1986. He has served as a technical staff member at the Sandia National Laboratory (1986–1987) and as professor in the Electrical and Computer Engineering Department at UNM (1987–1998), and is currently a technical staff member at the Los Alamos National Laboratory (from 1998 onward). He is a Senior Member of the IEEE and has served as an associate editor for the *IEEE Transactions on Neural Networks* and the *IEEE Signal Processing Magazine.*

1

Introduction

1.1 Digital Signal Processing

Digital signal processing (DSP) has become an established area of electrical and computer engineering in relatively recent times. In fact, because all types of signals, when they are processed, are now most often processed in digital form, scientists and engineers in all disciplines have come to at least a nodding acquaintance with the subject in order to understand what their instruments and displays are trying to tell them.

Compared with other areas of electrical engineering that share their analytic tools with DSP, such as fields and waves, communication theory, circuits, and control theory, DSP has a relatively short history, a relatively rapid growth, and astonishing diversification into nearly every branch of technology. This is primarily due to the remarkable growth and change in DSP electronic hardware, two aspects of which are illustrated in Figure 1.1.

The left plot* of the figure illustrates the exponential increase in the number of transistors on a processor chip over the years 1970–2010. Since the area of a chip has remained relatively constant, the plot also implies an exponential *decrease* in transistor size and spacing between transistors. The increase in number of transistors is called *exponential* because, when the ordinate scale is logarithmic as it is in Figure 1.1, the growth curve is a straight line. *Processing speeds*, in terms of operations per unit time, have also increased in proportion, due to faster switching rates as well as shorter delays caused by the decreased distance between transistors.

The right plot† of the figure illustrates a similar exponential increase in the number of computations per unit of energy, or in other words, an exponential *decrease* in the power and energy requirements of a chip designed to perform a given processing task. This has enabled the wide diversity in lightweight mobile and handheld DSP devices, as well as processors that run for long periods on low power in space and other remote locations.

* http://en.wikipedia.org/wiki/Moore's_law.

† Intel paper.

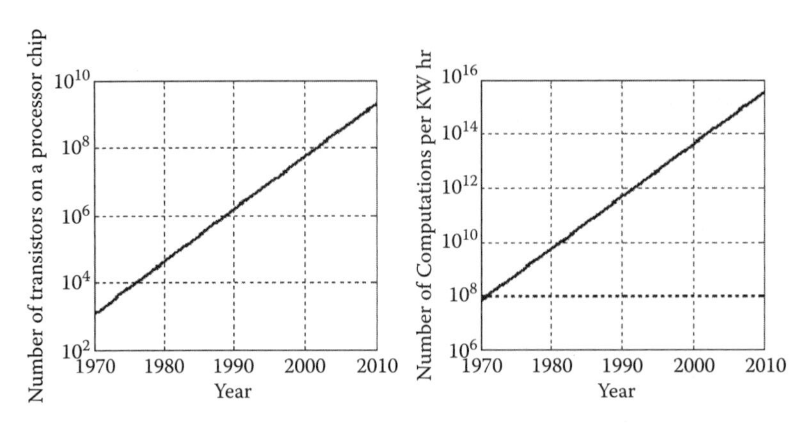

FIGURE 1.1
The remarkable growth and diversity in digital signal processing applications, especially since 1970, has been due, first, to the exponential increase in the number of transistors on a processor chip, and second, to the exponential increase in the number of computations possible per unit of energy.

Along with these rapid developments in DSP hardware, advances in DSP *software* have led to complex algorithms and devices that adapt to various inputs and seem to "think for themselves." The result is that in the last half-century or so the civilized world has, almost unknowingly, caused its very existence to depend on the reliable performance of signal processing hardware and software. And as long as the human race is left to fend for itself, there is no apparent way back to the simple life of the past.

The authors of this text have colleagues and friends who have built the foundations of DSP from the ground up, and many of them are still active contributors at the time of this writing. The list is too long to print here, but some of them can be found in the lists of "early" references at the end of this chapter[1–15] and subsequent chapters.

This book is the result of an attempt to organize the major contributions of these pioneers and to explain subjects that are basic to DSP as it exists currently, that is, subjects that one would find essential to or at least useful in the understanding of most current applications of DSP. There are now many special topics within DSP, such as image processing, digital communications, and signal coding and compression, so that no one is an expert in all areas. But hopefully, in this text, the reader will find subjects that are basic to (and useful in) all these different areas.

In other words, there is so much diversity in types of signals and types of processing and types of processors that we really only have two choices. One choice is to provide a complete text on a single subject, such as one of those just mentioned, or wavelet transforms or digital filters, for example. The second choice is to provide a connected knowledge base consisting of subjects basic to all kinds of digital signal processing, so that the reader may acquire at least the fundamentals but not all of the details.

It is this second option that we have chosen here. We hope that this text will introduce the reader to the basic areas of signal processing and also provide a reference for the reader who wishes to review the fundamentals of DSP.

1.2 How to Read This Text

There are two main ways to read this text. One way is to read it more or less in the order in which it is written and work some of the exercises as you go. Each chapter, especially Chapters 3–6, depends on subjects discussed in previous chapters. If the subjects are read and learned in order, the reader should end with a good foundation for work in most areas of signal processing.

The second way is to look in the text for a particular subject, such as filters, coding, or spectral estimation. The best way to do this is to look in the index. If your subject is not listed there, look in the table of contents for a related area. If you have at least some familiarity with DSP as well as the basic mathematics needed for analysis, there is no need to read all the text. The treatment of each individual subject is meant to be self-contained, although the text may refer to previous, more basic subjects.

With these two approaches in mind, the reader may now decide whether to read the remainder of this chapter or skip to a subject of interest. The rest of this chapter consists mostly of reviews of some basic mathematics and formulas useful in signal and system analysis, as well as the MATLAB® language, which has become a standard in signal processing analysis and system design. If you are familiar with the mathematics and are already a MATLAB user, these reviews may be skipped and treated as reference material.

1.3 Introduction to MATLAB®

MATLAB, a product of The MathWorks, Inc., is currently in use throughout the DSP community. Three principal reasons for this are (1) the syntax allows the user to do most DSP operations with very simple lines of code; (2) with MATLAB's graphics support, one can produce publishable plots with minimal effort; and (3) most importantly, you do not have to be an expert in MATLAB to use it. The syntax is easy to read and learn and use with almost no prior groundwork.

Most of the DSP examples in this text use the MATLAB language. This does not mean that the reader must own or operate MATLAB software, although having access to a computer running MATLAB gives a definite advantage. But the language is useful as a standard for describing DSP operations and algorithms. For readers unfamiliar with MATLAB notation, we begin here with some basics. If you do not own or operate MATLAB, you may wish to view the use of MATLAB in this text as a convenient and easy-to-read standard system for describing signal processing procedures. For more depth in MATLAB, the study by Etter[18] not only describes MATLAB but also addresses several of the topics in this text, with applications.

MATLAB uses single expressions called *commands*, which may be assembled into sets of commands called *m-files* (because the file extension is "m") or *functions*, which may be called by other m-files or functions. In this text, when you see lines inside a text box, these will usually be MATLAB expressions. For example,

```
>> x=4;
>> y=[2,3,4];
>> z=[1 2 3; 4 5 6];
```
(1.1)

Each of these lines is an individual expression (command) as indicated by the *command prompt* at the beginning of the line. Each line ends with a semicolon; if not, then MATLAB would echo the results of the line as in (1.2) below. *Row elements* of an array may be separated by commas (as in y) or by spaces (as in z). *Rows* are separated by semicolons (as in z). The results of (1.1) are shown in (1.2) when the MATLAB expression "x,y,z" is entered without a semicolon at the end.

```
>> x,y,z
x=
          4

y=
          2 3 4

z=
          1 2 3
          4 5 6
```
(1.2)

From basic expressions like these, we can proceed rapidly and simply in the following two sections to expressions that accomplish complicated DSP operations.

1.4 Signals, Vectors, and Arrays

High-level computer languages such as MATLAB are designed to process ordered sequences of elements, that is, *variables, vectors,* and *matrices.* As used in this text, these three terms form a hierarchy in the sense that:

- A variable is a single element (integer, real, or complex) [like x in (1.1)].
- A vector is an ordered sequence of variables [like y in (1.1)].
- A matrix is an ordered sequence of vectors [like z in (1.1)].

We usually use *array* as an inclusive term to designate a vector or a matrix. Sometimes our use of *vector* here is confusing at first, because we are used to vectors in electromagnetics or mechanics—in three-dimensional space—with exactly three components. But these are really just ordered sequences of three variables and are thus vectors in the sense used here, and we must now allow more dimensions in order to have vectors that represent signals in a "signal space," which we define as follows.

Figure 1.2 shows a sampled waveform and its corresponding signal vector. When we say a waveform is *sampled*, we mean that its amplitude is measured and recorded at different times. Usually, these different times are equally spaced in the time domain, but not always. We assume that the samples are equally spaced. The interval between samples is called the *sampling interval* or *time step*. In Figure 1.2, the time step is $\Delta t \equiv T = 1$. In this example, there are 10 sample values, and the *signal vector* consists of the ordered sequence of 10 integer samples. Because each sample is a variable, we may say that the *signal space* of the sampled waveform in Figure 1.2 has 10 dimensions, one for each sample. Thus, in this sense, the signal space of a sampled waveform has as many dimensions, or *degrees of freedom*, as there are samples in the waveform.

Figure 1.3 illustrates a (very small) grayscale image and its corresponding sample array. Each element of the image is called a *pixel* (picture element), and in this example, an element can take on only one of four values—0, 1, 2, or 3—and these values cover the range from black to white in the order given, that is, 0 = black, and 3 = white. The digital *image array* is shown next to the image, and in this case, we may say that the signal space associated with the image has 25 dimensions, one for each sample (variable) in the image.

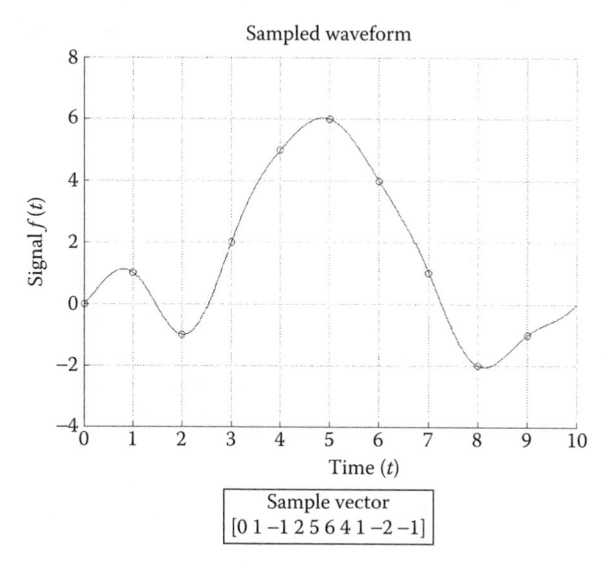

FIGURE 1.2
Sampled waveform and sample vector with time step $T = 1$.

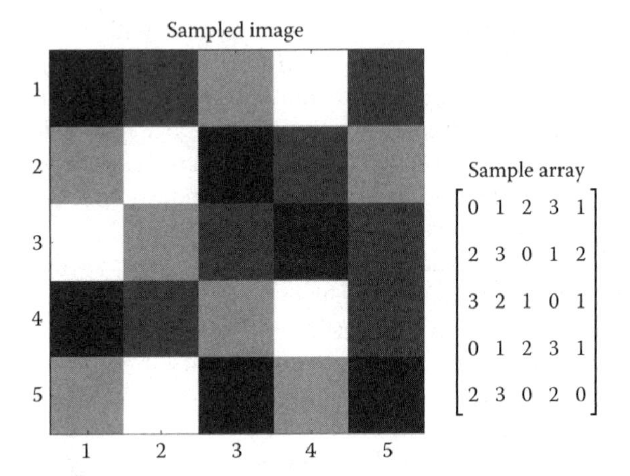

FIGURE 1.3
Sampled grayscale image and sample array.

Thus, through processes of sampling and analog-to-digital conversion that depend on the quantities being measured, signals and images in the real, continuous world are converted to vectors and matrices of integers in the discrete world of DSP. It is with these kinds of digital data that the subject of DSP begins. We should note that the examples in Figures 1.2 and 1.3 are simple for the sake of illustration. In real situations, waveform vectors and image arrays have large numbers of elements, as may be seen in some of the examples in this text. Real images, which consist of thousands or millions of pixels, typically have 256 gray levels, not just 4 as in Figure 1.3, and color images have many more levels to a pixel. Moreover, *video* signals, with the most data of all, consist of sequences of images.

1.5 Review of Vector and Matrix Algebra Using MATLAB® Notation

In the rest of this chapter, before we begin discussing DSP fundamentals in Chapter 2, we review some of the basic mathematics required to understand DSP systems, beginning in this section with vector and matrix algebra. We have seen how vectors and matrices are used to represent digitized signals and images. Vectors and matrices also appear in other ways in DSP, as we will see in Chapters 2 and 3. In modern DSP, and especially in languages like MATLAB, vector and matrix sums, products, and so on are the basic processing operations. These operations are mostly just demonstrated here. The development is not as formal as it would be in a text on linear algebra or matrix theory. Table 1.1 presents a summary of operations.

TABLE 1.1

Examples of Array Operations

Operation	Symbol	Example	Remarks
Sum or difference	+ or −	$\begin{bmatrix}1&2\\3&4\end{bmatrix}+\begin{bmatrix}0&-1\\1&2\end{bmatrix}=\begin{bmatrix}1&1\\4&6\end{bmatrix}$ $[1\ 2\ 3]-2=[-1\ 0\ 1]$	Dimensions must be the same unless one variable is a scalar.
Matrix product	*	$\begin{bmatrix}1&2&0\\3&4&2\end{bmatrix}*\begin{bmatrix}0&-1\\1&2\\-1&1\end{bmatrix}=\begin{bmatrix}2&3\\2&7\end{bmatrix}$ $2*[1\ 2\ 3]=[2\ 4\ 6]$	Inner dimensions must be the same unless one variable is a scalar.
Array product	.*	$\begin{bmatrix}1&2\\3&4\end{bmatrix}.*\begin{bmatrix}0&-1\\1&2\end{bmatrix}=\begin{bmatrix}0&-2\\3&8\end{bmatrix}$ $2.*[1\ 2\ 3]=[2\ 4\ 6]$	Dimensions must be the same unless one variable is a scalar.
Transposes	′	$\begin{bmatrix}1&2&3\\4&5&6\end{bmatrix}'=\begin{bmatrix}1&4\\2&5\\3&6\end{bmatrix};$ $[1+2j\ \ 3-4j]'=\begin{bmatrix}1-2j\\3+4j\end{bmatrix}$	$j=\sqrt{-1}$. Rows become columns. Elements are conjugated with ′ but not .′
	.′	$\begin{bmatrix}1&2&3\\4&5&6\end{bmatrix}.'=\begin{bmatrix}1&4\\2&5\\3&6\end{bmatrix};$ $[1+2j\ \ 3-4j].'=\begin{bmatrix}1+2j\\3-4j\end{bmatrix}$	
Exponentiation	.^	$2.\wedge[1\ 2\ 3]=[2\ 4\ 8]$ $[1\ 2\ 3].\wedge 2=[1\ 4\ 9]$ $\begin{bmatrix}0&1&2\\3&4&5\end{bmatrix}.\wedge\begin{bmatrix}0&1&2\\2&1&0\end{bmatrix}=\begin{bmatrix}0&1&4\\9&4&1\end{bmatrix}$	Dimensions must be the same unless one variable is a scalar.
Backslash	\	$a*b=c$, where $a=\begin{bmatrix}5&3\\2&2\end{bmatrix}$ and $c=\begin{bmatrix}1\\2\end{bmatrix}$ $b=a\backslash c=\begin{bmatrix}5&3\\2&2\end{bmatrix}\backslash\begin{bmatrix}1\\2\end{bmatrix}=\begin{bmatrix}-1\\2\end{bmatrix}$	This operation solves sets of linear equations.

(Quotients and products are similar.) If the reader is already familiar with the operations, the table should suffice as a convenient reference.

To begin, we note that a vector is an array with either one row (*row vector*) or one column (*column vector*). Thus, rules and statements about arrays are generally true for vectors.

The sum or difference of two arrays is obtained simply by adding or subtracting the individual elements. For example,

$$\begin{bmatrix}1\\2\\3\end{bmatrix}+\begin{bmatrix}4\\5\\6\end{bmatrix}=\begin{bmatrix}5\\7\\9\end{bmatrix};\ \begin{bmatrix}1 & 2 & 3\end{bmatrix}+\begin{bmatrix}0 & -1 & 4\end{bmatrix}=\begin{bmatrix}1 & 1 & 7\end{bmatrix};$$
$$\begin{bmatrix}1 & 2 & 0\\3 & 4 & 1\end{bmatrix}+\begin{bmatrix}-1 & 1 & 2\\2 & -2 & -1\end{bmatrix}=\begin{bmatrix}0 & 3 & 2\\5 & 2 & 0\end{bmatrix} \tag{1.3}$$

The MATLAB expressions for these operations are quite similar. For example,

```
>> [1,2,0; 3,4,1]+[-1,1,2; 2,-2,-1]
ans=
     0   3   2
     5   2   0
```
(1.4)

We do not really need the MATLAB versions here, where conventional expressions are more acceptable. Note that in order for an addition or subtraction to make sense, the two arrays must have the same dimensions, that is, 3×1, 1×3, and 2×3 as in (1.3). An exception occurs when one of the arrays is a scalar or single variable. In this case, the scalar or the variable is added to or subtracted from each element of the second array. Thus,

$$3+\begin{bmatrix}1 & 2\\3 & 4\end{bmatrix}=\begin{bmatrix}4 & 5\\6 & 7\end{bmatrix} \tag{1.5}$$

In this text, we use letters without subscripts to represent single vectors and arrays like those in (1.3) and (1.5), and letters with subscripts to represent the elements as in (1.6). For example,

$$a=\begin{bmatrix}a_{11} & a_{12} & a_{13}\\a_{21} & a_{22} & a_{23}\end{bmatrix} \tag{1.6}$$

The first index of each element designates the row, and the second the column. Indexes generally begin at one. If a is a vector, then only one index is required.

Two kinds of array *products* are used most often in DSP. The first is the ordinary matrix product with symbol $*$, formed by summing the products of row elements in the first array multiplied by column elements in the second. If a is an array with M rows and N columns, and if b is an array with N rows and K columns, then the elements of the array product $c = a * b$ are given by the following:

$$c_{ij} = \sum_{n=1}^{N} a_{in} b_{nj}; \quad 1 \le i \le M \quad \text{and} \quad 1 \le j \le K \tag{1.7}$$

For example, suppose a, b, and c are defined as follows:

$$a = \begin{bmatrix} 1 & -1 & 0 \\ 0 & -1 & 1 \\ 2 & 0 & 1 \end{bmatrix}; \quad b = \begin{bmatrix} b_1 \\ b_2 \\ b_3 \end{bmatrix}; \quad c = \begin{bmatrix} -1 \\ 1 \\ 2 \end{bmatrix} \tag{1.8}$$

Then, we can see from the matrix product definition in (1.7) that the array equation $a * b = c$ is an expression of three simultaneous equations. That is,

$$\begin{aligned} b_1 - b_2 &= -1 \\ -b_2 + b_3 &= 1 \qquad \text{and} \quad a * b = c \\ 2b_1 + b_3 &= 2 \end{aligned} \tag{1.9}$$

are equivalent expressions of three linear equations in three unknowns. Note how the matrix product $a * b$ is obtained by summing products along the rows of a and down columns of b.

In DSP, it is often necessary to solve simultaneous linear equations like (1.9), and we will return to this subject shortly.

The second kind of product, used just as often in DSP as the matrix product defined in (1.7), is obtained simply by multiplying corresponding elements together. To do this, the two arrays must have the same dimensions. This kind of operation is designated with the symbol .* (dot star) to distinguish it from the matrix product (*). Thus, if a and b are $M \times N$ arrays, an element of the *array product* (as opposed to matrix product), $c = a .* b$, is given by

$$c_{ij} = a_{ij} b_{ij}; \quad 1 \le i \le M \quad \text{and} \quad 1 \le j \le N \tag{1.10}$$

To illustrate the array product, suppose we wish to "fade" the image in Figure 1.3 by making the rows lighter from the bottom of the image to the top. Then we could construct a "fading matrix," Q, and use the array product to produce the faded version of the original sample array (S) in Figure 1.3. For example, let

$$Q = \begin{bmatrix} 5 & 5 & 5 & 5 & 5 \\ 4 & 4 & 4 & 4 & 4 \\ 3 & 3 & 3 & 3 & 3 \\ 2 & 2 & 2 & 2 & 2 \\ 1 & 1 & 1 & 1 & 1 \end{bmatrix} \tag{1.11}$$

The fading operation is then given by

$$Q.*S = \begin{bmatrix} 5 & 5 & 5 & 5 & 5 \\ 4 & 4 & 4 & 4 & 4 \\ 3 & 3 & 3 & 3 & 3 \\ 2 & 2 & 2 & 2 & 2 \\ 1 & 1 & 1 & 1 & 1 \end{bmatrix} .* \begin{bmatrix} 0 & 1 & 2 & 3 & 1 \\ 2 & 3 & 0 & 1 & 2 \\ 3 & 2 & 1 & 0 & 1 \\ 0 & 1 & 2 & 3 & 1 \\ 2 & 3 & 0 & 2 & 0 \end{bmatrix} = \begin{bmatrix} 0 & 5 & 10 & 15 & 5 \\ 8 & 12 & 0 & 4 & 8 \\ 9 & 6 & 3 & 0 & 3 \\ 0 & 2 & 4 & 6 & 2 \\ 2 & 3 & 0 & 2 & 0 \end{bmatrix} \tag{1.12}$$

Figure 1.4 shows the original and faded versions of the image in Figure 1.3. Note that there are 16 gray levels in the faded version, compared with 4 in the original.

The array product may be used in the same way to impose an "envelope" on a waveform. For example, the vector f consisting of 1000 samples of a unit sine wave with 50 samples per cycle is produced by the following MATLAB expressions:

```
>> n=[1:1000];
>> f=sin(2*pi* n/50);
```
(1.13)

Notice how the row vector n causes the sine function to produce f, also a row vector, with 1000 elements. The sine wave is then "dampened" to produce the vector g by imposing a decaying exponential envelope using the array product (.*), and then both f and g are plotted as follows:

```
>> g=exp(- n/250).*f;
>> plot(n,f,n,g)
```
(1.14)

In (1.14), each element (f_n) of f is multiplied by the exponential function $e^{-n/250}$ to produce an element (g_n) of g. Figure 1.5 shows the results of the MATLAB

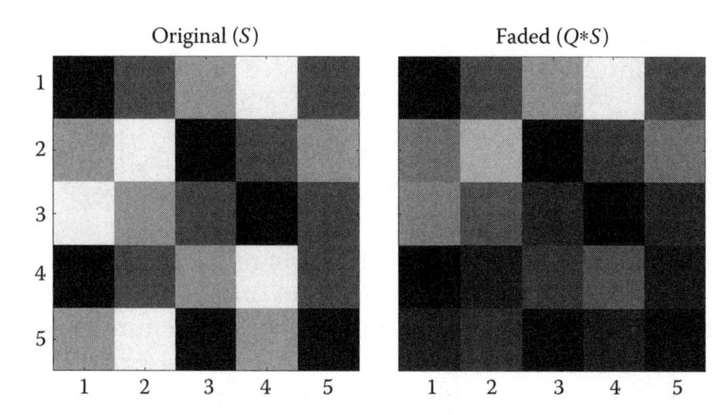

FIGURE 1.4
Original and faded images in equation (1.12).

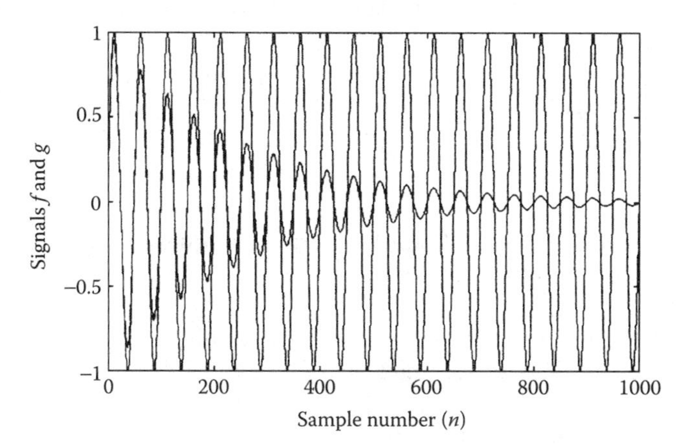

FIGURE 1.5
Plots of undamped and damped sine waves resulting from (1.13) and (1.14).

plot function, illustrating the ease of producing such a plot. It is also easy to add grid lines, labels, and so on, as we will later see in examples.

Having reviewed array sums and products, we focus on two operations that work with single arrays. First is the *transpose* with symbol ′ (prime), which changes rows to columns. Thus, each element a_{ij} in array a becomes a_{ji} in a'. The transpose is illustrated in Table 1.1. An important feature of the MATLAB transpose applies when the elements of the array are complex. Then, in a', the transposed elements are *conjugated*. If the elements are complex and conjugation is not desired, then .′ (dot prime) is used in the place of ′ (prime). Note the examples in Table 1.1.

The second operation is the matrix *inverse* with symbol ($^{-1}$) or *inv*. If a is a square matrix, *inv*(a) or a^{-1} is another square matrix such that $a * a^{-1} = I$, where I is the *identity matrix* with ones on the diagonal and zeros elsewhere. The inverse of any 2×2 matrix, which is needed often in DSP, is

$$a^{-1} = \begin{bmatrix} a_{11} & a_{12} \\ a_{21} & a_{22} \end{bmatrix}^{-1} = \frac{1}{a_{11}a_{22} - a_{12}a_{21}} \begin{bmatrix} a_{22} & -a_{12} \\ -a_{21} & a_{11} \end{bmatrix}; \quad a * a^{-1} = \begin{bmatrix} 1 & 0 \\ 0 & 1 \end{bmatrix} = I \qquad (1.15)$$

For a matrix to have an inverse, it must be square, because only then will the matrix product $a * a^{-1}$ be valid. The matrix must also be *nonsingular*, that is, no row may be a linear combination of other rows. This, in turn, means that the determinant[16,17] of the matrix must not vanish. In the case of (1.15), the determinant of a is $a_{11}a_{22} - a_{12}a_{21}$. In this text, we will prefer the *backslash* operation, which is described below in (1.17) to the inverse.

Array *exponentiation* is also useful in DSP and may be done with either scalars or arrays. In this text, we use only the MATLAB .^ (dot hat) operation, which is similar to .* (dot star) but produces powers instead of products. Thus,

$$\begin{bmatrix} 3 & 2 & 1 \\ -1 & 0 & 2 \end{bmatrix} .\wedge 2 = \begin{bmatrix} 9 & 4 & 1 \\ 1 & 0 & 4 \end{bmatrix} \qquad \begin{bmatrix} 3 & 2 & 1 \\ -1 & 0 & 2 \end{bmatrix} .\wedge \begin{bmatrix} 2 & -1 & 0 \\ -1 & 1 & 2 \end{bmatrix} = \begin{bmatrix} 9 & 0.5 & 1 \\ -1 & 0 & 4 \end{bmatrix} \qquad (1.16)$$

On the left, each element of the first array is squared, and on the right, each element of the first array is raised to the corresponding power in the second array.

The final array operation we review here is the backslash (\) operation, which is useful in DSP for solving linear equations like those in (1.9). We saw in connection with (1.9) that a set of linear equations may be expressed in the form $a * b = c$. The solution is found by multiplying on the left by a^{-1}, that is, $b = a^{-1} * c$. The backslash notation, $b = a \backslash c$, is equivalent, but in MATLAB, the latter is preferred, because it uses methods that produce the smallest round-off errors. Thus, in summary,

$$\text{If } a * b = c, \quad \text{then} \quad b = a \backslash c \qquad (1.17)$$

For example, the system in (1.9) is solved as follows:

$$b = a \backslash c = \begin{bmatrix} 1 & -1 & 0 \\ 0 & -1 & 1 \\ 2 & 0 & 1 \end{bmatrix} \backslash \begin{bmatrix} -1 \\ 1 \\ 2 \end{bmatrix} = \begin{bmatrix} 0 \\ 1 \\ 2 \end{bmatrix} \qquad (1.18)$$

The operations described in this section are the main algebraic operations that we use in DSP. We will also need relational and logical operations, but they will be introduced in terms of the MATLAB language as they are used.

This is the effortless way to learn MATLAB—learning necessary operations and functions as they are needed and memorizing those most often used. Table 1.1, consisting of examples of the operations discussed in this section, is provided as a reference.

1.6 Geometric Series and Other Formulas

The *geometric series* is used repeatedly to simplify expressions in DSP. The basic form of the series with N terms is

$$\sum_{n=0}^{N-1} x^n = 1 + x + x^2 + \cdots + x^{N-1} = \frac{1-x^N}{1-x} \tag{1.19}$$

The reader can prove this result by induction: It is obviously true for $N = 1$. And, you can easily show that if it is true for N, it is true for $N + 1$, and therefore true for all N. Moreover, if the magnitude of x is less than 1, the infinite version of (1.19) is as follows:

$$\sum_{n=0}^{\infty} x^n = \frac{1}{1-x}; \quad |x| < 1 \tag{1.20}$$

In DSP, x is often a complex exponential variable of the form e^{jk} in the geometric series, where $j = \sqrt{-1}$ and k is a real variable. Thus, for example,

$$\sum_{n=0}^{N-1} e^{j\frac{2\pi n}{N}} = \frac{1-e^{j2\pi}}{1-e^{j2\pi/N}} = 0 \tag{1.21}$$

Expressions in DSP are often simplified using the geometric series in this way. Furthermore, the *complex plane* is a useful aid for understanding why sums like this vanish. On the complex plane, $e^{j\theta}$ becomes a unit vector at angle θ measured counterclockwise from the positive real axis. For example, all the terms in (1.21) with $N = 8$ are plotted as unit vectors in Figure 1.6. That is, the vector at 3 o'clock is the term for $n = 0$, the vector at 12 o'clock is for $n = 2$, and so on. We can see from the symmetry of the plot that the sum of all eight vectors is zero, as in (1.21).

Some of the most useful *trigonometric identities* are provided in Table 1.2 as a reference for the remainder of the text. Trigonometric functions, especially sine and cosine functions, appear in different combinations in all kinds of harmonic analysis—Fourier series, Fourier transforms, and so on.

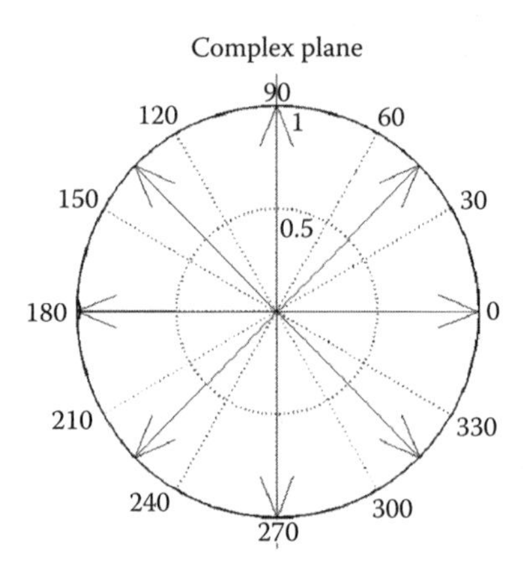

FIGURE 1.6
All the terms in (1.21) with $N = 8$ on the complex plane.

TABLE 1.2

Trigonometric Identities

$\sin\theta = \frac{1}{2j}(e^{j\theta} - e^{-j\theta})$	$\cos\theta = \frac{1}{2}(e^{j\theta} + e^{-j\theta})$
$e^{j\theta} = \cos\theta + j\sin\theta$	
$\sin(\theta+\alpha) = \sin\theta\cos\alpha + \cos\theta\sin\alpha$	$\cos(\theta+\alpha) = \cos\theta\cos\alpha - \sin\theta\sin\alpha$
$\sin\theta\sin\alpha = \frac{1}{2}(\cos(\theta-\alpha) - \cos(\theta+\alpha))$	$\cos\theta\cos\alpha = \frac{1}{2}(\cos(\theta+\alpha) + \cos(\theta-\alpha))$
$\sin\theta\cos\alpha = \frac{1}{2}(\sin(\theta+\alpha) + \sin(\theta-\alpha))$	
$\sin^2\theta + \cos^2\theta = 1$	$\cos^2\theta - \sin^2\theta = \cos 2\theta$
$\sin^2\theta = \frac{1}{2}(1 - \cos 2\theta)$	$\cos^2\theta = \frac{1}{2}(1 + \cos 2\theta)$
$\tan\theta = \frac{\sin\theta}{\cos\theta}$	$\cot\theta = \frac{\cos\theta}{\sin\theta}$

The identities that give sine and cosine functions in terms of exponentials are important, because they allow us to find sums of sines and cosines using the geometric series as in the example above. For example, with the exponential form of the sine or cosine function, we can easily show from (1.21) that when K is an integer,

$$\sum_{n=0}^{KN-1} \sin\left(\frac{2\pi n}{N}\right) = 0 \quad \text{and} \quad \sum_{n=0}^{KN-1} \cos\left(\frac{2\pi n}{N}\right) = 0 \tag{1.22}$$

TABLE 1.3

Additional Identities and Operations

If $x = a^y$, then $\log_a x = y$	$\log_{10} x = \log x \log_{10} e$, where "log" $\equiv \log_e$
$\sinh x = \frac{1}{2}(e^x - e^{-x})$	$\cosh x = \frac{1}{2}(e^x + e^{-x})$
$\sinh^{-1}(x) = \log(x + \sqrt{x^2+1})$	$\cosh^{-1}(x) = \log(x + \sqrt{x^2-1});\quad x > 1$
$\frac{d(uv)}{dx} = u\frac{dv}{dx} + v\frac{du}{dx}$	$\frac{d(u/v)}{dx} = \left[v\frac{du}{dx} - u\frac{dv}{dx}\right]\Big/ v^2$
$\int u\, dv = uv - \int v\, du$	

That is, a sum of equally spaced samples of any sine or cosine function is zero, provided the sum is over an integral number of cycles, or periods, of the function.

In addition to the trigonometric formulas in Table 1.2, some additional mathematical formulas and operations are summarized in Table 1.3. These are used frequently in DSP design and analysis as well as in the remainder of this text.

The last two expressions in Table 1.3, in which u and v are functions of an independent variable (say t), are especially useful with continuous transforms and transfer functions, because it is often necessary to differentiate or integrate products or quotients of functions when deriving transforms or transfer functions. For example, suppose we wish to integrate the product $\sin \alpha t\ e^{j\omega t}$ over an indefinite interval of t. If we set $u = \sin \alpha t$ and $dv = e^{j\omega t} dt$, then $v = (1/j\omega)e^{j\omega t}$ and $du = \alpha \cos \alpha t\ dt$, and the last expression in Table 1.3 becomes as follows:

$$\int \sin \alpha t e^{j\omega t} dt = \frac{1}{j\omega}\left(\sin \alpha t e^{j\omega t} - \alpha \int \cos \alpha t e^{j\omega t} dt\right) \tag{1.23}$$

At first, this result seems only to make the problem worse, but if we use the same method on the right-hand side integral in (1.23), this time with $u = \cos \alpha t$, the result is as follows:

$$\int \cos \alpha t e^{j\omega t} dt = \frac{1}{j\omega}\left(\cos \alpha t e^{j\omega t} + \alpha \int \sin \alpha t e^{j\omega t} dt\right) \tag{1.24}$$

Now, the integral on the right side in (1.24) is the same as the original integral on the left side in (1.23), and we can combine the two equations to give the desired result as follows:

$$\int \sin \alpha t e^{j\omega t} dt = \frac{e^{j\omega t}}{\alpha^2 - \omega^2}(j\omega \sin \alpha t - \alpha \cos \alpha t) \tag{1.25}$$

Thus, applications of the last expression in Table 1.3, which is called *integration by parts*, are often useful for integrating products of functions.

1.7 MATLAB® Functions in DSP

MATLAB *functions*, as we mentioned in Section 1.3, are m-files that can be called by other m-files. In the MATLAB system, these files are implemented in a way that is especially advantageous in developing DSP algorithms and systems. As you go through this text, there will be opportunities to write your own functions to use in exercises and then (hopefully) to apply in your own occupation. The MATLAB language thus helps people in DSP and other research and development areas work efficiently by stockpiling low-level operations, rather than having to redevelop them for each new project.

The *row_vec* function is a simple example. Suppose your MATLAB programs often require the conversion of unknown arrays to row vectors. Then, even though the conversion is simple, you can write the following function once and avoid repeating the task:

```
function v=row_vec(A)
%v=row_vec(A)
%
% Converts any array A into row vector v.
% If A is 1 × N, v=A. If A is N × 1, v=A.'.
% If A is a matrix, v=A scanned row by row.
AT=A';
v=AT(:)';
```
(1.26)

MATLAB allows you to store functions like this in your own system and then use them in other functions. If you are like most of us, you may eventually forget how to use the function, hence the comment (%) lines in (1.26). The comments allow one to request help at the MATLAB command prompt as in the following example:

```
>> help row_vec
   v=row_vec(A)
   Converts any array A into row vector v.
   If A is 1 × N, v=A. If A is N × 1, v=A.'.
   If A is a matrix, v=A scanned row by row.
```

That is, the *help* request at the command prompt causes the initial comment lines of the function to be printed. If you are new to MATLAB, it is a good idea to try the *help* command (you can even enter *help help*) and get instructions for some of the commonly used functions. It is also a good idea, as you read this text and work exercises, to form the habit of developing functions to do tasks that require repeating, and thus build your own personal library. In DSP, there is a real advantage in making a practice of doing this.

1.8 The Chapters Ahead

In the chapters ahead, we develop the fundamentals of DSP more or less from the ground up. As indicated in the Preface, we are attempting to cover the subject from an applied standpoint, assuming that the reader is planning to process digitized samples of real, continuous signals.

Chapter 2 is a review of least-squares analysis, orthogonality, and the Fourier series, all of which are needed for later chapters but may not be necessary for readers familiar with these subjects. Chapters 3 and 4 proceed through Fourier spectra, sampling and measurement of signals, the sampling theorem, linear algorithms, structures, and transfer functions. Chapters 2 through 4 are meant to be read successively, unless the reader is already familiar with the material.

The remaining chapters, from Chapter 5 on, cover different subjects basic to DSP—digital filters, statistical signal processing, least-squares system design, and so on. These chapters depend on Chapters 2–4 but are relatively independent of each other, and therefore not meant to be read in a particular order. The reader may choose to look only at chapters of current interest.

All of the chapters, except this one, have exercises at the end. This chapter is meant primarily for background and reference, but in each of the rest of the chapters, the exercises not only provide examples of the topics covered in the chapter, but they also introduce topics and applications not covered in the chapter. In this sense, the exercises are an essential component of the text.

References

Digital signal processing

1. Shannon, C. E. 1949. *The Mathematical Theory of Communication*. Urbana, IL: University of Illinois Press.
2. Blackman, R. B., and J. W. Tukey. 1958. *The Measurement of Power Spectra*. New York: Dover.
3. Hamming, R. W. 1962. *Numerical Methods for Scientists and Engineers*. New York: McGraw-Hill.
4. Kuo, F. K., and J. F. Kaiser, eds. 1967. *System Analysis by Digital Computer*. Chap. 7. New York: John Wiley & Sons.

5. Gold, B., C. M. Rader et al. 1969. *Digital Processing of Signals*. New York: McGraw-Hill.
6. Rabiner, L. R., and C. M. Rader, eds. 1972. *Digital Signal Processing*. New York: IEEE Press.
7. Oppenheim, A. V., and R. W. Schafer. 1975. *Digital Signal Processing*. Englewood Cliffs, NJ: Prentice Hall.
8. Rabiner, L. R., and B. Gold. 1975. *Theory and Application of Digital Signal Processing*. Englewood Cliffs, NJ: Prentice Hall.
9. Peled, A., and B. Liu. 1976. *Digital Signal Processing: Theory, Design, and Implementation*. New York: John Wiley & Sons.
10. Tretter, S. A. 1976. *Introduction to Discrete-Time Signal Processing*. New York: John Wiley & Sons.
11. Bellanger, M. 1984. *Digital Processing of Signals*. New York: John Wiley & Sons.
12. Orfanidis, S. 1985. *Optimum Signal Processing: An Introduction*. New York: MacMillan.
13. Ludeman, L. E. 1986. *Fundamentals of Digital Signal Processing*. New York: Harper & Row.
14. Roberts, R. A., and C. T. Mullis. 1987. *Digital Signal Processing*. Reading, MA: Addison-Wesley.
15. Oppenheim, A. V., and R. W. Schafer. 1989. *Discrete-Time Signal Processing*. Chaps. 4, 5. Englewood Cliffs, NJ: Prentice Hall.

Vectors, matrices, and linear algebra

16. Strang, G. 1988. *Linear Algebra and Its Applications*. 3rd ed. Orlando, FL: Harcourt Brace.
17. Lipshutz, S. 1991. *Schaum's Outline of Theory and Problems of Linear Algebra*. 2nd ed. New York: Schaum's Outline Series, McGraw-Hill.

MATLAB

18. Etter, D. M. 1993. *Engineering Problem Solving with MATLAB*. Englewood Cliffs, NJ: Prentice Hall.

Further Reading

Digital signal processing

Corinthios, M. 2004. *Signals, Systems, Transforms, and Digital Signal Processing with MATLAB*. Boca Raton, FL: CRC Press.

Hamdy, N. 2009. *Applied Signal Processing*. Boca Raton, FL: CRC Press.

Vectors, matrices, and linear algebra

Lyons, L. 1998. *All You Wanted to Know About Mathematics but Were Afraid to Ask*. Vols. 1 and 2. Cambridge, UK: Cambridge University Press.

MATLAB

The MathWorks, Inc., MATLAB & Simulink Student Version 2010.

2

Least Squares, Orthogonality, and the Fourier Series

2.1 Introduction

The three topics reviewed in this chapter (least squares, orthogonality, and the Fourier series) have been fundamental to digital signal processing (DSP) since its beginning. An understanding of these subjects provides insight into almost every area of spectral analysis and DSP system design. These three topics are related to each other. The use of orthogonal functions greatly simplifies the process of finding a *least-squares fit* of a linear function to a set of data. The Fourier series is an important example of a linear least-squares function, and the Fourier series is a series made up of orthogonal functions.

In this chapter, we provide a brief review of these topics as they are applied in DSP. References at the end of the chapter include more general texts on these subjects.

2.2 Least Squares

The principle of *least squares* is used often in DSP with signals and other functions of one or more variables. In this discussion, we assume functions of just one variable (time), that is, waveforms, because they are simplest, but the concepts apply to images as well.

To begin, suppose that we have two continuous functions, $f(t)$ and $\hat{f}(c,t)$, where c is a constant or an array of constants (i.e., c is not a function of t). The elements of c may then be selected to make $\hat{f}(c,t)$ a *least-squares approximation* to $f(t)$ with respect to a specified range of t, say from t_1 to t_2. If c is selected in this manner, then the *total squared error* (TSE),

$$\text{TSE} = \int_{t_1}^{t_2} [f(t) - \hat{f}(c,t)]^2 dt \tag{2.1}$$

is as small as possible.

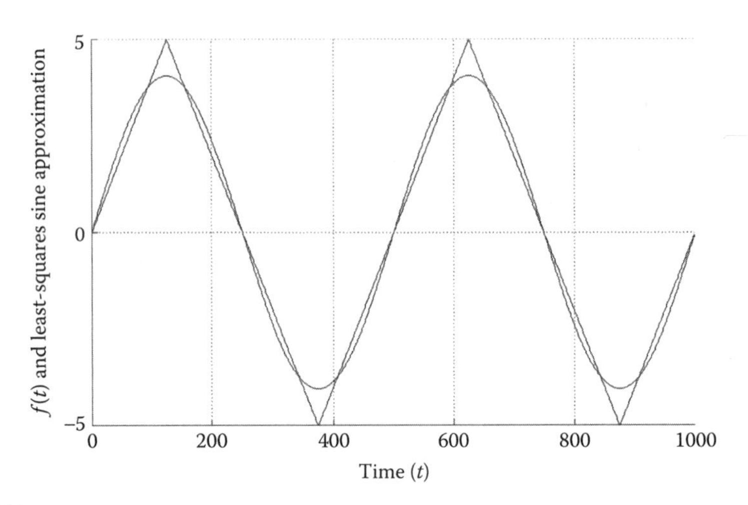

FIGURE 2.1
$f(t)$ and continuous least-squares approximation.

Figure 2.1 illustrates an example of a continuous least-squares fit. The first function, $f(t)$, is the triangular wave with period 500, plotted from $t_1 = 0$ to $t_2 = 1000$. In the second function, $\hat{f}(c,t) = c\sin(2\pi t/500)$, c is adjusted to minimize the TSE in (2.1), and so $\hat{f}(t)$ is a *least-squares fit* to $f(t)$ in the interval $[t_1, t_2] = [0, 1000]$. Note that there is only one constant to adjust in this example, that is, c is a scalar adjusted to minimize the TSE in this case.

In DSP, least-squares approximations are made more often to *discrete* (sampled) data rather than to continuous data. Instead of starting with a continuous function, $f(t)$, we start with a *sample vector* with N elements, $f = [f_1\ f_2\ \cdots\ f_N]$. If the approximating function is again $\hat{f}(c,t)$, then the TSE, similar to (2.1), is now given by the following equation:

$$\text{TSE} = \sum_{n=1}^{N} (f_n - \hat{f}(c, nT))^2 \tag{2.2}$$

where f_n is the nth element of f and T is the time step (interval between samples) described in Chapter 1, Section 1.4.

Figure 2.2 illustrates an example of a discrete least-squares fit. The functions are the same as shown in Figure 2.1, but here $f(t)$ is sampled, and c is adjusted to minimize the TSE in (2.2). The time step (sampling interval) in Figure 2.2 is $T = 35$, and the data vector, f, has $N = 28$ elements.

A *linear* least-squares approximation occurs when $\hat{f}(c,t)$ is a linear function of the elements of c. A *least-squares polynomial* of the form

$$\hat{f}(c,t) = c_1 + c_2 t + c_3 t^2 + \cdots + c_M t^{M-1} \tag{2.3}$$

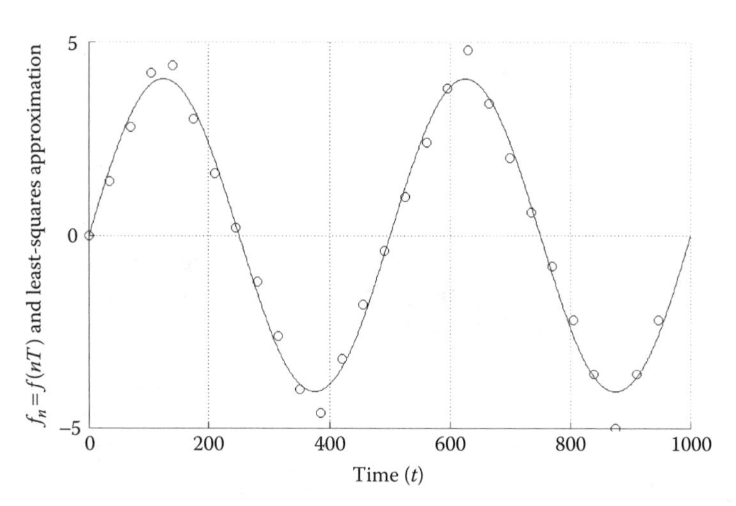

FIGURE 2.2
Least-squares approximation to sampled data with time step $T = 35$.

is an example of a linear least-squares function. In this case, there are M functions of the form t^m with m value starting at zero. More generally, suppose $g_m(t)$ is the mth function of t, then $g_{mn} = g_m(nT)$ is the value of $g_m(t)$ at $t = nT$, and there are M functions in all. Then, the linear least-squares function, beginning this time with $m = 1$, may be written as follows:

$$\hat{f}(c, nT) = c_1 g_{1n} + c_2 g_{2n} + c_3 g_{3n} + \cdots + c_M g_{Mn} \tag{2.4}$$

and the TSE in (2.2) becomes

$$\begin{aligned} \text{TSE} &= \sum_{n=1}^{N} (f_n - \hat{f}(c, nT))^2 \\ &= \sum_{n=1}^{N} \left(f_n - \sum_{m=1}^{M} c_m g_{mn} \right)^2 \end{aligned} \tag{2.5}$$

The TSE, considered a function of the vector c, is minimized by setting its gradient vector to zero, that is,

$$\nabla(\text{TSE}) = \left[\frac{\partial \text{TSE}}{\partial c_1} \;\; \frac{\partial \text{TSE}}{\partial c_2} \;\; \cdots \;\; \frac{\partial \text{TSE}}{\partial c_M} \right] = [0 \;\; 0 \;\; \ldots \;\; 0] \tag{2.6}$$

Using (2.5) in (2.6) produces M equations of the following form:

$$\begin{aligned} \frac{\partial \text{TSE}}{\partial c_m} &= -2 \sum_{n=1}^{N} g_{mn} \left[f_n - \sum_{k=1}^{M} c_k g_{kn} \right] \\ &= -2 \left[\sum_{n=1}^{N} g_{mn} f_n - \sum_{n=1}^{N} \sum_{k=1}^{M} c_k g_{mn} g_{kn} \right] = 0; \quad m = 1, 2, \ldots, M \end{aligned} \tag{2.7}$$

Dividing both sides of (2.7) by –2 and exchanging the summation order in the second term, we obtain the following expression:

$$\sum_{k=1}^{M} c_k \sum_{n=1}^{N} g_{mn} g_{kn} = \sum_{n=1}^{N} f_n g_{mn}; \quad m = 1, 2, \ldots, M \tag{2.8}$$

This is the set of linear equations to be solved for the least-squares coefficient vector, c. To simplify the notation, we use the following vector notation for the N samples of each function in (2.8):

$$G_m = [g_{m1} g_{m2} \cdots g_{mN}]; \quad f = [f_1 f_2 \ldots f_N]; \quad c = [c_1 c_2 \ldots c_M]' \tag{2.9}$$

Note that c is defined as a *column* vector. Then, using the MATLAB® operators described in Chapter 1 (see Table 1.1), all M equations in (2.8) are given by

$$\begin{bmatrix} G_1 * G_1' & G_1 * G_2' & \cdots & G_1 * G_M' \\ G_2 * G_1' & G_2 * G_2' & \cdots & G_2 * G_M' \\ \vdots & \vdots & \vdots & \vdots \\ G_M * G_1' & G_M * G_2' & \cdots & G_M * G_M' \end{bmatrix} * \begin{bmatrix} c_1 \\ c_2 \\ \vdots \\ c_M \end{bmatrix} = \begin{bmatrix} f * G_1' \\ f * G_2' \\ \vdots \\ f * G_M' \end{bmatrix} \tag{2.10}$$

This expression of (2.8) is an example of the use of the vector product (*) and transpose (′) operations in Table 1.1. An element $G_m * G_k'$ of the matrix on the left is seen to be $\sum_{n=1}^{N} g_{mn} g_{kn}$ in (2.8), an element $f * G_m'$ of the vector on the right is seen to be $\sum_{n=1}^{N} f_n g_{mn}$, and therefore, the mth row of (2.10) is the same as (2.8) for each value of m. Once we see that (2.10) expresses the M equations in (2.8), we can go one step further by defining

$$G = \left[G_1' \;\; G_2' \;\; \cdots \;\; G_M' \right] \tag{2.11}$$

so that there is a row of G for each *sample* (n) and a column for each *function* (m). Then, we may express (2.10) and, therefore, (2.8), as follows:

$$G' * G * c = (f * G)' \tag{2.12}$$

Equation 2.12 is a general expression of the discrete linear least-squares equations. According to Table 1.1 of Chapter 1, the solution is given by

$$c = (G' * G) \backslash (f * G)' \tag{2.13}$$

The best way to see how these results work is to do an example.

Suppose we have a signal for which the elements of the sample vector $f = [f_1\ f_2\ \cdots\ f_N]$ with $N = 12$ are the first 12 samples in Figure 2.2 (also shown in Figure 2.3) as follows:

$$f = [0.0, 1.4, 2.8, 4.2, 4.4, 3.0, 1.6, 0.2, -1.2, -2.6, -4.0, -4.6] \tag{2.14}$$

Instead of the single sine function shown in Figure 2.2, suppose we now wish to fit the data in f to the sum of three sine functions:

$$\begin{aligned} \hat{f}(C, nT) = c_1 \sin(2\pi nT/500) + c_2 \sin(6\pi nT/500) \\ + c_3 \sin(10\pi nT/500); \quad n = 0, 1, \ldots, 11 \end{aligned} \tag{2.15}$$

That is, g_{mn} in (2.5) is $g_{mn} = \sin(2\pi(2m-1)nT/500)$, with n beginning at zero here. The matrix G in (2.11) is now a matrix with $M = 3$ columns and $N = 12$ rows, with each element given by $g_{mn} = \sin(2\pi n(2m-1)T/500)$. (Note that the indexes are switched, because the G_m vectors in (2.11) are transposed.) The following MATLAB commands could be used to generate G:

```
m=[1:3];
n=[0:11];
T=35;
G=sin(2*pi*n'*(2*m-1)*T/500);
```
(2.16)

The key operation here is the vector product inside the sine argument that causes G to be a 12×3 matrix. Having found G, we now have all the elements needed to express the least-squares equations in (2.12). If you are using

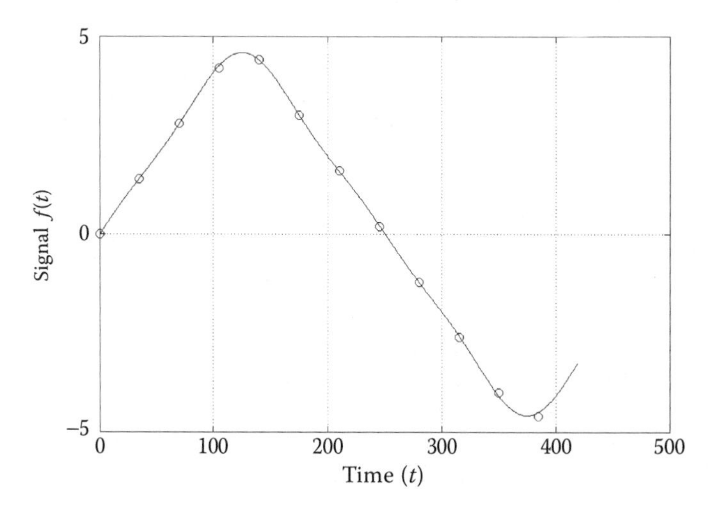

FIGURE 2.3
Least-squares fit plotted by the expression in (2.17).

MATLAB, the following expressions, along with (2.16) and row vector f, will solve the equations as in (2.13) and plot the data and the continuous approximation as shown in Figure 2.3.

```
c=(G'*G)\(f*G)';
n=[0:12*35-1]';
arg=2*pi*n/500;
fhat=[sin(arg), sin(3*arg), sin(5*arg)]*c;
plot(35*[0:11],f,'o',n,fhat);  grid;
```
(2.17)

Note that in (2.17) we have changed the time step from 35 to 1 to produce the continuous plot.

Another point may be made regarding the least-squares equations, (2.8) and (2.12): These equations may not be solvable. As with any set of M linear equations with M variables, a unique solution does not exist unless the equations are independent. This will not be the case if the number of data points, N, is less than M.[1] Hence, $N \geq M$ is always assumed in our discussions of least squares. Furthermore, if $N = M$, the function and the data have the same degrees of freedom, the fit is exact, and the TSE is zero. We will not discuss solvability further here, but the points just discussed are illustrated in Exercises 2.1(a) and 2.2 at the end of this chapter.

In summary, the *discrete least-squares process* is used in many areas of DSP. This procedure consists of fitting a linear combination of continuous functions with adjustable coefficients to a set of data samples in a way such that the TSE defined in (2.5) is minimized. This, in turn, involves solving the linear equations in (2.8) or in matrix form in (2.12) for the vector c of *least-squares coefficients.*

In Section 2.3, we will see that when the continuous functions are orthogonal with respect to the set of data samples (we will explain what this means), the process of finding the vector c is greatly simplified.

2.3 Orthogonality

In addition to the principle of least squares, orthogonality is another important concept often used in DSP. This word comes from a Greek word (*orthos*) implying a perpendicular or right-angled relationship. In mathematics, two row vectors a and b are said to be orthogonal if their inner product given by $a * b'$ is equal to zero. When a and b have two or three elements, then the original meaning of orthogonality holds and the vectors are perpendicular to each other, as illustrated in Figure 2.4. When vectors have more than three elements, we lose the spatial picture of orthogonality, but the definition, that

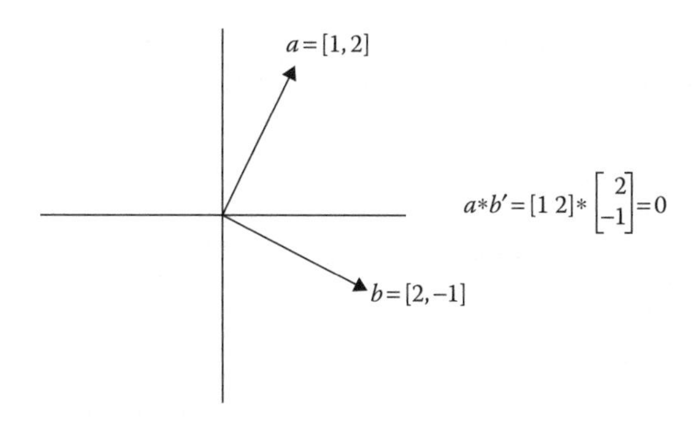

FIGURE 2.4
Orthogonal vectors a and b form a right angle at the origin.

is, $a * b' = 0$, holds for any number of elements, provided only that a and b have the same number of elements.

In Chapter 1, Section 1.4, we discussed the concept of *waveform vectors* in *signal space*. We say two such waveform vectors are *orthogonal* when their product vanishes as just described. Sometimes, these vectors occur from measurements, that is, analog-to-digital conversion, and sometimes they occur from the evaluation of functions. In the latter case, orthogonality must be evaluated *with respect to a specified set of samples*. That is, the vectors in this case consist of specified samples of the functions. For example, suppose we have N samples of two functions of t, $a(t)$ and $b(t)$, sampled at $t = 0, T, \ldots, (N-1)T$ to produce elements $a_0, \ldots, a_{N-1}$ and $b_0, \ldots, b_{N-1}$. These are then the elements of two vectors, a and b, and the condition for orthogonality with respect to the N samples is $a * b' = 0$, that is,

$$a * b' \equiv \sum_{n=0}^{N-1} a_n b_n = 0 \tag{2.18}$$

Suppose, for example, that $a_n = \sin(2\pi n/N)$ and $b_n = \cos(2\pi n/N)$. Then, using the result in Chapter 1, Equation 1.22, we can say that the two functions are orthogonal with respect to equally spaced samples over an integral number of cycles of the functions. But we can also show that the two functions are, in general, *not* orthogonal with respect to other sets of equally spaced samples.

Knowing when functions are orthogonal is important in the least-squares process, because when they are, the process is simplified. Look again at (2.4), in which the approximating function, $\hat{f}(c, nT)$, consists of a linear combination of M functions, $g_1, g_2, \ldots, g_M$. As in (2.9), each of these functions has a sample vector given by $G_m = [g_{m1}\ g_{m2}\ \cdots\ g_{mN}]$. Now, suppose that the M functions are *mutually orthogonal* with respect to the N samples, that is,

$$G_m * G_k' = 0; \quad m \neq k \tag{2.19}$$

Then, all the off-diagonal matrix elements in (2.10) are zero, and each of the equations in (2.10) becomes a simple expression of the following form:

$$G_m * G'_m * c_m = f * G'_m; \quad m = 1, 2, \ldots, M \tag{2.20}$$

Noting that c_m is a scalar and the two vector products are also scalars, we see that the solution for c_m in (2.20) is

$$c_m = \sum_{n=1}^{N} f_n g_{mn} \Big/ \sum_{n=1}^{N} g_{mn}^2; \quad m = 1, 2, \ldots, M \tag{2.21}$$

Thus, when the least-squares functions are orthogonal with respect to a given set of samples, we do not need to solve simultaneous equations to get the least-squares coefficient vector. The vector elements can be determined one by one as in (2.21).

In Section 2.4, we discuss the discrete Fourier series as an example of least-squares with orthogonal functions.

2.4 The Discrete Fourier Series

A discrete Fourier series[2,3] consists of combinations of sampled sine and cosine functions like those that we have been using as examples in this chapter. The Fourier series is named after J. B. J. Fourier (1768–1830). It is the basis of a branch of mathematics called *harmonic analysis*, which is applicable to the study of all kinds of natural phenomena, including the motion of stars, planets, and atoms, acoustic waves, and radio waves. The contributions of Fourier are a principal part of the foundation of DSP. Fourier's basic theorem, in terms of the DSP we are using here, states that any sample vector, regardless of its size or origin, can be exactly reproduced by summing a sufficiently long series of harmonic functions.

We must first describe what we mean by a *harmonic function*. To do this, we must first review a few basic properties of signals. Suppose we begin with a sample vector $x = [x_0, x_1, \ldots, x_{N-1}]$ (Note that here the indexes range from 0 to $N-1$ instead of 1 to N. This is admittedly annoying, but it cannot be helped, because sometimes we want the array index to start at 1 and sometimes we want the time of the first sample to be 0. We often need this convention while using MATLAB for DSP, for example, whenever the first element in an array is the sample at $t = 0$.)

We say the *fundamental period* of x is N samples. When we say this, we really imagine that the samples of x repeat, over and over again, in the time

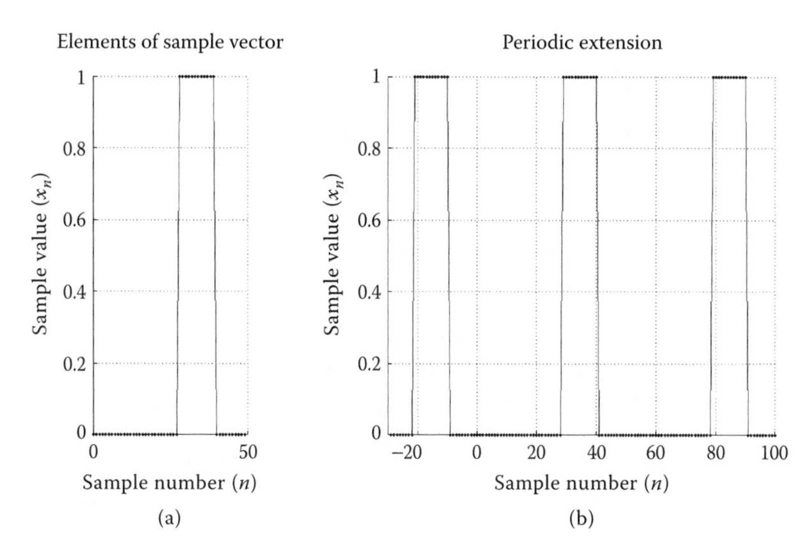

FIGURE 2.5
Sample vector and periodic extension, $N = 50$.

domain. In other words, we really think of a *periodic extension* of x, as illustrated in Figure 2.5. The elements of x are plotted on the left in Figure 2.5. They are shown connected with line segments to show approximately the original continuous function, $x(t)$, before sampling. The periodic extension of x on the right is formed by repeating x every N samples endlessly in both directions as shown in the figure.

Any function that repeats like the one on the right in Figure 2.5 is called a *periodic function*. The *fundamental period* is N samples, or NT seconds, where T is the time step in seconds. The *fundamental frequency* is the reciprocal of the fundamental period, or $f_0 = 1/(NT)$ Hertz (Hz). This, as we can see, is the number of times the function repeats (number of periods) in 1 second, that is, "Hertz" means "cycles per second." For example, suppose that $T = 1$ milliseconds (0.001 seconds) in Figure 2.5; then, the fundamental period is $NT = 50$ milliseconds, and the fundamental frequency is $f = 20$ Hz.

We also use two other measures of frequency other than f in this chapter. One of them is $\omega = 2\pi f$ radians per second. There are 2π radians in one cycle (rotation on the complex plane), and the factor 2π converts cycles per second to radians per second. We will introduce the third measure of frequency in Chapter 3.

We now introduce the *discrete Fourier series* as a *least-squares approximation* in the form of (2.4). We use two equivalent forms of the Fourier series in this chapter. These are given in (2.22) and the last line of (2.23).

$$\hat{x}_n \equiv \hat{x}(c, nT) = \sum_{m=-(M-1)}^{M-1} c_m e^{jmn\omega_0 T} \tag{2.22}$$

$$= c_0 + \sum_{m=1}^{M-1} c_m e^{jmn\omega_0 T} + c_{-m} e^{-jmn\omega_0 T}$$

$$= c_0 + \sum_{m=1}^{M-1} (c_m + c_{-m}) \cos mn\omega_0 T + j(c_m - c_{-m}) \sin mn\omega_0 T \tag{2.23}$$

$$= \frac{a_0}{2} + \sum_{m=1}^{M-1} a_m \cos mn\omega_0 T + b_m \sin mn\omega_0 T$$

In these expressions of the Fourier series, ω_0 is the fundamental frequency of the series in radians per second, $t = nT$ is the continuous time variable, and a_m, b_m, and c_m are coefficients similar to those in (2.4). Note that for c_m, m ranges from $-(M-1)$ to $M-1$ instead of 0 to $M-1$, and that $b_0 = 0$. Table 2.1 summarizes the relationships of the coefficients in the two forms of the series. They can be derived from the equivalence of (2.22) and (2.23). The prime (′) in Table 2.1 indicates the complex conjugate.

Neglecting the constant terms, a_0 and c_0, the Fourier series in its second form (2.23) is easily found to be periodic with period $2\pi/\omega_0$, because each sine and cosine function goes through m cycles during this period.

Furthermore, the Fourier series is a sum of *harmonic functions*. The terms for $m = 1$ in (2.22) and (2.23) are called *fundamental* terms. The terms for $m = 2$ are *second harmonic* terms, and so on. For each value of m, the frequency of the mth harmonic is $m\omega_0$ radians per second, that is, m times the fundamental frequency.

Because the Fourier series is a least-squares approximation, the coefficients in (2.22) and (2.23) must be least-squares coefficients. If we derive one set of these, we can find the other set using Table 2.1, so let us derive the a and b coefficients in (2.23). These may be determined using (2.21), because the harmonic functions are mutually orthogonal with respect to any set of regularly spaced samples over one period of the fundamental function. As shown in Figure 2.5, we may be thinking of a single sample vector, $x = [x_0, \ldots, x_{N-1}]$, or we may be analyzing a periodic function with repetitions of x. In either case, if the fundamental period, $2\pi/\omega_0$, covers N samples or NT seconds, then the fundamental frequency must be as follows:

$$\omega_0 = \frac{2\pi}{NT} \text{ rad/s} \tag{2.24}$$

TABLE 2.1

Equivalence of Fourier Series Coefficients

$[a,b]$ in terms of c	$a_0 = 2c_0;\quad a_m = c_m + c_{-m};\quad m > 0$
	$b_0 = 0;\quad b_m = j(c_m - c_{-m});\quad m > 0$
c in terms of $[a,b]$	$c_0 = a_0/2;\quad c_m = (a_m - jb_m)/2;\quad m > 0$
	$c_{-m} = c'_m$

With this substitution, to indicate sampling over exactly one fundamental period, (2.22) and (2.23) are now expressed as follows:

$$\hat{x}_n \equiv \hat{x}(c, nT) = \sum_{m=-(M-1)}^{M-1} c_m e^{j2\pi mn/N} \tag{2.25}$$

$$= \frac{a_0}{2} + \sum_{m=1}^{M-1} a_m \cos\left(\frac{2\pi mn}{N}\right) + b_m \sin\left(\frac{2\pi mn}{N}\right) \tag{2.26}$$

In this form, the harmonic functions in (2.26) are orthogonal with respect to the N samples of x. That is,

$$\begin{aligned} &\sum_{n=0}^{N-1} \cos\left(\frac{2\pi mn}{N}\right)\sin\left(\frac{2\pi kn}{N}\right) = 0; \quad m, k \geq 0 \\ &\sum_{n=0}^{N-1} \cos\left(\frac{2\pi mn}{N}\right)\cos\left(\frac{2\pi kn}{N}\right) = 0; \quad m, k \geq 0 \quad \text{and} \quad m \neq k \\ &\sum_{n=0}^{N-1} \sin\left(\frac{2\pi mn}{N}\right)\sin\left(\frac{2\pi kn}{N}\right) = 0; \quad m, k \geq 0 \quad \text{and} \quad m \neq k \end{aligned} \tag{2.27}$$

These results may be proved using the trigonometric identities in Table 1.2 and the geometric series application in (1.21).

Because the harmonic functions in (2.23) are all mutually orthogonal, we can apply (2.21) to solve for the least-squares coefficients. In the denominator of (2.21), if g_{mn} is one of the sine or cosine terms in (2.27), we can again use the identities in Table 1.2 for $\sin^2\theta$ and $\cos^2\theta$ and the geometric series formula to show that the denominator of (2.21) is $N/2$. For example,

$$\sum_{n=0}^{N-1} \sin^2\left(\frac{2\pi mn}{N}\right) = \sum_{n=0}^{N-1} \frac{\left(1 - \cos\left(\frac{4\pi mn}{N}\right)\right)}{2} = \sum_{n=0}^{N-1}\left(\frac{1}{2}\right) - 0 = \frac{N}{2} \tag{2.28}$$

Thus, using x_n for the sample data in (2.21), the least-squares Fourier coefficients in (2.26) are obtained:

$$\begin{aligned} a_m &= \frac{2}{N}\sum_{n=0}^{N-1} x_n \cos\left(\frac{2\pi mn}{N}\right); \quad 0 \leq m \leq M-1 \\ b_m &= \frac{2}{N}\sum_{n=0}^{N-1} x_n \sin\left(\frac{2\pi mn}{N}\right); \quad 1 \leq m \leq M-1 \end{aligned} \tag{2.29}$$

When using (2.21) to derive (2.29), we note that c in (2.21) stands for any coefficient so long as the orthogonality condition holds. When we are discussing

TABLE 2.2

Discrete Fourier Series and Coefficient Factors

$\hat{x}_n = \frac{a_0}{2} + \sum_{m=1}^{M-1} a_m \cos(2\pi mn/N) + b_m \sin(2\pi mn/N)$	$a_m = \frac{2}{N}\sum_{n=0}^{N-1} x_n \cos(2\pi mn/N);\quad 0 \le m \le M-1$ $b_m = \frac{2}{N}\sum_{n=0}^{N-1} x_n \sin(2\pi mn/N);\quad 1 \le m \le M-1$
$\hat{x}_n = \sum_{m=-(M-1)}^{M-1} c_m e^{j2\pi mn/N}$	$c_m = \frac{1}{N}\sum_{n=0}^{N-1} x_n e^{-j2\pi mn/N};\quad 0 \le m \le M-1$ $c_{-m} = c'_m$

the *Fourier coefficients* in (2.25), the elements of c, which are not all mutually orthogonal, are found using Table 2.1:

$$c_m = \frac{1}{N}\sum_{n=0}^{N-1} x_n e^{-j2\pi mn/N};\quad m \ge 0$$
$$c_{-m} = c'_m \tag{2.30}$$

A summary of the discrete Fourier series in (2.25) and (2.26) and the coefficient formulas we have just derived is presented in Table 2.2. We emphasize again that these formulas work only when the fundamental frequency of the series is $\omega_0 = 2\pi/NT$ radians per second. If we wish to apply the Fourier series to samples where this is not the case, as we did in the example using the samples in (2.14), then we must use the general least-squares solution in (2.13). We now consider an example in which the formulas in Table 2.2 can be applied.

Suppose a continuous signal is sampled, and the result is the vector with samples shown on the left in Figure 2.5. Then, the sine and cosine equations in Table 2.2 are applied to these samples with $M = 5$ in the following MATLAB file:

```
% Chapter 2: Discrete Fourier Series.
% Index vectors, n and m.
N=50;
n=[0:N-1];
M=5;
m=[0:M-1];                                                    (2.31)
% Sample vector, x.
x=[zeros(1,28),ones(1,12),zeros(1,N-40)];
% a and b coefficients. Note: b0=0 and a0/2 replaces a0.
a=(2/N)*x*cos(2*pi*n'*m/N);
a(1)=a(1)/2;
b=(2/N)*x*sin(2*pi*n'*m/N);
xhat=a*cos(2*pi*m'*n/N)+b*sin(2*pi*m'*n/N);
```

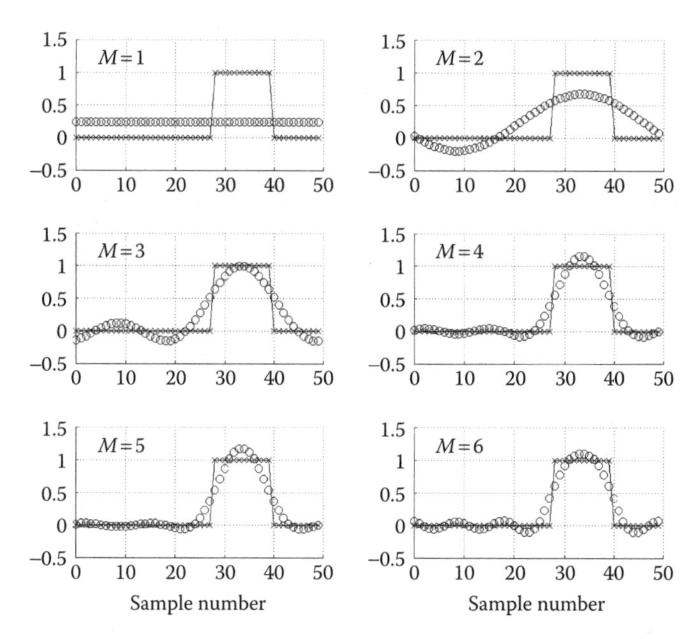

FIGURE 2.6
Discrete Fourier series approximations with increasing values of the series size, M. Waveforms are plotted versus the sample number.

There are several things to notice in the MATLAB code, especially in the array products. First, the index vectors (row vectors) are set to begin at zero. Then, the sample vector x in Figure 2.5 is specified as a row of zeros followed by ones followed by zeros, with a total of N elements. Next, in the computations of a and b, note how the product $n' * m$ produces an N by M array of sine or cosine values which, when multiplied by x, produce a and b row vectors. Note that $a(1)$, which is really a_0, is divided by 2 to satisfy the form of $\hat{x}_n$ in Table 2.2. Finally, note that $\hat{x}$ [$\hat{x}$ in (2.31)] is produced in a similar way, this time using the product $m' * n$, which produces an M by N array of the indexes. The result, $\hat{x}$, is plotted in Figure 2.6 for increasing values of M to show how the least-squares approximation improves with M.

As with any signal vector x, we may consider the vector elements, x_n and $\hat{x}_n$ in Table 2.2, as samples of a continuous waveform. If we do this, it is reasonable to make two substitutions in Table 2.2 to obtain an approximation, $\hat{x}(t)$, for the continuous waveform. First, we substitute the fundamental frequency, ω_0 in (2.24), and second, we substitute continuous time, t, for nT as in (2.4). The result is a continuous version of the discrete formulas in Table 2.2:

$$\hat{x}(t) = \frac{a_0}{2} + \sum_{m=1}^{M-1} a_m \cos(m\omega_0 t) + b_m \sin(m\omega_0 t) \tag{2.32}$$

$$\hat{x}(t) = \sum_{m=-(M-1)}^{M-1} c_m e^{jm\omega_0 t} \tag{2.33}$$

Because the coefficients here are derived from the samples as in Table 2.2, these results suggest a method for reconstructing a continuous function from a set of samples. We will discuss this idea further in Chapter 3.

The results in (2.32) and (2.33) also suggest a continuous form of the Fourier series, that is, a computation of the coefficients such that the reconstruction of $x(t)$ is exact. Instead of a formal development of the continuous form, we offer the following intuitive approach. In the formulas for a_m, b_m, and c_m in Table 2.2, we allow m to be unlimited and substitute the fundamental frequency, $\omega_0 = 2\pi/NT$ radians per second, so that the formula for c_m, for example, may be written as follows:

$$c_m = \frac{\omega_0}{2\pi} \sum_{n=0}^{N-1} x_n e^{-jm\omega_0 nT} T; \quad 0 \le m < \infty; \quad c_{-m} = c'_m \tag{2.34}$$

With c_m in this form, we can imagine decreasing the time step, T, toward zero, and at the same time increasing N proportionately so that the period, NT, remains constant. The samples (x_n) of $x(t)$ are thus packed more densely, so that in the limit, we have the Fourier series for a continuous periodic function, $x(t)$ as follows:

$$x(t) = \sum_{m=-\infty}^{\infty} c_m e^{jm\omega_0 t}; \quad c_m = \frac{\omega_0}{2\pi} \int_0^{2\pi/\omega_0} x(t) e^{-jm\omega_0 t} dt \tag{2.35}$$

Sometimes, for the sake of symmetry, c_m is given by an integral around $t = 0$ as follows:

$$c_m = \frac{\omega_0}{2\pi} \int_{-\pi/\omega_0}^{\pi/\omega_0} x(t) e^{-jm\omega_0 t} dt \tag{2.36}$$

By separating the integral in (2.35) into two integrals and substituting $u = t - 2\pi/\omega_0$ in the second integral, one can see that the two forms of c_m are equivalent. Derivation of formulas for a_m and b_m is left for Exercise 2.13 at the end of this chapter. In DSP, we are more generally concerned with sampled signals rather than continuous signals. The continuous forms of the Fourier series are, nevertheless, applicable to a wide range of natural periodic phenomena.

In this section, we have introduced two forms of the discrete Fourier series and have shown how to calculate the coefficients when the samples are taken over one fundamental period of the data, which is the usual case. In Chapter 3 we will see that harmonic analysis, and hence the Fourier series, forms the basis for the discrete Fourier transform and for spectral analysis in general.

Exercises

2.1 Find the parabola, that is, Equation 2.3 with $M = 3$, that forms a least-squares fit to each set of data points given in parts (a) through (d) below. In each case, assume that the time step, T, is one, so the samples in f are taken at $t = 0,1,\ldots,N-1$, where N is the length of f. Write a MATLAB file to perform the following steps:

- Compute the vectors G_1, G_2, and G_3 in (2.9), that is, the columns of G in (2.11).
- Form the array G in (2.11).
- Solve for the least-squares coefficients, c_1, c_2, and c_3, using (2.13).
- Make a continuous plot of the least-squares parabola, $\hat{f}(t)$, in the range $t = [0, N-1]$. Use 1000 values to produce a smooth plot.
- Finally, on the same graph, plot the elements of f as discrete points. Label the plot using functions *xlabel*, *ylabel*, and *title*, and observe whether the least-squares fit seems to be correct.

a. $f = [1, 3, 2]$

b. $f = [1, 3, 4, 2]$

c. $f = [1.1, 2.5, 3.2, 3.8, 3.7, 3.1, 2.0]$

d. $f = [8, 6, 4, 2.5, 1, 0, -0.5, -0.5, -0.5, 0, 1.5, 2.5, 4, 5, 7, 8]$

2.2 Try to execute the program in Exercise 2.1 using a vector $f = [-1, 3]$, that is, a vector with $N < M$.

a. Print the coefficient matrix and the vector in the right-hand side in (2.10).

b. Show from part (a) why the three linear equations are not independent, and why therefore, a unique solution for c does not exist.

c. Show what happens when the MATLAB backslash, (matrix)\(vector), is attempted.

d. Explain in simple geometric terms why the three equations do not have a single unique solution.

2.3 Some dates and closing values of the Dow Jones Industrial Average during 1960 are as follows:

Year	Month	Day	Close	Year	Month	Day	Close	Year	Month	Day	Close
1960	1	4	679.060	1960	5	3	607.730	1960	8	31	625.990
1960	1	22	645.850	1960	5	23	623.660	1960	9	21	594.260
1960	2	11	618.570	1960	6	13	655.850	1960	10	11	588.750
1960	3	3	612.050	1960	7	1	641.300	1960	10	31	580.360
1960	3	23	622.060	1960	7	22	609.870	1960	11	21	604.540
1960	4	12	626.500	1960	8	11	622.880	1960	12	12	611.940

a. Using steps similar to those in Exercise 2.1, find and print the coefficients of a least-squares straight line ("trend") using this data. (Hint: Change the dates to fractions of a year using a reasonable approximation.)

b. Plot the data points and the trend line on the same graph. Make the ordinate range [0, 1000] and extend the trend line beyond 1960 through the first half of 1961.

c. Also (on the same graph) plot the following closing values. Was the 1960 trend valid in the first half of 1961?

Year	Month	Day	Close	Year	Month	Day	Close
1961	1	3	610.250	1961	4	13	692.020
1961	1	23	639.820	1961	5	3	688.900
1961	2	10	639.670	1961	5	23	700.590
1961	3	3	671.570	1961	6	14	695.810
1961	3	23	675.450				

2.4 The intensity, I, of a certain extraterrestrial source is known to vary sinusoidally around a constant value, c_0, with a period of 1000 years. We have the following history:

Date	Intensity (I)	Date	Intensity (I)	Date	Intensity (I)
1/1/1900	167	1/1/1940	105	1/1/1980	42
1/1/1910	152	1/1/1950	89	1/1/1990	27
1/1/1920	136	1/1/1960	74	1/1/2000	11
1/1/1930	121	1/1/1970	58		

a. If time (t) is measured in years, what is the fundamental frequency (ω_0) of I in radians per year?

b. Using steps similar to those in Exercise 2.1, find c_0, c_1, and c_2, so that the function

$$\hat{f}(t) = c_0 + c_1 \sin \omega_0 t + c_2 \cos \omega_0 t$$

is a least-squares fit to the data.

c. Plot $\hat{f}(t)$ from the year 1900 through the year 2999, and plot the sample values on the same graph.

d. Explain why the Fourier coefficient formulas in Table 2.2 are not valid in this exercise.

2.5 Suppose we have two sample vectors, x and y, each with $N+1$ samples taken at $t = 0, T, 2T, \ldots, NT$ from the functions $x(t) = a(t - NT/2)$ and $y(t) = bx^2(t)$, where a and b are constants.

a. Write a program to show whether x and y are orthogonal for $N = 100$ and $N = 101$. For convenience, use $a = b = T = 1$.

b. Prove that x and y are orthogonal for any $N > 0$.

2.6 Prove that two different harmonic functions, $e^{-j2\pi mn/N}$ and $e^{-j2\pi kn/N}$, in the complex Fourier series are orthogonal with respect to the set of samples at $n = 0, 1, \ldots, N-1$.

2.7 Two theorems are important and helpful when applying the Fourier series to a sampled periodic function, say $x(t)$, when the samples are at $t = 0, T, \ldots, (N-1)T$. The first theorem is

> If $x(t)$ is an *even* function, that is, $x(nT) = x(-nT) = x\left((N-n)T\right)$, then the b_m coefficients are all zero, $c_m = -c_{-m}$ is real, and the Fourier series is a cosine series.

a. Assume N is even. In the formula for b_m in Table 2.2, show that the terms in the sum for $n = 0$ and $n = N/2$ are both zero.

b. Show that if $x(t)$ is even, $x_n = x_{N-n}$ for any value of n.

c. Using the condition given in part (b), separate the nonzero terms in the sum into two sums that cancel, thus proving the theorem.

2.8 The second theorem is similar to the first theorem in Exercise 2.7:

> If $x(t)$ is an *odd* function, that is, $x(nT) = -x(-nT) = -x\left((N-n)T\right)$, then the a_m coefficients are all zero, $c_m = -c_{-m}$ is imaginary, and the Fourier series is a sine series.

a. Assume N is even. In the formula for a_m in Table 2.2, show that, since $x_n = x_{-n}$, the terms in the sum for $n = 0$ and $n = N/2$ must both be zero.

b. Show that if $x(t)$ is odd, $x_n = -x_{N-n}$ for any value of n.

c. Using the condition given in part (b), separate the nonzero terms in the sum into two sums that cancel, thus proving the theorem.

2.9 The triangular waveform in Figure 2.1 is seen to have amplitude of 5.0 and period of 500.

a. Write a MATLAB expression to construct a row vector, x, of samples of the waveform $x(t)$ over exactly one period starting at $t = 0$ and using sampling interval $T = 0.5$.

b. Which of the theorems in Exercises 2.7 and 2.8 applies to this waveform?

c. Using the appropriate theorem, find and print the Fourier series coefficients for the first form of the series in Table 2.2 for $M = 6$.

d. Using the first form of the series in Table 2.2, plot exactly one cycle of the original waveform, $x(t)$, and the Fourier series approximation, $\hat{x}_n$ with $M = 6$, versus t on the same graph.

2.10 Repeat Exercise 2.9, but this time use $T = 25$ and do part (d) differently. Instead of plotting the original waveform, plot x as a set of discrete points. Then, plot the continuous approximation, $\hat{x}(t)$ in (2.32), using 2000 samples of t.

2.11 A periodic binary (two-valued) waveform has a fundamental period equal to 1 microsecond. One hundred samples are taken at $t = 0.00, 0.01, \ldots, 0.99$ microseconds. The sample vector x consists of 20 ones followed by 61 zeros followed by 19 ones.

a. Do either of the theorems in Exercises 2.7 and 2.8 apply to this vector? If so, which?

b. Using $M = 16$, compute the complex Fourier series coefficients, c_m in Table 2.2, for $m = 0, 1, \ldots, M-1$. Do the computed values agree with your answer to part (a)?

c. Plot x as a set of discrete points, and plot the continuous approximation, $\hat{x}(t)$ in (2.32), using 1000 samples of t.

2.12 The average monthly high and low temperatures (in Fahrenheit) in Albuquerque, NM, are as follows:

Month:	Jan	Feb	Mar	Apr	May	Jun	Jul	Aug	Sep	Oct	Nov	Dec
Avg. high:	46	53	61	70	79	90	92	89	81	71	57	47
Avg. low:	21	26	32	39	48	58	64	62	55	43	31	23

a. Do either of the theorems in Exercises 2.7 and 2.8 apply to this vector? If so, which?

b. Compute the Fourier series coefficients a_0, a_1, and b_1 in Table 2.2.

c. Plot each temperature vector as a set of discrete points, and plot each continuous approximation using (2.32) with 1000 samples of t. Make all plots on the same graph with the abscissa labeled in months, that is, Jan = 1, Feb = 2, and so on.

d. Comment on why the least-squares fit is as good when it is with $M = 2$. In other words, why is the average annual temperature variation nearly sinusoidal?

2.13 Complete the following:

a. Derive formulas similar to those in (2.35) for the continuous Fourier series in terms of sine and cosine functions. (Hint: Use the formula involving $e^{j\theta}$ in Chapter 1, Table 1.2.)

b. Give formulas similar to (2.36) for the Fourier series coefficients found in part (a).

References

Least squares

1. Giordano, A. A., and F. M. Hsu. 1985. *Least Square Estimation with Applications to Digital Signal Processing*. New York: John Wiley & Sons.

Fourier series

2. Bellanger, M. 1989. *Digital Processing of Signals*. 2nd ed. Chap. 1. New York: John Wiley & Sons.
3. Lyons, L. 1998. *All You Wanted to Know About Mathematics But Were Afraid to Ask*. Chap. 12. Cambridge, UK: Cambridge Press.

3

Correlation, Fourier Spectra, and the Sampling Theorem

3.1 Introduction

The word *spectrum* has several meanings. In signal processing, the spectrum of a signal is a particular type of mapping function. In the case of a waveform, the spectrum maps the signal from the *time domain* to the *frequency domain*, showing exactly how the signal content is distributed over frequency. In the case of an image, the spectrum maps the signal from the *spatial domain* of the image to a two-dimensional frequency domain. Sometimes, when we look at the signal in the frequency domain, the spectrum of a waveform or an image shows us things that we see less clearly or do not see at all in the time domain. It may, for example, bring out periodic components that are not clearly recognizable in the time domain.

In this chapter, we discuss the *Fourier spectrum* of a signal in terms of the *discrete Fourier transform* (DFT) and the *continuous Fourier transform*. (In Chapter 10 we discuss other transforms as well.) These two transforms allow us to obtain the spectral properties of continuous and discrete (sampled) signals.

The two Fourier transforms (FTs) also provide a means to prove the *Sampling Theorem*, which is basic to all digital signal processing (DSP) applications. The sampling theorem postulates how often a given continuous signal must be sampled to preserve all the information in the signal, so that we are able to compute the spectrum correctly and even reconstruct the continuous signal from the samples. The sampling theorem then allows us to discuss *aliasing*, a spectral phenomenon that occurs when a signal is not sampled often enough.

The final subject in this chapter is *waveform reconstruction*, which is a process of obtaining or estimating the original continuous waveform from a set of discrete samples by various methods.

Before we begin with these subjects, we introduce another operation known as *correlation*. Correlation has many uses in signal processing, but it is introduced here for one main reason: the discrete Fourier spectrum of a signal, $x(t)$, can be understood as a correlation of $x(t)$ with a set of sine and cosine functions.

3.2 Correlation

We begin the discussion of spectra by reviewing *correlation* as it applies to signal processing. Correlation is a measure used to tell how much two waveforms or images are like each other. We also use *autocorrelation* to tell how much a waveform at time t is like itself at some other time, $t + \tau$, or how much an image at location (x, y) is like itself at some other location, $(x + \sigma, y + \tau)$.

Suppose we have two waveforms, $x(t)$ and $y(t)$, defined in the range $t = [0, \infty)$. (This notation means the signals are defined from the lower limit to just before the upper limit.) Then, the correlation function of x and y is

$$\varphi_{xy}(\tau) = E\left[x(t)y(t+\tau)\right] = \lim_{L\to\infty}\frac{1}{L}\int_0^L x(t)y(t+\tau)dt; \quad \tau \geq 0 \tag{3.1}$$

where $E[\bullet]$ stands for the *expected value*, or *mean value*. The independent variable τ has the effect of shifting the time origin of y to an earlier time so that $y(t + \tau)$ coincides with $x(t)$, thus making the correlation function a continuous function of τ. For sampled waveforms, the correlation function is the discrete form of (3.1), that is,

$$\varphi_{xy}(m) = E[x_n y_{n+m}] = \lim_{L\to\infty}\frac{1}{L}\sum_{n=0}^{L-1} x_n y_{n+m}; \quad m \geq 0 \tag{3.2}$$

In this case, the time shift, m, is measured in terms of samples.

From (3.1) and (3.2), the correlation function, as defined here, is the *average product* of two signals. The product is averaged without an upper bound so that, as long as the time shift is positive, we do not need to be concerned with end effects. If the waveforms are defined only in the finite interval $[0, NT]$, then we usually make one of two assumptions:

1. The waveform values are zero outside the interval.
2. The waveform is periodic with period NT, as illustrated on the right in Figure 2.5.

An example of the correlation of two waveforms is given in Figure 3.1, in which $x(t)$ is the rectangular signal, and $y(t)$ is the triangular signal. Both signals are defined in the expressions for vectors x and y in the MATLAB® code (3.3). The MATLAB code computes the correlation function $\varphi_{xy}(m)$ using (3.2) with periodic extension, recognizing that the average product in (3.2) can be computed by averaging over one period. Note how the use of yy, the one-period extension of y, is used to prevent y_{n+m} in (3.2) from extending beyond the end of the vector.

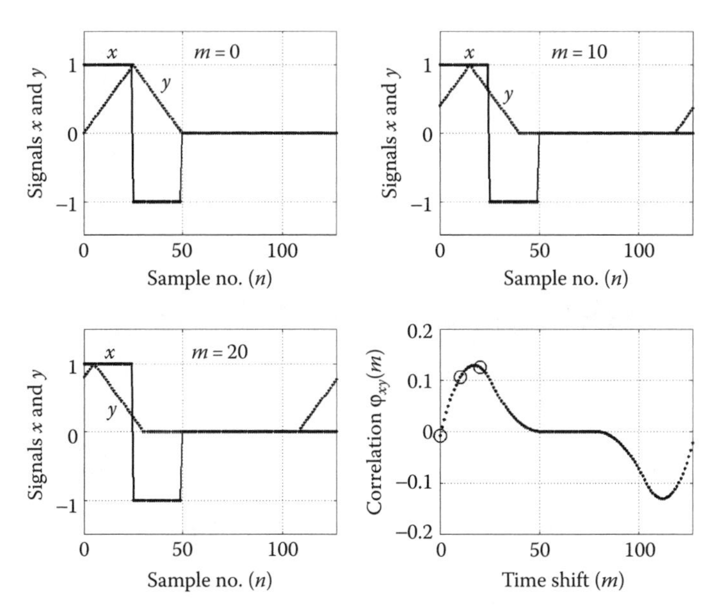

FIGURE 3.1
Signals $x(t)$ and $y(t + mT)$, with $m = 0$, 10, and 20, and correlation function $\varphi_{xy}(mT)$. Each circle on the correlation plot corresponds to one of the signal plots.

```
n=0:127;
% Waveform values with periodic extension.
x=[ones(1,25),-ones(1,25),zeros(1,78)];
y=[0:24,25:-1:1,zeros(1,78)]/25;
yy=[y,y];
% Correlation function.
for m=0:127
   phi(m+1)=(1/128)*x*yy(m+1:m+128)';
end
```
(3.3)

Figure 3.1 illustrates this operation by showing the two signals aligned first with $m = 0$, next with $m = 10$, and finally with $m = 20$. In each picture, x_n remains stationary, while y_{n+m} (with periodic extension) moves to the left. These three cases are circled in the plot of $\varphi_{xy}(m)$ in the lower right corner of Figure 3.1. In each case, you can observe how the value of $\varphi_{xy}(m)$ is the sum of the product of the two curves on the corresponding plot.

Correlation has important uses of its own in signal processing. For example, a *correlation detector* is a system that correlates an incoming signal in real time with a stored "target," the objective being to detect the presence of the target in the incoming signal. An example of correlation detection is given in Figure 3.2. The target signal is shown in the upper plot. The target signal appears again in additive random noise at sample numbers 1000 and 3700 in the center plot. The signal can be discerned in the center plot but is more easily detected in the lower plot, which is the correlation of the target signal with the center waveform.

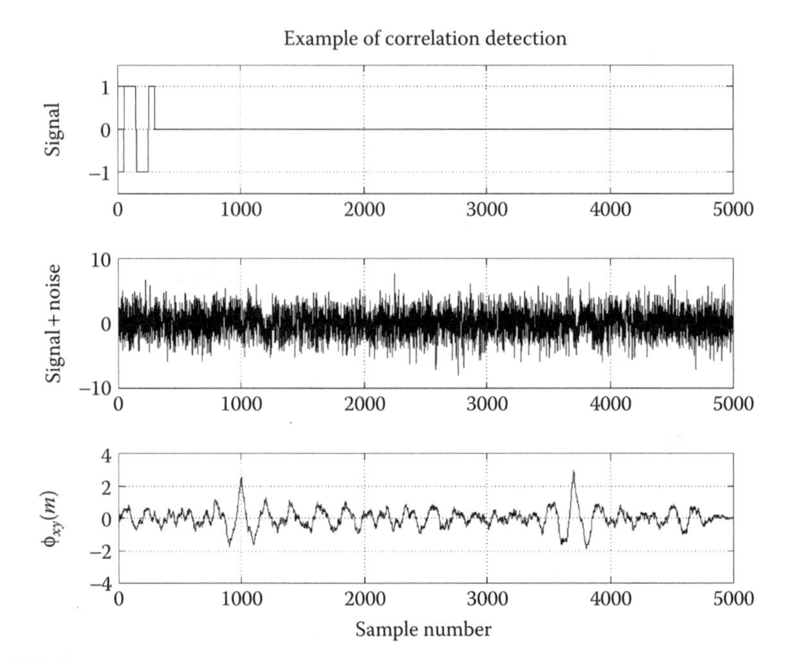

FIGURE 3.2
An example of correlation detection. The signal is evident more clearly in the correlation function (lower plot) than in the raw data (center plot).

3.3 The Discrete Fourier Transform (DFT)

The DFTs transform the sampled waveform or image data into spectral information about the respective waveform or image. We begin our discussion of DFTs with the complex Fourier series coefficients presented in Table 2.2, represented as follows:

$$c_m = c'_{-m} = \frac{1}{N}\sum_{n=0}^{N-1} x_n e^{-j2\pi mn/N}; \quad 0 \le m \le M-1 \tag{3.4}$$

Because $e^{-j2\pi mn/N} = \cos(2\pi mn/N) - j\sin(2\pi mn/N)$, and because the sum in (3.4) goes over n, we can observe that the complex Fourier coefficient in (3.4) resembles the finite correlation function in (3.2). In fact, if we define u_n and v_n as $u_n = \cos(2\pi mn/N)$ and $v_n = -\sin(2\pi mn/N)$, then, using the definition in (3.2), c_m is the sum of the two correlation functions and is written as

$$c_m = \varphi_{xu}(0) + j\varphi_{xv}(0) \tag{3.5}$$

Thus, recalling from (2.24) that the frequency corresponding to index m is

$$\omega_m = m\omega_0 = \frac{2\pi m}{NT} \tag{3.6}$$

we conclude that c_m is a measure of the correlation of the vector $x = [x_0\, x_1 \ldots x_{N-1}]$ with the unshifted sine and cosine vectors at the frequency ω_m. In this sense, c_m is a measure of the *spectral content* of x at ω_m radians per second.

The DFT is a scaled version of c_m, and is, therefore, a measure of the spectral content of x. However, the DFT has exactly N components; that is, M in (3.4) is equal to N. The N components of the DFT are

$$\boxed{X_m = \sum_{n=0}^{N-1} x_n e^{-j2\pi mn/N}; \quad \text{where } m = 0, 1, \ldots, N-1} \tag{3.7}$$

Thus,

$$X_m = Nc_m; \quad m = 0, 1, \ldots, M-1 \tag{3.8}$$

We will discuss the relative values of M and N later in Section 3.10, dealing with the sampling theorem.

As in Chapters 1 and 2, we use X without a subscript to represent the DFT vector with components $X_0 \; X_1 \; \ldots \; X_{N-1}$. Thus,

$$X = \begin{bmatrix} X_0 & X_1 & \cdots & X_{N-1} \end{bmatrix} \text{ is the DFT of } x = \begin{bmatrix} x_0 & x_1 & \cdots & x_{N-1} \end{bmatrix} \tag{3.9}$$

3.4 Redundancy in the DFT

It is easy to show that the indexes $m = 0, 1, \ldots, N-1$ in (3.7) are sufficient to supply *all possible* values of the DFT; in other words, values of the DFT outside this range of m are repetitive. Let K be any integer. Then, because $e^{\pm j2\pi nK} = 1$,

$$X_{m \pm KN} = \sum_{n=0}^{N-1} x_n e^{-j2\pi(m \pm KN)n/N} = \sum_{n=0}^{N-1} x_n e^{-j2\pi m/N} e^{\pm j2\pi nK} = X_m \tag{3.10}$$

Thus, the DFT is a periodic complex function of m with a period N, and its values simply repeat outside the foregoing range of m.

Furthermore, when all the elements of the sample vector are real, essentially only half of the components of the DFT are independent. Again, because $e^{-j2\pi n} = 1$ for any value of n, we have

$$X_{N-m} = \sum_{n=0}^{N-1} x_n e^{-j2\pi(N-m)/N} = \sum_{n=0}^{N-1} x_n e^{j2\pi m/N} = X'_m \tag{3.11}$$

where X'_m is the complex conjugate of X_m, as in Chapter 2, Table 2.1. Thus, when the samples (x_n) are real, all the independent values of X_m are in the following range:

$$m = 0, 1, \ldots, \frac{N}{2} \tag{3.12}$$

To illustrate the DFT and the properties we have discussed so far, our next example uses the sample vector x in the previously applied example of a Fourier series on page 30, which is shown on the left in Figure 2.5 and again on the left in Figure 3.3. The waveform vector, x, and the DFT vector, X, are computed as follows:

```
% Chapter 3 Example: Discrete Fourier Transform.
% Index vectors, n and m.
N=50;
n=[0:N-1];
m=[0:N-1];
% Sample vector, x.
x=[zeros(1,28),ones(1,12),zeros(1,N-40)];
% DFT, X.
X=x*exp(-j*2*pi*m'*n/N);
```
(3.13)

Notice again, as in (2.31) of Chapter 2, the index vectors (n and m) are multiplied to produce a square matrix, so that the computation of "X" produces all the N values of X_m shown in (3.7).

The signal vector, x, is plotted on the left in Figure 3.3. The magnitude of the DFT, $|X_m|$, is plotted on the right. As shown in (3.10), the DFT is periodic outside the range shown. Furthermore, the result in (3.11) is illustrated, because we are plotting the *magnitude* of the DFT, $|X_{N-m}| = |X'_m| = |X_m|$, and so the left and the right halves of the DFT plot are mirror images.

A final item to note in the plot of $|X_m|$ in Figure 3.3 is the value of X_0, which, according to (3.7), is the sum of all the elements in the sample vector x. Because there are 12 ones in x and all the remaining elements are zeros, we have $X_0 = 12$.

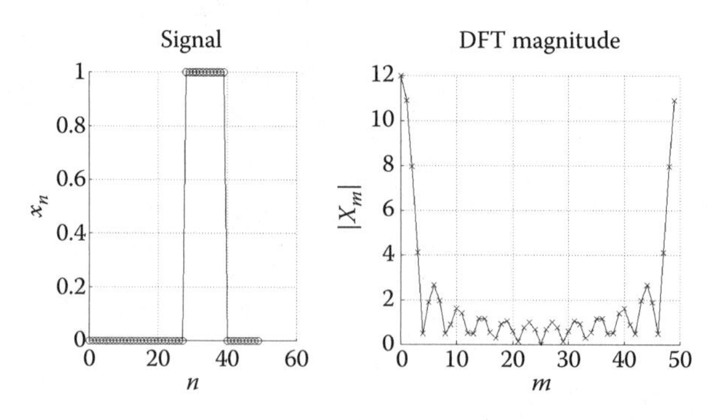

FIGURE 3.3
Signal and discrete Fourier transform magnitude in (3.13).

3.5 The FFT Algorithm

The fast Fourier transform (FFT) is not a new transform. It is an algorithm for producing the DFT exactly as defined in (3.7), but with fewer complex products. The FFT appears in different forms in DSP hardware and software. In the literature,[1–8] the operation and structure of the FFT are described, in addition to the interesting history of its development. Our purpose here is only to observe that the computation of the DFT, as expressed in (3.7), contains redundant complex products (products that must be essentially repeated during the computation of the DFT) and that repetitions of these products may be eliminated to produce a faster computation.

If (3.7) is used to compute the DFT in the form shown herein, there are N complex products in the sum and N sums in the transform; accordingly, there are N^2 products in all. Let us now express the DFT in the following form:

$$X_m = \sum_{n=0}^{N-1} x_n W_N^{mn}; \quad m = 0, 1, \ldots, N-1, \quad \text{where } W_N = e^{-j2\pi/N} \tag{3.14}$$

Thus, W_N is a unit vector on the complex plane, and W_N^k is periodic over k with a period N; that is, in MATLAB notation, $W_N^k = W_N^{\text{rem}(k,N)}$, where rem($k$, N) is the remainder after dividing k by N.

Redundant values of W_N^k are illustrated with $N = 8$ in Figure 3.4. With $N = 8$, the product $k = mn$ in the DFT represented in (3.7) ranges from 0 through 49. However, only eight of these products are unique, as shown in the figure. Using this redundancy in W_N^k, the construction of a fast DFT algorithm begins by decomposing the sample vector. Suppose N is a multiple of 2. Then, we decompose the samples into two vectors containing even- and odd-numbered samples, and we write (3.14) as follows:

$$X_m = \sum_{n=0}^{N/2-1} x_{2n} W_N^{2mn} + W_N^m \sum_{n=0}^{N/2-1} x_{2n+1} W_N^{2mn}; \quad m = 0, 1, \ldots, N-1 \tag{3.15}$$

In the definition of W_N^k in (3.14), $W_{N/2}^k = W_N^{2k}$ for any value of k. Thus, (3.15) becomes

$$X_m = \sum_{n=0}^{N/2-1} x_{2n} W_{N/2}^{mn} + W_N^m \sum_{n=0}^{N/2-1} x_{2n+1} W_{N/2}^{mn}; \quad m = 0, 1, \ldots, N-1 \tag{3.16}$$

Although this result appears a rather complicated way to express the DFT, it forms the basis for fast algorithms, because each sum in (3.16) is a DFT with size $N/2$ rather than N, requiring $(N/2)^2$ instead of N^2 products. Each smaller DFT is periodic in m with period $N/2$, and hence, two periods are contained in the range $0 \le m < N$. In addition, there are $N/2$ unique outer products in

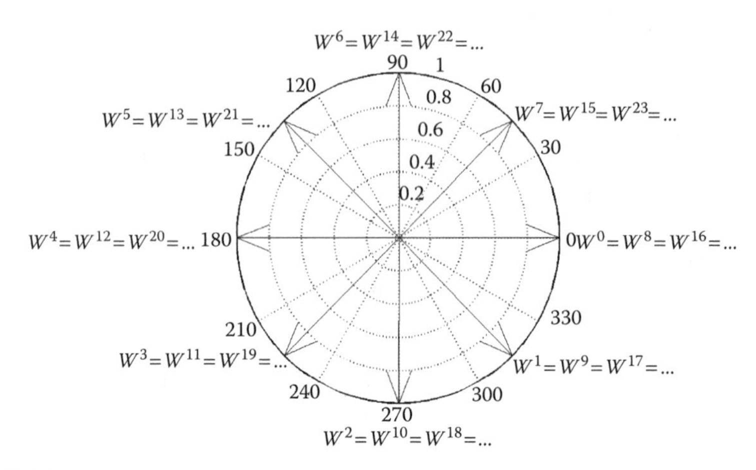

FIGURE 3.4
Repeating values of W^N with $N = 8$, showing that only N values are unique.

the right-hand term, because, as illustrated in Figure 3.4, $W_N^k = -W_N^{k-N/2}$ when $k \geq N/2$. Thus, (3.16) expresses a total of $2(N/2)^2 + N/2 = N^2/2 + N/2$ complex products, which are fewer than the N^2 products in (3.14).

Suppose further that N is not only a multiple of 2 but also a *power* of 2, say $N = 2^K$. Then, each of the sums in (3.16) can be decomposed as we have just described, and (3.16) can be written as the sum of four sums, and so on, until after K such steps, there are N sums with only one term in each sum. In this process, we have added $N/2$ complex products at each stage and finally produced N sums that contribute $N/2$ additional nonredundant products. Thus, if $N = 2^K$ and (3.14) is decomposed iteratively using (3.16) to produce the FFT, the DFT computation is reduced as follows:

$$\begin{aligned} &\text{Complex products in DFT} = N^2 \\ &\text{Complex products in FFT} = \frac{N}{2}K = \frac{N}{2}\log_2 N \end{aligned} \tag{3.17}$$

FFT algorithms are most efficient when N is a power of 2, which is assumed in (3.17), because then the number of decompositions ($\log_2 N$) is the maximum. However, the process just described works when N has any prime factors other than itself. For example, suppose N is a multiple of three. Then the equivalent version of (3.16) is as follows:

$$X_m = \sum_{n=0}^{N/3-1} x_{3n} W_{N/3}^{mn} + W_N^m \sum_{n=0}^{N/3-1} x_{3n+1} W_{N/3}^{mn} + W_N^{2m} \sum_{n=0}^{N/3-1} x_{3n+2} W_{N/3}^{mn}; \tag{3.18}$$

$$m = 0, 1, \ldots, N-1$$

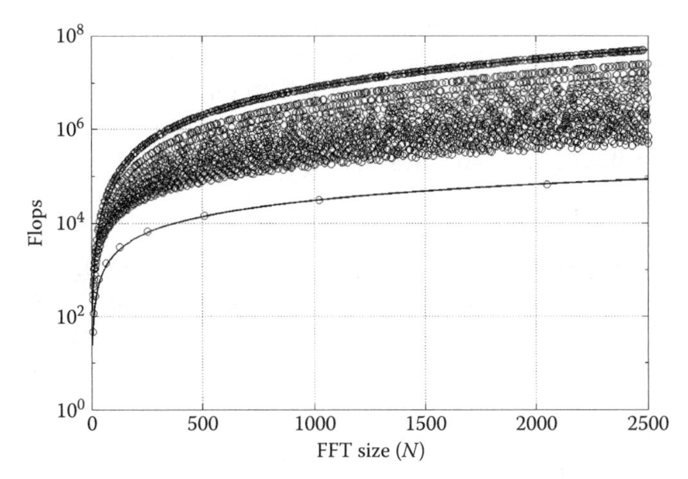

FIGURE 3.5
Floating-point operations (flops) as a function of the fast Fourier transform size (N). Lower boundary plot is proportional to $(N/2)\log_2 N$; upper boundary is proportional to N^2.

Here, we see that there are $(N/3)^2$ products in each reduced DFT, plus N outer products in the second and third terms. Therefore, the original N^2 products are reduced to $3(N/3)^2 + 2N$ (that is, $N^2/3 + 2N$) products in this case.

FFT algorithms have different specific forms in software and hardware, but they are all based on decomposing the DFT iteratively into successively smaller DFTs, as illustrated in (3.16) and (3.18). Robust FFT algorithms, such as the MATLAB *fft* function, take advantage of all prime factors of the signal vector size (N) to reduce the number of complex products.

Figure 3.5 illustrates the results of computation of $f(N)$, the number of floating-point operations required for an FFT of size N, for a range of values of N. All values of $f(N)$ for $4 \le N \le 2500$ are plotted as circles. In addition, two boundary curves are plotted. The upper plot is of a constant times N^2, and the lower one is of the same constant times $(N/2)\log_2 N$.

The measure (number of floating-point operations) involves more than just complex products; nevertheless, the values of $f(N)$ conform to the theory, as shown in Figure 3.5. Notice that when N is a power of 2, $f(N)$ lies close to the lower bound. On the contrary, when N is a prime number and has no factors other than itself, $f(N)$ is on the upper bound.

3.6 Amplitude and Phase Spectra

We turn now from the computation of the DFT to its use in spectral analysis. We introduce the subject in this chapter and continue it in Chapter 7. We begin with a discussion of the *amplitude spectrum* and the *phase spectrum* of a signal vector. These are easily defined. The amplitude spectrum is the vector with elements that are amplitudes of the DFT components, and the phase spectrum is the vector of DFT-component phase angles in radian units. Thus, if $X = [X_0, X_1, \ldots, X_{N-1}]$ is the DFT of a signal vector $x = [x_0, x_1, \ldots, x_{N-1}]$, then

$$\begin{aligned} \text{Amplitude spectrum} &= \text{abs}\,(X) = \begin{bmatrix} |X_0| & |X_1| & \cdots & |X_{N-1}| \end{bmatrix} \\ \text{Phase spectrum} &= \text{angle}\,(X) = \begin{bmatrix} \angle X_0 & \angle X_1 & \cdots & \angle X_{N-1} \end{bmatrix} \end{aligned} \quad (3.19)$$

When these DFT functions are applied to measured data, that is, samples of a continuous signal, each component is associated with a corresponding frequency. Frequency is usually measured in one of four systems of units. It is important to get these units right in one's mind, because they are all used in the literature, and they cause a surprising amount of confusion among students and even engineers, who should know them.

All the four frequency scales are illustrated below the DFT index (m) in the example in Figure 3.6, which is a plot of the amplitude spectrum of

$$x = \left[\frac{\sin[2\pi(n - 50.5)/5]}{n - 50.5}; \quad n = 0, 1, \ldots, 99 \right] \quad (3.20)$$

It is interesting to note that the spectral content of x is mostly below the index $m = 20$; however, we leave this point for the present and concentrate on the frequency scales below the figure.

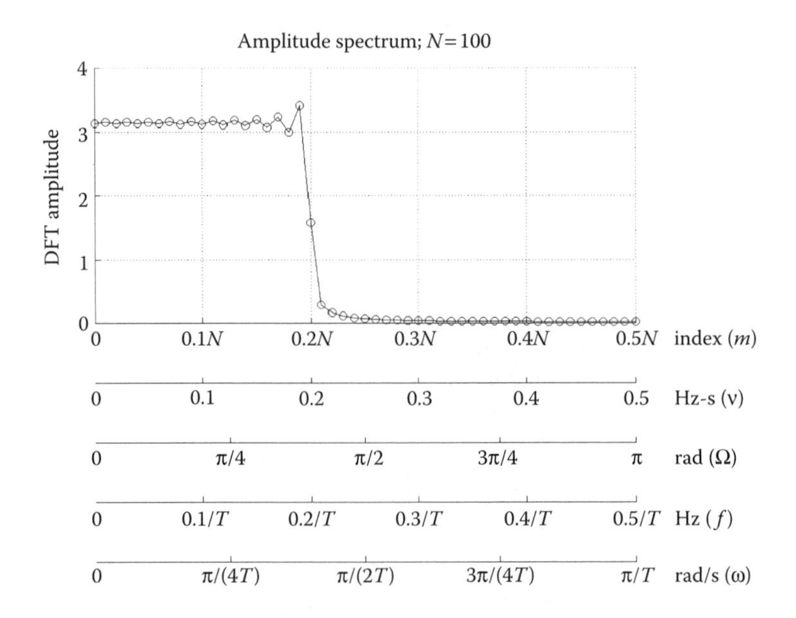

FIGURE 3.6
Amplitude spectrum of $X = \sin[2\pi(n - 50.5)/5]/(n - 50.5)$; $n = 0,1,\ldots,99$, illustrating different frequency scales.

First, note that the range of the DFT index (m) is $[0, N/2]$. As shown in (3.10) and (3.11), the amplitude spectrum simply repeats outside this range. That is, from $m = N/2$ to N, the amplitude spectrum is the mirror image of Figure 3.6, and outside the range $m = [0, N-1]$, the DFT is periodic. Thus, in the range $m = [0, N/2]$, we have the complete amplitude spectrum.

The first two frequency scales below the index are dimensionless. These are based on one rotation of the unit vector $e^{j2\pi m/N}$ on the complex plane, as shown previously. The Hertz-second (Hz-s) scale is defined as m/N, and the radian (rad) scale is defined as $2\pi m/N$.

The last two frequency scales in Figure 3.6 are used with real data, where the time step T between samples is given in seconds (s). The Hertz (Hz) scale is defined as m/NT, and the radians per second (rad/s) scale is defined as $2\pi m/NT$. Parentheses in these expressions as well as others in Tables 3.1 and 3.2 are omitted to simplify them; for example, "m/NT" stands for "$m/(NT)$", etc.

The symbol used commonly to represent each frequency measure is shown at the right of each scale in Figure 3.6: ν for Hertz-second, Ω for radian, and so on. All five scales are used in practice to describe frequency in whatever units appear convenient to the situation; hence, the DSP engineer must be able to move easily from one scale to another. Table 3.1 relates the scales by providing factors for all possible conversions. Note that each pair of symmetric table elements above and below the main diagonal represents a reverse conversion, and the factors are, therefore, inverses.

Two other important features of the frequency scales—the *sampling frequency* and the DFT *frequency increment*—are given in Table 3.2 for the frequency index and for each frequency scale. The DFT index N designates the sampling frequency in the sense that, with $m = N$ in the DFT formula (3.7), the complex sinusoid $e^{-j2\pi mn/N}$ goes through a cycle each time the sample number (n) is increased by one. Given $m = N$ at the sampling frequency, the remaining sampling frequencies are found by using the top row of conversions in

TABLE 3.1

Frequency-Conversion Factors

	To				
From	**Index (m)**	**Hz-s (ν)**	**rad (Ω)**	**Hz (f)**	**rad/s (ω)**
Index (m)	M	m/N	$2\pi m/N$	m/NT	$2\pi m/NT$
Hz-s (ν)	νN	ν	$2\pi\nu$	ν/T	$2\pi\nu/T$
rad (Ω)	$N\Omega/2\pi$	$\Omega/2\pi$	Ω	$\Omega/2\pi T$	Ω/T
Hz (f)	fNT	fT	$2\pi fT$	f	$2\pi f$
rad/s (ω)	$N\omega T/2\pi$	$\omega T/2\pi$	ωT	$\omega/2\pi$	ω

Example: Ω (Frequency in radians) $= 2\pi f T$, where f = frequency in Hertz.

TABLE 3.2
Sampling Frequencies and Increments in Discrete Fourier Transform Frequencies

	Index (m)	Hz-s (ν)	rad (Ω)	Hz (f)	rad/s (ω)
Sampling frequency	N	1	2π	$1/T$	$2\pi/T$
Frequency increment	1	$1/N$	$2\pi/N$	$1/NT$	$2\pi/NT$

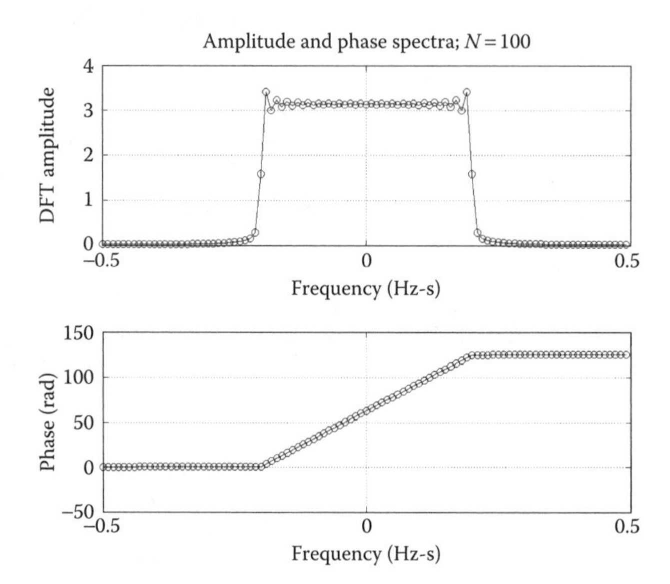

FIGURE 3.7
Amplitude and phase spectra of $X = \sin[2\pi(n-50.5)/5]/(n-50.5)$; $n = 0,1,\ldots,99$.

Table 3.1. The frequency increments in Table 3.2 are determined simply by dividing each sampling frequency by N, the number of frequency increments from zero up to the sampling frequency. Because the index increment is one, the frequency increments are identical with the conversion factors in the first row of Table 3.1.

Finally, Figure 3.7 is provided to illustrate the phase spectrum of the signal vector in (3.20), together with the amplitude spectrum. Here, the complete frequency range from −0.5 to + 0.5 Hz-s (one period of the DFT) is shown to illustrate the redundancy in (3.11), that is, $X_{N-m} = X'_m$, which implies that the amplitude spectrum is evenly symmetric about the origin and that the phase spectrum has an odd symmetry about the origin. Amplitude and phase plots are often shown in this form.

The amplitude and phase spectra in Figure 3.7 may be computed using the MATLAB code shown below. The expressions are self-explanatory, with the exception of the *fftshift* and *unwrap* functions. The *fftshift* function swaps the upper and lower halves of the DFT vector (X). Thus (assuming N is even), the first $N/2$ DFT components cover [0.5, 1) Hz-s or, equivalently, [–0.5, 0] Hz-s (refer to Equation 3.10), and the complete frequency range is therefore [–0.5, 0.5), as shown.

The *unwrap* function is used to "unwrap" the output of the *angle* function. *Angle* computes the phase in radians of each DFT component using the *atan2* arctangent function, which resolves all angles into the range [–π, π] radians. The *unwrap* function adds multiples of 2π to eliminate "jumps" from –π to π or vice versa in the progression of phase elements.

```
n=0:99;
x=sin(2*pi*(n-50.5)/5)./(n-50.5);
X=fftshift(fft(x));
amplitude=abs(X);
phase=unwrap(angle(X));
```
(3.21)

3.7 The Inverse DFT

The DFT in the form of (3.7) may be viewed as a mapping in N dimensions of a signal vector, x, into a DFT vector, X. Each vector has N components. Only half plus one of the DFT components are independent, but half minus one are complex and have two independent components each (X_0 and $X_{N/2}$ are the exceptions). Therefore, x and X have equal degrees of freedom, and an inverse mapping is possible.

The inverse mapping may be found by multiplying X_m in (3.7) by $e^{j2\pi mn/N}$ and summing over m. [Note that because n appears in the left-hand sum, we use k instead of n on the right when (3.7) is substituted for X_m.]

$$\sum_{m=0}^{N-1} X_m e^{j2\pi mn/N} = \sum_{m=0}^{N-1}\sum_{k=0}^{N-1} x_k e^{j2\pi m(n-k)/N} = \sum_{k=0}^{N-1} x_k \sum_{m=0}^{N-1} e^{j2\pi m(n-k)/N} = Nx_n \quad (3.22)$$

Once again, this result is an application of (1.21), where the geometric series on the right in (3.22) adds up to zero for all values of k, except $k = n$, where the sum is N. Therefore, we can now present the DFT as a symmetric pair as follows:

$$\boxed{\begin{aligned} X_m &= \sum_{n=0}^{N-1} x_n e^{-j2\pi mn/N}; \quad m = 0,1,\ldots,N-1 \\ x_n &= \frac{1}{N}\sum_{m=0}^{N-1} X_m e^{j2\pi mn/N}; \quad n = 0,1,\ldots,N-1 \end{aligned}} \quad (3.23)$$

Thus, the inverse transformation from X back to x is the same as the forward transform, except for a change of sign in the exponent and the scaling factor $1/N$. The MATLAB *ifft* function implements the inverse transformation as shown, again using the fast algorithm described in Section 3.5, with the scaling factor included, so that, for example, the expression "*y* = *ifft*(*fft*(*x*))" would (neglecting roundoff errors) be the same as "*y* = *x*."

3.8 Properties of the DFT

Besides the inverse transform, the DFT has other properties worth mentioning. In addition to the properties described in Sections 3.4 and 3.6, four additional properties are summarized in Table 3.3. In this section, we discuss these properties. If the reader is already familiar with them, he or she may wish to skip the discussion.

The *periodicity* and *redundancy* properties of the DFT are given by (3.10) and (3.11). Note that if the signal vector x is complex, the redundancy property does not hold.

The *linearity* property follows from the DFT definition expressed in (3.7). If $ax_n + by_n$ is substituted for x_n in (3.7), we can write the sum in (3.7) as the sum of a times the DFT of x plus b times the DFT of y, thus proving the property. We can apply the linearity property to the DFT of a complex signal vector as follows. If $x = u + jv$, then the DFT of x must be the complex sum of two DFTs of real vectors, that is, $X = U + jV$. Because this is true, it is usually sufficient to use real signal vectors in our discussions of the DFT.

Property 4 states that when x is real, X_0 is the sum of the elements of x, and $X_{N/2}$ is the alternating sum of the elements. As discussed in Section 3.7, this property results in the DFT of a real signal vector having exactly N degrees of freedom in the inclusive index range $[0, N/2]$.

Properties 5 through 7 in Table 3.3 are interesting and useful characteristics of the DFT. Properties 5 and 7 are, in a sense, duals of each other. Each causes a form of *interpolation* in its corresponding transform domain.

TABLE 3.3

Properties of Discrete Fourier Transforms (DFTs):
$x = [x_0, x_1, \ldots, x_{N-1}]$; $X = \text{DFT}(x) = [X_0, X_1, \ldots, X_{N-1}]$

1. Periodicity	$X_{m \pm KN} = X_m$
2. Redundancy	$X_{N-m} = X'_m$, that is, $X_{-m,} = X'_m$, if x is real
3. Linearity	If $s = ax + by$, then $S = aX + bY$
4. Two real elements	$X_0 = x_0 + x_1 + \cdots + x_{N-1}$ and $X_{N/2} = x_0 - x_1 + x_2 - \cdots \pm x_{N-1}$ are real, if x is real
5. Zero extension	If K zeros are appended to x, $\Delta\nu$ decreases to $\Delta\nu = 1/(N+K)$ Hz-s
6. Zero insertion	If K zeros are inserted after each sample in x, the discrete Fourier transform repeats K times
7. Resampling	By inserting zeros properly into X, the time step in x can be changed
8. Linear phase shift	If x_{n-k} is substituted for x_n, $2\pi km/N$ radians are subtracted from angle (X_m)

First, we consider *zero extension in the time domain* (Property 5). Suppose we form a vector y by appending K zeros to a row vector, x. The MATLAB expression for doing this would be $y = [x, zeros(1, K)]$. (The *zeros* function in this case creates a row of K zeros as in (3.3), etc.) We now find the DFT of y using (3.7). Noting that the last K elements of y are zeros, the result is

$$Y_m = \sum_{n=0}^{N+K-1} y_n e^{-j2\pi mn/(N+K)} = \sum_{n=0}^{N-1} x_n e^{-j2\pi mn/(N+K)}; \quad m = 0, 1, \ldots, N+K-1 \tag{3.24}$$

One way to interpret this result is to say that the DFT of y is the same as the DFT of the original vector x, but interpolated in frequency, that is, given at frequencies that are more closely spaced. For example, suppose $K = N$ so that the length of y is twice the length of x. Then, it is easy to conclude from (3.24) that $Y_{2m} = X_m$, that is, every other value in Y is an original DFT component, and the in-between values are "interpolated" DFT components.

An example of this DFT with $K = 3N$ is given in Figure 3.8. The two signals and DFTs are produced with the following MATLAB expressions:

```
N1=32; N2=128;
x=[ones(1,N1/2),-ones(1,N1/2)];
X=fft(x,N1);
Y=fft(x,N2);
```

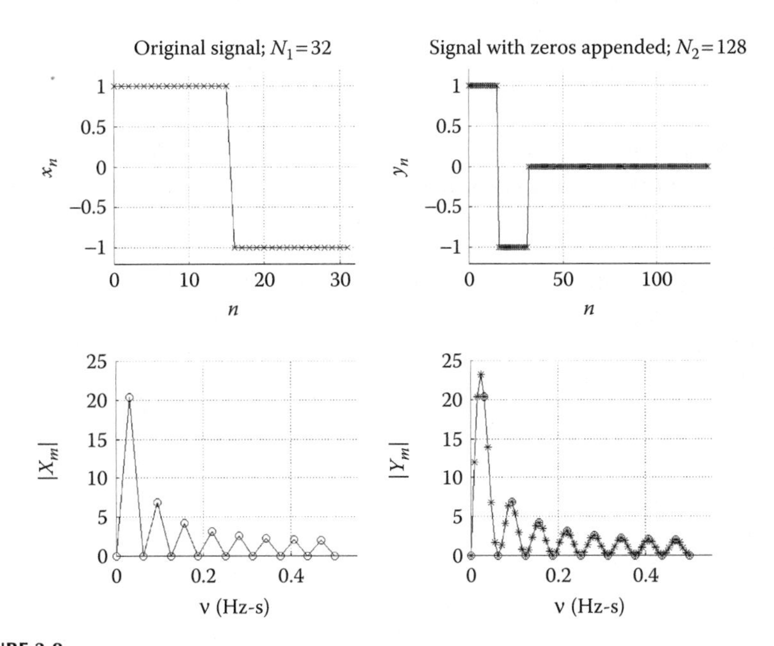

FIGURE 3.8
Zero extension in time resulting in interpolation in frequency. With $N_2 = 4N_1$, there are four times as many points in the interpolated spectrum.

The original signal is shown at the top left with $N_1 = 32$, and again with 96 zeros appended ($N_2 = 128$) at the top right. The original amplitude spectrum is plotted at the bottom left, and the interpolated version is plotted at the bottom right. In the latter, note that the original points (circled) coincide with the left-hand spectrum and that three interpolated points are present between each pair of original points. Notice also that the "Hz-s" frequency scale is used to emphasize that the frequency *range* does not change.

This effect of zero extension in the time domain leads to a *continuous* version of the DFT, because there is theoretically no limit to the number of appended zeros and, therefore, no limit to the amount of interpolation. That is, if we substitute ω (radians per second) into (3.7) in accordance with Table 3.1, we obtain a continuous form of the DFT:

$$\text{Continuous DFT:}\quad X_m = \sum_{n=0}^{N-1} x_n e^{-j\omega nT};\quad 0 \le \omega < 2\pi/T;\quad m = \frac{N\omega T}{2\pi} \tag{3.25}$$

The question of whether the continuous DFT gives a true representation of the spectrum of a sampled continuous signal, $x(t)$, must remain until we discuss the *sampling theorem* (Section 3.10) in this chapter.

Insertion of zeros between the samples of a signal vector in the time domain also has an interesting effect on the DFT. Suppose K zeros are inserted after each sample in x to produce y, so that $y_k = x_n$ when $k = (K+1)n$; and $y_k = 0$ when k is not a multiple of $K+1$. Then, the DFT of y is

$$Y_m = \sum_{k=0}^{(K+1)N-1} y_k e^{-j\frac{2\pi mk}{(K+1)N}} = \sum_{n=0}^{N-1} x_n e^{-j\frac{2\pi mn}{N}} = X_m;\quad m = 0,1,\ldots,(K+1)N-1 \tag{3.26}$$

We get the second sum by using the expression $y_k = x_k$, when $k = (K+1)n$; and $y_k = 0$, otherwise. Thus, the DFT of x with zeros inserted is a succession of $K+1$ DFTs of x. Now, suppose x is a sample vector with $T_1 = T$ seconds between samples. With zeros inserted in x, the interval is then reduced to $T_2 = T/(K+1)$, and the corresponding sampling frequencies are $f_1 = 1/T$ and $f_2 = (K+1)/T$ Hz. This, together with the result in (3.26), suggests the *time-domain interpolation* scheme illustrated in Figure 3.9, which will in turn take us to Property 7 in Table 3.3.

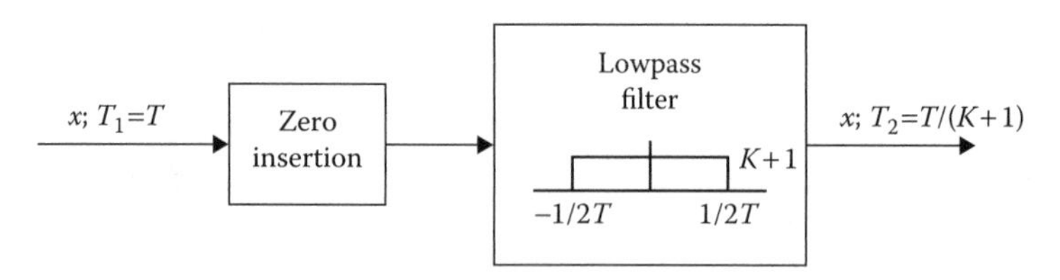

FIGURE 3.9
Interpolation between samples of vector x by inserting K zeros between the samples and then removing all spectral content above $1/(2T)$ Hz.

In Figure 3.9, an ideal lowpass filter is used following the insertion of K zeros after each sample in x. The filter has the effect of limiting the DFT of y, so that it contains only the components of the first of the sequence of DFTs in (3.26), thus creating an *interpolated* version of x; in other words, this vector has $K+1$ times as many samples of x and a proportionately smaller time step (T_2), and yet, it has the same spectrum as the original x and the same values as at the original sample points of x.

The entire process in Figure 3.9 can be accomplished in the frequency domain simply by expanding X, the DFT of x, with zeros, and effectively creating Y in (3.26) without any frequencies above $1/(2T)$ Hz, that is, without any nonzero components at indexes above $m = N/2$. This we do by inserting zeros in the *middle* of X, as in the following MATLAB expressions:

```
X=fft(x);
Y=[X(1:(N+1)/2),zeros(1,K*N),X((N+1)/2+1:N)];
y=(K+1)*ifft(Y);
```
(3.27)

Here, we assume that N (the number of elements in x) is odd; so, $X(2{:}N)$ can be split into equal halves, and there is no component at $m = N/2$. (If N is even, $X_{N/2}$ must be split into two halves on either side of the zeros and the number of zeros reduced to KN–1.) Notice that Y, the DFT of y, now has $(K+1)N$ components and no frequencies above $1/(2T)$ Hz. Thus, when the inverse transform is taken, y contains K interpolated samples after each original sample in x, no frequencies above $1/(2T)$ Hz, and a smaller time step, $T_2 = T/(K+1)$. The inverse transform is multiplied by $K+1$, because the factor ahead of the sum in (3.23) has changed from $1/N$ to $1/((K+1)N)$.

An example of this interpolation process with $N = 17$ and $K = 2$ is shown in Figure 3.10. The original signal vector (x) is at the upper left with time step at $T_1 = 3$ milliseconds. The amplitude spectrum, $|X|$, is shown on the upper right and on the lower left with zeros inserted. The spectra are shown using the *fftshift* function described previously to emphasize the frequency content, but the computations are done exactly as in (3.27). Finally, the interpolated vector (y) is plotted on the lower right, with x overplotted to show that the samples of x and y coincide. Again, the properties of (y) are that it occupies essentially the same time frame as the original vector, the time step has been decreased to $T_2 = 1$ milliseconds, and the spectral content has not changed from that of the original vector.

At this point, it is natural to ask whether increasing K indefinitely in the process just described would allow one to recover the original continuous waveform from which the samples in x were taken. And the answer, as we shall find, is "yes," provided the *sampling theorem* (Section 3.10) holds for the vector x.

The final DFT property, Property 8 in Table 3.3, is called *linear phase shift*. The simplest way to illustrate this property is to assume that there is a *signal component* given by

$$x = [x_n; \quad 0 \le n < N]$$

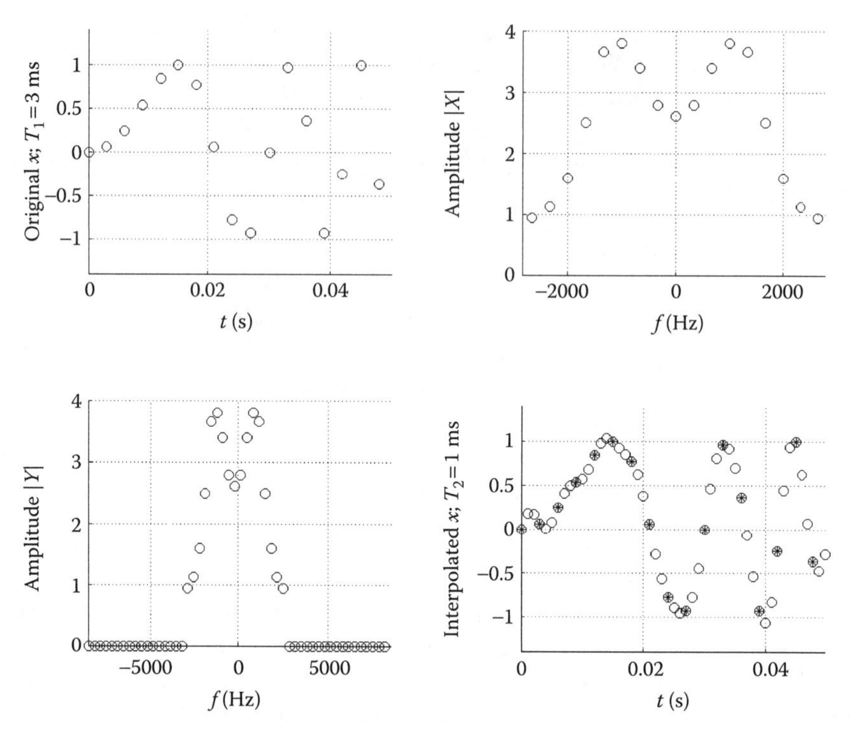

FIGURE 3.10
Interpolation in the time domain by expanding the discrete Fourier transform with zeros in the frequency domain. $N = 17$ and $K = 2$ in this example. The process is equivalent to that of Figure 3.9.

within a long vector of length $L >> N$; and a time-shifted version of x given by

$$y = [y_n;\quad 0 \le n < N] = [x_{n-k};\quad k \le n < N + k],$$

such that x and y are entirely within the long vector. We now examine the contribution of y to the DFT of the long vector using (3.7).

$$Y_m = \sum_{n=0}^{L-1} x_{n-k} e^{-j2\pi mn/L} = \sum_{i=0}^{L-1} x_i e^{-j2\pi mi/L} e^{-j2\pi mk/L} = X_m e^{-j2\pi mk/L};\quad 0 \le m < L \quad (3.28)$$

The second sum is found by substituting the expression $i = n - k$ and maintaining the same limits, because the indexes of x and y are, by definition, within the limits $0 \le i < L$. Thus, we have shown that delaying a signal component by k samples causes no change in the component's amplitude spectrum and subtracts a *linear phase* component given by the term $2\pi mk/L$, where L is the size of the transform, from the phase spectrum.

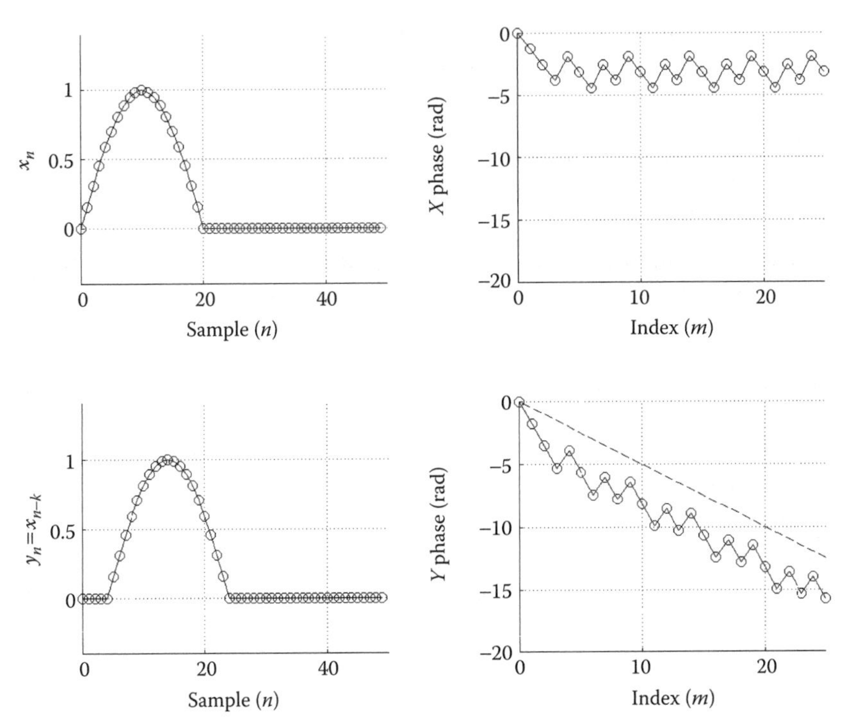

FIGURE 3.11
Illustration of linear phase shift. Signal samples $[x_n]$ are delayed to produce $[y_n]$ with $y_n = x_{n-k}$; $0 \le n < N$. The delay does not alter the amplitude spectrum but subtracts $2\pi mk/N$ rad from the phase spectrum of x. The dashed line shows this linear phase component. $N = 50$ and $k = 4$ in this illustration.

A linear phase shift is illustrated in Figure 3.11 using a transform size of $L = 50$ and a delay of $k = 4$ samples. The signal component, x, is on the upper left, and its shifted version, y, is on the lower left. The respective phase spectra are on the right. In this example, the linear phase component, plotted at the lower right as a dashed line, is $-2\pi km/L = -0.50m$ rad.

3.9 Continuous Transforms, Linear Systems, and Convolution

The focus of DSP is on vectors and arrays containing elements taken from discrete sample spaces. However, to keep in touch with reality, engineers must have ways to relate these vectors and arrays of discrete measured samples to their origin in the continuous real world, that is, to their corresponding continuous waveforms and images. This leads first to a discussion of the FT, which we introduce as a continuous, limiting version of the DFT. The FT of a continuous waveform, $x(t)$, is represented as

$$\boxed{\text{Fourier Transform:} \quad \text{FT}\{x\} \equiv X(j\omega) = \int_{-\infty}^{\infty} x(t)e^{-j\omega t}dt} \tag{3.29}$$

where ω is the continuous frequency in radians per second, and t is the time in seconds. (The use of X here for the FT raises the possibility of confusion with its use to represent the DFT vector, but this is conventional; hence, we must remember that the argument $(j\omega)$ implies the FT rather than the DFT. Usually, the implication comes from the context in which X is used anyway.)

To relate the FT to the DFT, we define another continuous function of time, $\bar{x}(t)$, as a semi-infinite sequence of impulses equal to $x(t)$ at $t = 0,T,2T,\ldots$ given by

$$\bar{x}(t) = \sum_{n=0}^{\infty} x(t)\delta(t - nT); \quad 0 \le t < \infty \tag{3.30}$$

where $\delta(t - nT)$ is the unit impulse at $t = nT$. Then, the FT of $\bar{x}(t)$ is

$$\text{FT}\{\bar{x}\} = \int_{0}^{\infty} \sum_{n=0}^{\infty} x(t)\delta(t - nT)e^{-j\omega t}dt = \sum_{n=0}^{\infty} x(nT)e^{-j\omega nT} \tag{3.31}$$

The sum on the right results from exchanging the summation and integration operations and noting that in general $\int_{-\infty}^{\infty} f(t)\delta(t-\tau)\,dt$ is equal to $f(\tau)$ for any value of τ. Now, if we assume $x(nT) = x_n$ is a discrete sample of $x(t)$ and is nonzero only for $0 \le n < N$, (3.31) is equivalent to

$$\text{FT}\{\bar{x}\} \equiv \sum_{n=0}^{N-1} x_n e^{-j\omega nT} \tag{3.32}$$

We recognize this as the DFT with continuous frequency, ω, that is, X_m in (3.25). Thus, we have the following interesting relationship between the FT and the DFT:

> The Fourier transform (FT) of a sequence of impulse functions, each of which is an element of the sample vector x, is equal to the DFT of x at each frequecncy at which the DFT is measured. (3.33)

In addition, the FT has many other applications. The FT of a continuous signal provides the amplitude and phase spectra of the signal, just as the DFT provides the amplitude and phase in (3.19). In other words, the amplitude spectrum is the magnitude of the FT, and the phase spectrum is the angle of the FT.

There is also an inverse FT, just as there is an inverse DFT. The inverse FT is given by

$$\boxed{\text{Inverse FT:}\quad x(t) = \frac{1}{2\pi}\int_{-\infty}^{\infty} X(j\omega)e^{j\omega t}d\omega} \tag{3.34}$$

The inverse FT is verified when we substitute (3.29) for $X(j\omega)$ and switch the order of integration. Thus, in the following expression, the term in square brackets is equal to one when $t = \tau$ and zero when $t \neq \tau$, that is, equal to $\delta(t-\tau)$:

$$x(t) = \int_{-\infty}^{\infty} x(\tau)\left[\frac{1}{2\pi}\int_{-\infty}^{\infty} e^{j\omega(t-\tau)}d\omega\right]d\tau = \int_{-\infty}^{\infty} x(\tau)\delta(t-\tau)d\tau = x(t) \tag{3.35}$$

In some applications used hereafter in this text, the use of the FT is limited, because even though the integral in (3.29) is infinite, the energy in the waveform must, in general, be finite for the integral to converge; that is, for convergence of the FT,

$$\int_{-\infty}^{\infty} |x^2(t)|\,dt < \infty \tag{3.36}$$

Thus, for example, the FT of any periodic function does not exist according to the definition in (3.29), although functions of ω that "work" as transforms of periodic functions in some sense are sometimes included in transform tables.

The *Laplace transform* is a modification of the FT that converges for periodic functions and, in fact, for any finite-valued waveform. The Laplace transform is an FT with frequency (ω) replaced by a complex variable, $s = \alpha + j\omega$. In this case, we can use $X(s)$ without ambiguity to represent the Laplace transform of $x(t)$, and thus we define the Laplace transform as

$$\boxed{\text{Laplace transform:}\quad X(s) = \int_{-\infty}^{\infty} x(t)e^{-st}dt} \tag{3.37}$$

Similar to the FT, the Laplace transform has an inverse resembling (3.34); however, the integration is on the complex plane, being carried out over s

instead of over $j\omega$. The Appendix contains a table of various Laplace transforms. More extensive transform tables are available in the literature.[9,10]

The continuous transforms are useful mainly with continuous signals, and therefore they have limited use in this text, which deals with DSP as the main subject. However, the Fourier and Laplace transforms are useful in the discussion of the sampling theorem in the next section (Section 3.10) and in the design of optimal infinite impulse response filters, which we discuss in Chapter 6. The main purpose of this section is to introduce the two transforms and to show how they are related to each other and to the DFT. The principal relationships are expressed in (3.29), (3.33), (3.34), and (3.37).

Before we complete this discussion, it is also helpful to introduce the concept of *convolution* in terms of continuous transforms and linear systems.[6–17] The convolution concept is discussed in Chapter 4 in terms of discrete transforms and is the basis of all linear filtering operations, both analog and digital.

Convolution is closely related to the *transfer function* concept. In the frequency domain, a system is defined to be *linear* if its response to a pure sinusoidal signal at any given frequency is also sinusoidal at the same frequency. Thus, we may characterize the response of any linear system in terms of a Fourier or Laplace transform, or *transfer function,* because the complex value of the transform at each frequency (ω) gives us the amplitude gain (amplitude spectrum) and phase shift (phase spectrum) at that frequency; that is, if $H(j\omega)$ (or $H(s)$ with $s = j\omega$) is the transfer function, then

$$\begin{aligned} \text{Amplitude gain} &= \left|H(j\omega)\right|; \\ \text{Phase shift} &= \angle H(j\omega) \end{aligned} \tag{3.38}$$

Thus, each component of $H(s)$ indicates how the amplitude and phase of a signal are modified at a particular frequency; therefore, if $X(s)$ is the transform of an input signal, $x(t)$, to a system with transfer function $H(s)$, the spectrum of the output signal, $y(t)$, must be

$$Y(s) = X(s)H(s) \tag{3.39}$$

This concept is illustrated in Figure 3.12.

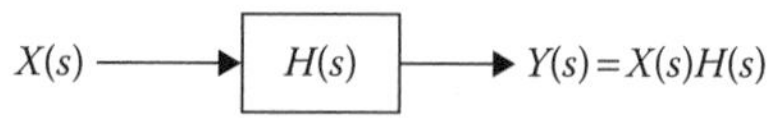

$$X(s) \longrightarrow \boxed{H(s)} \longrightarrow Y(s) = X(s)H(s)$$

FIGURE 3.12
Illustration of how a linear transfer function, $H(s)$, alters the spectrum of an input, $X(s)$, to produce the output, $Y(s)$.

Taken together, all the components of $H(s)$ describe how the linear system "transfers" the input signal to the output. Thus, if the amplitude gain is, say, A at frequency ω_0, then the system's response to a unit sine wave at frequency ω_0 is a sinusoidal signal at the same frequency with amplitude A. And if the phase shift at ω_0 happens to be, for example, $\pi/2$ radians, the response to an input sine wave is a cosine wave at ω_0.

The inverse transform of $H(s)$, that is, $h(t)$, is called the *impulse response* of the system because it is the response of $H(s)$ if the input signal is a unit impulse, that is, to the function $\delta(t)$ introduced in Section 3.9. For discrete systems, the unit impulse vector is one at $t = 0$ and zero elsewhere, as described in (3.30), in Chapter 4, and in line 100 of the Appendix. For continuous systems, it is convenient to define the unit impulse as the rectangular pulse with unit area shown in Figure 3.13. As T approaches zero, the rectangular pulse approaches the delta function, $\delta(t)$, while the area of the pulse remains constant at one. Using this definition of the continuous impulse function, we can derive its FT, $\Delta(j\omega)$, using (3.29), as follows:

$$\begin{aligned}\Delta(j\omega) &= \int_{-\infty}^{\infty} \delta(t)e^{-j\omega t}dt = \lim_{T\to 0}\frac{1}{T}\int_{-T/2}^{T/2} e^{-j\omega t}dt \\ &= \lim_{T\to 0}\left(-\frac{1}{j\omega T}\left(e^{-j\omega T/2} - e^{-j\omega T/2}\right)\right) = \lim_{T\to 0}\frac{\sin\left(\frac{\omega T}{2}\right)}{\left(\frac{\omega T}{2}\right)} = 1\end{aligned} \tag{3.40}$$

That is, *the* FT *(spectrum) of* $\delta(t)$ *is one at all frequencies.* Thus, the inverse transform of $H(s)$, $h(t)$, is called the "impulse response" because it is the output, $y(t)$, when $x(t)$ is the unit impulse function.

The *convolution integral* can be derived from the inverse transform of the transfer function, $H(s)$, by substituting the FT formula, (3.29), into the right side of (3.39):

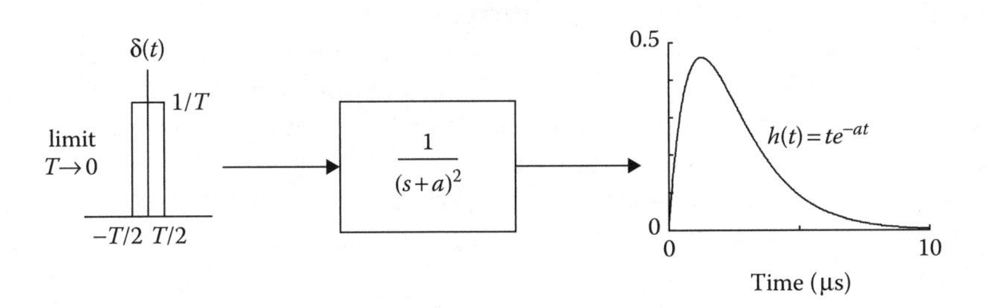

FIGURE 3.13
Response of the linear system, given by $H(s) = 1/(s + a)^2$, to a continuous unit impulse function.

$$\begin{aligned} Y(j\omega) &= \int_{-\infty}^{\infty} x(\tau)e^{-j\omega\tau}d\tau \int_{-\infty}^{\infty} h(v)e^{-j\omega v}dv = \int_{-\infty}^{\infty}\int_{-\infty}^{\infty} x(\tau)h(v)e^{-j\omega(\tau+v)}d\tau dv \\ &= \int_{-\infty}^{\infty}\left(\int_{-\infty}^{\infty} x(\tau)h(t-\tau)d\tau\right)e^{-j\omega t}dt; \quad v = t-\tau \end{aligned} \tag{3.41}$$

In this result, we have $Y(j\omega)$ itself in the form of (3.29); therefore, the function within parentheses must be $y(t)$; that is,

$$\boxed{y(t) = \int_{-\infty}^{\infty} x(\tau)h(t-\tau)d\tau} \tag{3.42}$$

This is the *convolution integral* for continuous functions. Because $y(t)$ above and $X(s)H(s)$ in (3.39) are a transform pair, we have shown that *the transform of the convolution of two signals is the product of the transforms of the two signals.*

3.10 The Sampling Theorem

Having discussed discrete and continuous transforms, which are used to map time-varying waveform and image data into the frequency domain, we are now in a position to discuss and apply the well-known *sampling theorem.* The sampling theorem can be stated in different ways, but its main use in DSP is to tell us the rate at which we must sample a continuous waveform to convey all the information in the waveform, that is, to be able to reconstruct the waveform from its samples.

Such a reconstruction is, of course, impossible in general, because any set of samples "belongs" to an infinite set of continuous waveforms, as illustrated in Figure 3.14, which shows three different waveforms with the same sample values at $t = 0,2,4,\ldots$. In other words, given a finite set of sample points, it is always possible to construct an infinite number of continuous functions through these points. If this is the case, then how can any continuous waveform ever be reconstructed uniquely from a set of sample points?

The answer lies in allowing only certain frequency components to exist in the continuous waveform. The sampling theorem may be stated simply as follows:

If a continuous signal is sampled at a rate greater than twice its highest-frequency component, then it is possible to recover the signal from its samples.	(3.43)

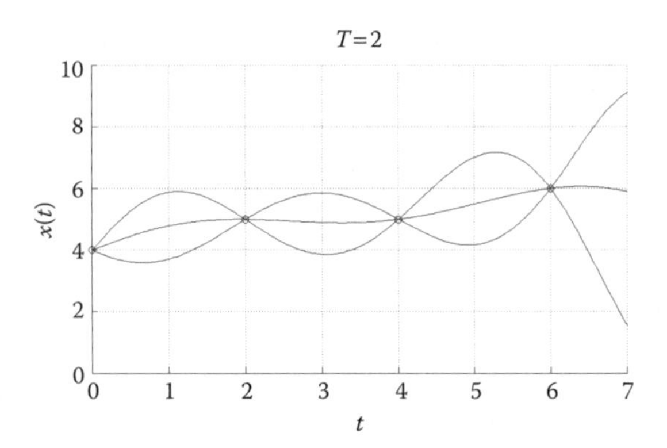

FIGURE 3.14
Different waveforms with the same samples at $t = 0,2,4,6,\ldots$.

Notice that if the signal can be recovered from the samples, then all the information in the signal can be recovered from the samples. For example, if the signal is a speech signal conveying news about events, then all the news is also conveyed in the individual samples, provided the sampling theorem is satisfied. In addition, the samples in (3.43) are assumed accurate. In some cases, they are noticeably inaccurate, and *quantizing errors* must additionally be taken into account. Quantizing errors are treated as statistical or random errors and are discussed briefly in Chapter 10.

The sampling theorem is easy to prove, and in fact, might be anticipated from the previous results presented in Sections 3.3 and 3.4. Suppose we have accurate samples of $x(t)$ taken at $t = 0,T,\ldots,(N-1)T$ and we wish to recover $x(t)$ in the range $t = [0, NT)$. Let $\bar{x}(t)$ represent the corresponding set of impulse samples, as in (3.30). Then, according to (3.33), the FT of $\bar{x}(t)$ and the DFT of the vector $x = \left[x_0, x_1, \ldots, x_{N-1}\right]$ are the same; that is, $\text{DFT}\{x\} = \text{FT}\{\bar{x}\} = \bar{X}(j\omega)$. Now, suppose we express the impulse sequence in (3.30) as $\bar{x}(t) = x(t)d(t)$, where

$$d(t) = \sum_{n=-\infty}^{\infty} \delta(t - nT) \tag{3.44}$$

The Fourier series coefficients for this function, which is obviously periodic—with period T and fundamental frequency $\omega_0 = 2\pi/T$, may be computed using (2.35), as follows:

$$c_m = \frac{1}{T}\int_{-T/2}^{T/2} \sum_{-\infty}^{\infty} \delta(t - nT)e^{-j2\pi mt/T}dt = \frac{1}{T}\sum_{-\infty}^{\infty}\int_{-T/2}^{T/2} \delta(t - nT)e^{-j2\pi mt/T}dt = \frac{1}{T} \tag{3.45}$$

In this result, the integrand on the right equals one at $t = 0$ and zero elsewhere. Using (2.34), the Fourier series for $\bar{x}(t) = x(t)d(t)$ is

$$\bar{x}(t) = \frac{1}{T}\sum_{m=-\infty}^{\infty} x(t)e^{j2\pi mt/T} \tag{3.46}$$

The final step is to derive the FT of this version of $\bar{x}(t)$, as follows:

$$\begin{aligned}\bar{X}(j\omega) &= \frac{1}{T}\int_{-\infty}^{\infty}\sum_{m=-\infty}^{\infty} x(t)e^{j2\pi mt/T}e^{-j\omega t}dt \\ &= \frac{1}{T}\sum_{m=-\infty}^{\infty}\int_{-\infty}^{\infty} x(t)e^{-j(\omega-2\pi m/T)t}dt = \frac{1}{T}\sum_{m=-\infty}^{\infty} X\big(j(\omega - 2\pi m/T)\big)\end{aligned} \tag{3.47}$$

Thus, $\bar{X}(j\omega)$, which is the FT of the sequence of impulse samples of $x(t)$, is $1/T$ times an infinite superposition of the FTs of $x(t)$. But, if $X(j\omega)$, the FT of $x(t)$, is zero at and beyond half the sampling rate, that is, for $|\omega| \geq \pi/T$, then the terms in the sum in (3.47) do not overlap, and $X(j\omega) = T\bar{X}(j\omega)$ in this frequency range.

Thus, the sampling theorem is proved, because if we can compute $X(j\omega)$ from the DFT [that is, by transforming the sample sequence to produce $\bar{X}(j\omega)$], then we can recover $x(t)$ as the inverse transform of $X(j\omega)$. Two useful recovery methods that effectively accomplish these operations when the sampling theorem is satisfied are described next in Section 3.11.

3.11 Waveform Reconstruction and Aliasing

Our first waveform-reconstruction method is based on the inverse DFT in (3.23) and on the equivalence of the DFT components with the Fourier series coefficients in (3.8). These suggest that writing the inverse DFT as a continuous function of t provides a continuous Fourier series for a function that passes through all the sample points and has no frequencies beyond half the sampling rate. Because (as per the sampling theorem) the only function having these properties is $x(t)$, we conclude that the reconstruction, which we may call the *Fourier series reconstruction*, is exact when the sampling theorem holds. Furthermore, even when the sampling theorem does not hold, the Fourier series reconstruction is exact at all the sample points and is guaranteed to contain no frequencies at or above half the sampling rate.

Using the redundancy in (3.11), we write the inverse DFT (3.23) as follows, so that the frequency index stays at or below half the sampling rate index, that is, $\le N/2$:

$$\begin{aligned} x_n &= \frac{1}{N}\sum_{m=0}^{N-1} X_m e^{j2\pi mn/N} \\ &= \frac{1}{N}\left[X_0 + 2\sum_{m=1}^{m<N/2} \mathrm{Re}\{X_m\}\cos\left(\frac{2\pi mn}{N}\right) - \mathrm{Im}\{X_m\}\sin\left(\frac{2\pi mn}{N}\right)\right] \end{aligned} \tag{3.48}$$

The second line in (3.48) is obtained from the first by separating the first sum into two halves and applying (3.10) and (3.11). If N is even, $X_{N/2}$ must be zero for the sampling theorem to hold. Next, we substitute $\omega_0 = 2\pi/NT$ as in (2.24), and we let $t = nT$ to obtain the Fourier series reconstruction from (3.48) at any point in time as follows:

$$x(t) = \frac{1}{N}\left(X_0 + 2\sum_{m=1}^{m<N/2} \mathrm{Re}\{X_m\}\cos m\omega_0 t - \mathrm{Im}\{X_m\}\sin m\omega_0 t\right) \tag{3.49}$$

Two examples of the use of (3.49) are shown in Figure 3.15. The original waveform, $x(t)$, is shown in the upper plot. The spectrum of $x(t)$ is zero at and above 1.0 Hz. In the center plot, $x(t)$ is sampled at 2 samples/s, and $x(t)$ is reconstructed using (3.49). Because the sampling theorem is satisfied in this case, the reconstruction is exact. In the lower plot, $x(t)$ is sampled at 1.5 samples/s, and $x(t)$ is again reconstructed using (3.49). Because the sampling theorem is not satisfied in this case, the reconstruction (solid curve) differs from $x(t)$ (dashed curve), both of which are shown. However, the reconstruction passes through the sample points and interpolates smoothly between the sample points. One would expect the latter, because, in this case, the reconstruction (using Equation 3.49) contains only frequencies below 0.75 Hz.

The second reconstruction method was developed previously in Section 3.8. It is based on the *resampling* method on line 7 of Table 3.3, in which zeros are inserted into the DFT, resulting in interpolation between samples in the time domain. Because this method produces a more densely packed sample sequence that includes the original sample points and has an unchanged spectrum, we can now use the sampling theorem to say that, if the sampling theorem is satisfied and if an increasingly large number of zeros is inserted into the DFT in the manner described [(3.27), for example], the resampled function will, in the limit, become the original continuous function.

The resampling method is especially easy to implement using MATLAB expressions. The *resamp* function on the publisher's website for this text is one such implementation. The expressions in *resamp* are similar to the

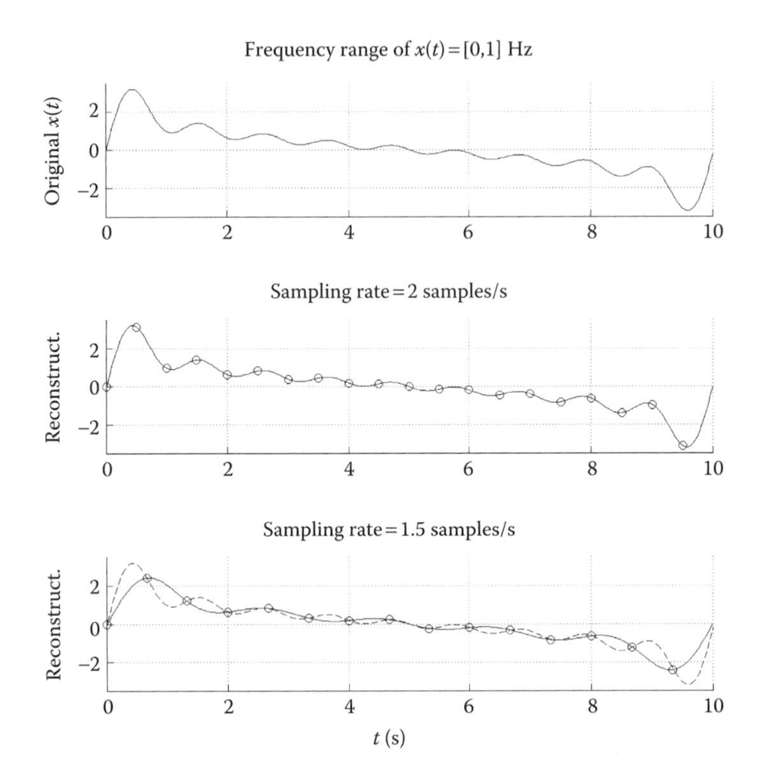

FIGURE 3.15
Two examples of Fourier series reconstruction using (3.49). In the center plot, the sampling theorem is satisfied, and reconstruction is exact. In the lower plot, the sampling rate is too low, and the reconstruction (solid curve) is not exact.

expressions in (3.27), but modified as described in Section 3.12 to improve its performance with short waveform segments. An example of the use of *resamp* is provided in Figure 3.14. The implementing MATLAB expression is

$$y = resamp(x, L_y) \tag{3.50}$$

in which x is the sample vector (with length $L_x = 18$ in Figure 3.16), y is the interpolated version of x, and L_y is the specified length of y, which was set at 500 for this example. With 500 points, the reconstruction appears continuous.

Both reconstruction methods—Fourier series and resampling—produce *band-limited* approximations of the original signal from which the samples were taken. That is, the spectra of the reconstructions are limited to frequencies below half the sampling rate. These methods differ from other interpolation methods, such as linear interpolation, splines, and so on, which are not band limited.

A method used often in the design of DSP systems is shown in Figure 3.17. To prevent errors in the reconstruction (in addition to other types of

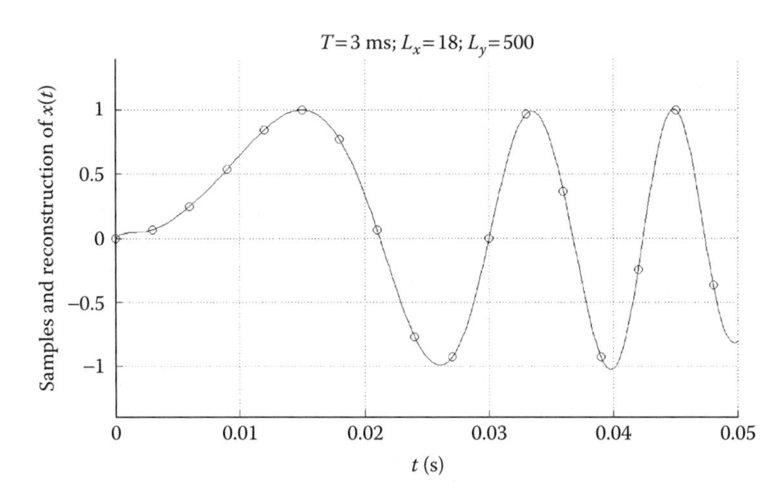

FIGURE 3.16
Samples and reconstruction of a continuous signal, $x(t)$, using the "resampling" method of interpolation between samples. The reconstructed signal has no spectral content above half the sampling rate (166 Hz, in this example).

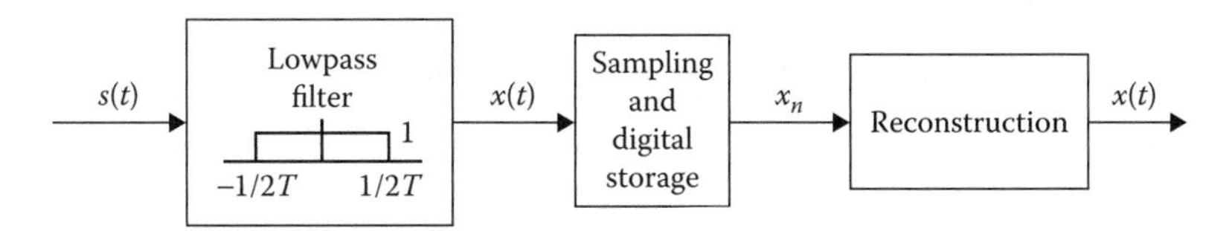

FIGURE 3.17
Reconstruction of a signal, $x(t)$, that has been prefiltered to remove all spectral content over half the sampling rate.

processing such as filtering), a signal is "prefiltered" to remove all frequency components over half the sampling rate. Then, as suggested in the illustration, the prefiltered signal can be sampled, stored, and later reconstructed in its original form.

The concept in Figure 3.17, although often used, cannot be done ideally as shown. Ideal filters do not exist, as we will find later in this text, beginning in Chapter 5. So the question arises: How is a signal distorted, or *aliased*, when it is reconstructed after having been sampled at a rate that is somewhat less than twice its highest spectral content? The answer lies essentially in (3.33) and (3.47). Each of the reconstruction methods produces a signal with a spectrum that is the continuous DFT of the samples. But, from (3.33) and (3.47), the spectrum of the samples is identical with $1/T$ times the superposition of shifted FTs of the original signal in the range $|\omega| < \pi/T$, and is zero outside this range.

This phenomenon, called *aliasing*, is illustrated in Figure 3.18. The upper plot shows the spectrum of the original signal, $x(t)$. The center plot shows the two essential components in the sum in (3.47) (with $m = 0$ and $m = 1$) and the

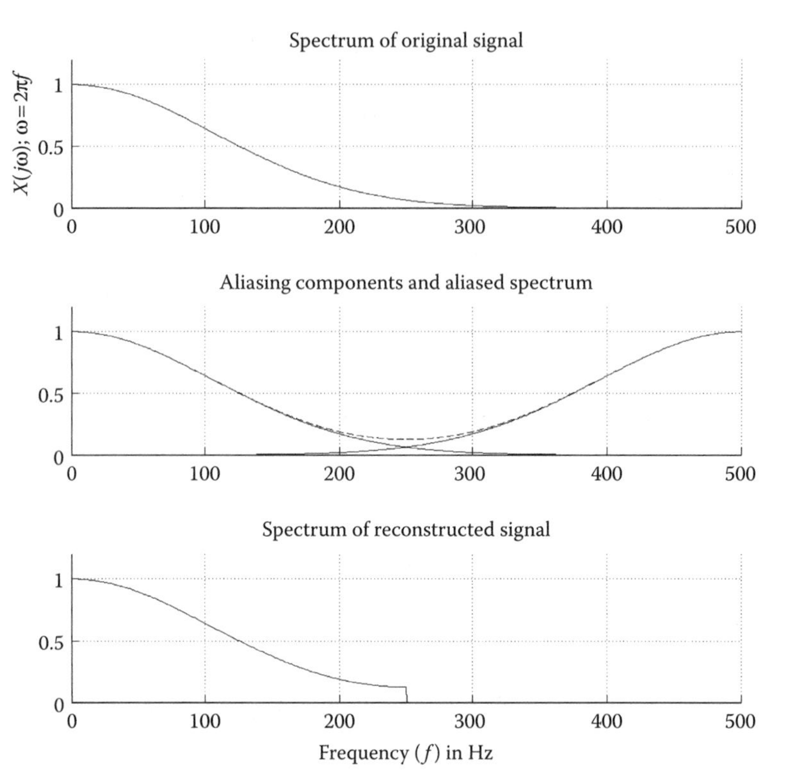

FIGURE 3.18
Illustration of aliasing in accordance with (3.45); sampling rate = 500 Hz. Upper: Spectrum of original signal. Center: Components of the sum in (3.45) and the sum (dashed curve). Lower: Spectrum of the reconstructed signal.

sum, which is $\bar{X}(j\omega)$ in (3.47), assuming that $x(t)$ is sampled at 500 samples/s. The lower plot shows the reconstructed spectrum, which is distorted above approximately 150 Hz and is zero above 250 Hz (or half the sampling rate).

A further illustration of the situation in Figure 3.18 is provided in the two plots of $\bar{X}(j\omega)$ versus f, that is, $\omega/2\pi$, in Figure 3.19, which shows several terms in (3.47)—frequencies ranging from zero to several times the sampling frequency, and illustrates how the terms do not overlap when the sampling frequency is 1000 Hz, but overlap and cause aliasing when the sampling frequency is reduced to 500 Hz.

Finally, no discussion of waveform reconstruction would be complete without mentioning the early work of E. T. Whittaker, who published the sampling theorem in a remarkable paper[11] that established much of the groundwork for modern DSP. His paper included the now-famous formula for reconstructing a waveform from its sample set. Whittaker developed a formula that employs the "cardinal function" for interpolating a function between its sample points. It is an important and fundamental formula and

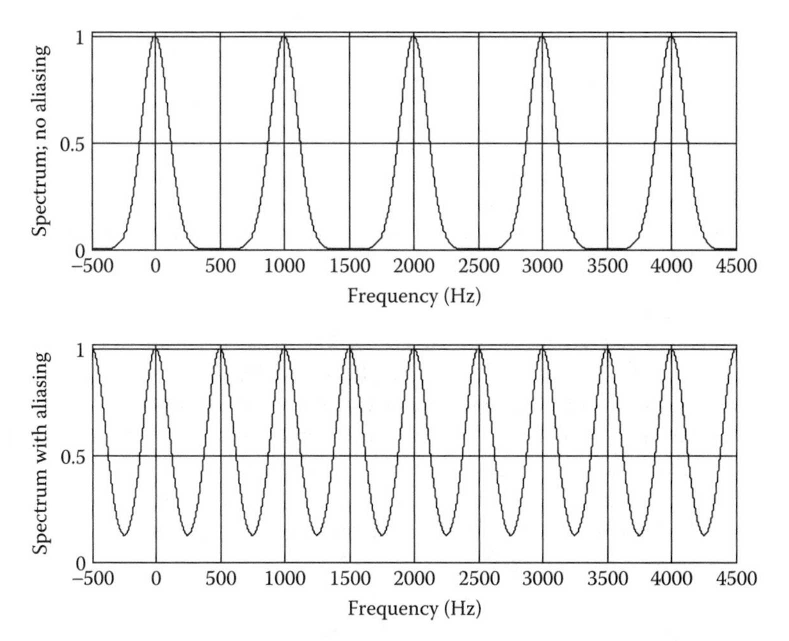

FIGURE 3.19
The upper plot shows the terms in (3.47) for $m = 0$ through $m = 4$ when $X(j\omega)$, which is made real for the sake of the illustration, is the spectrum in Figure 3.16 and the sampling frequency is 1000 Hz. The lower plot shows how the terms in (3.47) overlap when the sampling frequency is reduced to 500 Hz and aliasing occurs.

can be developed as presented in the next paragraph based on the foregoing discussions of continuous linear systems presented in Section 3.9 and the sampling theorem described in Section 3.10.

Suppose we use (3.30) with $t = nT$ to express the sample vector, x, as a function of time as follows:

$$\bar{x}(nT) = \sum_{n=0}^{\infty} x_n \delta(t - nT); \quad 0 \le n < \infty \tag{3.51}$$

From the result in (3.47), which proved the sampling theorem, we know that if we could reduce the sum of transforms to just the term with $m = 0$, we would have

$$\bar{X}(j\omega) = \frac{1}{T} X(j\omega) \tag{3.52}$$

This result suggests that, if the sampling theorem is satisfied so that the term with $m = 0$ in (3.47) is separate from the other terms and $\bar{X}(j\omega)$ is processed

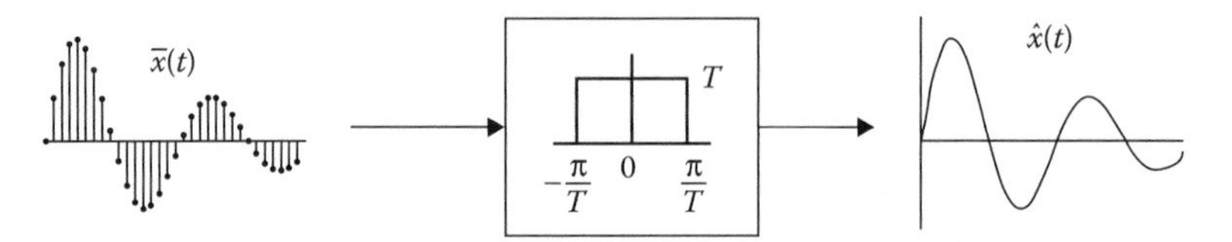

FIGURE 3.20
Filtering $\bar{x}(t)$ in the time domain to recover the original signal as the inverse transform of $X(j\omega)$, that is, the term with $m = 0$ in (3.47).

to extract the spectral content in the range $|\omega| < \pi/T$, we would have an exact recovery of $X(j\omega)$. In the upper plot in Figure 3.19, for example, this would mean processing $\bar{x}(t)$ using a lowpass transfer function with gain equal to one in the range $|f| < 500$ Hz, and zero elsewhere. The process is illustrated in Figure 3.20, which shows $\bar{x}(t)$ being processed by a lowpass filter with gain $= T$ to produce a recovered version of $x(t)$. Using (3.42) to represent the process in Figure 3.20 and noting that $\bar{x}(t)$ is nonzero only at the sample points, we have

$$\hat{x}(t) = \int_{-\infty}^{\infty} \bar{x}(\tau)h(t-\tau)d\tau = \sum_{n=-\infty}^{\infty} x_n h(t-nT) \tag{3.53}$$

In this case, h is the impulse response of the filter in Figure 3.20, which is the inverse transform of the transfer function, that is,

$$\begin{aligned} h(t) &= \frac{1}{2\pi}\int_{-\pi/T}^{\pi/T} Te^{j\omega t}d\omega = \frac{T}{2\pi}\left(\frac{e^{j\pi t/T} - e^{-j\pi t/T}}{jt}\right) \\ &= \frac{1}{2j}\left(\frac{e^{j\pi t/T} - e^{-j\pi t/T}}{\left(\frac{\pi t}{T}\right)}\right) = \frac{\sin\left(\frac{\pi t}{T}\right)}{\left(\frac{\pi t}{T}\right)} \end{aligned} \tag{3.54}$$

Using this result in (3.53) gives us the reconstruction formula for recovering $x(t)$ from the sample vector, x:

$$\boxed{\hat{x}(t) = \sum_{n=-\infty}^{\infty} x_n \frac{\sin\left[\frac{\pi(t-nT)}{T}\right]}{\pi\left(\frac{t-nT}{T}\right)}} \tag{3.55}$$

Whittaker[11] was able to show that his reconstruction of $x(t)$, that is, $\hat{x}(t)$ in (3.55), contains no frequencies above half the sampling rate, which is obvious here from the way the reconstruction is developed in Figure 3.20. His formula is not generally used in modern applications to reconstruct signals due to a couple of practical considerations, but is worth noting for its simplicity and universal applicability as a unique interpolation formula that preserves the sampling theorem in the reconstruction of $\hat{x}(t)$.

One practical consideration involves "end effects." When a segment of a long waveform is reconstructed using (3.55), terms beyond the ends of the segment are needed to reconstruct the segment accurately near its endpoints.

Another practical consideration occurs when we try to compute (3.55) for a value of t at one of the sample points, say at $t = kT$. Then, all the sine arguments in the sum are multiples of π, so that all the terms in the sum, except the term with $n = k$, are zero. Furthermore, because

$$\lim_{t \to kT} \frac{\sin\left[\dfrac{\pi(t-kT)}{T}\right]}{\pi\left(\dfrac{t-kT}{T}\right)} = 1 \tag{3.56}$$

we can recognize that (3.55) reduces to

$$\hat{x}(kT) = x_k \tag{3.57}$$

Thus, the reconstruction is exact at the sample points but must be evaluated separately at these points to avoid division by zero.

A further practical consideration concerns the number of operations needed to compute (3.55), compared with the number of computations required with the transform-reconstruction method described in Section 3.8. This can be alleviated somewhat by noting that

$$\begin{aligned} \sin\left[\frac{\pi(t-nT)}{T}\right] &= \sin\left(\frac{\pi t}{T}\right)\cos(n\pi) - \cos\left(\frac{\pi t}{T}\right)\sin(n\pi) \\ &= (-1)^n \sin\left(\frac{\pi t}{T}\right) \end{aligned} \tag{3.58}$$

Using this in (3.55) and moving the sine function outside the sum allows fewer computations of the reconstruction for each value of t, as shown in (3.59):

$$\hat{x}(t) = \frac{\sin\left(\dfrac{\pi t}{T}\right)}{\left(\dfrac{\pi}{T}\right)} \sum_{n=0}^{N-1} x_n \frac{(-1)^n}{t-nT} \tag{3.59}$$

Here, we have set finite limits on the sum to indicate reconstruction using a sample vector of finite length (N).

3.12 Resampling

The *resamp* function described in Section 3.11 incorporates additional steps, which reduce the "end effects" caused by the zero-insertion method illustrated in (3.27). These were hardly noticeable in the example in Figure 3.10; but they can be objectionable when a short waveform or a small segment of an image is being resampled. These effects occur when the sample values differ at the beginning and end of the segment, because the FFT method presented in (3.27) assumes that the segment is periodic and therefore jumps between the first and last samples at the beginning and the end.

An example is shown in Figures 3.21 and 3.22. We desire to reconstruct a segment of the continuous signal in Figure 3.21, sampled at 1 GHz and known to contain no significant content above half the sampling rate, that is, 500 MHz. The plot on the left shows a reconstruction using the method in (3.27), with $K = 9$ samples between each pair of original samples. The plot on the right shows a similar reconstruction with $K = 9$ using the *resamp* function.

The two reconstructions are nearly the same, except near the ends of the segment. The problem is that the waveform from which the segment was taken is continuous at the segment ends, whereas a periodic extension of the segment, which (3.27) effectively uses, is not continuous. In this case, there is a single large step from x_{N-1} to x_0 which is not present in the continuous

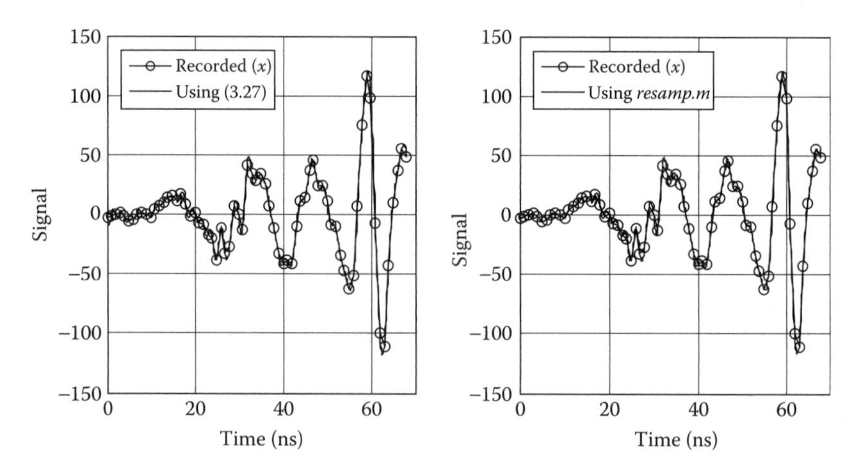

FIGURE 3.21
Sampled waveform segments with $T = 1$ ns, reconstructed with $K = 9$ samples between each original pair of samples, using the method in (3.27) (left plot) and using *resamp.m* (right plot).

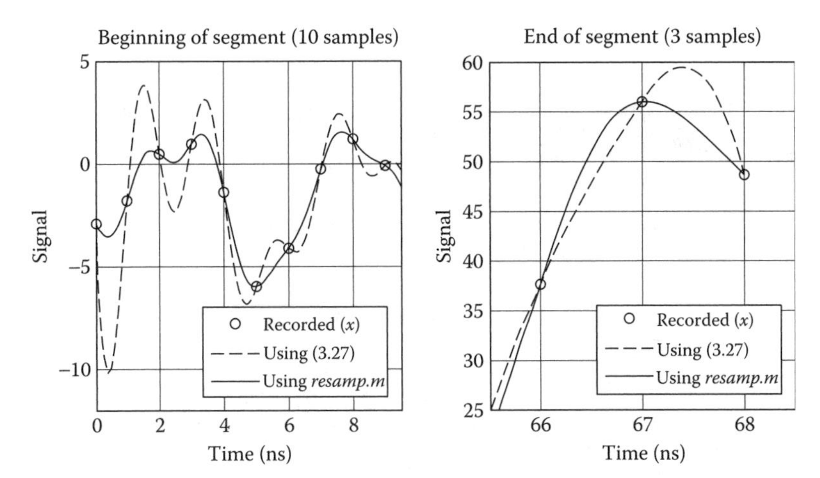

FIGURE 3.22
Beginnings (left) and ends (right) of the two plots in Figure 3.21, showing the end effects caused by using (3.27) in the reconstruction, which are not found when the *resamp* function is used.

waveform, and this step causes the spurious oscillations in the reconstruction shown clearly in Figure 3.22.

The *resamp* function creates an altered version, v, of the original segment vector, x, in two simple steps, with the result that the periodic extension of v is continuous and has a continuous first derivative at the endpoints. The procedure is illustrated in Figure 3.23. First, the elements of a ramp function (straight line) from x_0 to x_{N-1} are subtracted from the corresponding elements of x. The resulting reduced version, u, has $N-2$ degrees of freedom because $u_0 = u_{N-1} = 0$. The latter is then used to form the vector v of length $2N-2$ as follows:

$$v = \left[u_0 \; u_1 \cdots u_{N-2} \; u_{N-1} \; -u_{N-2} \cdots -u_2 \; -u_1 \right] \tag{3.60}$$

We can understand why v, as opposed to x, might be useful for interpolation. If v in (3.60) is viewed as one cycle of a periodic waveform, a continuous version of the waveform would then be continuous and have a continuous first derivative at the middle and end of each period. Furthermore, because v is an odd function, its Fourier series is a sine series (Chapter 2, Exercise 2.8). When the sampling theorem is satisfied and there are no components above π Hz-s, (2.23) gives the following Fourier series for v, noting that the period in this case is $2N-2$ rather than N:

$$v_n = \sum_{m=1}^{N-1} b_m \sin \frac{\pi mn}{N-1}; \quad 0 \le n \le N-1 \tag{3.61}$$

We limit n to the range shown, so that the Fourier series includes just the original range of x. The sine coefficients are given by the expression

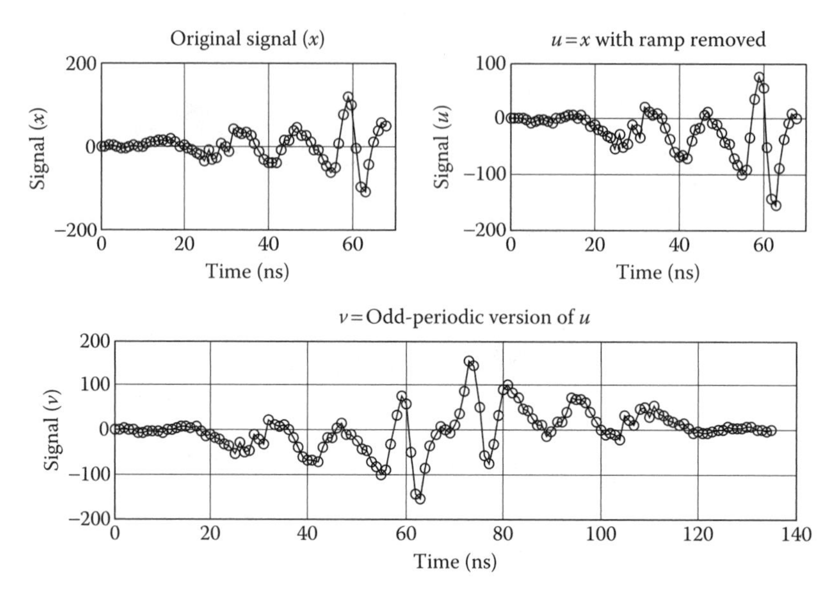

FIGURE 3.23
Procedure for creating the odd-periodic version of the original signal, x. First the straight line from x_0 to x_{N-1} is removed to produce u, then u is appended to itself, as described in the text, to produce v.

$$b_m = \frac{1}{N-1}\sum_{n=0}^{2N-3} v_n \sin\frac{\pi mn}{N-1}; \quad 1 \le m \le N-1 \tag{3.62}$$

We can change v_n in (3.61) into a continuous function of t, again with no frequency content beyond half the sampling rate, simply by letting $t = nT$, where t is continuous time and T is the time interval between the samples. Then, in effect, we have a method for *resampling*, that is, changing the original time step to any new value as follows:

$$v(t) = \sum_{m=1}^{N-1} b_m \sin\frac{\pi mt}{(N-1)T}; \quad 0 \le t \le (N-1)T \tag{3.63}$$

To implement this procedure in a MATLAB function, we use the fast-transform procedure described in Chapter 10, Section 10.6, in the derivation of the Discrete Sine Transform, which is discussed further there. Here, we only note that the extension of v is odd-periodic and the DFT, V, is therefore imaginary. When the zero-insertion method (3.27) is applied, using V instead of X, the resulting reconstruction is generally more accurate near the endpoints of x; that is, near the beginning and center of v. The interpolated version of x is found in *resamp* by inserting the appropriate number of zeros in the middle of V and taking a scaled version of essentially the first half of the inverse FFT of the result.

As mentioned in (3.46), the implementing MATLAB expression for *resamp* is

$$y = resamp(x, L_y) \tag{3.64}$$

in which L_y is the desired length of the interpolated signal, y. The zero-insertion method guarantees that y will have no frequency components above half the sampling rate of the original signal, x, and that whenever two samples, x_k and y_n, happen to occur at the same time, they are equal.

The *resamp* function also allows *decimation*, that is, it allows L_y to be less than L_x, the length of x. The procedure is similar to that in (3.27) except that, instead of inserting zeros in the middle of V to make V longer, we *remove* components from the center of V to make V shorter. That is, when *resamp* is used to decimate a signal vector x, it deletes the high-frequency content of x, leaving a possibly altered version of the original signal. The usual application with $L_y < L_x$ would therefore be *compression* rather than reconstruction, especially in cases where x has negligible content at frequencies above half the sampling rate of y; that is, above $1/(2T_y)$.

In either case, because any vector of length N has $N-1$ steps from beginning to end, the time step in x, T_x, is reduced or increased in the interpolated version, y, according to the following rule:

$$T_y = T_x \frac{L_x - 1}{L_y - 1} \tag{3.65}$$

where L_x and L_y are the lengths of x and y, respectively. An example is shown in Figure 3.24, in which a small segment of the waveform in Figure 3.21 is resampled using *resamp*, with L_y adjusted to produce exactly three samples

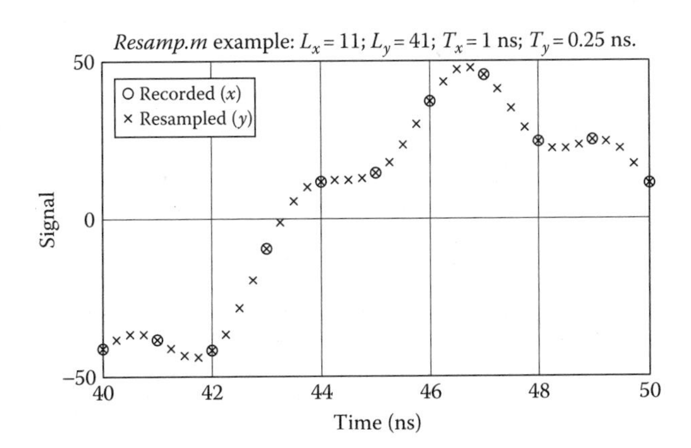

FIGURE 3.24
Example of the use of *resamp.m* on a small segment of the waveform in Figure 3.21. In this case, L_y in (3.62) is chosen to create three samples in y between each pair of samples in x; in other words, $T_y/T_x = 1/4$.

between each pair of samples in x, that is, L_y is adjusted to make $T_y/T_x = 1/4$. Note that when $L_y \geq L_x$ and T_y/T_x is a ratio of integers, as is the case here, the output vector, y, includes all the elements of the input vector, x.

3.13 Nonuniform and Log-Spaced Sampling

As we discussed in Chapter 1, Section 1.4, when signals or images are digitized, they are usually sampled at regular intervals with a constant time step (T). Nearly all analog-to-digital converters work in this manner to digitize analog waveforms coming from instruments and transducers, such as microphones, accelerometers, seismometers, and so on. Cameras produce images of two-dimensional space using uniformly spaced pixels, video scanners digitize a constant-rate scan signal at regular intervals, and so on.

Nevertheless, there are situations in which measurements of phenomena we wish to analyze in the frequency domain are taken at irregular times. Examples are shown in Figures 3.25, 3.26, and 3.27. The first example in Figure 3.25 is a series of geological measurements taken over a period of twenty years, with the later measurements taken less often. The exact sample times are shown along the time axis.

The second example in Figure 3.26 illustrates a phenomenon known as "jitter" in analog-to-digital (A–D) conversion, which causes samples to be taken at times slightly different from the prescribed regular intervals. Jitter, which can be a problem especially in high-rate A–D conversion, may be due

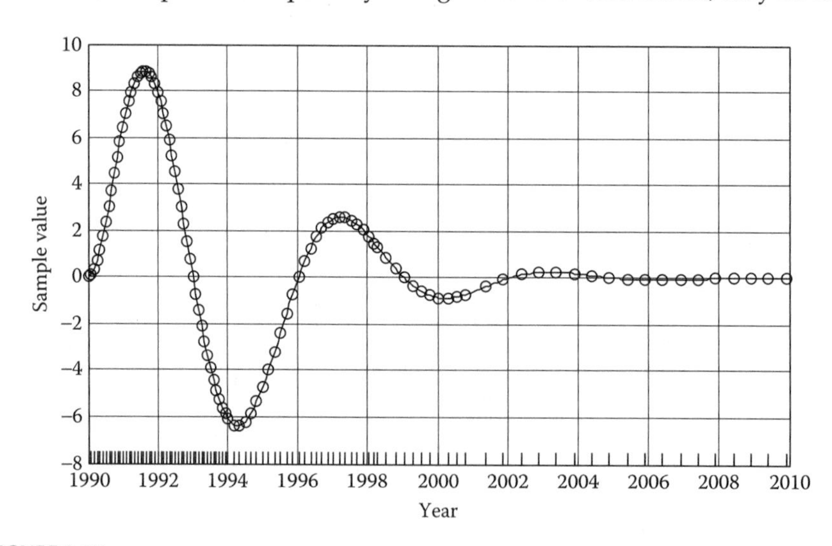

FIGURE 3.25
Samples taken at irregular intervals over the years 1990–2010. The sampling rate changed in 1994, in 1998, and in 2001, as shown by the marks along the time axis.

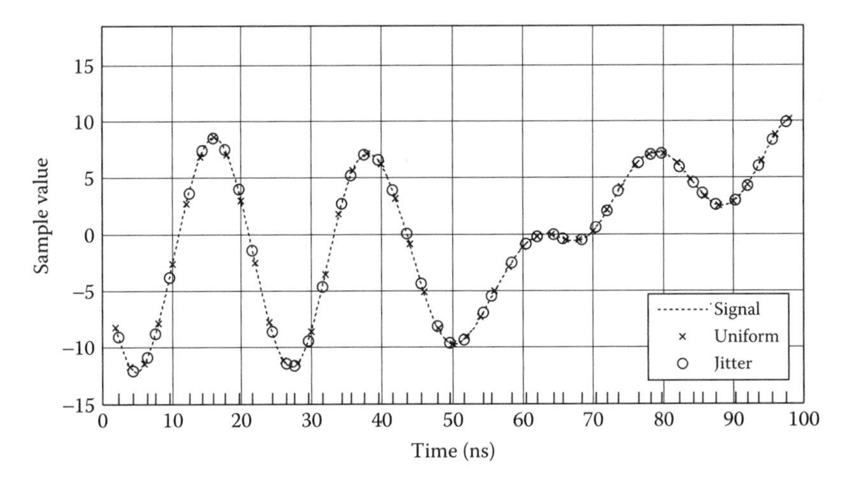

FIGURE 3.26
Sampling with jitter. In this illustration the uniformly-spaced samples of x at intervals of 2 ns are shown together with a corresponding set (o) of samples spaced irregularly due to time jitter. Both sets are assumed to be accurate, but the samples in the second set are not uniformly spaced in time. The time tics at the bottom of the figure indicate the irregular sample times due to jitter.

to inaccuracies in the system clock, inconsistent triggering of the converter by the clock, or in the manner in which the converter itself acquires each sample. Figure 3.26 illustrates the type of jitter in which the samples are accurate but not necessarily taken at the correct time steps. Again, the exact sample times, shown along the time axis, are observed to "jitter" within their corresponding interval of length $T = 2$ ns.

A number of articles and textbooks[1] address various aspects of nonuniform sampling. Our discussion here is limited to the problem of processing nonuniformly sampled data in the frequency domain; that is, reconstructing the signal or image in a way that allows us, as far as possible, to preserve the original spectral content in a uniformly sampled waveform or image. Having the latter, we can then go on to filter the uniformly sampled signal, analyze its spectrum, and so on, using standard DSP algorithms, which do not work with nonuniformly sampled data.

It is possible, as in the examples above, to have signals that have been nonuniformly or even randomly sampled without aliasing, that is, at rates that may be said to preserve the "local spectral content" throughout the signal. There are cases in mechanics and shock physics in which the waveform changes rapidly and has broad spectral content initially and then loses high frequencies in an exponentially decaying fashion. In these cases, *logarithmic sampling* is sometimes used to reduce the sampling rate where only lower frequencies are present and thereby reduce the length of the stored signal vector.

In logarithmic sampling, we space the samples uniformly in log time, which means, of course, that the samples become more widely separated as linear time increases. With uniform spacing in log time, a MATLAB *semilogx* plot, that is, a plot with a logarithmic abscissa, will have uniformly spaced samples and can therefore be easily resampled and processed in the log-time domain.

Suppose we have a digitizer running with time step T between samples. We wish to select samples of a signal, $x(t)$, which are uniformly spaced in the log-time domain over a total time interval from kT to $(k+K)T$ seconds, where K is a very large number. We define a time vector of length N spanning this interval as follows:

$$t = \begin{bmatrix} \delta & \delta^2 & \delta^3 & \delta^4 & \cdots & \delta^{N-1} & \delta^N \end{bmatrix} T \tag{3.66}$$

If sampling is done in accordance with (3.66) over the total interval of KT seconds, then we have

$$(\delta^N - \delta)T = KT, \text{ or } \delta^N = K + \delta \tag{3.67}$$

For values of N greater than 100 and K less than 10^{30}, it is easy to show that δ is a number in the interval [1, 2]. Thus, (3.66) requires interpolation to determine samples of x among the original samples as they arrive from the A–D converter. In practice, $x(t)$ is often oversampled; therefore, linear interpolation can be used.

We can solve (3.67) approximately for δ by extending the original period (K) by two samples and defining

$$\delta \triangleq (K+2)^{1/N} \tag{3.68}$$

Then, because $\delta < 2$, when we use $K + 2$ in place of K in (3.67), we have

$$\delta^N - \delta = K + 2 - \delta; \quad \text{thus, } K < \delta^N - \delta < K + 1 \tag{3.69}$$

That is, with δ computed as in (3.68), the resulting log-spaced vector in (3.66) extends just slightly beyond the original interval, KT.

In the log time domain, the uniformly spaced samples given by (3.66) are

$$\begin{aligned} \tau = \log t &= \log\begin{bmatrix} \delta & \delta^2 & \delta^3 & \delta^4 & \cdots & \delta^{N-1} & \delta^N \end{bmatrix} + \log T \\ &= \begin{bmatrix} 1 & 2 & 3 & 4 & \cdots & N-1 & N \end{bmatrix} \log\delta + \log T \end{aligned} \tag{3.70}$$

That is, the step size in log time is log δ. Any base may be used for the logarithms in these results. In engineering applications, it is usually advantageous to use the base 10.

To illustrate logarithmic sampling, we will use what is typically called a *chirping* signal, that is, a sinusoidal signal whose frequency changes monotonically with time, either up or down. The simplest such signal may be described as follows:

$$x_n = \sin[2\pi(\nu_0 + \alpha n)n]; \quad n = 0, 1, \ldots, N - 1 \tag{3.71}$$

The frequency is initially ν_0 and then changes either up or down at a linear rate given by α. The "instantaneous frequency" in the vector at index k is found by using the period, P, from $k - P/2$ to $k + P/2$, over which x goes through one cycle; that is,

$$[\nu_0 + \beta(k + P/2)](k + P/2) - [\nu_0 + \beta(k - P/2)](k - P/2) = 1 \text{ period} \tag{3.72}$$

From this, we find that $P = 1/(\nu_0 + 2\alpha k)$, and thus the instantaneous frequency at point k is

$$\nu_k = \nu_0 + 2\alpha k \text{ Hz-s}, \quad \text{or} \quad f_k = f_0 + 2\alpha k/T \text{ Hz} \tag{3.73}$$

That is, the "frequency" of x begins at f_0 and increases linearly with k at rate 2α.

Linear chirping is discussed further in Chapter 4, Section 4.5, in connection with the chirp z-transform. Here, we desire a signal similar to the example in Figure 3.27, with the instantaneous frequency decaying exponentially. Signals from explosions, in addition to signals from other geophysical and biological phenomena, behave in this manner. Logarithmic sampling may be used with these types of signals because the necessary sampling rate decreases with time.

In Figure 3.27, the signal is "chirping" because the *frequency*, not the amplitude, decreases with time. A signal that is essentially the response to an

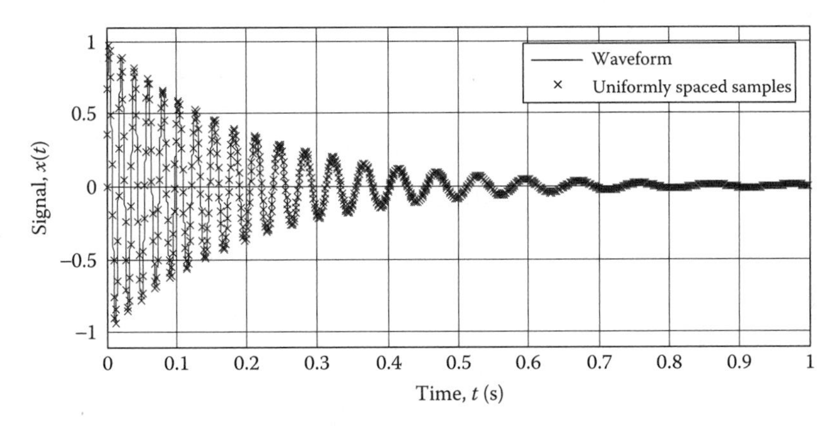

FIGURE 3.27
Chirping signal for logarithmic sampling with 1000 uniformly spaced samples.

impulse, such as an impact or explosion, typically displays a "$1/f$" characteristic; but, this description refers to the signal's *power spectrum* decreasing with frequency in inverse proportion to f, not to the fact that the signal is chirping.

A log-sampled version of Figure 3.27 is shown in Figure 3.28. The sample times are given in (3.66), with $N = 200$ and δ chosen such that the log-spaced sample vector spans approximately one second, that is, approximately the span of the signal, x. The plot is constructed over linear time; thus, the samples are observed to become less dense as time increases, yet dense enough to accomodate the lower frequencies occurring at later time points. The signal, x, is also plotted versus log time in Figure 3.29 to illustrate how the log-spaced samples are, of course, spaced uniformly over log time.

When nonuniformly sampled signals, such as those discussed in the previous paragraph, are subjected to spectral analysis or just about any type of

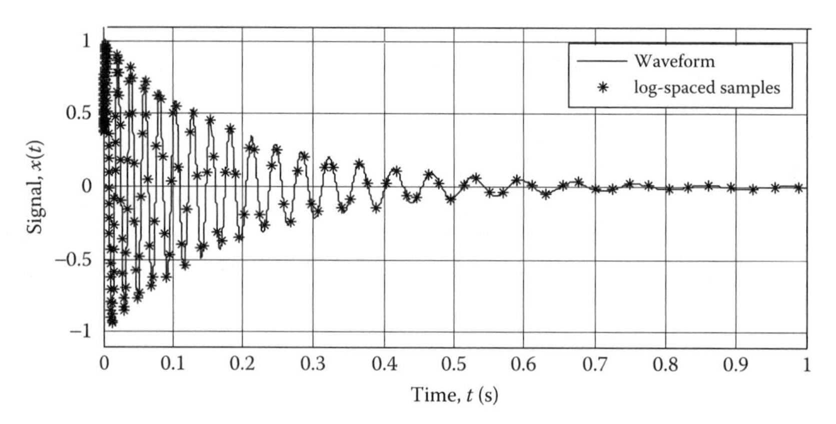

FIGURE 3.28
Chirping signal in Figure 3.27, with $N = 200$ log-spaced samples.

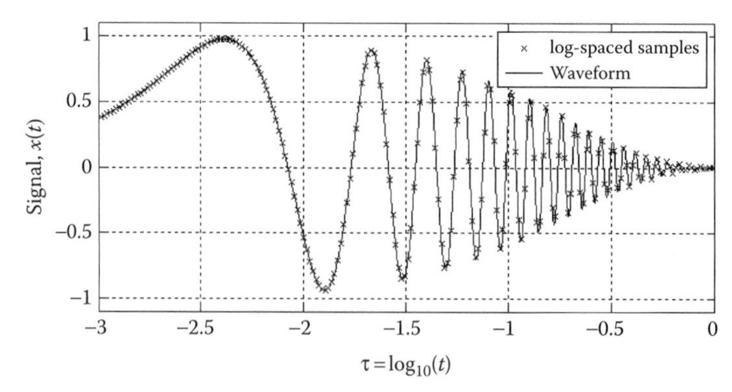

FIGURE 3.29
Chirping signal in Figure 3.28, with 200 uniformly spaced samples over log time.

linear processing, the most practical approach is to first produce a reasonable uniformly sampled version of the signal and then carry out the analysis or processing. No method for producing a uniform sample vector can be optimal for all nonuniformly sampled signals. Here, we describe one such method using the function *resamp_r*, which is intended for resampling a variety of different nonuniformly sampled signals, including those discussed in the foregoing text.

First, if the samples are uniformly sampled over $\tau = \log_{10} t$, as shown in Figure 3.28, or over any other continuous function of time, the uniformly sampled vector can be resampled *in the nonuniform time domain* using the function *resamp* described in Section 3.12 or any similar frequency-preserving method. Doing so allows us to either (1) represent the waveform as a continuous function of t in the form of (3.63) and, hence, at any set of points evenly spaced in t, or (2) to apply *resamp_r* to an arbitrarily dense set of nonuniform samples and, thus, obtain evenly spaced samples in linear time.

The *resamp_r* function works in the same manner as *resamp*, described in Section 3.12, except that instead of inserting zeros in the spectrum of the sample vector, since the samples are not spaced uniformly in this case, we must make a least-squares fit of a Fourier sine series to the odd-periodic vector similar to the one in Figure 3.27, but with nonuniformly spaced samples. The process in this case is computationally much more intensive than in *resamp*. Referring to Chapter 2, (2.8), the function g in this case is

$$g_{mn} = \sin\left(\frac{2\pi m t_n}{P}\right); \quad 0 \le n < N; \quad 1 \le m \le M \tag{3.74}$$

The time vector, $[t_n]$, spans the time domain, and m spans the frequency domain up to M/P Hz-s where P is the period of the odd-periodic vector. The vector G_m is then formed as in (2.9), and the M-by-M matrix, G in (2.11), is inverted to solve for the M coefficient vector, c, as in (2.13). The m components of c then become the coefficients in the continuous sine series for the interpolated waveform, y, as follows:

$$y(t) = \sum_{m=1}^{M} c_m \sin\left(\frac{2\pi m t}{P}\right) \tag{3.75}$$

From this result, we can define t to be any vector of uniformly spaced samples spanning the original time domain and thereby producing y as the vector of uniformly spaced samples approximating the original signal, x.

In this type of reconstruction, or interpolation, when the samples of x are not uniformly spaced and we desire to preserve the original frequency content, there is no general rule for choosing an optimal value for M, the number of sine components in (3.75). This number is limited by (1) the number

(N) of the nonuniform samples, (2) the speed and capacity of the processor, which must invert the M-by-M matrix G, and (3) the precision of the processor, which affects the condition of G.

The *resamp_r* function works by first partitioning x into segments of essentially equal length (N), which, in a modern computer using MATLAB, can be made at least on the order of 100, then setting the initial value of M equal to N. After forming G using the settings, the function tests the condition of G using the MATLAB function *cond.m* and reduces M, if necessary, until the condition is less than another fixed parameter on the order of 10. Having determined M in this manner, *resamp_r* then proceeds to compute the coefficient vector, c, in (3.75).

The results of applying *resamp_r* to the waveforms in Figures 3.25 and 3.26 are shown respectively in Figures 3.30 and 3.31. In the first of these, the original waveform vector consisted of 103 nonuniform samples, and the function limited M to 39 sine components due to the condition of G, which in turn is affected mainly by the frequency content of the original waveform. If higher frequencies had been present in the latter, the function would have chosen a larger value of M.

In the second example in Figure 3.26, there were only 49 nonuniform samples, and *resamp_r* used $M = 49$, that is, the maximum number of independent sine coefficients. In both examples, the original waveform was recovered with negligible error. The *resamp_r* function will not always work this well, especially if the portions of the signal are not sampled at least at twice the highest significant frequency, and oversampling always guarantees a more accurate reconstruction.

The latter principle, oversampling, can be applied easily in the example of the chirping signal in Figure 3.28, because the log-spaced samples are spaced *uniformly* in log time, as in Figure 3.29. The *resamp* (not *resamp_r*) function

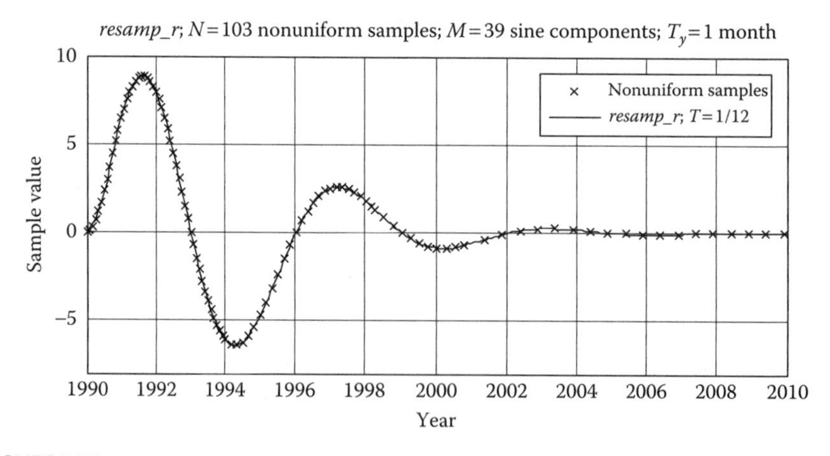

FIGURE 3.30
Reconstruction of Figure 3.25 using the function *resamp_r.m*.

discussed in Section 3.12 may be applied to "oversample" the signal in the log-time domain, as illustrated in Figure 3.32, in which five times the number of original samples are now present, uniformly spaced in log time. These are now used by *resamp_r* to resample the signal in linear time, and the result is shown in Figure 3.33. In this case, there is nearly zero error in the reconstruction of the chirping signal in Figure 3.27.

Thus, the procedure in *resamp_r*, that is, of fitting a Fourier sine series to an odd function constructed from the original vector of nonuniformly spaced samples, is a simple and practical way to reconstruct a signal from a non-

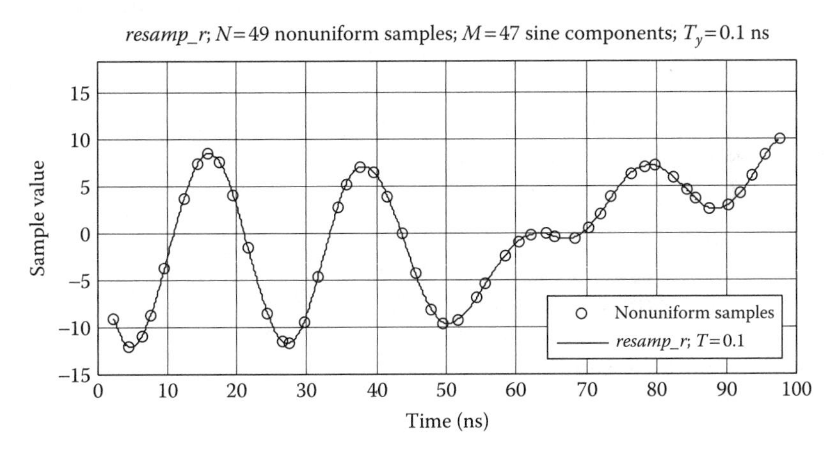

FIGURE 3.31
Reconstruction of Figure 3.26 using the function *resamp_r.m*.

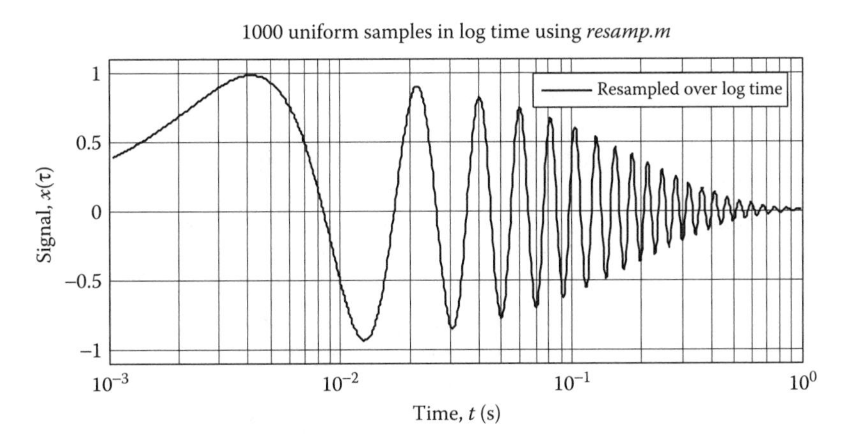

FIGURE 3.32
Resampled version of the uniformly sampled waveform in Figure 3.29 using the function *resamp.m*.

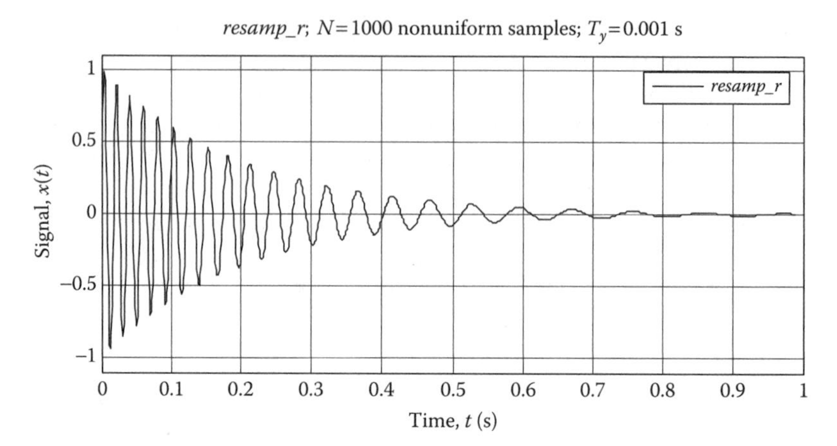

FIGURE 3.33
Resampled version of the chirping waveform in Figure 3.27 obtained by using the function *resamp_r.m* on the log-spaced samples in Figure 3.32.

uniform set of samples. Such a procedure works well, provided (1) the original waveform segments can be represented using a Fourier series with a reasonable number of terms, (2) the nonuniform sampling has not resulted in aliasing in portions of the signal, and (3) the original sample vector is not too long for the processor to produce a uniformly sampled version of x in a reasonable amount of time.

Exercises

General instructions: (1) Whenever you make a plot in any of the exercises, be sure to label both axes, so we can tell exactly what the units mean. (2) Construct continuous plots unless otherwise specified. "Discrete plot" means to plot discrete symbols, unconnected unless otherwise specified. (3) In plots, and especially in subplots, use the *axis* function for best use of the plotting area.

3.1 Write a program similar to (3.13), and use the result to produce a figure similar to Figure 3.3, illustrating the signal and the magnitude of its DFT. In place of the signal in (3.13), use

$$x_n = 2\sin\left(\frac{2\pi n}{10}\right); \quad n = 0, 1, \ldots, 49$$

3.2 Repeat Exercise 3.1, this time using the following signal with $N = 500$:

$$x_n = 2\sin\left(\frac{2\pi n}{50}\right)e^{-n/100}; \quad n = 0, 1, \ldots, 499$$

3.3 Create a 2×2 array of plots using *subplot*(2,2,*). On the upper left, plot the signal vector x, where

$$x_n = \sin\left(\frac{2\pi n}{20}\right); \quad n = 0, 1, \ldots, 99$$

versus time, assuming time step T = 0.05 seconds. On the lower left, make a discrete plot of the DFT magnitude, $|X_m|$, versus m for m = 0,1,...,50. On the upper right, make a discrete plot of the DFT magnitude versus frequency in the range [0, 0.5] Hz-s. On the lower right, make a discrete plot of the DFT magnitude versus frequency (in Hertz) from zero to half the sampling rate.

3.4 In this exercise, we examine the phenomenon known as *leakage*. The term refers to the spectrum as represented by the DFT, in which the "true" spectrum of a periodic signal or component is spread out as a consequence of transforming a signal vector of finite length that does not contain an integral number of cycles of the periodic component.

Create a 2×2 array of subplots. On the upper left, plot the signal vector with $N = 92$:

$$x_n = \sin\left(\frac{2\pi n}{20}\right); \quad n = 0, 1, \ldots, 91$$

Below this, make a discrete plot of the amplitude spectrum, $|X_m|$, versus frequency in the range [0, 0.5] Hz-s and note the leakage around the signal frequency, 0.05 Hz-s. Repeat these two plots on the right with the signal vector length increased to $N = 292$, and note the effect on leakage of the increased vector length.

3.5 Make two subplots, side by side. On the left, plot the following sample vector:

$$x_n = \sin\left(\frac{2\pi n}{20}\right)e^{-n/75}; \quad n = 0, 1, \ldots, 299$$

On the right, make a connected discrete plot of the amplitude spectrum in terms of the frequency in radians. Use the correct symbol in Table 3.1 to label the frequency axis.

3.6 Carry out Exercise 3.5, assuming an interval of 0.1 microseconds between samples; label the time and frequency axes appropriately. Use frequency units f (MHz).

3.7 Complete the following:

a. Do Exercise 3.5, assuming an interval of 0.1 microseconds between samples, but instead of the amplitude spectrum, plot the unwrapped phase spectrum in terms of degrees versus frequency in MHz.

b. Prove that for any real vector x, the DFT at $\omega T = 0$ and $\omega T = \pi$ must be real, and therefore, the phase must be either zero or ±180°.

3.8 Beginning at $t = 0$, 2048 samples of $f(t) = te^{-t/3}$ are collected with time step $T = 0.015$ seconds. In the upper of two subplots, make a connected discrete plot of the amplitude spectrum of $f(t)$ over the range [0, 0.6] Hz. In the lower subplot, make a connected discrete plot over the same frequency range of the interpolated amplitude spectrum using Property 5 in Table 3.3, such that the density of points is four times the density in the upper plot.

3.9 A function given by

$$x(t) = e^{-t/4} \sin\left(\frac{2\pi t}{5}\right); \quad t \geq 0$$

is sampled once per second from $t = 0$ through $t = 16$ seconds. In the upper of two subplots, make a discrete plot of x, the sample vector. In the lower subplot, make a discrete plot of the resampled version of x, with three samples between each pair of original samples. Use the resampling method (Table 3.3, Property 7) illustrated in (3.27).

3.10 Starting with the definition of the FT in (3.29), derive the FT of the continuous signal, $x(t)$, in Exercise 3.9. (You may wish to use the example of integration by parts in Chapter 1.) Plot the amplitude spectrum, $|X(j\omega)|$ versus ω in the range [0, 10] rad/s. Is the sampling interval in Exercise 3.9 adequate to prevent aliasing? Why, or why not? How would you select an "adequate" sampling rate for $x(t)$?

3.11 The spectrum of $x(t)$ is given by the expression $X(j\omega) = X(j2\pi f) = 10e^{-|f|/2}$, where f is in Hertz. Note that the spectrum is real in this example. In each of two subplots, plot the continuous DFT of $x(t)$, that is, $\bar{X}(j\omega)$ in (3.47), over the frequency range [0, 40] Hz. Use sampling intervals $T = 0.04$ and 0.07 seconds in the upper and lower subplots, respectively. Use as many terms as necessary from the sum in 3.47 to complete the plot.

3.12 A continuous waveform is described for $t \geq 0$ as follows:

$$x(t) = 100te^{-150|t-0.02|} \sin(200\pi t)$$

Create a sample vector, x, of $x(t)$ in the range $t = [0, 0.07]$ seconds using time step $T = 0.1$ milliseconds. In the upper of two subplots, plot $x(t)$ as a continuous curve through the points in x. In the lower subplot, make a connected discrete plot of the amplitude spectrum in terms of the complete DFT of x versus frequency in the range [0, 10] KHz. Based on the lower plot, estimate the lowest sampling frequency for $x(t)$ that would not violate the sampling theorem.

3.13 Compare the DFTs of two sample vectors taken from $x(t)$ in Exercise 3.12, both over the same range t = [0, 0.07] seconds, but this time, with T = 3.0 and 2.0 milliseconds. Plot the DFT amplitudes in two subplots, one above the other, both versus frequency in KHz over the range from zero to the sampling frequency. Observe the values of the DFT amplitudes around half the sampling rate and comment on whether either or both of the two time steps are small enough to satisfy the conditions of the sampling theorem.

3.14 This is a reconstruction exercise using the resampling method described in Table 3.3, Property 7. In the center of three subplots, plot the sample vector of the waveform in Exercise 3.12 with $T = 0.1$ milliseconds. For the upper subplot, create a sample vector over the same 70-milliseconds time interval using T = 3.0 milliseconds, and plot the reconstructed continuous $x(t)$ using Property 7. Finally, do the same in the lower plot with T = 2.0 milliseconds. Construct all plots with time range [0, 70] milliseconds and amplitude range [–2, 2]. Comment on the adequacy of the sampling frequencies for subplots 1 and 3 by comparing both with subplot 2.

3.15 Repeat Exercise 3.14 using the Fourier series reconstruction instead of the resampling method. Comment on which reconstruction method is preferable for this particular signal and why it is preferable.

3.16 Show that the statement concerning the time kernel, δ, in Section 3.13 is correct; that is, to satisfy the relation $\delta^N = K + \delta$, with $N > 100$ and $K < 10^{30}$, δ must be a number in the interval from 1 to 2.

References

1. Proakis, J. G., C. M. Rader, F. Ling, C. L. Nikias, M. Moonen, and I. K. Proudler. 2002. *Algorithms for Statistical Signal Processing*. Chap. 2. Upper Saddle River, NJ: Prentice Hall.
2. Brigham, E. O. 1974. *The Fast Fourier Transform*. Chap. 10. Englewood Cliffs, NJ: Prentice Hall.
3. Ahmed, N., and K. R. Rao. 1975. *Orthogonal Transforms for Digital Signal Processing*. New York: Springer-Verlag.
4. Elliott, D. F., and K. R. Rao. 1983. *Fast Transforms*. New York: Academic Press.
5. Bergland, G. D. 1969. A guided tour of the fast Fourier transform. *IEEE Spectr* 6(7):41–52.
6. *IEEE Transactions on Audio and Electroacoustics* (Special Issues on the Fast Fourier Transform), AU-15, June 1967 and AU-17, June 1969.

7. Cooley, J. W., and J. W. Tukey. 1965. An algorithm for the machine calculation of complex Fourier series. *Math Comput* 19(90):297–301.
8. Good, I. J. 1958 and 1960. The interaction algorithm and practical Fourier series. *J R Stat Soc Ser B* 20:361 and 22:372.
9. Nixon, F. E. 1965. *Handbook of Laplace Transformation*. Englewood Cliffs, NJ: Prentice Hall.
10. Holbrook, J. G. 1966. *Laplace Transforms for Electronic Engineers*. Oxford: Pergamon Press.
11. Whittaker, E. T. 1915. Expansions of the interpolation-theory. *Proc R Soc Edinb* 35:181–94.
12. Marvasti, F., ed. 2001. *Nonuniform Sampling*. New York: Springer.
13. Higgins, J. R. 1996. *Sampling Theory in Fourier Signal Analysis*. Oxford: Oxford University Press.
14. Wikepedia, *Discrete Fourier Transform*, http://en.wikipedia.org/wiki/Discrete_Fourier_transform.
15. Kailath, T., ed. 1980. *Linear Systems*. Englewood Cliffs, NJ: Prentice Hall.
16. Oppenheim, A. V., and A. S. Wilsky. 1983. *Signals and Systems*. 2nd ed. Englewood Cliffs, NJ: Prentice Hall.
17. Roberts, M. J. 2004. *Signals and Systems, Analysis Using Transform Methods and MATLAB*. New York: McGraw-Hill.

Further Reading

Belhaouane, A., et al. 1996. Reconstruction method for jitter tolerant data acquisition system. *J Elect Testing* 9(1–2):177–85.

Cartinhour, J. 2000. *Digital Signal Processing: An Overview of Basic Principles*. Chap. 8. Upper Saddle River, NJ: Prentice Hall.

Corinthios, M. 2009. *Signals, Systems, Transforms, and Digital Signal Processing with MATLAB*. Boca Raton, FL: CRC Press.

Lyons, R. G. 1997. *Understanding Digital Signal Processing*. Chaps. 3 and 4. Reading, MA: Addison-Wesley.

Marvasti, F., ed. 2001. *Nonuniform Sampling Theory and Practice*. New York: Kluwer/Plenum.

Oppenheim, A. V., and R. W. Schafer. 1989. *Discrete-Time Signal Processing*. Chap. 8. Englewood Cliffs, NJ: Prentice Hall.

Orfanidis, S. J. 1996. *Introduction to Digital Signal Processing*. Chap. 9. Upper Saddle River, NJ: Prentice Hall.

Papoulis, A. 1966. Error analysis in sampling theory. *Proc IEEE* 54(7):947–55.

4

Linear Systems and Transfer Functions

4.1 Continuous and Discrete Linear Systems

Digital signal processing (DSP) technology has its roots in the analysis of continuous linear systems. We saw in the previous chapter how continuous signals may be regarded as limiting forms of corresponding discrete sample vectors, and the same idea holds for continuous and discrete processing systems. Continuous linear systems[7,8] are described with *linear differential equations*, and discrete linear systems are described with *linear difference equations*.

There are many reasons to keep these "continuous roots" in our thoughts as we proceed to develop the subject of DSP. In this chapter particularly, the time and frequency domains are similar for digital and continuous systems. Convolution in one domain (e.g., time) is equivalent to multiplication in the other domain (frequency) in the case of either discrete or continuous signals. Most of the other notions in continuous linear systems, such as resonance, stability, impulse response, step response, transfer function, filtering, and so on are carried over and applied in the analysis of discrete systems.

Our discussion of linear systems begins with a description of the discrete linear difference equation and goes on to develop the linear transfer function as well as some of the rest of the concepts just mentioned. The development will be complete and (we hope) understandable in the terms given; nevertheless, it will be helpful, as it has been helpful in previous chapters, always to think of the sample vectors being derived from continuous signals.

4.2 Properties of Discrete Linear Systems

A *linear* equation is an equation in which the variable terms appear only in the first degree. Take for example the following equation:

$$y_k = bx_k + ay_{k-1} \tag{4.1}$$

In this example, discrete values of the *dependent* variable, y, are related to discrete values of the *independent* variable, x. In the terminology of signal processing, when (4.1) describes a *linear system* (4.1), x is called the *input signal* and y is called the *output signal.*

The two constants in (4.1), b and a, are called *coefficients* or *weights.* We say the equation is *linear* because the elements of vectors x and y, that is, x_k, y_k, and y_{k-1}, appear only in the first degree and are not multiplied together. But would this be true if y were expressed *nonrecursively,* that is, not in terms of itself as it is in (4.1)? To answer, we must "solve" the equation so that y is given explicitly in terms of x.

Before we solve (4.1), we should note that the subscript k denotes the position in an ordered sequence (vector) as in previous chapters, and this, in turn, implies an *order of processing* in equations like (4.1). Beginning at $k = 0$ (for convenience), y_0 is computed (usually with y_{-1} assumed to be zero) as $y_0 = bx_0$, then, $y_1 = bx_1 + ay_0$, $y_2 = bx_2 + ay_1$, and so on. We also associate continuous time with the order of processing, that is, $t = kT$ as in Chapters 1–3, T being the time step between any two adjacent samples.

Several other properties of linear equations are related to the association of k with the time domain. The first is *realizability.* The linear operation in (4.1) is *realizable,* because the current value, y_k, of the output signal is computed in terms of its own past values (y_{k-1} in this case), and not in terms of any future values of y, such as y_{k+1}, y_{k+2}, and so on.

The second property is *causality.* The linear operation in (4.1) is *causal,* because y_k is computed in terms of present and past input signal values (x_k in this case), and not in terms of any future values of x. The term comes from the idea that in nature, results in the present are not affected by events that have not yet happened. For example, if x_k in (4.1) were changed to x_{k+1}, then (4.1) would be a *noncausal* operation. Examples of causality and realizability are given at the end of this section.

The linear equation (4.1) is not difficult to solve by induction, that is, by substituting repeatedly for the sample of y on the right. The result (assuming $x_k = 0$ for $k < 0$) becomes

$$\begin{aligned} y_k &= bx_k + a(bx_{k-1} + ay_{k-2}) \\ &= bx_k + a(bx_{k-1} + a(bx_{k-2} + a(bx_{k-3} + \cdots))\cdots) \\ &= b\sum_{n=0}^{k} a^n x_{k-n} \end{aligned} \tag{4.2}$$

Now we see (4.1) in a nonrecursive form, which is again linear, because all the variables, that is, the elements of vectors x and y, are in the first degree.

A linear equation such as (4.1) in which the output, y, is given in terms of its own past values as well as values of the input, is called *recursive*. If past values of y do not appear on the right, as in (4.2), then the linear equation is *nonrecursive*, that is, recursion is not needed to express y_k in terms of samples of x. Note that (4.1), which is recursive, becomes nonrecursive when expressed in the form of (4.2). However, (4.2) indicates that y_k, in response to a single nonzero input sample x_0, will continue forever, because $a^k x_0$ is always the final term in the sum. Thus, a recursive linear system is also known generally as an *infinite impulse response* (*IIR*) system. Conversely, a nonrecursive system is known as a *finite impulse response* (*FIR*) system.

Another property related to the time-domain behavior of (4.1) is *stability*. A linear system is *stable* if the magnitude of its response to a finite input does not grow without bound. In the case of (4.1), we can see from its nonrecursive form in (4.2) that (4.1) describes a stable system when $|a| < 1$ and an unstable system when $|a| > 1$, because, with $|a| > 1$, the final term in the sum in (4.2), which is $a^k x_0$, increases in magnitude with k for any nonzero value of x_0. Stability is an important property of linear systems and is discussed in Section 4.6 in terms of poles of the transfer function.

In most of our discussions, the weights in the linear equation, that is, b and a in (4.1), are constant real numbers. However, sometimes these weights are equal to, or related to, measured properties of a system, that is, *parameters* of the system, and it is possible that these parameters may change or drift with time, so that the weights are not constant but are functions of k instead. In this case, the linear equation is describing a *time-varying* or *adaptive* signal processing system. Adaptive signal processing, in which the weights change with time and adapt according to a specified purpose, has become an important part of DSP and is introduced in Chapter 9.

The foregoing properties of discrete linear equations are summarized in Table 4.1. For each property, there is a simple example where the property exists and a simple example where the property does not exist.

With these properties in mind, we are ready to discuss linear transfer functions, which, as we shall see, lead to better ways to solve recursive linear equations and, more importantly, provide useful insights into the behavior of linear systems.

TABLE 4.1

Properties of Discrete Linear Systems

Property	Exists	Does not Exist
Linear	$y_k = bx_k + ay_{k-1}$	$y_k = b\|x_k\| + ay_{k-1}$
Realizable	$y_k = bx_k + ay_{k-1}$	$y_k = bx_k + ay_{k+1}$
Causal	$y_k = bx_k + ay_{k-1}$	$y_k = bx_{k+1} + ay_{k-1}$
Recursive (IIR)	$y_k = bx_k + ay_{k-1}$	$y_k = bx_k + ax_{k-1}$
Stable	$y_k = bx_k + 0.9y_{k-1}$	$y_k = bx_k + 1.1y_{k-1}$
Adaptive	$y_k = b_k x_k + ay_{k-1}$	$y_k = bx_k + ay_{k-1}$

4.3 Discrete Convolution

We begin this section with the general form of (4.1), known as the *direct form*, that describes any causal, realizable linear system with constant weights:

$$y_k = \sum_{n=0}^{N-1} b_n x_{k-n} - \sum_{m=1}^{M-1} a_m y_{k-m}; \quad k = 0, 1, \ldots \tag{4.3}$$

Causality ($n \geq 0$) is not essential in this discussion, but it simplifies the index notation, and it also enables the system to operate in "real time," which is often required. Also, we can usually define the time index, k, so that the signals are zero for $k < 0$, and we will assume this unless we state otherwise. Thus, we have a *startup* sequence that begins as follows:

$$\begin{aligned} y_0 &= b_0 x_0 \\ y_1 &= b_0 x_1 + b_1 x_0 - a_1 y_0 \\ y_2 &= b_0 x_2 + b_1 x_1 + b_2 x_0 - (a_1 y_1 + a_2 y_0), \quad \ldots\text{etc.} \end{aligned} \tag{4.4}$$

The startup sequence continues until $k = M - 1$ or $k = N - 1$, whichever is later, that is, until the final terms in the sums in (4.3) involve samples of x and y at and above $k = 0$. From this point, the computation in (4.3) involves a constant number of $(N + M - 1)$ products.

Each of the sums in (4.3) is a *convolution* of two vectors, a weight vector and a signal vector. Convolution is a fundamental operation in continuous as well as digital signal processing. An illustration of the left-hand convolution in (4.3), with $N = 4$, is given in Figure 4.1. There is nothing really new in the illustration, but it is often helpful to picture the convolution as shown, with

$$k=0:\begin{bmatrix} & & & & b_0 & b_1 & b_2 & b_3 \\ \cdots\, x_4 & x_3 & x_2 & x_1 & x_0 & & & \end{bmatrix} \quad y_0 = b_0 x_0$$

$$k=1:\begin{bmatrix} & & & b_0 & b_1 & b_2 & b_3 & \\ \cdots\, x_4 & x_3 & x_2 & x_1 & x_0 & & & \end{bmatrix} \quad y_1 = b_0 x_1 + b_1 x_0$$

$$k=2:\begin{bmatrix} & & b_0 & b_1 & b_2 & b_3 & & \\ \cdots\, x_4 & x_3 & x_2 & x_1 & x_0 & & & \end{bmatrix} \quad y_2 = b_0 x_2 + b_1 x_1 + b_2 x_0$$

$$k\geq 3:\begin{bmatrix} & & b_0 & b_1 & b_2 & b_3 \\ \cdots & x_{k+1} & x_k & x_{k-1} & x_{k-2} & x_{k-3} \end{bmatrix} \quad y_3 = b_0 x_k + b_1 x_{k-1} + b_2 x_{k-2} + b_3 x_{k-3}$$

FIGURE 4.1
Discrete convolution of signal vector x with weight vector b, showing the alignment of the vectors at each time step to produce the sum of products on the right side of the figure.

the signal vector, x, reversed and aligned with the weight vector, b, in accordance with the time index, k, so that each product in the sum is evident.

4.4 The z-Transform and Linear Transfer Functions

Just as we used the discrete Fourier transform (DFT) to transform signals from the time domain to the frequency domain, we may also use the DFT to transform the convolutions in (4.3) from the time domain to the frequency domain. To do this, we define the *z-transform*, which is really just the DFT expressed with modified notation using the substitution

$$z = e^{j\omega T} \tag{4.5}$$

in which ωT is the frequency in radians as described in Section 3.6. The z-transform of a vector x, which we denote $X(z)$, is the continuous DFT (3.25) with the substitution (4.5), that is,

$$\boxed{z\text{-transform:}\quad X(z) = \sum_{n=0}^{N-1} x_n z^{-n}} \tag{4.6}$$

The limits of the sum in this definition are not fixed; they are set, in each case, to include all the samples in the vector x, or whatever set of samples we may wish to include, in the transform. Either limit may be infinite if so desired. For example, suppose $x_k = e^{-\alpha k}$ for $k = [0:\infty]$. Then

$$z\text{-transform:}\quad X(z) = \sum_{k=0}^{\infty} e^{-\alpha k} z^{-k} = \frac{z}{z - e^{-\alpha}};\quad \text{DFT:}\quad X(e^{j\omega T}) = \frac{e^{j\omega T}}{e^{j\omega T} - e^{-\alpha}} \tag{4.7}$$

The closed expression for $X(z)$ is found using the geometric series formula (1.20), which, together with (1.19), are the key formulas used for expressing z-transforms of functions of this sort in closed form. In modern DSP, we deal mostly with sample vectors that are not usually expressible as simple discrete functions such as x_k in (4.7). Thus, in this text, we do not (except in a few special cases) discuss the theory of forward and inverse z-transforms of functions. References Oppenheim and Schafer,[1] Orfanidis,[2] and Stearns and Hush[3] are recommended for this purpose.

We now consider an application of z-transforms, namely, the z-transform of a convolution. The transform of a convolution such as those in (4.3) is found as follows:

$$\text{Let } u_k = \sum_{n=0}^{N-1} b_n x_{k-n}, \quad 0 \le k < \infty$$

$$\text{Then } U(z) = \sum_{k=0}^{\infty}\sum_{n=0}^{N-1} b_n x_{k-n} z^{-k} = \sum_{n=0}^{N-1} b_n z^{-n} \sum_{k=0}^{\infty} x_{k-n} z^{-(k-n)} \tag{4.8}$$

$$= \sum_{n=0}^{N-1} b_n z^{-n} \sum_{i=0}^{\infty} x_i z^{-i} = B(z)X(z)$$

In the final sum, we note that $i = k - n$, and i can begin at zero, because k begins at zero, and x_k is zero for $k < 0$. Thus, the z-transform of the convolution shown here is the product of the z-transforms of the two vectors being convolved. This is true in all cases in which one of the vectors has infinite extent, as x has in this case. Cases in which both vectors are finite will be discussed later.

With the result in (4.8), we are now able to transform the direct form equation in (4.3). Let $a_0 = 1$. Then, (4.3) and its transform become

$$\sum_{m=0}^{M-1} a_m y_{k-m} = \sum_{n=0}^{N-1} b_n x_{k-n}; \quad 0 \le k \le \infty$$

$$A(z)Y(z) = B(z)X(z) \tag{4.9}$$

$$\frac{Y(z)}{X(z)} = \frac{B(z)}{A(z)} = \frac{b_0 + b_1 z^{-1} + \cdots + b_{N-1} z^{-(N-1)}}{1 + a_1 z^{-1} + \cdots + a_{M-1} z^{-(M-1)}} \equiv H(z)$$

When the transforms are written in this form, the quotient $H(z) = B(z)/A(z)$ is called a *transfer function*, because it contains all the information on how to relate the input signal to the output signal, that is, how to "transfer" the signal through the system. Furthermore, the transfer function supplies this information in the *frequency domain* because of the association of z with frequency in (4.5). In other words, equivalent descriptions of a discrete linear system in the time and frequency domains are as follows:

$$\text{Time domain: } y_k = \sum_{n=0}^{N-1} b_n x_{k-n} - \sum_{m=1}^{M-1} a_m y_{k-m} \tag{4.10}$$

$$\text{Transfer function: } H(z) = \frac{Y(z)}{X(z)} = \frac{b_0 + b_1 z^{-1} + \cdots + b_{N-1} z^{-(N-1)}}{1 + a_1 z^{-1} + \cdots + a_{M-1} z^{-(M-1)}} \tag{4.11}$$

$$\text{Frequency response: } H(e^{j\omega T}) = \frac{b_0 + b_1 e^{-j\omega T} + \cdots + b_{N-1} e^{-j(N-1)\omega T}}{1 + a_1 e^{-j\omega T} + \cdots + a_{M-1} e^{-j(M-1)\omega T}} \tag{4.12}$$

Now, $H(e^{j\omega T})$ is, as we have said, the ratio of the DFTs of the vectors a and b. Thus, just as the amplitude and phase spectra of a signal vector are given in (3.18) as the magnitude and angle of the DFT of the vector, so the *amplitude response* (or *amplitude gain*) and *phase response* (or *phase shift*) of a linear system are given by the magnitude and angle of $H(e^{j\omega T})$ in (4.12); that is,

$$\begin{aligned} \text{Amplitude gain} &= \text{abs}(H(e^{j\omega T})) = \frac{|B(e^{j\omega T})|}{|A(e^{j\omega T})|} \\ \text{Phase shift} &= \text{angle}\,(H(e^{j\omega T})) = \angle B(e^{j\omega T}) - \angle A(e^{j\omega T}) \end{aligned} \tag{4.13}$$

An example expressed in MATLAB® notation of the computation of (4.13) is given next in (4.14). The weight vectors, a and b, with $M = N = 5$ in (4.3), are copied into the first two expressions. These were generated by a function to be discussed in Chapter 6. So, for the present, we take b and a as given:

```
b=[0.2021  0.3328  0.4517  0.3328  0.2021];
a=[1.0000 -0.2251  0.6568  0.0349  0.0548];
H=fft(b,1000)./fft(a,1000);
subplot(1,2,1);
plot(linspace(0,.5,501),abs(H(1:501)));
subplot(1,2,2);
plot(linspace(0,.5,501),unwrap(angle(H(1:501))));
```

(4.14)

Next, $H(e^{j\omega T})$ is computed as the ratio of DFTs in (4.12). Both DFTs are extended with zeros to length 1000 to produce a more densely packed response vector and thus allow continuous amplitude gain and phase plots. (See "Zero Extension" in Table 3.3.) Amplitude and phase response vectors are then computed as in (4.13). [The operation of *unwrap* was explained in connection with (3.21)] The response curves are plotted in the two subplots in Figure 4.2.

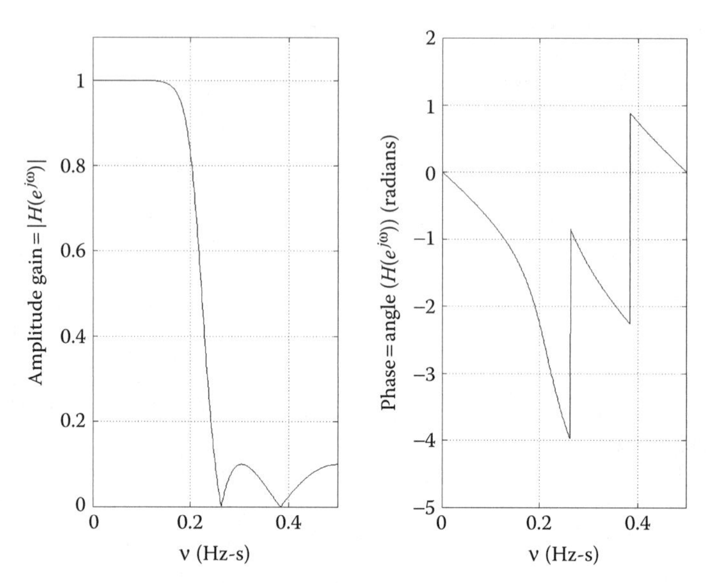

FIGURE 4.2
Amplitude gain and phase response of the linear system with weights given in (4.14). The system is called a "lowpass filter," because it passes the low-frequency components of a signal and blocks the high-frequency components.

Because, as explained in Chapter 3, the second half of a DFT is a mirror image of the first half, it is conventional to plot responses only up to half the sampling rate, that is, only for $\omega = [0, \pi/T]$ rad/s, or $\nu = [0, 0.5]$ Hz-s. Expressions for labels and so forth, in the figure are not shown in (4.14) for the sake of simplicity. The complete m-file is included in the software available on the website for this text.

As a matter of interest, the amplitude gain curve in Figure 4.2 is called a "lowpass" characteristic, because it describes a system that passes low-frequency components and blocks high-frequency components of a signal.

4.5 The Complex *z*-Plane and the Chirp *z*-Transform

The variable z in (4.5) is called a *complex variable* because it has in general a real part and an imaginary part; that is,

$$z = e^{j\omega T} = \cos \omega T + j \sin \omega T \tag{4.15}$$

Thus, z is complex unless ωT is a multiple of $\pi/2$; that is, z is real if ωT is an even multiple of $\pi/2$, imaginary if z is an odd multiple of $\pi/2$, and complex otherwise. Therefore, z must be represented as a point on a plane rather than a point on a line. Furthermore, as long as z is restricted to this definition and ω and T are real variables, the point on the z-plane must be located on the unit circle, $|z| = 1$, since

$$|z| = |\cos \omega T + j \sin \omega T| = \sqrt{\cos^2 \omega T + \sin^2 \omega T} = 1 \tag{4.16}$$

The concept is illustrated in Figure 4.3 in which a particular value of z is shown as a point on the unit circle. The real part (real component) of z is $\cos \omega T$ and the imaginary part is $\sin \omega T$, and of course, the magnitude of z is one in accordance with (4.16). Thus, as the frequency, ω, increases from zero to half the sampling rate, that is, from zero to π/T, the value of z begins at $z = 1$, moves around the upper half of the unit circle, and ends at $z = -1$.

We will see that the complex z-plane is useful in several ways in the analysis of linear systems. For example, in the next section of this chapter, when the poles and zeros of the transfer function, that is, the roots of the polynomials in the denominator and numerator of $H(z)$ in (4.11), are plotted on the z-plane, they provide a geometrical interpretation of the amplitude gain and phase shift characteristics of a linear system.

In Chapter 6, we will see how the z-plane is useful in the understanding and analysis of methods for designing IIR digital filters. The design methods

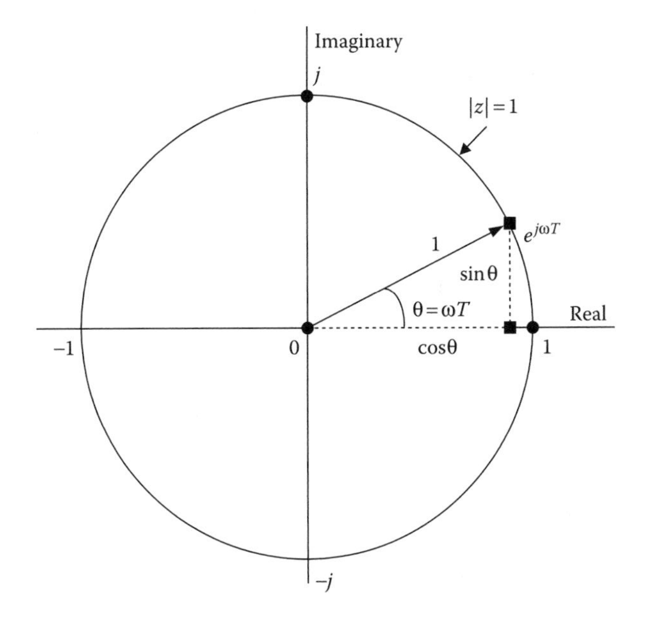

FIGURE 4.3
Illustration of the vector drawn from $z = 0$ to $z = e^{j\omega T}$ on the z-plane, with magnitude $|z| = 1$, angle θ, and components $\text{Re}\{z\} = \cos\theta$ and $\text{Im}\{z\} = \sin\theta$.

require essentially the mapping of poles and zeros of analog filters on the s-plane to digital poles and zeros on the z-plane.

In Chapter 11, which concerns digital models of continuous systems, the digital model is viewed as a mapping of the analog poles and zeros on the s-plane to digital poles and zeros on the z-plane.

As our first example of the geometry of the complex plane, we consider a modification of the DFT in Chapter 3 known as the *chirp-z transform*. We have seen in (4.13) how the DFT is used to produce the amplitude gain and phase shift of a linear system at evenly spaced points in the frequency range $\omega = [0, \pi/T]$. That is, the DFT of any vector, x, is the complex spectrum of x at each of these frequency points, and there are $N/2+1$ points spaced at equal intervals $\Delta\omega = 2\pi/NT$, as indicated in Table 3.2.

Suppose we wish to examine the gain of a linear system in a particular part of the spectrum, rather than over the entire range from $\omega = 0$ to $\omega = \pi/T$. This might be the case if, for example, the system is expected to be processing signals with power only in this particular part of the spectrum.

In other words, in computing the gain of a linear system, we may wish to compute the spectrum of a vector, x, (1) in more detail than that given by the DFT, but (2) only in part of the overall frequency range from zero to half the sampling rate.

In Section 3.8, we saw that (1) could be accomplished simply by extending x with zeros before performing the DFT (or fast Fourier transform [FFT]). But then, instead of (2), we would have computed the detailed spectrum over the entire frequency range, rather than just the desired part of the overall range. We also saw in Chapter 3 that the continuous form of the DFT of x,

$$X(\omega) = \sum_{n=0}^{N-1} x_n e^{-jn\omega T}; \quad \omega = \omega_1, \omega_2, \ldots, \omega_M \tag{4.17}$$

offers a way to accomplish both (1) and (2), but also requires MN complex products.

The *chirp-z transform*[1] is a clever method that accomplishes (2) as well as (1) and does so generally with fewer than the MN products required by (4.17). The latter is accomplished by using the FFT to perform the major computational steps in the transformation. Since we are interested in acquiring samples of the Fourier transform here, the z-plane contour is restricted to an arc along the unit circle. The transform algorithm, however, is not limited to such contours. A more general class of contours is made possible by allowing z to move off the unit circle.

To begin our description of the chirp-z transform, let x be the N-point sequence whose spectrum is of interest. The z-transform of x is given by

$$X(z) = \sum_{n=0}^{N-1} x_n z^{-n} \tag{4.18}$$

The Fourier transform of x is obtained by substituting $z = e^{j\omega T} = e^{j\theta}$, where $\theta = \omega T$ is the angular frequency in radians, that is, the angle shown in Figure 4.3 of the vector drawn on the z-plane from $z = 0$ to $z = e^{j\omega T}$.

Now, as suggested above, suppose we wish to sample the Fourier transform at M evenly spaced discrete frequencies, beginning at angular frequency θ_0, and with spacing $\Delta\theta$, that is, at angular frequencies

$$[\theta_n] = [\theta_0,\ \theta_0 + \Delta\theta,\ \ldots,\ \theta_0 + (M-1)\Delta\theta] \tag{4.19}$$

In Figure 4.3 we can see how these angular frequencies are represented by a set of points along an arc on the unit circle, as illustrated in Figure 4.4. Thus, we now have z in (4.18) restricted to this set of points, that is,

$$z_m = e^{j\theta_m} = e^{j\theta_0} e^{jm\Delta\theta}; \quad m = 0, 1, \ldots, M-1 \tag{4.20}$$

The chirp-z transform is developed by letting $A = e^{j\theta_0}$ and $B = e^{-j\Delta\theta}$. Using the definitions in (4.20) and substituting z_m into (4.18), we have $z_m = AB^{-m}$ and

$$X(z_m) = \sum_{n=0}^{N-1} x_n (AB^{-m})^{-n} = \sum_{n=0}^{N-1} x_n A^{-n} B^{mn}; \quad m = 0, 1, \ldots, M-1 \tag{4.21}$$

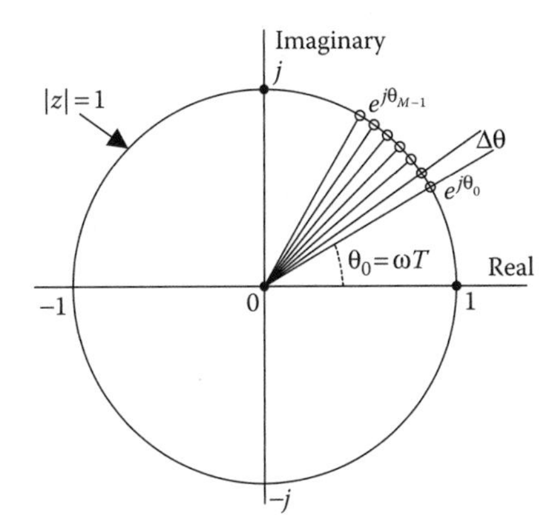

FIGURE 4.4
Illustration of points on the unit circle given by (4.17) with $\theta_0 = \pi/6$, $\Delta\theta = \pi/36$, and $M = 7$, representing seven samples of the portion of the spectrum from $\omega = \pi/6T$ through $\omega = \pi/3T$.

The next step in our development, due to Bluestein,[10] appears to complicate rather than simplify this expression, but it is the key step in arriving at an efficient version of (4.21). This step uses the following identity:

$$2nm = n^2 + m^2 - (m-n)^2 \tag{4.22}$$

With this substitution, (4.21) becomes

$$\begin{aligned} X(z_m) &= \sum_{n=0}^{N-1} x_n A^{-n} B^{n^2/2} B^{m^2/2} B^{-(m-n)^2/2} \\ &= B^{m^2/2} \sum_{n=0}^{N-1} x_n A^{-n} B^{n^2/2} B^{-(m-n)^2/2} \end{aligned} \tag{4.23}$$

Now we define two vectors, v and c, as follows:

$$v_n = x_n A^{-n} B^{n^2/2}; \quad c_n = B^{-n^2/2}; \quad n = 0,1,\ldots,N-1 \tag{4.24}$$

Using these definitions, (4.23) becomes

$$X_m \equiv X(z_m) = B^{m^2/2} \sum_{n=0}^{N-1} v_n c_{m-n}; \quad m = 0,1,\ldots,M-1 \tag{4.25}$$

Thus, with Bluestein's substitution, the spectral computation in (4.17) has been formulated in (4.25) in terms of a convolution, and the key to an efficient implementation is to perform the convolution using a product of transforms as explained in Section 4.4. The algorithm for this implementation is illustrated in Figure 4.5 and is implemented in the function *chirp_z.m* included in the *functions* folder of the software available on the publisher's website.

The name "chirp" stems from the nature of the vector c in (4.24). Using the definition $B = e^{-j\Delta\theta}$ in (4.24) results in

$$c_n = (e^{-j\Delta\theta})^{-n^2/2} = e^{j(\Delta\theta/2)n^2} \tag{4.26}$$

Thus, c_n is a unit vector like the vector in Figure 4.3, which moves at an increasing rate (accelerates) around the unit circle as n increases. Such a

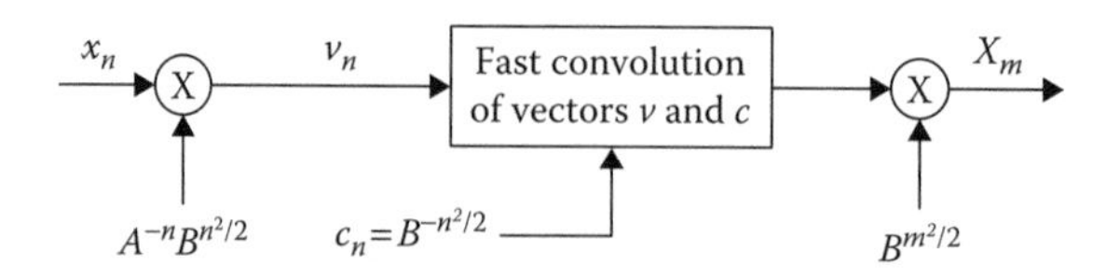

FIGURE 4.5
Illustration of the chirp-z transform algorithm, which is implemented in the function *chirp_z.m*.

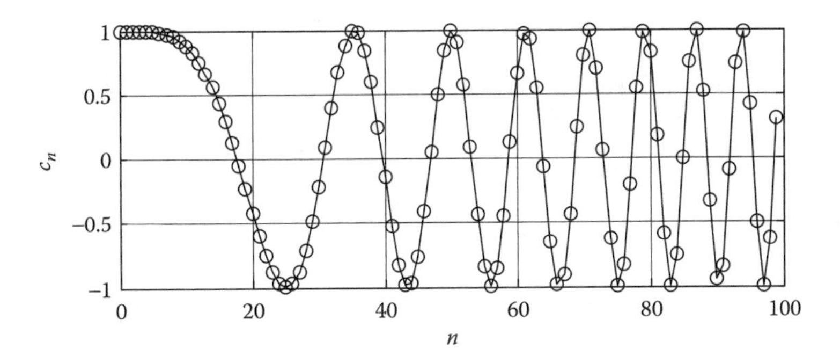

FIGURE 4.6
Real part of the "chirping" vector c in (4.24) with $\Delta\theta = 0.01$ and $N = 100$.

signal is said to "chirp" because its frequency increases with n, as illustrated in Figure 4.6, in which the real part of the vector c is plotted. In the figure, $\Delta\theta$ has been reduced (from its value 4.16) to 0.01, and N has been increased to 100, to achieve more closely spaced samples of the chirping signal.

The fast convolution block in Figure 4.5 amounts essentially to the inverse FFT of the product of the FFTs of vectors v and c, as in (4.8). However, each vector must be modified prior to its FFT to account for the finite lengths of both vectors, which was not the case in (4.8), where one of the vectors is infinite in length. These modifications are discussed later in this chapter under "FFT Algorithms" in Section 4.13 (specifically in Equation 4.80), and they may also be seen in the code of the function *chirp_z.m*. The use of *chirp_z* is demonstrated in Example 4.1.

EXAMPLE 4.1

An example in which the chirp-z transform is used to approximate the spectrum of a band-limited vector given in Figure 4.7. The vector itself could represent any signal, including the impulse response of a digital filter. In the latter case, the spectrum of the signal becomes the complex gain of the filter, as discussed in Section 4.7. The vector in this case is a segment of a band-limited random waveform, and it serves to illustrate the use of the chirp-z transform in approximating the continuous spectrum. We can see in Figure 4.7 that the DFT magnitude (that is, the amplitude spectrum) in the center plot gives a rough estimate of the continuous amplitude spectrum, and that the chirp-z transform, with a higher density of points in the frequency band of interest, produces a more detailed estimate.

4.6 Poles and Zeros

The geometrical interpretation of the z-transform described in Section 4.5 and illustrated in Figure 4.3, when applied to any transfer function, $H(z)$, becomes a useful aid in understanding the steady-state frequency response

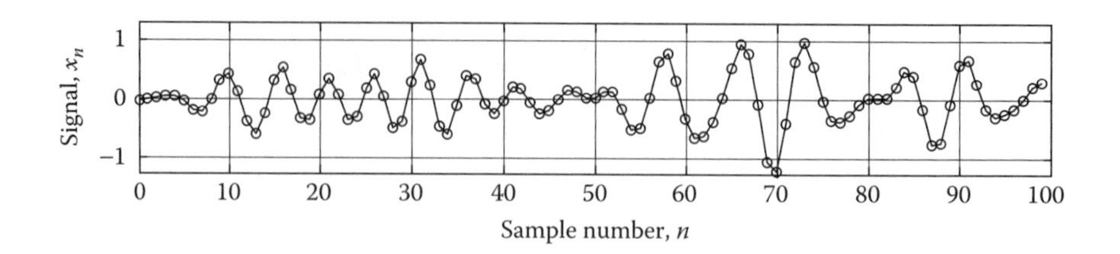

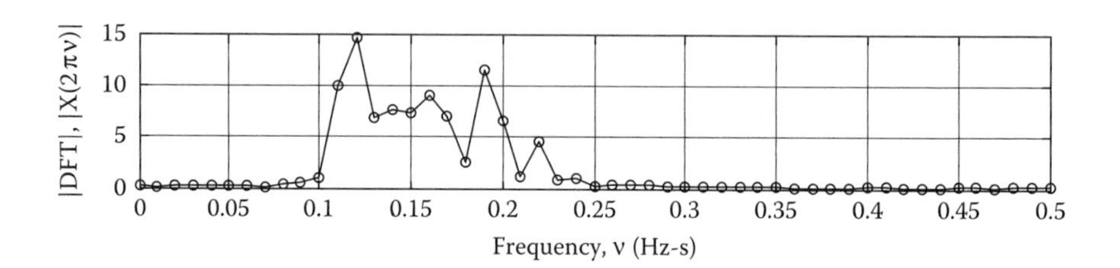

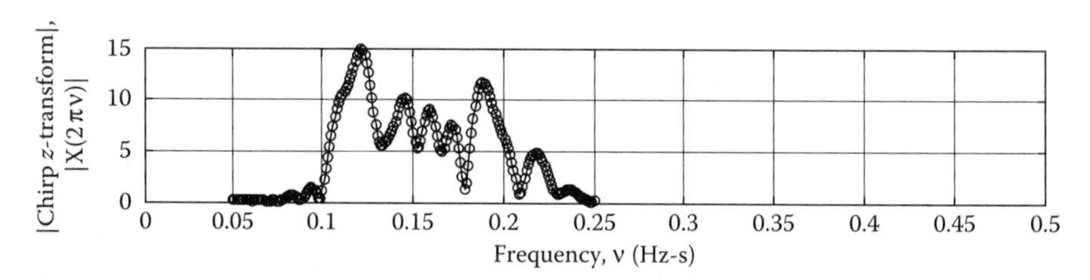

FIGURE 4.7
Chirp-z transform used to approximate the continuous amplitude spectrum of a band-limited signal.

and also the transient response of the corresponding linear system. The interpretation involves the *z-plane,* with Cartesian coordinates that are the real and imaginary parts of z, and the *poles* and *zeros* of $H(z)$. The latter are the roots, respectively, of the denominator and numerator of $H(z)$ in (4.11).

We can best explain this transfer function geometry with a simple example. Suppose the linear transfer function, corresponding with $y_k = x_k + x_{k-1} + 0.9y_{k-1}$, is as follows:

$$H(z) = \frac{1+z^{-1}}{1-0.9z^{-1}} = \frac{z+1}{z-0.9} \tag{4.27}$$

The pole at $z = 0.9$ and the zero at $z = -1$ are plotted on the z-plane in the left half of Figure 4.8. The numerator and denominator of the complex gain, $H(e^{j\omega T})$, are also shown as vectors on the z-plane. First, we see that the *amplitude gain* is a ratio of *distances,* α/β, that is,

$$|H(e^{j\omega T})| = \frac{|e^{j\omega T}+1|}{|e^{j\omega T}-0.9|} = \frac{\alpha}{\beta} \tag{4.28}$$

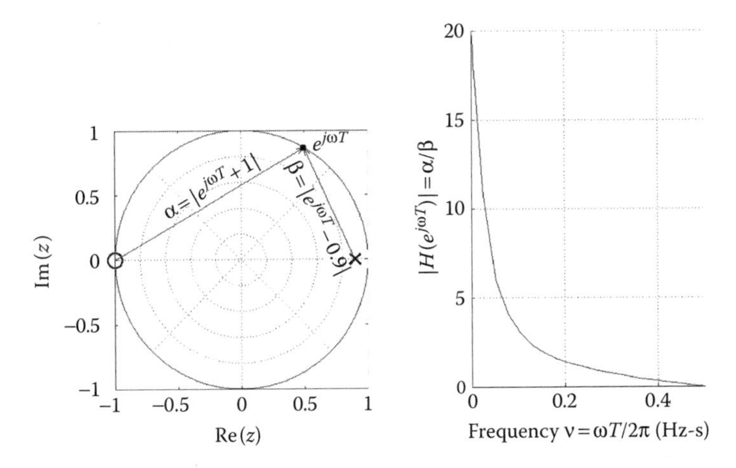

FIGURE 4.8
Left: Pole–zero plot of $H(z) = (z+1)/(z-0.9)$, showing how the amplitude gain at ω radians per second may be visualized as a quotient of lengths, that is, $|H(e^{j\omega T})| = \alpha/\beta$. Right: Plot of $|H(e^{j\omega T})|$ versus frequency in Hz-s.

The point on the unit circle labeled $e^{j\omega T}$ designates the frequency at which the gain is being observed and is called the *operating point*. As the frequency, ω rad/s, moves from zero to half the sampling rate, the operating point moves from 0° to 180° around the upper half of the unit circle. As it moves, we can see, in this case, that α decreases from 2.0 to 0, and β increases from 0.1 to 2.0. Thus, we are able to see, from the locations of the pole and zero, the form of the amplitude gain plot, without actually calculating values in (4.28). The amplitude gain plot is shown in the right half of Figure 4.8.

A similar plot of the phase response of the linear system in (4.27) is shown on the left in Figure 4.9. As in (4.13), the phase response is the algebraic sum of angles of the poles and zeros of $H(e^{j\omega T})$. In Figure 4.9, the zero angle is labeled θ, and the pole angle is labeled φ, so the phase response is as follows:

$$\angle H(e^{j\omega T}) = \theta - \varphi \tag{4.29}$$

Note that as ω increases from 0 to π, $[\theta, \varphi]$ increase from [0, 0] to $[\pi/2, \pi]$. Thus, the phase response in (4.29) decreases from 0 to $-\pi/2$, as seen on the right in Figure 4.9.

More generally, to find the poles and zeros of a linear system in the form of (4.11), the first step is to write $H(z)$ as a ratio of polynomials in z:

$$H(z) = \frac{b_0 + b_1 z^{-1} + \cdots + b_{N-1} z^{-(N-1)}}{1 + a_1 z^{-1} + \cdots + a_M z^{-(M-1)}} = z^{M-N} \frac{b_0 z^{N-1} + b_1 z^{N-2} + \cdots + b_{N-1}}{z^{M-1} + a_1 z^{M-2} + \cdots + a_{M-1}} \tag{4.30}$$

Then, the poles and zeros are the roots of the denominator and numerator polynomials in (4.30). Notice that, if $N \neq M$, there is also a pole or zero of order $|M - N|$ at $z = 0$.

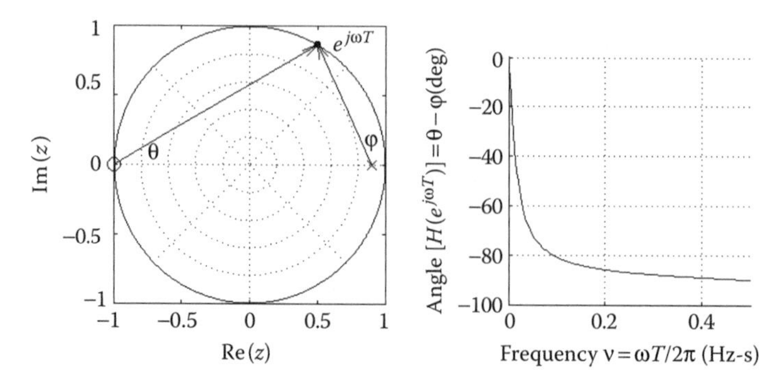

FIGURE 4.9
Left: Pole–zero plot of $H(z) = (z+1)/(z-0.9)$, showing how the phase shift at ω radians per second may be visualized as a difference of angles, that is, $\angle H(e^{j\omega T}) = \theta - \varphi$. Right: Plot of $\angle H(e^{j\omega T})$ versus frequency in Hz-s.

In MATLAB, the roots may be determined with the *roots* function. The syntax is r = *roots*(w), in which w is a vector of weights and r is a vector of roots. The corresponding polynomial form for r = *roots*(w) is as follows:

$$p(z) = w_1 z^{N-1} + w_2 z^{N-2} + \cdots + w_N = w_1(z - r_1)(z - r_2)\cdots(z - r_{N-1}) \tag{4.31}$$

The *roots* function allows us to make pole–zero plots easily using MATLAB. For example, the system in (4.14) resulted in the gain and phase plots in Figure 4.2. The poles and zeros of the transfer function may be obtained with the following expressions:

```
b=[0.2021   0.3328   0.4517   0.3328   0.2021];
a=[1.0000  -0.2251   0.6568   0.0349   0.0548];
zeros=roots(b)
poles=roots(a)
```
(4.32)

When these expressions in Equation 4.32 are executed, the zeros and poles are printed in the MATLAB command window as follows:

```
zeros =
   -0.7444 + 0.6677i
   -0.7444 - 0.6677i
   -0.0789 + 0.9969i
   -0.0789 - 0.9969i

poles =
    0.1668 + 0.7577i
    0.1668 - 0.7577i
   -0.0542 + 0.2968i
```
(4.33)

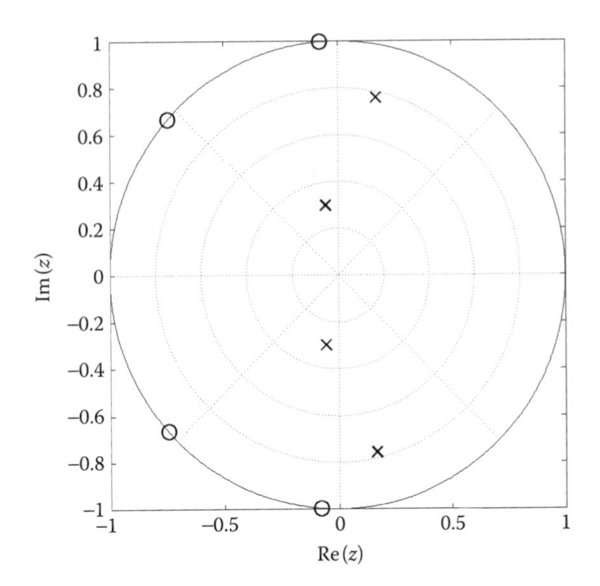

FIGURE 4.10
Poles (×) and zeros (o) of the linear transfer function with coefficients given in (4.14) and amplitude and phase responses plotted in Figure 4.2.

When roots are complex, as these are, they always appear in *conjugate pairs*. Otherwise, as one can easily see by reconstructing the polynomial from the roots, some of the weights would have to be complex.

The poles and zeros in (4.33) are plotted on the z-plane in Figure 4.10. Imagine moving the operating point around the unit circle from $\omega T = 0$ to $\omega T = \pi$, and note how the ratio of distances from the operating point to all the poles and zeros produces the amplitude gain plot in Figure 4.2. Also note that the 180°-phase jumps produced by the zeros on the unit circle in Figure 4.10 appear on the phase plot in Figure 4.2. In this way, the pole–zero plot illustrates the response of a linear system at all frequencies.

The function *pz_plot(b, a)* is included in the *functions* folder on the publisher's website for this text. It uses the *roots* function internally to produce the actual plot in a MATLAB figure window. The plot in Figure 4.10 was produced by executing *pz_plot(b, a)*, where b and a were specified as in (4.32).

4.7 Transient Response and Stability

Having the poles and zeros of a linear system allows us to express the direct form transfer function in (4.11) as a *sum of partial fractions*. Even with the poles and zeros given, the partial-fraction form is not easy to construct. Another MATLAB function called *residue* is useful in the construction, and it is, in fact, used in the function *part_frac_exp* provided on the publisher's website for this text, which computes the pole–zero expansion of the direct-form transfer function (4.11). The operation of these functions is complicated when there are *multiple poles* [a multiple pole is due to a term of the form $(z-p)^n$ with $n > 1$ in the denominator of $H(z)$]. We discuss the partial fraction form only briefly in this section, because normally, in DSP, partial fraction expansions of transfer functions are needed only in the special situations discussed here.

For an example of the use of *part_frac_exp*, we again use the linear system with weights given in (4.14) as follows:

```
b=[0.2021   0.3328   0.4517   0.3328   0.2021];
a=[1.0000  -0.2251   0.6568   0.0349   0.0548];
[r,p,c]=part_frac_exp(b,a)
```
(4.34)

When these expressions are executed, the result is

```
zeros =
   -0.7444 + 0.6677i
   -0.7444 - 0.6677i
   -0.0789 + 0.9969i
   -0.0789 - 0.9969i

poles =
    0.1668 + 0.7577i
    0.1668 - 0.7577i
   -0.0542 + 0.2968i
   -0.0542 - 0.2968i
```
(4.35)

and the corresponding partial-fraction form of $H(z)$ is

$$H(z) = c + \frac{r_1 p_1 z}{z - p_1} + \frac{r_2 p_2 z}{z - p_2} + \frac{r_3 p_3 z}{z - p_3} + \frac{r_4 p_4 z}{z - p_4} \tag{4.36}$$

Further aspects of partial-fraction expansions may be found in the references[1,2,9], but we wish to turn to a more important topic, namely, *transient response* and the question of *stability*.

The *transient response* of a system is the output signal resulting from a single transient input signal, such as an impulse or a step function. In particular, the *impulse response* of a linear system is the response to a single unit sample at $t = 0$, that is, the response to the *unit impulse*:

$$\text{Unit impulse:} \quad i = [1\ 0\ 0\ \ldots] \tag{4.37}$$

In definition (4.6), we can see that the z-transform of the unit impulse is $I(z) = 1$, and therefore, the impulse response of any linear system is just the inverse transform of the transfer function, that is, $Z^{-1}\{H(z)\}$.

Consider now the impulse response of $H(z)$ in (4.36), which is a constant, c, plus the sum of responses contributed by each additional term. Any of these terms describes a simple linear system of the form given in (4.1), with $b = r_n p_n$ and $a = p_n$ and the corresponding "solution" for the output in (4.2). When applying this solution with $x = i$ in (4.37), the impulse response contribution from this nth term is found to be of the following form:

$$y_k = r_n p_n (p_n)^k = r_n p_n^{k+1}; \quad k = 0, 1, \ldots \tag{4.38}$$

Here, we can see that the impulse response will increase with k, provided $|p| > 1$, that is, provided the pole is outside the unit circle on the z-plane, $|z| = 1$. From this result, we can easily reach the following general conclusion concerning the stability of linear systems:

A system is *unstable* if its response to a transient input increases without bound. The linear system described by $H(z)$ is *stable* if and only if all poles of $H(z)$ are inside the unit circle on the z-plane.	(4.39)

This is the usual criterion for stability. If $H(z)$ has pole(s) *on* the unit circle, the system is said to be *conditionally stable.*

4.8 System Response via the Inverse z-Transform

We have just seen a situation where the inverse z-transform was useful in determining the response of a linear system. In general, given analytic descriptions of the input signal, x, and the transfer function, $H(z)$, we can obtain the output signal, y, as an inverse z-transform, that is,

$$y = Z^{-1}\{X(z)H(z)\} \tag{4.40}$$

The uses of this approach are limited, because signals are not usually known in analytic form, but sometimes, the impulse or step-function response of a system, or the transient response to a sinusoidal input, is required, and the inverse transform approach is useful in these cases.

In general, inverse z-transforms and inverse Laplace transforms are derived with the help of the Residue Theorem, which is part of complex variable theory.[4] Because this subject is not a normal prerequisite for a first course in DSP, we will not pursue it here. Instead, a short list of z-transforms is included in Table 4.2, and a longer list in the Appendix will be of use later in this text. Table 4.2 is arranged in conventional form, with the transforms on the right.

TABLE 4.2

A Short Table of z-Transforms

Line	Function of k	Function of z
A	$x = [x_0, x_1, \ldots]$	$X(z) = \sum_{k=0}^{\infty} x_k z^{-k}$
B	$\alpha x + \beta y$	$\alpha X(z) + \beta Y(z)$
C	$y_k = \begin{cases} 0; & k < \lambda \\ x_{k-\lambda}; & k \geq \lambda \end{cases}$	$Y(z) = z^{-\lambda} X(z)$
D	$y_k = e^{-\alpha k} x_k; \quad k \geq 0$	$Y(z) = X(ze^{\alpha})$
E	$y_k = k\, x_k; \quad k \geq 0$	$Y(z) = -z \dfrac{\partial}{\partial z} X(z)$
1	[1, 0, 0, ...] (impulse function)	1
2	[1, 1, 1, ...] (step function)	$\dfrac{z}{z-1}$
3	$\alpha^k; \quad k \geq 0$ and $\lvert\alpha\rvert < 1$	$\dfrac{z}{z-\alpha}$
4	$\sin(\alpha k); \quad k \geq 0$	$\dfrac{z \sin\alpha}{z^2 - 2z\cos\alpha + 1}$
5	$\cos(\alpha k); \quad k \geq 0$	$\dfrac{z(z - \cos\alpha)}{z^2 - 2z\cos\alpha + 1}$
6	$\dfrac{\alpha^k - \beta^k}{\alpha - \beta}; \quad \lvert\alpha\rvert, \lvert\beta\rvert < 1$	$\dfrac{z}{(z-\alpha)(z-\beta)}$
7	$\dfrac{R^k}{\beta}\sin(k\theta); \quad R = \sqrt{\alpha^2 + \beta^2},\ \theta = \tan^{-1}\left(\dfrac{\beta}{\alpha}\right)$	$\dfrac{z}{(z-\alpha)^2 + \beta^2}$

By combining lines in the table, one can find transforms and inverse transforms of a wide variety of functions. The lettered lines, A through E, are functional relations that may be proved from the definition of the z-transform, which appears on line A. Notice that lines B and C are equivalent to the linearity and phase shift properties (lines 2 and 8) of the DFT in Table 3.3.

The numbered lines, 1 through 7, then provide the transforms of specific functions. These lines may be combined or modified using the relations on the lettered lines to produce other transform pairs. For example, we would produce the transform of $e^{-\lambda k}\sin(\alpha k)$ using lines D and 4 (Table 4.2) as follows:

$$Z\{e^{-\lambda k}\sin(\alpha k)\} = \frac{ze^{\lambda}\sin\alpha}{z^2 e^{2\lambda} - 2ze^{\lambda}\cos\alpha + 1} \tag{4.41}$$

Table 4.2 may be used in a number of ways to determine transient response characteristics of linear systems. For example, we saw in Section 4.7 that a linear transfer function may be expressed as a sum of partial fractions, as in (4.36). The conjugate pairs of fractions may then be combined to produce rational quadratic forms with real coefficients. We will discuss this form of the transfer function (called the *parallel* form) in the next section, but here we note that with $H(z)$ written in the parallel form, lines 6 and 7 of Table 4.2 may be used to determine the impulse response of each term, and thus (using line B), the impulse response of $H(z)$. There are exercises on this and other topics relating to Table 4.2 in the "Exercises" section at the end of this chapter.

4.9 Cascade, Parallel, and Feedback Structures

We have seen how the linear transfer function, $H(z)$ in (4.11), may be factored so that its numerator and denominator are in the form of (4.31) or are written as a sum of partial fractions in the form of (4.36). The terms in the factored version may be combined to form a *cascade filter structure*, and the terms in the partial fraction version may be combined to form a *parallel* structure. Both structures are illustrated in Figure 4.11. The corresponding relationships are as follows:

$$\text{Cascade:}\quad H(z)=\prod_{n=1}^{N}H_n(z);\quad \text{Parallel:}\quad H(z)=\sum_{n=1}^{N}H_n(z) \qquad (4.42)$$

When the two forms in (4.42) are used to implement a given linear transfer function, $H(z)$, the overall transfer function is the same, but the individual terms, $H_n(z)$, are not the same. They imply different computations in the time domain, and sometimes, especially in DSP hardware design, one form may be preferred over the other.

A third form, the *feedback structure*, is illustrated in Figure 4.12. The output signal, y, is processed by $H_2(z)$ and added to the input signal, x. The sum is

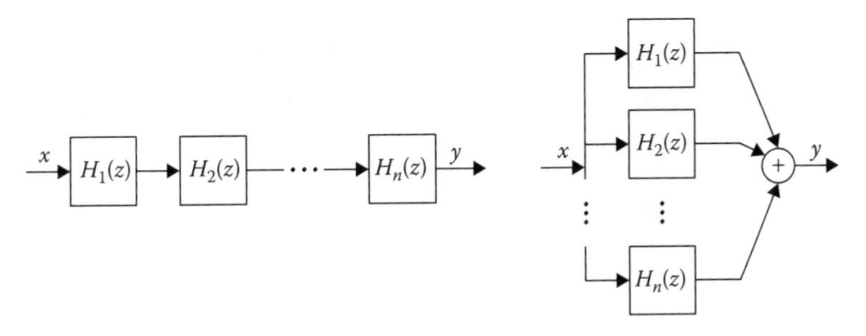

FIGURE 4.11
Cascade (left) and parallel (right) filter structures.

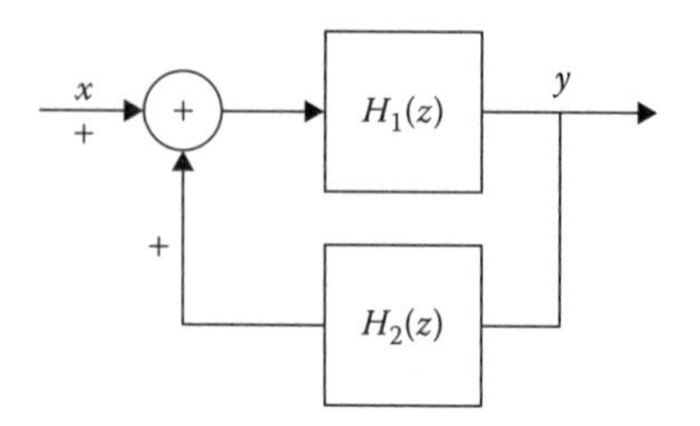

FIGURE 4.12
Linear feedback. The output, y, is filtered by $H_2(z)$, and the result is added to the input, x. Then, the sum is filtered by $H_1(z)$ to produce y. For realizability, the path through $H_1(z)$ and $H_2(z)$ must include a delay.

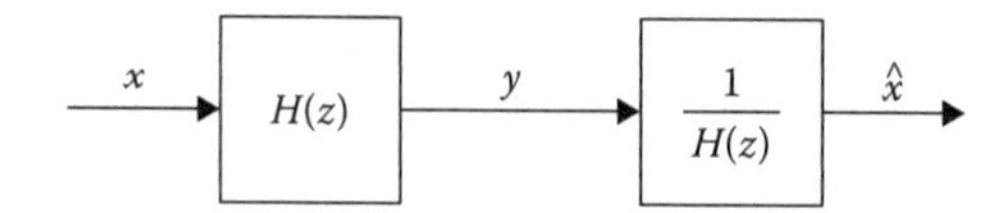

FIGURE 4.13
Inverse operation to recover x from y.

then processed by $H_1(z)$ to produce y. For realizability, element y_k obviously cannot be computed in terms of itself. Therefore, we must include a delay in the path through $H_1(z)$ and $H_2(z)$. Assuming this is true and the structure is realizable, we can apply the transform relation (4.11) and obtain the following:

$$Y(z) = (X(z) + Y(z)H_2(z))H_1(z)$$

Solving this for the overall transfer function, $H(z)$, we obtain

$$H(z) = \frac{Y(z)}{X(z)} = \frac{H_1(z)}{1 - H_1(z)H_2(z)} \tag{4.43}$$

The stability of the linear feedback form does not depend on the individual stabilities of $H_1(z)$ and $H_2(z)$ but instead relies on the location of the poles in (4.43). Notice also, that for $H(z)$ to be *realizable*, the combination $H_1(z)H_2(z)$ must delay the input by at least one sample.

The final form we wish to discuss in this section, illustrated in Figure 4.13, involves a simple concept—the *inversion* of a transfer function. In the figure, we can see that if the second system is the inverse of the first, then, except for delays and possible round-off errors, vectors $\hat{x}$ and x should be identical. Two questions arise about such a system: (1) Is it ever useful? and (2) When is the inverse operation realizable?

Regarding the first question, there are at least two useful applications—encryption and compression. If x is a secret signal, then $H(z)$ could amount to an "encryption key" that allows one to send y in the clear and allows a receiver with knowledge of $H(z)$ to recover x from y. In the compression application, $H(z)$ could be a filter that reduces the amount of information in x in a manner such that y requires a shorter description and is thus a *compressed* version of x. There is an example of signal compression and recovery in Chapter 8.

The answer to the second question is that, at least with signal vectors of finite length, inversion is essentially always possible. This can be seen simply by solving (4.9) for x_k:

$$x_k = \frac{1}{b_0}\left[y_k + \sum_{m=1}^{M-1} a_m y_{k-m} - \sum_{n=1}^{N-1} b_n x_{k-n} \right]; \quad k \geq 0 \tag{4.44}$$

In accordance with Table 4.1, this result is both realizable and causal. One may argue further that because y was computed originally from x, and y is finite, instability cannot become a problem in this case.

As a final point, (4.44) is obviously not valid if $b_0 = 0$. If such is the case, then (4.9) can be solved for x_{k-i}, where b_i is the first nonzero weight. Then, the inverse operation involves a delay but is still realizable.

The structures reviewed in this section are representative of a variety of electromechanical systems and processes, as well as other physical phenomena. The feedback concept is applicable to *feedback control systems*, that is, control systems that measure their own performance. Many control applications that were formerly accomplished with continuous electromechanical systems are now implemented with digital processors, essentially in the form of Figure 4.12.

4.10 Direct Algorithms

Each of the structures just described is an assembly of linear transfer functions; each is in the direct form of (4.11). The direct form of the transfer function implies a computational procedure, or *algorithm*, in the time domain, just as (4.11) implies (4.10). But, (4.10) represents only one of several algorithms, all of which produce (4.11) as the transfer function. The algorithms are all equivalent in that they produce the same output signal in response to a given input signal, but each implies a different set of computations, or, if implemented in hardware, a different chip design.

First, three *direct* forms of (4.10) are shown in Figures 4.14 through 4.16. These are all equivalent and produce the same output for a given input, but the algorithms differ and imply different realizations in hardware. For example, the first diagram would require twice as many storage units compared with the other two. In languages like MATLAB, any of the three forms may easily be implemented. The MATLAB *filter* function accepts data in the form of Figure 4.14.

The diagram in Figure 4.14 is seen to represent explicitly the direct form in (4.10). The "u" symbols in Figures 4.15 and 4.16 each stand for intermediate signal vectors like x and y. To prove the form in Figure 4.15 is equivalent to

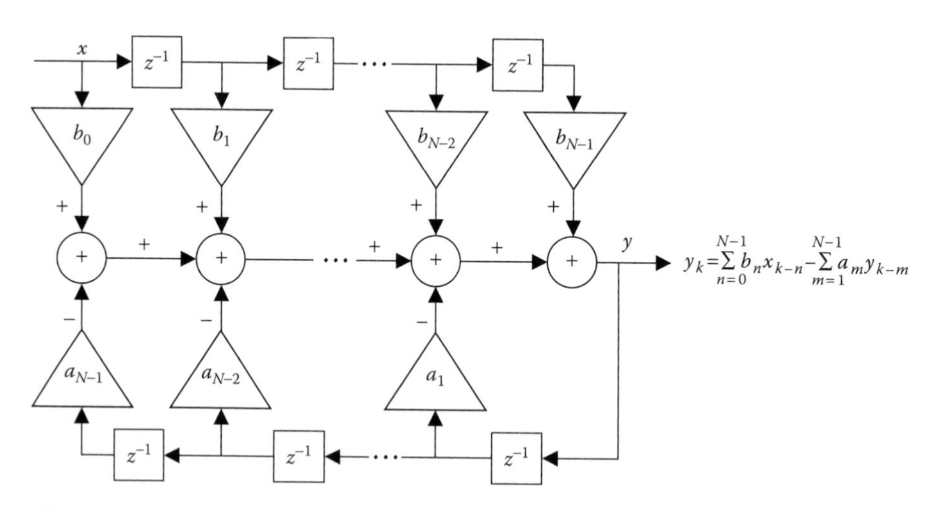

FIGURE 4.14
Linear system diagram 1. The diagram illustrates the direct form in (4.10), with $N = M$. Any of the weights may be zero.

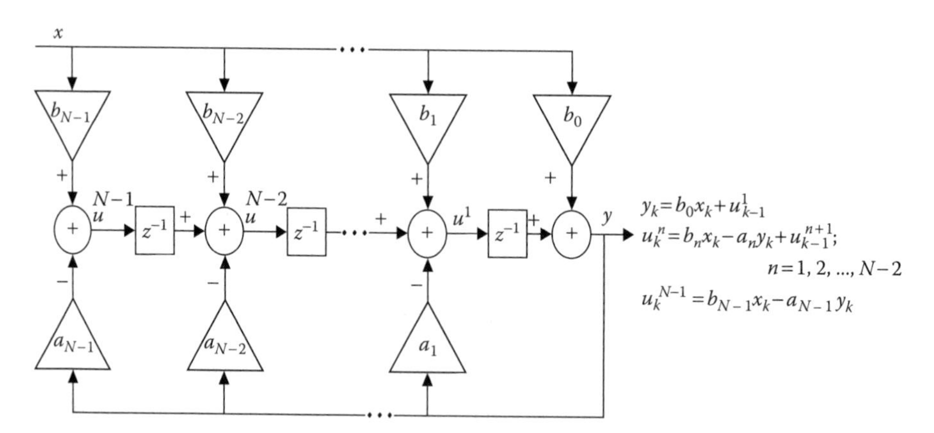

FIGURE 4.15
Linear system diagram 2. The diagram illustrates the direct form with $N = M$, but the form of the algorithm differs from (4.10). Any of the weights may be zero.

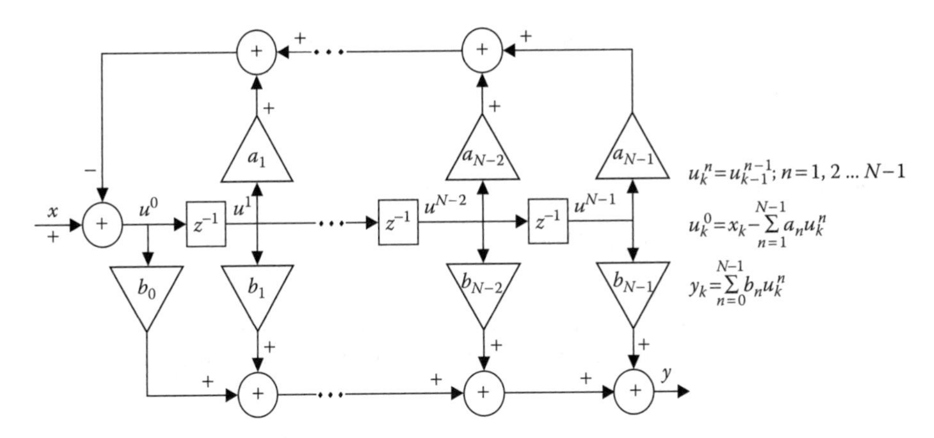

FIGURE 4.16
Linear system diagram 3. The diagram illustrates the direct form with $N = M$, in a form that differs from the two diagrams shown in Figures 4.14 and 4.15. Any of the weights may be zero.

(4.10), substitute the second line of the equation into the first line recursively for $n = 1, 2, \ldots, N - 2$, and substitute the last line for $n = N - 1$, as follows:

$$\begin{aligned} y_k &= b_0 x_k + (b_1 x_{k-1} - a_1 y_{k-1} + (b_2 x_{k-2} - a_2 y_{k-2} + \cdots \\ &\qquad + (b_{N-1} x_{k-N+1} - a_{N-1} y_{k-N+1}) \cdots)) \\ &= \sum_{n=0}^{N-1} b_n x_{k-n} - \sum_{m=1}^{N-1} a_m y_{k-m} \end{aligned} \tag{4.45}$$

Proving that Figure 4.16 is equivalent to (4.10) is more difficult. The first equation recurses by self-substitution to $u_k^n = u_{k-n}^0$. Substituting this result into the second equation, we obtain the following:

$$u_k^0 = x_k - \sum_{n=1}^{N-1} a_n u_{k-n}^0, \quad \text{or} \quad x_k = \sum_{n=0}^{N-1} a_n u_{k-n}^0; \quad a_0 \equiv 1 \tag{4.46}$$

Similarly, substituting $u_k^n = u_{k-n}^0$ into the third equation produces

$$y_k = \sum_{n=0}^{N-1} b_n u_{k-n}^0 \tag{4.47}$$

Next, as in (4.8), we take the z-transforms of the convolutions in (4.46) and (4.47) and eliminate the transform of u^0 as follows:

$$X(z) = A(z)U^0(z); \quad Y(z) = BU^0(z); \quad \therefore \frac{Y(z)}{X(z)} = \frac{B(z)}{A(z)} \tag{4.48}$$

This is the transfer function in (4.11), and therefore, Figures 4.16 and 4.14 are equivalent.

4.11 State-Space Algorithms

In addition to the three direct forms in Figures 4.14 through 4.16, we have the *state-space* or *state-vector* form, which is closely related to the third form in Figure 4.16. In this form, the internal *state* of the system at time step k is specified by the *state vector*,

$$u_k = [u_k^{N-1} \ u_k^{N-2} \ \cdots \ u_k^1]' \tag{4.49}$$

Each of the elements of the state vector is called a *state variable* of the system. Note that the state vector is a column vector and does not include the input state variable, u^0. Next, we define the following arrays, each having $N-1$ rows:

$$A = \begin{bmatrix} 0 & 1 & 0 & 0 & \cdots & 0 & 0 \\ 0 & 0 & 1 & 0 & \cdots & 0 & 0 \\ \vdots & & & & & & \\ 0 & 0 & 0 & 0 & \cdots & 0 & 1 \\ -a_{N-1} & -a_{N-2} & -a_{N-3} & -a_{N-4} & \cdots & -a_2 & -a_1 \end{bmatrix}$$

$$B = \begin{bmatrix} 0 \\ 0 \\ \vdots \\ 0 \\ 1 \end{bmatrix} \quad C = \begin{bmatrix} b_{N-1} - b_0 a_{N-1} \\ b_{N-2} - b_0 a_{N-2} \\ \vdots \\ b_2 - b_0 a_2 \\ b_1 - b_0 a_1 \end{bmatrix} \tag{4.50}$$

With these definitions, the state-space equations for Figure 4.16 are as follows:

$$\boxed{\begin{aligned} u_{k+1} &= A * u_k + Bx_k \\ y_k &= C' * u_k + b_0 x_k \end{aligned}} \tag{4.51}$$

These are the state-space equations for a linear discrete system, with array products indicated with "*". The state-space form is used often in control literature and with feedback systems in general.

To show that the state-space equations represent the diagram in Figure 4.16 and therefore any linear system, we provide the following demonstration with $N = 4$, which should be easier to follow and understand than a formal proof. With $N = 4$ and the definitions just given, we have the following arrays:

$$u_k = \begin{bmatrix} u_k^3 \\ u_k^2 \\ u_k^1 \end{bmatrix} \quad A = \begin{bmatrix} 0 & 1 & 0 \\ 0 & 0 & 1 \\ -a_3 & -a_2 & -a_1 \end{bmatrix} \quad B = \begin{bmatrix} 0 \\ 0 \\ 1 \end{bmatrix} \quad C = \begin{bmatrix} b_3 - b_0 a_3 \\ b_2 - b_0 a_2 \\ b_1 - b_0 a_1 \end{bmatrix} \tag{4.52}$$

Substituting (4.52) into the first equation in (4.51), we obtain

$$u_{k+1} = \begin{bmatrix} u_{k+1}^3 \\ u_{k+1}^2 \\ u_{k+1}^1 \end{bmatrix} = \begin{bmatrix} 0 & 1 & 0 \\ 0 & 0 & 1 \\ -a_3 & -a_2 & -a_1 \end{bmatrix} * \begin{bmatrix} u_k^3 \\ u_k^2 \\ u_k^1 \end{bmatrix} + \begin{bmatrix} 0 \\ 0 \\ 1 \end{bmatrix}; \quad x_k = \begin{bmatrix} u_k^2 \\ u_k^1 \\ x_k - \sum_{n=1}^{3} a_n u_k^n \end{bmatrix} = \begin{bmatrix} u_k^2 \\ u_k^1 \\ u_k^0 \end{bmatrix} \tag{4.53}$$

Thus, in (4.53), we have the first two equations shown in Figure 4.16 with $N = 4$. Substituting (4.52) into the second equation in (4.51), we obtain

$$\begin{aligned} y_k &= [b_3 - b_0 a_3 \quad b_2 - b_0 a_2 \quad b_1 - b_0 a_1] * \begin{bmatrix} u_k^3 \\ u_k^2 \\ u_k^1 \end{bmatrix} + b_0 x_k \\ &= b_0 \left(x_k - \sum_{n=1}^{3} a_n u_k^n \right) + b_1 u_k^1 + b_2 u_k^2 + b_3 u_k^3 = \sum_{n=0}^{3} b_n u_k^n \end{aligned} \tag{4.54}$$

This completes the demonstration with $N = 4$, because the result in (4.54) is the last line in Figure 4.16. The demonstration works in the same way for any value of N.

The state-space equations (4.51) may also be written in a form that allows an output signal element, y_k, to be computed at any time step nonrecursively, that is, in terms only of the initial state vector and the history of the input signal. This nonrecursive form of (4.51) is as follows:

$$\boxed{\begin{aligned} u_k &= A^k * u_0 + \sum_{m=0}^{k-1} A^m * B x_{k-1-m} \\ y_k &= C' * u_k + b_0 x_k \end{aligned}} \tag{4.55}$$

The second lines in (4.51) and (4.55) are identical. The validity of the first line of (4.55) may be proved by induction. First, with $k = 1$, and writing the first line of (4.51) for k instead of $k + 1$, the two equations for u_k are the same.

Now, suppose (4.55) is true for time step k. Then, at step $k+1$, the first line of (4.55) is as follows:

$$\begin{aligned} u_{k+1} &= A*\left(A^k * u_0 + \sum_{n=1}^{k} A^{n-1} * Bx_{k-n}\right) + Bx_k \\ &= A*\left(A^k * u_0 + \sum_{m=0}^{k-1} A^{m} * Bx_{k-m-1}\right) + Bx_k \end{aligned} \tag{4.56}$$

But, if (4.55) is true for k, then the quantity in brackets in (4.56) must be u_k, and so (4.56) is the same as the first line in (4.51). Thus, we have shown that (4.55) is true for $k=1$. Also, we have shown that if (4.55) is true for k, it must be true for $k+1$. So, the validity of (4.55) is proved by induction.

4.12 Lattice Algorithms and Structures

The dictionary defines a lattice as an open framework of strips of wood or metal interwoven to form a regular pattern. The lattice form of a linear system gets its name, because if you substitute signal paths for the wood and metal strips, its diagram fits the definition.

Lattice algorithms are like direct algorithms in that both provide implementations of any linear system. As we shall see, converting from direct to lattice form is complicated, and the reader may ask whether it is worth the effort, considering that the lattice does nothing really new. The answer is that it may not be worth the effort, and you may wish to skip this section, unless (1) you may be interested in DSP operations that use the lattice structure for greater stability or a preferable hardware layout, or (2) you are a student and your instructor thinks this subject is important. In either case, the result of your study may be beneficial.

A single *symmetric two-multiplier lattice stage* is shown in Figure 4.17. These stages are connected to form lattice structures. The signal vectors, u^n, v^n, and so on are indicated at each stage, with the stages being numbered from right to left. (Note that the superscripts are *not* exponents.) In z-transform notation, we can see the following relationships in Figure 4.17:

$$\begin{aligned} U^n(z) &= U^{n+1}(z) - \kappa_n z^{-1} V^n(z) \\ V^{n+1}(z) &= \kappa_n U^n(z) + z^{-1} V^n(z) \end{aligned} \tag{4.57}$$

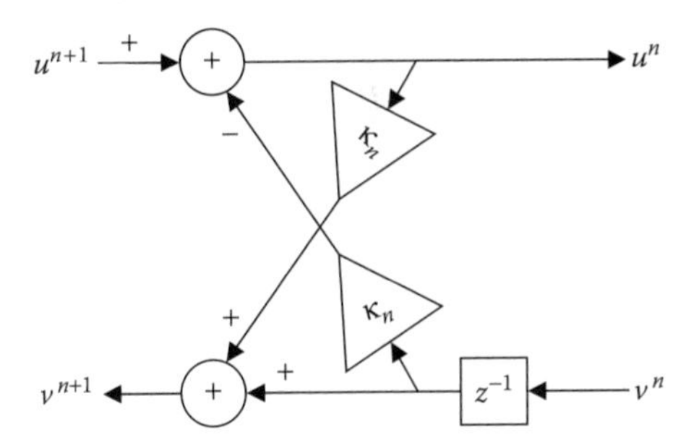

FIGURE 4.17
Symmetric two-multiplier lattice stage.

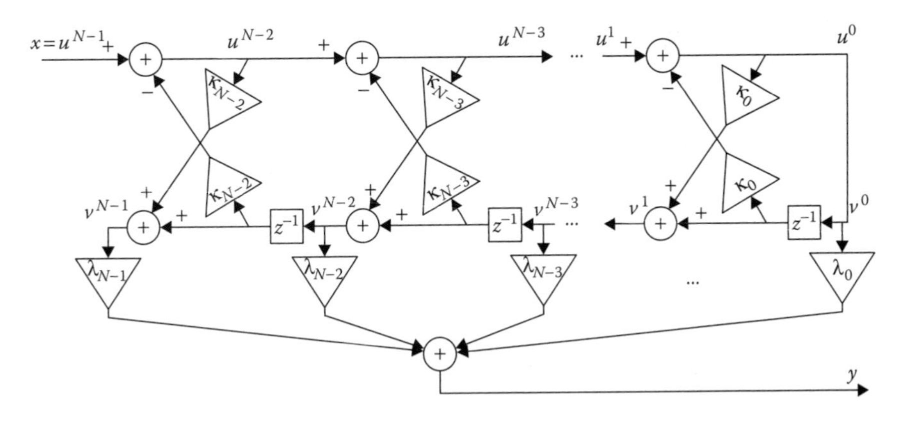

FIGURE 4.18
Symmetric recursive lattice structure.

A form of (4.57) more useful for analysis allows us to move from stage n to stage $n+1$:

$$\begin{bmatrix} U^{n+1}(z) \\ V^{n+1}(z) \end{bmatrix} = \begin{bmatrix} 1 & \kappa_n z^{-1} \\ \kappa_n & z^{-1} \end{bmatrix} * \begin{bmatrix} U^n(z) \\ V^n(z) \end{bmatrix} \tag{4.58}$$

A lattice may be formed by connecting stages like this in cascade. A *symmetric recursive lattice* equivalent to the direct form is shown in Figure 4.18, following Gray and Markel.[5] There are $N-1$ "κ" weights and N "λ" weights; hence, there are the same degrees of freedom as in Figures 4.14 through 4.16. To prove that Figure 4.18 is an implementation of Figure 4.14, and to show how to convert the weights in Figure 4.18 to direct weights, we first normalize (4.58) as follows:

$$P^n(z) = \frac{U^n(z)}{U^0(z)};\quad Q^n(z) = \frac{V^n(z)}{V^0(z)};\quad \begin{bmatrix} P^n(z) \\ Q^n(z) \end{bmatrix} = \begin{bmatrix} 1 & \kappa_{n-1} z^{-1} \\ \kappa_{n-1} & z^{-1} \end{bmatrix} * \begin{bmatrix} P^{n-1}(z) \\ Q^{n-1}(z) \end{bmatrix} \tag{4.59}$$

Noting that $u^0 = v^0$ at the right end of Figure 4.18, and therefore, $U^0(z) = V^0(z)$, we can express the transfer function, $H(z)$, in terms of these new variables and the weights in the diagram:

$$\frac{X(z)}{U^0(z)} = P^{N-1}(z); \quad \frac{Y(z)}{U^0(z)} = \sum_{n=0}^{N-1} \lambda_n V^n(z); \quad \therefore H(z) = \frac{Y(z)}{X(z)} = \frac{\sum_{n=0}^{N-1} \lambda_n Q^n(z)}{P^{N-1}(z)} \tag{4.60}$$

Noting that $P^0(z) = Q^0(z) = 1$ by definition, we now have a method for converting the lattice structure in Figure 4.18 to a direct algorithm. The method is summarized as follows:

Lattice-to-direct conversion:

$$P^0(z) = Q^0(z) = 1$$

$$\begin{bmatrix} P^n(z) \\ Q^n(z) \end{bmatrix} = \begin{bmatrix} 1 & \kappa_{n-1} z^{-1} \\ \kappa_{n-1} & z^{-1} \end{bmatrix} * \begin{bmatrix} P^{n-1}(z) \\ Q^{n-1}(z) \end{bmatrix}; \quad n = 1, 2, \ldots, N-1 \tag{4.61}$$

$$H(z) = \frac{\sum_{n=0}^{N-1} \lambda_n Q^n(z)}{P^{N-1}(z)} = \frac{B(z)}{A(z)}$$

The lattice-to-direct conversion in (4.61) is easy to implement using MATLAB. The function $[b, a] = lat\_to\_dir(kappa, lambda)$ is included with the functions on the publisher's website for this text. The conversion is accomplished with the following expressions within the function:

```
% Initialize P(1:N), q(1:N), b(1:N) at stage 0.
p=[1,zeros(1,N-1)];
q=[1,zeros(1,N-1)];
b=lambda(1)*q;
% Update p, q, and b at stages 1 thru N-1.
for n=2:N
        p1=p;
        p=p1+kappa(n-1)*[0,q(1:N-1)];
        q=kappa(n-1)*p1+[0,q(1:N-1)];
        b=b+lambda(n)*q;
end
% Set a equal to final value of p(1:N).
a=p;
```
(4.62)

In (4.62), p and q are coefficient vectors of $P^n(z)$ and $Q^n(z)$ in (4.61), and these keep changing in the "for" loop in (4.62). At the nth stage, the vectors are p^n

and q^n, and $\lambda_n q^n$ is added to the b vector, that is, the numerator of $H(z)$ in (4.61). At the final stage, $A(z)$ is set equal to $P^{N-1}(z)$, the denominator of $H(z)$ in (4.61).

Having seen equivalent lattice and direct forms, it is reasonable to ask how *stability* translates from direct to lattice. A stable direct system has all poles inside the unit circle, and in (4.61), we can see that the poles of $P^{N-1}(z)$ must also lie inside the unit circle for stability. Suppose $N = 2$, that is, the lattice in Figure 4.18 has only a single stage. Then, in (4.61), $P^{N-1}(z)$ would become $P^1(2)(z) = 1 + \kappa_0 z^{-1}$, and the requirement for stability would be $|\kappa_0| < 1$. Because the lattice stages in Figure 4.18 are cascaded, it is not surprising that, as Jury[6] has proved, the weight in each stage must satisfy $|\kappa_n| < 1$ for stability. Thus,

For stablility in a symmetric lattice:
$\lvert\kappa_n\rvert < 1$ in each stage of the lattice

We turn now to the question of an inverse for (4.61), that is, a *direct-to-lattice* conversion. First, we prove an interesting property of the symmetric lattice. By repeated use of the second equation in (4.61), we can write

$$\begin{bmatrix} P^n(z) \\ Q^n(z) \end{bmatrix} = \begin{bmatrix} 1 & \kappa_{n-1} z^{-1} \\ \kappa_{n-1} & z^{-1} \end{bmatrix} * \cdots * \begin{bmatrix} 1 & \kappa_0 z^{-1} \\ \kappa_0 & z^{-1} \end{bmatrix} * \begin{bmatrix} 1 \\ 1 \end{bmatrix} \tag{4.63}$$

[Here, we can see that $p_n^n = \kappa_{n-1}$ and $q_n^n = 1$ are the coefficients of z^{-n} in $P^n(z)$ and $Q^n(z)$, respectively. We will use this property below following (4.68).] In the first equation in (4.63), we substitute z^{-1} for z and multiply by z^{-n} to obtain

$$\begin{bmatrix} z^{-n} P^n(z^{-1}) \\ Q^n(z) \end{bmatrix} = \begin{bmatrix} z^{-1} & \kappa_{n-1} \\ \kappa_{n-1} & z^{-1} \end{bmatrix} * \cdots * \begin{bmatrix} z^{-1} & \kappa_0 \\ \kappa_0 & z^{-1} \end{bmatrix} * \begin{bmatrix} 1 \\ 1 \end{bmatrix} \tag{4.64}$$

If we carry out the product in (4.64) from right to left, the elements of the right-hand column vector are always equal. Thus, we have this important relationship at each stage of a symmetric lattice: The coefficient vector q^n is the reverse of the coefficient vector p^n, that is,

$$\boxed{Q^n(z) = z^{-n} P^n(z^{-1})} \tag{4.65}$$

With this relationship (4.65), we can construct a direct-to-lattice algorithm. We rewrite the center equation in (4.61) as follows:

$$\begin{bmatrix} P^n(z) \\ Q^n(z) \end{bmatrix} = \begin{bmatrix} 1 & \kappa_{n-1} \\ \kappa_{n-1} & 1 \end{bmatrix} * \begin{bmatrix} P^{n-1}(z) \\ z^{-1} Q^{n-1}(z) \end{bmatrix} \tag{4.66}$$

Next, we invert the weight matrix in accordance with (1.15) and rewrite (4.66) as follows:

$$\begin{bmatrix} P^{n-1}(z) \\ z^{-1}Q^{n-1}(z) \end{bmatrix} = \frac{1}{1-\kappa_{n-1}^2} \begin{bmatrix} 1 & -\kappa_{n-1} \\ -\kappa_{n-1} & 1 \end{bmatrix} * \begin{bmatrix} P^n(z) \\ Q^n(z) \end{bmatrix} \tag{4.67}$$

Equation (4.67) gives us the formula for $P^{n-1}(z)$ in terms of $P^n(z)$:

$$P^{n-1}(z) = \frac{P^n(z) - \kappa_{n-1}Q^n(z)}{1-\kappa_{n-1}^2}; \quad n = N-1, N-2, \ldots, 1 \tag{4.68}$$

Because, as seen in (4.63), $p_n^n = \kappa_{n-1}$ is the coefficient of z^{-n} in $P^n(z)$, we have in (4.68) a method for regressing from left to right in Figure 4.13 and computing κ_{n-1}, p^n, and q^n at each stage. Furthermore, in the numerator of $H(z)$ in (4.61), we can see that λ_{N-1} must be the coefficient of $z^{-(N-1)}$ in the sum, because, again as in (4.63), the coefficient q_{N-1}^{N-1} equals 1. Therefore, $\lambda_{N-1} = b_{N-1}$, and we can remove the final term in the sum and thus move to the left in Figure 4.18, computing λ_n similarly at each stage. All of this is summarized in the algorithm (4.69), in which $S^n(z)$ is the reduced numerator of (4.61) at the nth stage.

As before, the conversion in (4.69) from direct to lattice weights is easy to accomplish in MATLAB. The function [*kappa*, *lambda*] = *dir_to_lat*(*b*, *a*), which is the inverse of *lat_to_dir*, is included with the functions on the publisher's website for this text. Within the function, the conversion is accomplished with the expressions in (4.70). The expressions are essentially an implementation of (4.69). The *p*, *q*, and *s* vectors are coefficient vectors of $P^n(z)$, $Q^n(z)$, and $S^n(z)$, and these change during execution. The only difference is that in the m-file, array indices must begin at one instead of zero.

Direct-to-lattice conversion:

$$P^{N-1}(z) = A(z); \quad S^{N-1}(z) = B(z); \quad \lambda_{N-1} = b_{N-1}$$

$$\left.\begin{array}{l} \kappa_{n-1} = p_n^n \\ Q^n(z) = z^{-n}P^n(z) \\ P^{n-1}(z) = \dfrac{P^n(z) - \kappa_{n-1}Q^n(z)}{1-\kappa_{n-1}^2} \\ S^{n-1}(z) = S^n(z) - \lambda_n Q^n(z) \\ \lambda_{n-1} = s_{n-1}^{n-1} \end{array}\right\}; \quad n = N-1, N-2, \ldots, 1 \tag{4.69}$$

```
% Initialize p, s, kappa, and lambda.
p=a;
s=b;
kappa=zeros(1,N-1);
lambda=[zeros(1,N-1),b(N)];
% Update kappa, q, p, s, and lambda recursively.
for n=N:-1:2
    kappa(n-1)=p(n);
    q=p(n:-1:1);
    p=(p(1:n)-kappa(n-1)*q)/(1-kappa(n-1)^2);
    s=s(1:n)=lambda(n)*q;
    lambda(n-1)=s(n-1);
end
```
(4.70)

Thus, we have algorithms for transforming from direct to lattice form and vice versa. Lattice filtering is accomplished simply by implementing Figure 4.13, by setting the appropriate initial conditions, including $u^{N-1} = x$, where x is the input signal vector, computing successive values of $U^n(z)$ and $V^n(z)$ at each time step in accordance with (4.57) going from left to right in Figure 4.18, and computing y_k at each time step as the sum shown in Figure 4.18. The lattice algorithm may be summarized in z-transform form as follows, with zero initial conditions assumed. The input and output signal vectors in Figure 4.18 are x_k and y_k with $k = 0,1,\ldots,K-1$:

Lattice filtering:

$$U^{N-1}(z) = X(z)$$

$$\left.\begin{aligned} U^n(z) &= U^{n+1}(z) - \kappa_n V^n(z) \\ V^{n+1}(z) &= \kappa_n U^n + V^n(z) \end{aligned}\right\}; \quad n = N-2, N-3, \ldots, 0 \tag{4.71}$$

$$Y(z) = \sum_{n=0}^{N-1} \lambda_n V^n(z)$$

The *lat_filter* function is an implementation of (4.71). The reader may examine the m-file, which is quite short, to see the procedure in detail.

We have seen that a *nonrecursive direct* form could be constructed simply by setting all the "a" weights in (4.10) to zero, so $H(z) = B(z)$, a polynomial in z. A *nonrecursive symmetric lattice* form is also possible but is not attained quite as simply. In (4.61), if we set $\lambda_0 = 1$ and the rest of the λ's to zero, then, because $Q^0(z) = 1$, we have

$$H(z) = \frac{Y(z)}{X(z)} = \frac{1}{P^{N-1}(z)} = \frac{U^0(z)}{U^{N-1}(z)} \tag{4.72}$$

Now suppose in (4.72) that we could reverse the signal flow and thus invert the transfer function. Then, we would have a nonrecursive filter with transfer function:

$$H(z) = P^{N-1}(z) = \frac{U^{N-1}(z)}{U^0(z)} = B(z) = 1 + b_1 z^{-1} + \cdots + b_{N-1} z^{-(N-1)} \tag{4.73}$$

That is, a polynomial in z with $b_0 = 1$. But, this requires $X(z) = U^0(z)$ and $Y(z) = U^{N-1}(z)$, which means that the signal in each lattice stage must flow from u^n to u^{n+1}, *without changing the relationships* in (4.57). Thus, the operation of the lattice stages must be described by (4.58) instead (4.57). A lattice of this form is shown in Figure 4.19. The lattice-to-direct conversion is the same as in (4.61), except now $H(z) = P^{N-1}(z)$:

Nonrecursive lattice-to-direct conversion:

$$P^0(z) = Q^0(z) = 1$$

$$\begin{bmatrix} P^n(z) \\ Q^n(z) \end{bmatrix} = \begin{bmatrix} 1 & \kappa_{n-1} z^{-1} \\ \kappa_{n-1} & z^{-1} \end{bmatrix} * \begin{bmatrix} P^{n-1}(z) \\ Q^{n-1}(z) \end{bmatrix}; \quad n = 1, 2, \ldots, N-1 \tag{4.74}$$

$$H(z) = P^{N-1}(z)$$

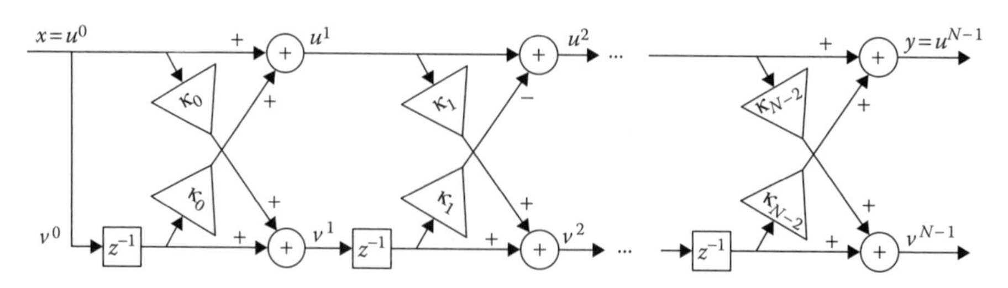

FIGURE 4.19
Symmetric nonrecursive lattice structure.

Similarly, the direct-to-lattice conversion algorithm follows from (4.69) with $P^{N-1}(z)$ initialized to $B(z)$ to conform with (4.73):

Nonrecursive direct-to-lattice conversion:

$$P^{N-1}(z) = B(z) = 1 + b_1 z^{-1} + \cdots + b_{N-1} z^{-(N-1)}$$

$$\left.\begin{aligned} \kappa_{n-1} &= p_n^n \\ Q^n(z) &= z^{-n} P^n(z) \\ P^{n-1}(z) &= \frac{P^n(z) - \kappa_{n-1} Q^n(z)}{1 - \kappa_{n-1}^2} \end{aligned}\right\}; \quad n = N-1, N-2, \ldots, 1 \tag{4.75}$$

Note that the direct weight b_0 must equal one. If this is not so, that is, $b_0 \neq 1$ in (4.75), one might first think of dividing $B(z)$ by b_0, but if $b_{N-1} = b_0$ (and this is often the case as we will see in Chapter 5), then dividing by b_0 would result in $\kappa_{N-1} = 1$, and again, (4.75) would not work. So, the preferred approach when $b_0 \neq 1$ is to convert $H(z) = 1 + B(z)$ to a lattice using (4.75), then subtract one from the transfer function by subtracting the input from the output as in Figure 4.20.

A MATLAB function called *nr_dir_to_lat* converts any causal nonrecursive system from direct to lattice form by implementing (4.75) and, in cases where $b_0 \neq 1$, modifies the system as in Figure 4.20, and issues a warning that the input signal must be subtracted from the lattice output. The reader may wish to compare the expressions in *nr_dir_to_lat* with (4.75).

A companion function, *nr_lat_to_dir*, converts a nonrecursive lattice to direct form in accordance with (4.74). The output of *nr_lat_to_dir* is a direct weight vector with length one greater than the length of the input lattice weight vector, and with weight b_0 always equal to one.

Another function, *nr_lat_filter*, implements the nonrecursive lattice filter in the form of Figure 4.19. The reader may examine the m-file to see the procedure, which follows essentially from (4.58), in detail.

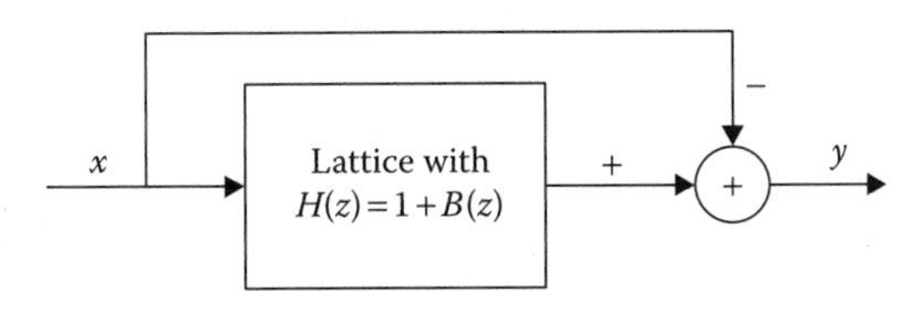

FIGURE 4.20
Conversion to nonrecursive lattice when $b_0 \neq 1$. The overall transfer function is $Y(z)/X(z) = B(z)$.

In this section, we discussed recursive and nonrecursive lattice algorithms and structures. The development was rather long, so we omitted examples, and the reader is encouraged to look at some of the exercises at the end of this chapter on conversion to lattice forms and lattice filtering. One of the best ways to begin with lattices is to convert a simple linear system from direct to lattice form, then process a given signal vector with both forms and see that, although the structures are very different, they produce the same output signal vector.

One final point concerning lattice structures is worth considering: We have seen that in the recursive lattice, the "κ" weights must be less than one for stability. In the nonrecursive lattice, if the κ weights are all less than one, all the zeros of $H(z)$ will be inside the unit circle, resulting in a *minimum-phase* system,[1] which is often desirable. Thus, generally speaking, lattice structures have κ weights less than one; that is, each κ weight is an *attenuator* rather than an amplifier of the signal samples. This property has been, and probably will be, an advantage in DSP systems that operate at the highest possible sampling rates, because, with proper scaling, "hybrid" lattice filters can be made with attenuators that perform analog rather than digital weight multiplication. Hybrid applications continue to exist, because as processor speeds increase, requirements for greater speeds somehow manage to stay ahead, just out of reach.

4.13 FFT Algorithms

The previous remarks on processor speed requirements bring us to this discussion of *FFT algorithms,* which may also increase the processing speed of a linear system. FFT algorithms are based essentially on (4.8), that is, on the fact that the DFT (or FFT) of the convolution of two vectors is the product of the two individual DFTs (or FFTs).

In any discrete linear system, we have the equivalent relationships in (4.10) through (4.12), that is, $Y(z) = H(z)X(z)$. Also, in connection with (4.37), we have seen that the impulse response, h, is the inverse transform of $H(z)$. If $H(z)$ is nonrecursive, then h is a finite vector equal to b, the weight vector. If $H(z)$ is recursive, h is an infinite vector with elements equal to coefficients of the polynomial resulting from the division of $B(z)$ by $A(z)$. The point here is that, in any case, if $H(z)$ is expressed as a polynomial in z, and if $Y(z) = H(z)X(z)$ in which $x = [x_0\ x_1\ \dots\ x_{K-1}]$ is a signal vector, then

$$y_k = \sum_{n=0}^{\infty} h_n x_{k-n}; \quad k = 0, 1, \dots, K-1 \tag{4.76}$$

Furthermore, if we assume as before that x_k is 0 for $k < 0$, then the products in the sum are all zero for $n \geq K$, and

$$y_k = \sum_{n=0}^{K-1} h_n x_{k-n}; \quad k = 0, 1, \ldots, K-1 \tag{4.77}$$

Thus, we have the convolution of two finite vectors of equal length (K). (In the nonrecursive system with $h = [b_0 \, b_1 \, \ldots \, b_{N-1}]$, we simply append $K - N$ zeros to b in order to obtain h with length K.) We must now ask whether $Y(z)$ can be expressed as a product of transforms in this finite case, as it was for the infinite case in (4.8). To answer, we examine once again the inverse DFT of the product of two finite DFTs. We use the DFT now instead of the z-transform, because the DFT has discrete components, and the inverse DFT is given in terms of these components. From definition (3.23), the inverse of the DFT product is as follows:

$$\begin{aligned}
\mathrm{DFT}^{-1}\{H_m X_m\} &= \frac{1}{K} \sum_{m=0}^{K-1} H_m X_m e^{j2\pi mk/K} \\
&= \frac{1}{K} \sum_{m=0}^{K-1} \sum_{n=0}^{K-1} h_n e^{-j2\pi mn/K} \sum_{i=0}^{K-1} x_i e^{-j2\pi mi/K} e^{j2\pi mk/K} \\
&= \frac{1}{K} \sum_{m=0}^{K-1} \sum_{n=0}^{K-1} h_n x_i \sum_{m=0}^{K-1} e^{j2\pi m(k-i-n)/K}; \quad k = 0, 1, \ldots, K-1
\end{aligned} \tag{4.78}$$

In the final expression of (4.78), the sum on the right is equal to K when $i = k - n \pm \eta K$, where η is any integer, and is equal to zero otherwise. Therefore, the expression becomes

$$\boxed{\mathrm{DFT}^{-1}\{H_m X_m\} = \sum_{n=0}^{K-1} h_n x_{k-n\pm\eta K}; \quad k = 0, 1, \ldots, K-1; \quad \eta = 0, 1, \ldots \infty} \tag{4.79}$$

Thus, the inverse DFT of a DFT product is essentially the same as in (4.8), except that when the signal vectors h and x are finite, as they are here, h is convolved with the *periodic extension* of x, that is, an infinite vector with the elements of x repeating with period K. [Naturally, in (4.8), we did not notice the periodicity of x, because the period was infinite.] An example of (4.79) similar to Figure 4.1, showing the alignment of h and x with $K = 4$, is shown in Figure 4.21. Alignments are shown in the figure for the computation of y_k with all K values of k.

Comparing Figures 4.21 and 4.1, it is obvious that periodic convolution differs from ordinary convolution, and in this sense, does not represent a linear

$$\begin{bmatrix} h_0 & h_1 & h_2 & h_3 \\ x_0 & x_3 & x_2 & x_1 \end{bmatrix} \quad y_0 = h_0x_0 + h_1x_3 + h_2x_2 + h_3x_1$$

$$\begin{bmatrix} h_0 & h_1 & h_2 & h_3 \\ x_1 & x_0 & x_3 & x_2 \end{bmatrix} \quad y_1 = h_0x_1 + h_1x_0 + h_2x_3 + h_3x_2$$

$$\begin{bmatrix} h_0 & h_1 & h_2 & h_3 \\ x_2 & x_1 & x_0 & x_3 \end{bmatrix} \quad y_2 = h_0x_2 + h_1x_1 + h_2x_0 + h_3x_3$$

$$\begin{bmatrix} h_0 & h_1 & h_2 & h_3 \\ x_3 & x_2 & x_1 & x_0 \end{bmatrix} \quad y_3 = h_0x_3 + h_1x_2 + h_2x_1 + h_3x_0$$

FIGURE 4.21
Illustration of the periodic convolution of vectors h and x, each of length $K = 4$, showing the computation of y_0 through y_3. Compare with Figure 4.1.

$$\begin{bmatrix} h_0 & h_1 & h_2 & h_3 & 0 & 0 & 0 & 0 \\ x_0 & 0 & 0 & 0 & 0 & x_3 & x_2 & x_1 \end{bmatrix} \quad y_0 = h_0x_0$$

$$\begin{bmatrix} h_0 & h_1 & h_2 & h_3 & 0 & 0 & 0 & 0 \\ x_1 & x_0 & 0 & 0 & 0 & 0 & x_3 & x_2 \end{bmatrix} \quad y_1 = h_0x_1 + h_1x_0$$

$$\begin{bmatrix} h_0 & h_1 & h_2 & h_3 & 0 & 0 & 0 & 0 \\ x_2 & x_1 & x_0 & 0 & 0 & 0 & 0 & x_3 \end{bmatrix} \quad y_2 = h_0x_2 + h_1x_1 + h_2x_0$$

$$\begin{bmatrix} h_0 & h_1 & h_2 & h_3 & 0 & 0 & 0 & 0 \\ x_3 & x_2 & x_1 & x_0 & 0 & 0 & 0 & 0 \end{bmatrix} \quad y_3 = h_0x_3 + h_1x_2 + h_2x_1 + h_3x_0$$

FIGURE 4.22
Illustration of the periodic convolution of vectors h and x as in Figure 4.21, but with zero extension to prevent the periodic effect, with K now increased from 4 to 8.

transfer function. To apply the FFT, we must make periodic convolution look like ordinary convolution. This may be done by appending K zeros to the h and x vectors. Then, again using vectors of length four as an illustration and extending to $K = 8$, the periodic convolution looks like Figure 4.22 instead of Figure 4.21, and the values of y_k in Figure 4.22 now represent the operation of a linear system.

Thus, in general, linear processing may be done using products of FFTs, provided the signal vectors are extended with zeros. In MATLAB, the process takes the following form, in which $x(1{:}K)$ is the signal vector and $h(1{:}K)$ is the system impulse response vector.

$$\begin{aligned} y &= \text{IFFT}\left[\text{FFT}(h, 2*K).*\text{FFT}(x, 2*K)\right] \\ y &= y(1:K) \end{aligned} \qquad (4.80)$$

This expression describes, for example, the "fast convolution of v and c" block in Figure 4.5. Before going further with this concept, we may ask whether the FFT method will really save time, because the vectors are all twice as large. The answer depends on the size and value of K, as suggested in Figure 3.5. Using the number of complex products as a measure of computing time, we can see in (4.77) that there are K products per sum and K sums, therefore, K^2 products in the convolution. Assuming the best case where K (and hence, $2K$) is a power of two, (3.17) with $N = 2K$ gives $K\log_2(2K)$ complex products per FFT. A complex product in Cartesian form requires four products, so we use $4K\log_2(2K)$ as the number of products per FFT. In (4.80), there are two FFTs and one inverse FFT, for a total of $12K\log_2(2K)$ products. Therefore,

$$\frac{\#\text{ products using (4.77)}}{\#\text{ products using (4.80)}} = \frac{K^2}{12K\log_2(2K)} = \frac{K}{12(1+\log_2 K)} \tag{4.81}$$

This ratio is plotted in Figure 4.23 for cases where K is a power of 2. When the ratio exceeds one, the FFT method becomes profitable in terms of computing time. Thus, the FFT method is worth considering for $K \geq 128$, and definitely preferable for very large values of K.

Seeing that the FFT method can be useful for large values of K, we look in more detail first at the *nonrecursive* case, where (4.77) describes the linear system as a single convolution with $h = b$, the nonrecursive weight vector. In practical linear operations where K is large, which we assume here, the number of weights is usually less than K. Suppose this is true. To see the effect

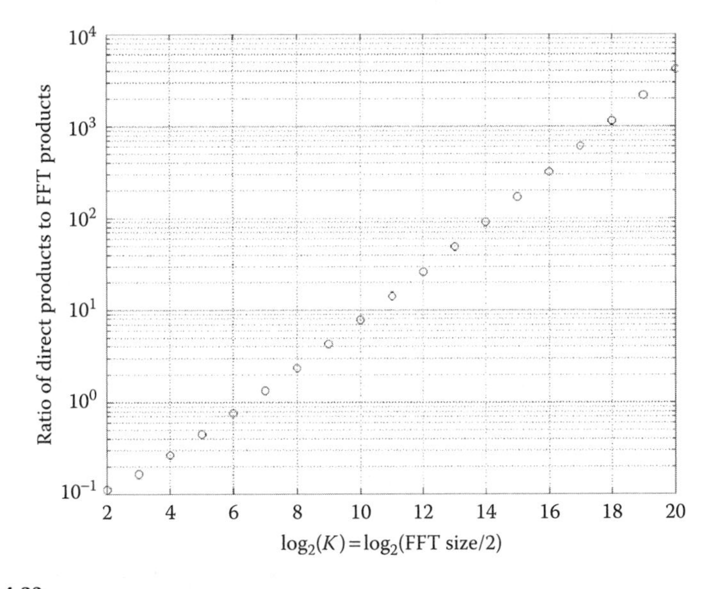

FIGURE 4.23
Ratio of products required for convolution using the direct method (4.77) to products required using the FFT method (4.80), when the vector length, K, is a power of 2. The FFT method becomes profitable as K increases from 64 to 128.

on periodic convolution, suppose $h_2 = h_3 = 0$ in Figure 4.22. Then, we would not need as many zeros to avoid circular convolution. In fact, we only need to extend x with N zeros, where N is the number of nonrecursive weights. So, assuming we still require the FFT size to be a power of two, we can modify (4.80) as follows:

$$\boxed{\begin{array}{l} \text{Nonrecursive FFT filtering;} \quad K \gg N \\ N = \text{length}\,(b) \\ K = \text{length}\,(x) \\ L = \text{next power of } 2 \geq K + N \\ y = \text{IFFT}\left[\text{FFT}(b, L) .* \text{FFT}(x, L)\right] \\ y = \text{real}\left(y(1:K)\right) \end{array}} \tag{4.82}$$

In (4.82), the FFTs are smaller than in (4.80), but the convolution in (4.77) now has N products per sum instead of K products per sum. So Figure 4.23, while no longer valid for this case, is still an approximation. The last line of (4.82) is written to adjust y to its correct length and to remove small imaginary components that are the result of rounding errors.

In the *recursive* case, direct processing as in (4.10) consists of two convolutions. Assume $M = N$ so there are $2N$ products per sum, or a total of $2NK$ products in the direct form. In the FFT version, we must first find h as the inverse transform of $B(z)/A(z)$. Division of FFTs is not a good idea, because even assuming the system is stable, $A(z)$ could have zeros near the unit circle at points where the FFTs are computed, resulting in an inaccurate computation of h.

The preferred method for finding h with a recursive system is *polynomial division*, in which the division of $B(z)$ by $A(z)$ is simply carried out by long division to the desired length of h. The process is easy to encode in languages like MATLAB that operate on vectors. A function called *imp_resp* is included in the *functions* folder on the publisher's website to illustrate the process. Once $h(1{:}K)$ is found, it may be used as the nonrecursive weight vector in place of b in (4.82) to implement recursive filtering using FFTs.

If MATLAB is being used to do the processing, the MATLAB *conv, deconv,* and *filter* functions, which use compiled algorithms, should be used, because, with long signal vectors, they are at least an order of magnitude faster than *imp_resp* and the expressions in (4.82). (See also the *filters* function described at the end of Section 6.7.)

On the other hand, if a processor is using a primitive language, and particularly if a 2^K-element FFT chip is available, then *block FFT filtering* using the algorithm in (4.82) should allow processing at the highest possible sample rates. In *block FFT filtering*, as the name suggests, the input signal is processed in successive blocks (vectors) of length 2^K in order to use the FFT chip most efficiently.

The exercises include applications of FFT filtering with recursive and non-recursive filters. The MATLAB language provides an excellent set of tools to test such applications, as we saw in (4.82), which is almost a set of MATLAB expressions. The main purpose of this section has been to show that, essentially as the result of the relationship in (4.8) (that is, convolution in the time domain implies multiplication in the frequency domain), the efficiency of the FFT algorithm may be applied to digital filtering.

4.14 Discrete Linear Systems and Digital Filters

The names *digital filter* and *discrete linear system* refer to signal processing systems like those we have been discussing in this chapter. But when we use *filter*, we usually are thinking of a process analogous to filtering in chemistry, where some of the ingredients of a mixture are removed by a filter to produce a mixture of greater purity. In signal processing, the mixture is a signal mixed with noise. The filter is a discrete processor, and the desired output is the signal with less noise.

In Chapters 5 and 6, we discuss filters in which the discrete processor is a linear nonrecursive (FIR) or recursive (IIR) system of the type discussed in this chapter. When a linear processor of this type is used as a filter, its effect is most evident in the frequency domain—particularly in the amplitude response characteristic. Certain frequency components of the input signal are allowed to pass through the filter, while other components are attenuated by the filter. *Power gain* characteristics of the four types of filters used most commonly in DSP are illustrated in Figure 4.24. Power gain is the square of amplitude gain and is usually measured in *decibels* (dB), that is,

$$\begin{aligned}\text{Power gain in dB} &= 10\log_{10}\left|H\left(j\frac{2\pi\nu}{T}\right)\right|^2 \\ &= 20\log_{10}\left|H\left(j\frac{2\pi\nu}{T}\right)\right|\end{aligned} \tag{4.83}$$

where $|H(j\omega)|$ is the amplitude gain in (4.13), and $\nu = \omega T/2\pi$ Hz-s. Power gain plots are commonly used in the literature to illustrate filter gain properties.

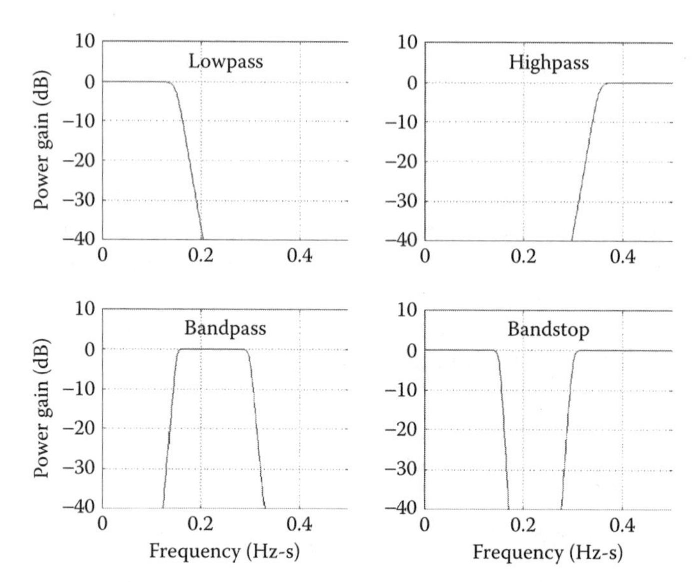

FIGURE 4.24
Power gain characteristics of four common types of digital filters. The filter weights are determined using algorithms described in Chapter 6.

In each case, the name of the filter in Figure 4.24 describes the gain property (the processing effect of the filter).

The filter weights used for the plots in Figure 4.24 were computed using methods described in Chapter 6; thus, the plots are for typical recursive digital filters. From (4.83), we can see that a power gain equal to 0 dB translates to an amplitude gain equal to one; a power gain equal to −40 dB translates to an amplitude gain equal to 0.01.

There are also digital filters that are not linear processors. Nonlinear filters such as neural networks, support vector machines, and time-varying or adaptive filters are examples. The discussion of these is left to Chapter 9 in the case of adaptive filters and to Chapter 12 in the case of support vector machines.

4.15 Functions Used in This Chapter

Several functions, some (but not all) of which have been described, are used for the gain and phase computations, transfer function analysis, filter structures, filtering, and so on discussed in this chapter. In the following list, all the functions listed, with the exception of *filter*, are included in the *functions* folder on the publisher's website.

Function	Used to Compute
[X,v]=chirp_z(x,v1,v2,N)	Chirp-z transform of a vector
y=filter(b,a,x)	y = filtered version of x; filter in direct form
y=filters(b,a,x)	y = filtered version of x; filter in cascade form
[H,v]=gain(b,a,N)	Complex gain of a linear system
H=gain_f(b,a,freq)	Complex gain at one or more specific frequencies
[r,p,k]=part_frac_exp(b,a)	Partial fraction expansion of a linear transfer function
pz_plot(b,a)	Poles and zeros of a linear system, plotted on z-plane
[dB,v]=power_gain(b,a,N)	Power gain in dB of a linear system
h=imp_resp(b,a,K)	Impulse response of a linear system
[kappa,lambda]=dir_to_lat(b,a)	Lattice weights from given direct weights
y=lat_filter(kappa,lambda,x)	y = lattice-filtered version of x
[b,a]=lat_to_dir(kappa,lambda)	Direct weights from given lattice weights
kappa=nr_dir_to_lat(b)	FIR lattice weights from given direct weights
y=nr_lat_filter(kappa,x)	y = nonrecursive lattice-filtered version of x
b=nr_lat_to_dir(kappa)	Direct FIR weights from nonrecursive lattice weights

In addition to *filter*, the basic MATLAB language abounds with functions, some of which we have used in this chapter, that are useful in processing signals, analyzing linear systems, and doing all sorts of other DSP operations. MATLAB's *Signal Processing Toolbox*, which is an add-on to the basic MATLAB language, contains additional functions especially designed for DSP operations.

Exercises

General instructions: (1) Whenever you make a plot in any of the exercises, be sure to label both axes, so we can tell exactly what the units mean. (2) Make continuous plots unless otherwise specified. "Discrete plot" means to plot discrete symbols, unconnected unless otherwise specified. (3) In plots, and especially in subplots, use the *axis* function for best use of the plotting area.

4.1 Find a nonrecursive expression of the linear system described by $y_k = x_k + x_{k-1} + y_{k-1}$. Assume zero initial conditions at $k = 0$. Then, present a clear inductive proof of your result.

4.2 Decide whether the first four properties in Table 4.1 exist for the following systems:

a. $y_k = b \sin x_{k-1}$

b. $y_k = bx_{k-1} - ay_{k-2}$

c. $y_k = b_1x_{k-1} + b_2x_k + b_3x_{k+1}$

d. $y_k = x_k + bx_{k-1}y_{k-1}$

4.3 Suppose the sample vector x, which begins at $k = 0$, is $x = [1\ 1\ -1\ 0\ 0\ \ldots]$. Show the startup sequences for linear systems (c) and (d) in Exercise 4.2.

4.4 Write the transfer function, $H(z)$, for linear systems (b) and (c) in Exercise 4.2. For (b), express the transfer function as a ratio of polynomials in z.

4.5 The transfer function of a digital filter is given by

$$H(z) = \frac{0.0675 + 0.1350z^{-1} + 0.0675z^{-2}}{1 - 1.1430z^{-1} + 0.4128z^{-2}}$$

a. Using (4.12), show that the filter gain is one at $\omega T = 0$ and zero at $\omega T = \pi$.

b. Plot the amplitude gain versus frequency in rad, and verify that your answer to (a) is correct. Use zero extension with an FFT size of 1000 in a manner similar to (4.14).

4.6 For the filter in Exercise 4.5, plot phase shift in degrees versus frequency in Hz-s again in a manner similar to (4.14), but this time, make a plot of connected discrete points using an FFT size of 100. Note that the phase shift at 0.5 Hz-s is indeterminate, because the gain at this frequency is zero.

4.7 Make a connected discrete plot of the complete impulse response of the filter in Exercise 4.5, from its beginning until it becomes negligible. Normalize the plot so its maximum value is 1.0. Also, print the normalized impulse response and identify the point where its magnitude stays below 1%.

4.8 The weights of a digital filter in the form of (4.11) are as follows:

$$b = [0.0675 \quad 0.0000 \quad -0.1349 \quad 0.0000 \quad 0.0675]$$

$$a = [1.0000 \quad -1.9425 \quad 2.1192 \quad -1.2167 \quad 0.4128]$$

Plot the continuous amplitude response versus frequency in terms of KHz, assuming a sampling frequency of 40 KHz.

4.9 Make a continuous plot of the phase shift of the filter in Exercise 4.8, in terms of degrees versus frequency in KHz, assuming the filter is working with a time step of 0.025 ms.

4.10 Make a connected discrete plot of the complete impulse response of the filter in Exercise 4.8, from its beginning until it becomes negligible. Normalize the plot so its maximum *magnitude* is 1.0. Also, print the normalized impulse response and identify the point where its magnitude stays below 1%.

4.11 Plot the poles and zeros of the digital filter in Exercise 4.8 using the *pz_plot* function. Explain the following:

a. How did the pole and zero locations indicate a band-pass gain characteristic?

b. What is the approximate frequency range of the pass band?

c. Why is the phase shift zero at $\omega = 0$ and then jumps to 180° when ω is just above zero?

4.12 Write a MATLAB function $[H, \nu] = my\_gain(b, a)$ to compute the complex gain of any linear system, recursive or nonrecursive, in terms of the b and a vectors in (4.12). Have the function return N values of H and ν, evenly spaced from zero through half the sampling rate, including the endpoints, with N set internally to 1024. Verify that the function works (1) by duplicating the left-hand plot in Figure 4.2 and (2) comparing its result with that of the *gain* function included in the *functions* folder on the website. If the function works, you may use it as desired in subsequent applications.

4.13 Write a MATLAB function $[dB, \nu] = my\_power\_gain(b, a)$ to compute power gain in decibels versus frequency in Hz-s for any linear system. Have the function return N values of dB and ν, with N set internally to 1024. Write the function so that it limits the range of dB to 200. Test the function on the following vectors:

$$b = [0.0675 \quad 0.1350 \quad 0.0675]$$

$$a = [1.0000 \quad -1.1430 \quad 0.4128]$$

Plot dB versus ν, and use *axis*([0 0.5 –80 0]) to set the ν and dB axis limits. If your plot shows 0 dB at $\nu = 0$ and –25 dB at $\nu = 0.3$, your function is probably correct, and you may use it to make power gain plots in subsequent chapters.

4.14 A linear transfer function is given as $H(z) = (z^2 + 1)/(z^2 - 1.93z + 1.21)$. Make two subplots. On the left, plot the poles and zeros of $H(z)$. On the right, plot the first 60 samples of the impulse response. Discuss the stability of $H(z)$ in terms of each plot.

4.15 Using lines B, 4, and 5 in Table 4.2, find the z-transform, $X(z)$, of $x_k = 2\sin(\omega k + \pi/6)$.

4.16 Two linear systems, $H_1(z) = 1 + 0.4z^{-1}$ and $H_2(z) = -bz^{-1}$, are connected together to form $H(z)$ in the feedback configuration of Figure 4.12.

a. On the complex z-plane, make a pole–zero plot showing the trajectory of the poles as b varies from 0 to 1.7.

b. For what range of b is $H(z)$ stable?

c. For what range of b are the poles of $H(z)$ real?

4.17 The direct form of a particular linear system is given by the following:

$$H(z) = \frac{2 + 1.1z^{-1} - 0.5z^{-2}}{1 + 0.1z^{-1} - 0.3z^{-2}}$$

a. Express $H(z)$ in cascade form with two sections.
b. Express $H(z)$ in parallel form with two sections.

4.18 A linear system has the transfer function $H(z) = (0.5 + 0.6z^{-2})/(1 - 0.4z^{-2})$.

a. Is the system stable?
b. Write the difference equation for the system and make a diagram similar to Figure 4.14.
c. Make a diagram similar to Figure 4.15, including the difference equations.
d. Make a diagram similar to Figure 4.16, including the difference equations.

4.19 Define the necessary arrays and write the state-space equations for the linear system in Exercise 4.18.

4.20 Do the following:

a. Find the weights of the lattice form of $(1 + z^{-1})/(1 - 0.8z^{-1})$.
b. Make diagrams of the direct and lattice forms.
c. Check for equivalence by computing the output vectors of each form when the input signal vector is [0 1 3 –1 0 0 0 0].

4.21 Complete the following:

a. Convert

$$H(z) = \frac{1 + 1.1z^{-1} - 0.5z^{-2}}{6(1 + 0.1z^{-1} - 0.8z^{-2})}$$

to a lattice filter, and draw the lattice diagram, showing all weights.

b. Make connected discrete plots of the direct and lattice impulse responses in two subplots, one above the other, to show the equivalence of the two forms.

4.22 Complete the following:

a. Convert $H(z) = 1 + z^{-3} - z^{-4}$ to a nonrecursive lattice filter, and draw the direct and lattice diagrams, showing all weights.
b. Compute and print the direct and lattice impulse responses to show the equivalence of the two forms.

4.23 Plot the power gain in dB versus Hz-s of $H(z)$ in Exercise 4.21, which is the transfer function of a certain cable. Describe the effect of the cable when it is used to transmit signals.

4.24 The impulse response of a filter is given as follows:

$$h_k = \frac{\sin(1.25(k-25))}{3(k-25)}; \quad k = 0, 1, \ldots, 50$$

a. In the upper plot of three subplots, plot the following signal:

$$x_k = 10\sin(0.5\pi k) + \cos(0.1\pi k); \quad k = 0, 1, \ldots, 499$$

b. Use the FFT filtering algorithm in (4.82) to produce y, the result of filtering x with the filter just described. Plot y below x in the center plot.

c. Plot the amplitude gain of the filter in the lower plot, and explain what the filter did to x.

4.25 Do Exercise 4.24 again, but this time use the following IIR filter:

$$H(z) = \frac{(0.2533z^2 + 0.5066z + 0.2533)(0.1839z^2 + 0.3678z + 0.1839)}{(z^2 - 0.4531z + 0.4663)(z^2 - 0.3290z + 0.0646)}$$

In the FFT product, use an impulse response having the same length as x.

4.26 Use the chirp-z transform to analyze the *speech* signal in the *data* folder on the website. Make four subplots. Plot the speech signal, $s(t)$, in the first subplot. The sampling rate is 40 KHz. Label the time axis in ms.

In the second subplot, plot the power spectrum in dB versus KHz using the absolute value of the FFT of $s(t)$.

In the third subplot, plot the same power spectrum, but only in the range from 0 to 100 Hz. How many points does this plot contain?

In the fourth subplot, plot the same power spectrum from 0 to 100 Hz, but use the chirp-z transform this time to get a continuous plot with 1000 points. Comment on the differences and similarities between the last two plots.

References

1. Oppenheim, A. V., and R. W. Schafer. 1989. *Discrete-Time Signal Processing*. Chaps. 4, 5. Englewood Cliffs, NJ: Prentice Hall.
2. Orfanidis, S. J. 1996. *Introduction to Digital Signal Processing*. Chap. 5. Upper Saddle River, NJ: Prentice Hall.

3. Stearns, S. D., and D. R. Hush. 1989. *Digital Signal Analysis*. Chap. 7. Englewood Cliffs, NJ: Prentice Hall.
4. Churchill, R. V., J. W. Brown, and R. F. Verhey. 1976. *Complex Variables and Applications*. New York: McGraw-Hill.
5. Gray, A. H., and J. D. Markel. 1973. Digital lattice filter synthesis. *IEEE Trans Audio Electroacoust* 21(6):491–500.
6. Jury, E. I. 1964. A note on the reciprocal zeros of a real polynomial with respect to the unit circle. *IEEE Trans Commun Tech* 11(2):292–4.
7. Oppenheim, A. V., and A. S. Willsky. 1983. *Signals and Systems*. Englewood Cliffs, NJ: Prentice Hall.
8. Schwarz, R. J., and B. Friedland. 1965. *Linear Systems*. New York: McGraw-Hill.
9. Moon, T. K., and W. C. Stirling. 2000. *Mathematical Methods and Algorithms for Signal Processing*. Upper Saddle River, NJ: Prentice Hall.
10. Bluestein, L. I. 1970. A linear filtering approach to the computation of the discrete fourier transform. *IEEE Trans Audioelectroacoust* 18(4):451–5.

Further Reading

Itakura, F., and S. Saito. 1971. Digital filtering techniques for speech analysis and synthesis. *Proceedings of 7th International Congress on Acoustics*, vol 3. p. 261. Budapest.

Kumar, B. P. 2010. *Digital Signal Processing Laboratory*. 2nd ed. Boca Raton, FL: CRC Press.

Makhoul, J. 1977. Stable and efficient lattice methods for linear prediction. *IEEE Trans Acoust Speech Signal Process* 25(5):423–8.

Mitra, S. K., P. S. Kamat, and D. C. Huey. 1977. Cascaded lattice realization of digital filters. *IEEE Trans Circuit Theory Appl* 5(1):3–11.

Tan, L. 2007. *Digital Signal Processing—Fundamentals and Applications*. New York: Elsevier.

5

FIR Filter Design

5.1 Introduction

In Chapter 4, we saw that the transfer function of a linear nonrecursive or finite impulse response (FIR) filter is the transform of the filter's impulse response vector, the latter being equal to the filter's weight vector. In this chapter, we discuss the simplest FIR design techniques, which are based simply on starting with a desired transfer function in the form of a discrete Fourier transform (DFT) and then obtaining the filter weights via the inverse DFT. Because many different applications take the form of one of the four filters in Figure 4.24—lowpass, highpass, bandpass, and bandstop—we focus in this chapter on these four basic types of filters.

Compared with IIR filters, FIR filters have advantages and disadvantages. The major disadvantage is the result of the FIR transfer function being a polynomial in z and z^{-1} rather than a ratio of polynomials and thus having zeros anywhere but poles only at the origin. Without poles away from the origin, more weights are generally required to provide a given transition from passband to stopband.

One major advantage of FIR filters comes in a property called *linear phase,* which we defined in Chapter 3 in terms of a shift in the time domain. We showed that when a signal x is delayed n time steps, $n\omega T$ radians are added to the phase of x, that is, added to the angle of the transform of x. This means that if x is expressed as a Fourier series, each component of x is adjusted in phase to produce a delay of exactly n time steps. So then, if the transfer function of a linear digital filter is given by

$$H(e^{j\omega T}) = \left|H(e^{j\omega T})\right| e^{-jn\omega T} \tag{5.1}$$

then, in a like manner, the effect of the filter will be to shift each component of the input signal by n time steps while modifying its amplitude. Thus, a linear-phase digital filter always has a transfer function in the form of (5.1). As we shall see, it is easy to impose this requirement on the design of FIR filters. We will also see in Chapter 6 that linear phase shift is not possible in stable IIR filters that operate in real time; therefore, linear phase is a major consideration in choosing the FIR construction.

FIR filters have other advantages as well. Having only zeros, they are not unstable as long as the weights are finite. Also, the finite impulse response property is often an advantage, because we can always specify the exact *startup time*, that is, the duration of the filter's response to a transient signal.

5.2 An Ideal Lowpass Filter

To derive the filter weights in the manner just described, that is, as the inverse transform of the filter transfer function, we begin with the ideal transfer function shown in Figure 5.1, which is ideal in two ways: (1) the gain is exactly one at frequencies $|\nu| \leq \nu_c$ Hz-s and exactly zero for $|\nu| > \nu_c$ Hz-s, and (2) the transfer function is real, implying zero phase shift.

The units in Figure 5.1 require some discussion. The frequency units are Hz-s. (It is helpful here to recall the frequency conversions in Table 3.1, especially $\omega = 2\pi\nu/T$.) The filter gain, which is real, is expressed (for convenience) as a continuous Fourier transform. The cutoff frequency, which is illustrated here at $\nu_c = 0.1$ Hz-s, is below half the sampling rate, that is, $\nu_c < 0.5$, and as long as this is true, the sampling theorem holds, and (3.47) gives a simple relationship between the Fourier transform and the DFT of the impulse samples, that is,

$$\bar{H}_{ideal}(j\omega) = \frac{1}{T} H_{ideal}(j\omega)$$

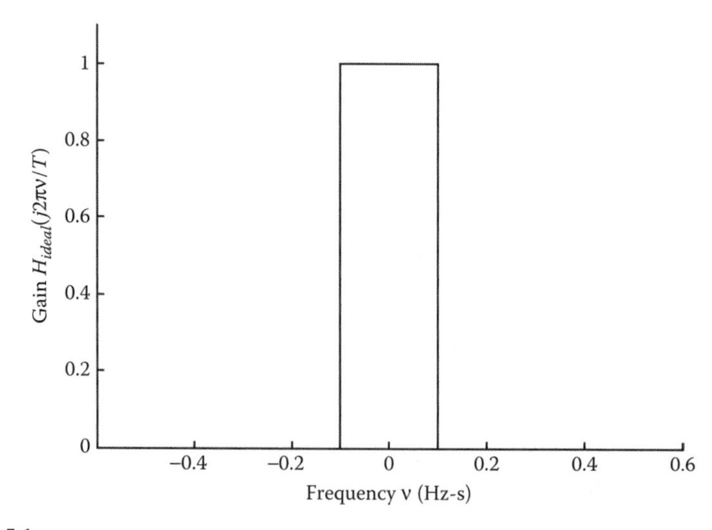

FIGURE 5.1
Gain of an ideal lowpass filter. The filter passes all signal components below $\nu_c = 0.1$ Hz-s without alteration and stops all components above ν_c Hz-s.

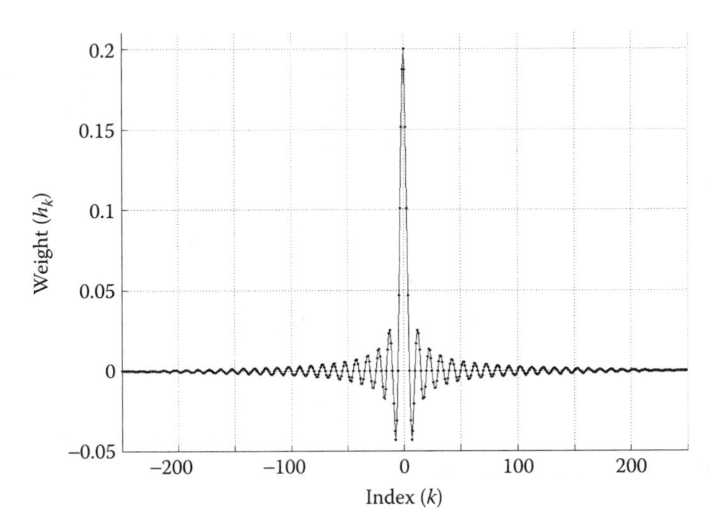

FIGURE 5.2
Weight vector of the digital version of the ideal lowpass filter in Figure 5.1. Weights extend to $\pm\infty$.

Accordingly, because we wish $\bar{H}_{ideal}$ rather than H_{ideal} to be one at frequencies below ω_c, the discrete samples of the impulse response of this ideal filter must be T times the inverse Fourier transform of the transfer function; that is, according to (3.34),

$$h_{ideal}(kT) = \frac{T}{2\pi}\int_{-\infty}^{\infty} H(j\omega)e^{j\omega kT}d\omega = \frac{T}{2\pi}\int_{-2\pi\nu_c/T}^{2\pi\nu_c/T}(1)e^{j\omega kT}d\omega$$
$$= \begin{cases} 2\nu_c; & k=0 \\ \dfrac{\sin(2\pi\nu_c k)}{\pi k}; & 1 \le |k| < \infty \end{cases} \tag{5.2}$$

The central part of the ideal weight vector corresponding with Figure 5.1, $b_{ideal}(k) = h_{ideal}(kT)$, is plotted in Figure 5.2. Note that the maximum weight is $b_0 = 2\nu_c$ and that the weights extend infinitely in both directions.

5.3 The Realizable Version

The simplest way to make a realizable FIR filter out of the infinite weight vector in (5.2) is simply to truncate the vector at, say, $\pm L$, so there are $N = 2L + 1$ weights in the realizable version. We can see from (3.47) that the phase of the truncated vector is still zero as long as the weights remain symmetric about zero, because each term in the vector's DFT sum is added to its complex conjugate, resulting in a real transfer function. Furthermore, if the entire weight vector is shifted L steps so that h_{-L} becomes h_0, the result is a *causal* digital filter with linear phase shift, in accordance with Table 3.3, Property 8. In other words,

$$\boxed{\begin{array}{l}\text{Weights of any casual FIR filter with linear phase shift:}\\ \qquad b=[b_0\, b_1 \ldots b_{2L}] \text{ with}\\ \qquad b_{L-n}=b_{L+n};\quad n=1,2,\ldots,L\\ \text{Total number of weights: } N=2L+1.\end{array}} \tag{5.3}$$

Using (5.2), the impulse response (or weights) of our realizable lowpass causal, linear-phase filter may be given by

$$h_k=\begin{cases} 2\nu_c; & k=L \\ \dfrac{\sin(2\pi\nu_c(k-L))}{\pi(k-L)}; & k=0,1,\ldots,L-1,L+1,\ldots,2L \end{cases} \tag{5.4}$$

For weights computed in this manner and with $\nu_c = 0.1$ Hz-s, the power gain characteristic is plotted in Figure 5.3 for $L = 4$, 8, and 128. The plots illustrate the effects of truncating the weight vector. First, the gain curve becomes less rectangular as L is decreased. Second, there is significant *passband ripple* and

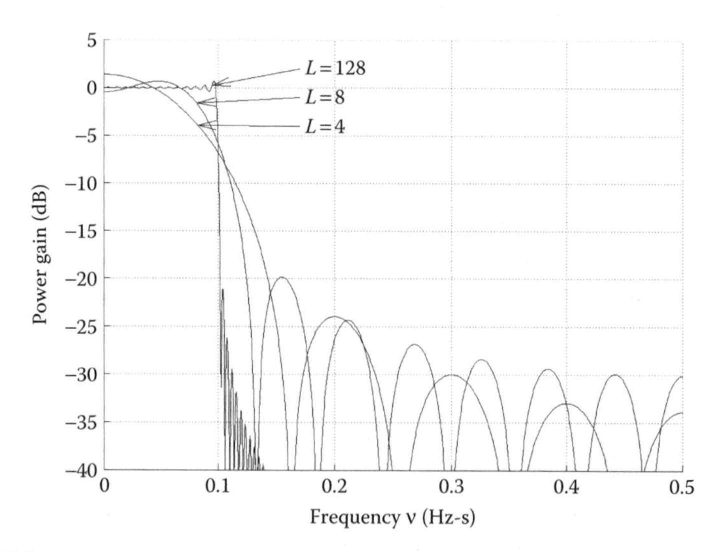

FIGURE 5.3
Power gain of three lowpass FIR filters, each with cutoff at 0.1 Hz-s. The number of weights in each case is $N = 2L + 1$.

stopband ripple, both of which are undesirable. Third, the power gain at cutoff is ideally –6 dB, corresponding with ideal amplitude gain 0.5, that is, halfway from one to zero in Figure 5.1. This also may be altered by truncation, that is, by the undesirable ripple in the gain characteristic.

The first effect, the nonrectangular gain curve, can obviously be improved by increasing L, but a larger L means more hardware and more computation per sample.

There is also another less obvious way to improve the shape of the gain curve, if the sampling rate can be adjusted. Suppose you are examining a digital processing system and notice that the analog-digital converter is set to convert at 10^8 samples per second, that is, the sampling frequency is $f_s = 100$ MHz. Suppose you also know there is never any content above 5 MHz in the analog input signal. Then, obviously, f_s is about 10 times higher than required, and the filter gain, which is the DFT of the weights, is spread over 10 times the necessary range. Thus, the gain may be made more rectangular in cases like this by reducing the sampling frequency to a more reasonable value, say 10 MHz, which still avoids aliasing. An example is given in Figure 5.4. The upper plot is the power gain with $f_s = 100$ MHz, and the lower plot is the power gain with $f_s = 10$ MHz.

Even if the analog-to-digital conversion rate cannot be changed, f_s can be reduced simply by *decimation*, or *down-sampling*, provided that aliasing is not introduced. In the example just discussed, the process is illustrated in Figure 5.5. Down-sampling is represented by a box with a down-arrow and

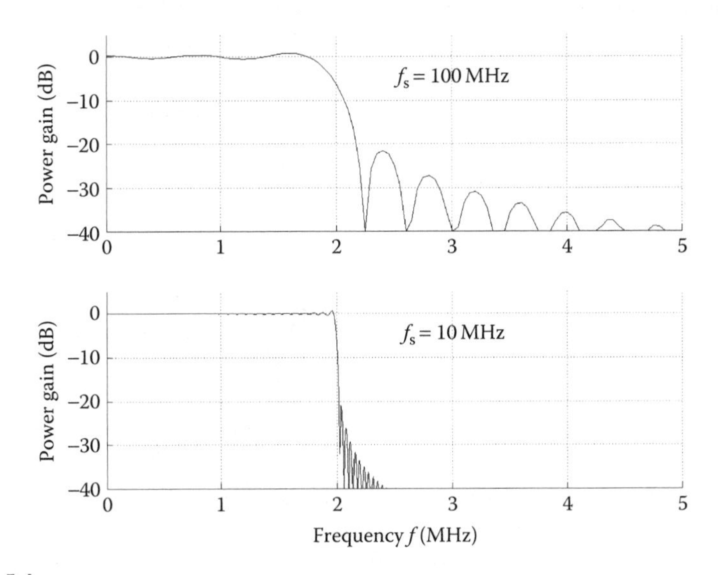

FIGURE 5.4
Power gain of two lowpass FIR filters, each with $N = 257$ weights and cutoff frequency $f_c = 2$ MHz. The upper filter is designed for sampling frequency $f_s = 100$ MHz and the lower for $f_s = 10$ MHz.

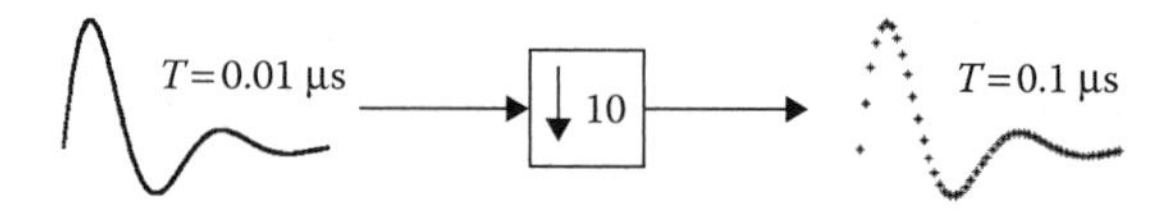

FIGURE 5.5
Down-sampling. The sampling rate is reduced from 100 to 10 MHz.

a number indicating the output/input time-step ratio. Aliasing is not introduced in this case, and the result is a waveform that we can filter successfully with fewer weights, as suggested in Figure 5.4.

When down-sampling with time-step ratio K is accomplished by decimation, $K - 1$ out of K samples of the input are ignored in each output time step. An obvious alternative to decimation is *averaging*, in which each output sample is the average of K input samples. Averaging is preferable to decimation, because it usually reduces the quantizing error of the analog-to-digital converter. (See also the discussion of *resampling* in Chapter 3, Section 3.12.)

The second adverse effect caused by truncating the weight vector (the passband and stopband ripple evident in Figure 5.3) cannot be improved by the methods just discussed, but may be improved and almost eliminated with the use of *window functions*, which is the subject of Section 5.4.

5.4 Improving an FIR Filter with Window Functions

The name *window function* comes from the idea of taking an infinite impulse response function such as $[h_k]$ in Figure 5.2 and superimposing a rectangular frame, like a window frame, through which you can only see impulse response values from index $k = -L$ to L. A *rectangular window* with length $2L + 1$ has the effect of multiplying $h(-L{:}L)$ by one and the rest of the h's by zero. There are also *nonrectangular* windows, and these have the effect of scaling the weights within the window frame.

We have already seen how the rectangular window is used to obtain a realizable lowpass FIR filter with noticeable passband and stopband ripple. Our goal in this section is to apply nonrectangular windows to the ideal impulse response function and obtain a smoother gain characteristic with less ripple.

The theory we need for this application is similar to the convolution result in (4.8), that is, the DFT of the convolution of periodic functions in the time domain is the product of the two individual DFTs. Suppose we now take the DFT of the product of the truncated ideal impulse response, $[h_k;\ k = 0,1,\ldots, N-1]$, and any window function, $[w_k;\ k = 0,1,\ldots,N-1]$:

$$
\begin{aligned}
\mathrm{DFT}\{h_k w_k\} &= \sum_{k=0}^{N-1} h_k w_k e^{-j2\pi mk/N} \\
&= \frac{1}{N^2}\sum_{k=0}^{N-1}\sum_{i=0}^{N-1} H_i e^{j2\pi ik/N}\sum_{n=0}^{N-1} W_n e^{j2\pi nk/N} e^{-j2\pi mk/N} \\
&= \frac{1}{N^2}\sum_{k=0}^{N-1} H_i \sum_{n=0}^{N-1} W_n \sum_{k=0}^{N-1} e^{j2\pi(i+n-m)k/N} \\
&= \frac{1}{N}\sum_{i=0}^{N-1} H_i W_{m-i}; \quad m = 0,1,\ldots,N-1
\end{aligned}
\tag{5.5}
$$

On the second line, inverse DFTs are substituted for the time functions. On the third line, the final sum is N when $n = m - i$ and zero otherwise. Thus, the DFT of a time-domain product is the convolution of individual DFTs, and taken together, (4.8) and (5.5) illustrate the duality of convolutions and transforms in the time and frequency domains. If x and y are vectors of length N,

Convolution in the time and frequency domains:

$$
\mathrm{DFT}^{-1}\{X_m Y_m\} = \sum_{n=0}^{N-1} x_n y_{k-n\pm\eta N}; \quad k = 0,1,\ldots,N-1; \quad \eta = 0,1,\ldots \tag{5.6}
$$

$$
\mathrm{DFT}^{-1}\{x_k y_k\} = \frac{1}{N}\sum_{n=0}^{N-1} X_n Y_{m-n}; \quad m = 0,1,\ldots,N-1 \tag{5.7}
$$

Notice that both convolutions involve periodic functions. In the convolution in (5.7), the DFT is always periodic as stated in Table 3.3, Property 1.

Thus, we see in (5.5) that the gain of the FIR filter is the convolution of the DFT of the truncated ideal impulse response, $h(0{:}N{-}1)$, with the DFT of the window vector, $w(0{:}N{-}1)$. The DFT of the rectangular window has high side lobes, hence the ripple in the filter gain function.

We illustrate this effect on the FIR lowpass power gain function using the four window functions illustrated with $N = 31$ in Figures 5.6 and 5.7. The relatively low value of N is used so the structures of the power spectra may be observed easily. The windows are in Figure 5.6, and their power spectra are in Figure 5.7. We use *boxcar*, the MATLAB® term which is descriptive and easy to remember, for *rectangular* in Figure 5.6. In Figure 5.7, each of the corresponding window functions was scaled, so the power gain was 0 dB at $\nu = 0$.

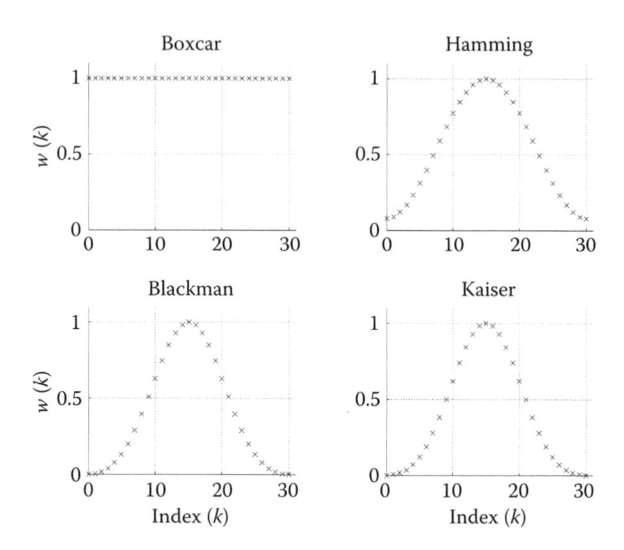

FIGURE 5.6
Rectangular (boxcar), Hamming, Blackman, and Kaiser window functions; $N = 31$ weights.

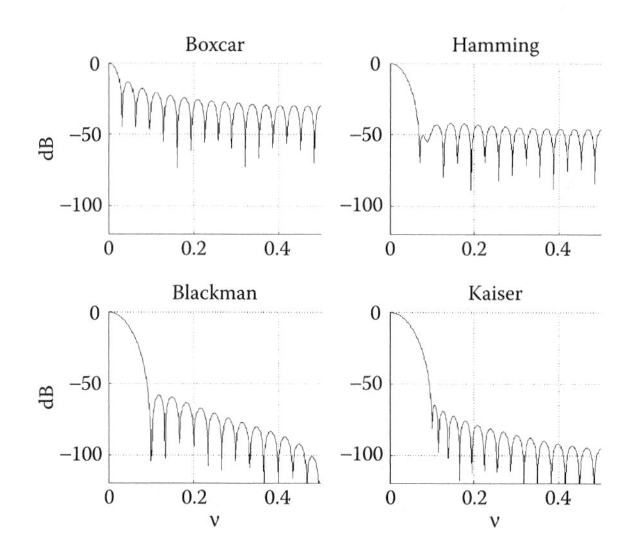

FIGURE 5.7
Continuous power gain spectra of the rectangular (boxcar) window and three other window functions, each scaled to produce 0 dB at $\nu = 0$; $N = 31$ weights.

Before discussing the other three windows in detail, imagine convolutions in (5.7) using the power spectra shown in Figure 5.7, convolving each of these from right to left over the ideal lowpass function in Figure 5.1 (with the gain in dB instead of amplitude units). As the main lobe at the left end of the window spectrum passes over the cutoff point of the ideal filter spectrum, the convolution causes the resulting gain to spread or "roll off" around the cutoff frequency. Then, as the window spectrum moves further to the left, the side lobes interact with the ideal spectrum to produce the ripple we have seen in the resulting passband and stopband. In brief, we may summarize these effects as in Table 5.1.

Having seen how windows operate to smooth the ripple in the FIR power gain, we turn now to formulas for the windows. The subject of windows is extensive in the literature on signal processing and statistical analysis of time series.[1–7] Here we present, besides the boxcar, just the three windows in Figure 5.6 that are well known and have certain optimal properties. Algebraic and MATLAB expressions for these windows are given in Table 5.2. The expressions are easy to understand and implement, except for the Kaiser expressions, which require explanation.

The *Hamming* window is the raised cosine function illustrated in Figure 5.6 and is usually the preferred "two-term" function in filter design. Another window, the *Hanning* window, named after J. von Hann, is similar but is zero at the end points. The *Blackman* window is a modulated version of the Hanning window. As seen in Figure 5.7, the Blackman window has a relatively wide main spectral lobe and lower side lobes, resulting in a smoother FIR power gain function with longer transition between passband and stopband.

The Kaiser window requires more discussion, but it is worth the trouble, because its properties lead to superior FIR filter designs. Unlike the other

TABLE 5.1

Effects of Window Spectrum Properties on FIR Filter Power Gain

Window Spectrum Property	Effect on Power Gain of FIR Filter
Wider main lobe	Wider transition from passband to stopband
Higher side lobes	Higher ripple in passband and stopband

TABLE 5.2

Window Formulas—Algebraic and MATLAB

Window	Formulas for Weights: $k = 0,1,\ldots, N-1$ (Algebraic), or $k = [0{:}N-1]$ (MATLAB)
Hamming	$w_k = 0.54 - 0.46\cos\left(\frac{2\pi k}{N-1}\right)$
	`w=0.54-0.46*cos(2*pi*k/(N-1))`
Blackman	$w_k = 0.42 - 0.5\cos\left(\frac{2\pi k}{N-1}\right) + 0.08\cos\left(\frac{4\pi k}{N-1}\right)$
	`w=0.42-0.5*cos(2*pi*k/(N-1))+0.08*cos(4*pi*k/(N-1))`
Kaiser	$w_k = \dfrac{I_0\left(\beta\sqrt{1-\left(\frac{2k}{N-1}-1\right)^2}\right)}{I_0(\beta)}$
	`w=bessel_0(beta*sqrt(1-(2*k/(N-1)-1).^2))/bessel_0(beta)`

windows, it has a parameter β, which has been set in Figures 5.6 and 5.7 to produce a main spectral lobe about the same as that of the Blackman window. Typically, the value of β is in the range from 4 to 9. Values of β that give main lobe widths comparable to those of the Hamming and Blackman windows are

$$\boxed{\begin{array}{ll} \beta = \sqrt{3}\pi & \text{Kaiser and Hamming main lobe widths about equal} \\ \beta = 2\sqrt{2}\pi & \text{Kaiser and Blackman main lobe widths about equal} \end{array}} \tag{5.8}$$

Power spectra of windows with these β values are illustrated in Figure 5.8 with N = 127 weights. In the upper half of the figure, for example, with $\beta = \sqrt{3}\pi$, the Kaiser and Hamming windows have similar main lobes. However, the Kaiser window has lower side lobes, resulting in less ripple in an FIR filter gain function.

The Kaiser window, as presented in Table 5.2, also requires computation of the zero-order modified Bessel function of the first kind, given by

$$I_0(x) = 1 + \sum_{n=1}^{\infty} \left(\frac{(x/2)^n}{n!} \right)^2 \tag{5.9}$$

Fortunately, this sum converges rapidly due to the factorial in the denominator, and is easily approximated in various ways. The MATLAB function

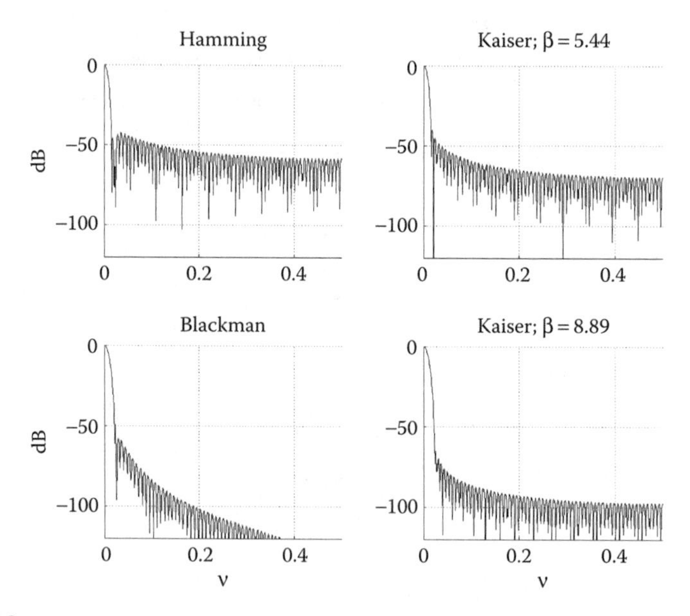

FIGURE 5.8
Power spectra of Kaiser windows with β set to give main lobe widths approximately the same as Hamming and Blackman main lobe widths.

besseli (0, *x*) may be used to compute (5.9). If MATLAB is not available, a simple function called *bessel_0* is included in the *functions* folder on the publisher's website for this text. The use of *bessel_0* is illustrated in Table 5.2.

There is an important point about using windows in the manner just described to design lowpass FIR filters: due to the truncation of the original ideal impulse response followed by the application of a window, the windowed weights may require *scaling* to adjust the lowpass gain to 0 dB at $\nu = 0$. (We noted that this was done for the plots in Figure 5.7.) For lowpass filters, because the amplitude gain at $\nu = 0$ is just the sum of the weights, the adjustment (in MATLAB notation) after the weights are computed is simply

$$b = b/sum(b)$$

For highpass, bandpass, and bandstop filters, the gain may be adjusted to one at $\nu = 0.5$, $\nu = (\nu_1 + \nu_2)/2$, and $\nu = 0$, respectively. These adjustments are made in the *fir_weights* function described in Section 5.5.

To conclude this section, Figure 5.9 illustrates power gain spectra for four lowpass FIR filters, each with $N = 127$ weights and cutoff at $\nu_c = 0.1$ Hz-s. In each case, the lowpass power gain results from the use of the window with a power spectrum in the corresponding position in Figure 5.8. The relative effects on passband ripple and passband-to-stopband transition are not easy to see in Figure 5.9, but the relative effects on stopband ripple are evident.

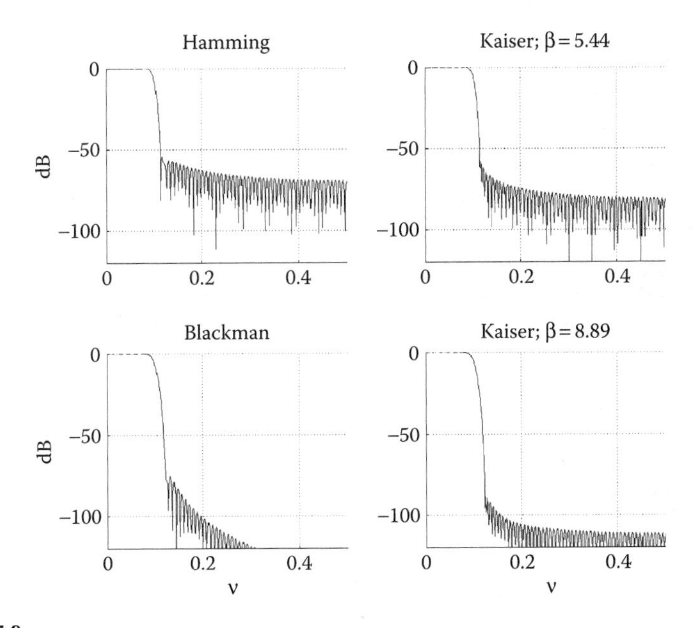

FIGURE 5.9
Lowpass FIR filter power gain; $L = 63$ and $\nu_c = 0.1$ Hz-s. Each power gain function results from the use of the window with corresponding power gain in Figure 5.8.

5.5 Highpass, Bandpass, and Bandstop Filters

In Section 5.4, we saw how to apply a window to a truncated ideal impulse response to achieve a lowpass FIR filter. This concept applies equally well to impulse responses of other "ideal" FIR filters with rectangular gain functions, and in particular, to highpass, bandpass, and bandstop FIR filters. Hence, all we need for these filters is the impulse response functions, or weight vectors, of the filter types to use in place of the lowpass vector, *h*, in (5.4).

Weight vectors for the other filter types can be derived by subtracting lowpass weight vectors in the form of (5.4). For example, Figure 5.10 illustrates how to produce an ideal highpass filter gain by taking the difference between two ideal lowpass filter gains. We use the notation $H_{lp}(\nu,\nu_c)$ to represent the gain of an ideal lowpass filter with passband from zero to ν_c Hz-s. The filter on the left in Figure 5.10 with gain $H_{lp}(\nu,0.5)$ is a lowpass filter with passband extending to half the sampling frequency, that is, an allpass filter. The second filter is an ideal lowpass filter with cutoff at ν_c Hz-s. The difference between these two gains, which are both one in the passband, is the ideal highpass gain, $H_{hp}(\nu,\nu_c)$, on the right.

Ideal bandpass and bandstop gains are produced similarly. These are denoted $H_{bp}(\nu,\nu_1,\nu_2)$ and $H_{bs}(\nu,\nu_1,\nu_2)$, respectively, where ν_1 and ν_2 are the band-edge frequencies. We summarize all three conversions as follows:

$$\begin{array}{|l|}\hline \text{Conversion of ideal lowpass FIR filter gain to} \\ \text{Highpass: } H_{hp}(\nu,\nu_c)=H_{lp}(\nu,0.5)-H_{lp}(\nu,\nu_c) \\ \text{Bandpass: } H_{bp}(\nu,\nu_1,\nu_2)=H_{lp}(\nu,\nu_2)-H_{lp}(\nu,\nu_1) \\ \text{Bandstop: } H_{bs}(\nu,\nu_1,\nu_2)=H_{lp}(\nu,0.5)-H_{bp}(\nu,\nu_1,\nu_2) \\ \hline \end{array} \tag{5.10}$$

Because the DFT is a linear operation (Table 3.3, Property 3), the implication of (5.10) is that we may construct highpass, bandpass, and bandstop FIR filters using lowpass weight vectors given by (5.4). First, we note that the impulse response of the all-pass filter is the discrete impulse function. Using $\nu_c = 0.5$ in (5.4) and denoting the all-pass weight vector b_{ap}, we have the following:

$$b_{ap} = [\text{zeros}(1,L),1,\text{zeros}(1,L)] \tag{5.11}$$

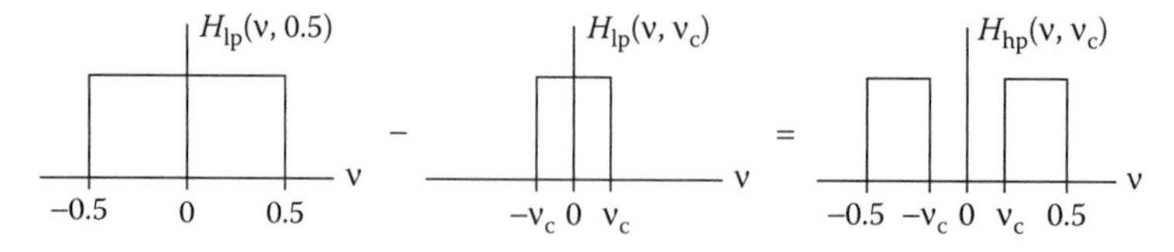

FIGURE 5.10
All-pass gain minus lowpass gain equals highpass gain, assuming the H_{lp}'s are of equal amplitude.

Using similar notation, that is, using $b_{hp}(\nu_c)$ for the weight vector corresponding with $H_{hp}(\nu, \nu_c)$, and so on, and the relationships in (5.10), the highpass, bandpass, and bandstop weight vectors may be formulated as follows:

$$\boxed{\begin{array}{l}\text{Conversion of ideal lowpass FIR filter weights to}\\ \text{Highpass: } b_{hp}(\nu_c) = b_{ap} - b_{lp}(\nu_c)\\ \text{Bandpass: } b_{bp}(\nu_1, \nu_2) = b_{lp}(\nu_2) - b_{lp}(\nu_1)\\ \text{Bandstop: } b_{bs}(\nu_1, \nu_2) = b_{ap} - b_{bp}(\nu_1, \nu_2)\end{array}} \quad (5.12)$$

Thus, in (5.4), (5.11), and (5.12), we have weight-vector formulas for the four basic types of FIR digital filters. A function b = *fir_weights(N, band, windo,* ν1, ν2), in which N is the number of weights, *band* designates the band (lowpass, highpass, etc.), *windo* designates the window, and [ν1, ν2] are the critical frequencies, is included with the functions on the publisher's website for this text.

Any of the windows discussed in the previous section may be applied to the weight vectors in (5.12), either before or after they are combined, with the same effect. Another function, w = *window(N, type, beta)*, generates several different nonnormalized windows and is used internally by *fir_weights*. For example, the weight vector for the FIR filter with $N = 127$ weights, *band* = 1 for lowpass, *window* = [7, 8.89] for the Kaiser window with $\beta = 8.89$, and $\nu_c = 0.1$ Hz-s on the lower right in Figure 5.9 is produced by the following expression:

```
b = fir_weights(127,1,[7,8.89],.1);                    (5.13)
```

To conclude this section, four examples of FIR filter power gain are presented in Figure 5.11. In each case, the number of weights is $N = 127$. The four windowed weight vectors were computed with the following expressions:

```
b1 = fir_weights(127,1,[7,2*sqrt(2)*pi],.1);
bh = fir_weights(127,2,[7,2*sqrt(2)*pi],.4);           (5.14)
bp = fir_weights(127,3,[7,2*sqrt(2)*pi],.1,.2);
bs = fir_weights(127,4,[7,2*sqrt(2)*pi],.1,.2);
```

Let b stand for any of the four weight vectors above. The filter gain in each case in Figure 5.11 is as follows:

```
H = fft(b,1000);                                       (5.15)
```

Hopefully, the reader is familiar by now with the effect of appending zeros to the windowed weight vector before taking the FFT. The result in this case is a DFT with 1000 points and a 501-point gain function in the range [0, 0.5] Hz-s,

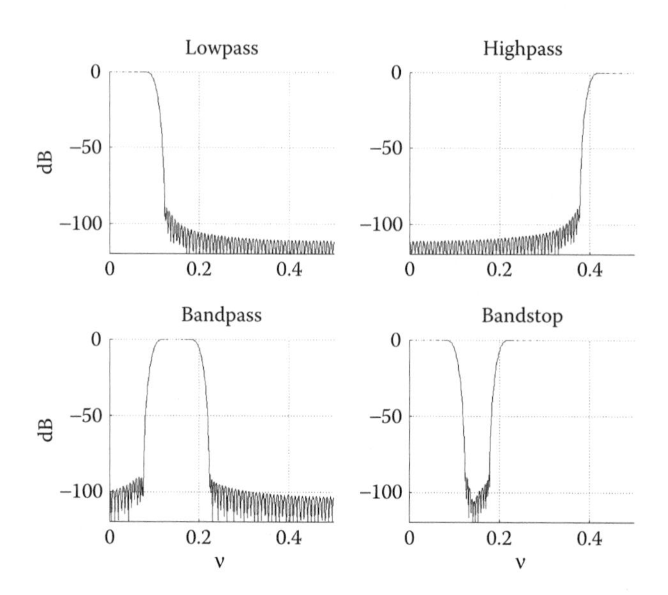

FIGURE 5.11
Power gain of four typical FIR filters, each with $N = 127$ weights and using the Kaiser window with $\beta = 8.89$.

that is, from zero through half the sampling rate. The final step is to convert the amplitude of Hz to decibels:

```
dB = 20*log10(max(abs(H(1:501)),1.e-6));                    (5.16)
```

The resulting power gain in decibels is plotted in Figure 5.11 for each of the four filters.

5.6 A Complete FIR Filtering Example

We now have all the tools necessary for the application of any of the four common types of FIR filters to a digitized signal vector. There are a few practical matters worth noting when this is done, such as conversion of units, phase shift, and so on. To illustrate some of these, we provide the following example in which an input vector consisting of a signal with additive high-frequency random noise is filtered to improve the signal-to-noise ratio. We should note at the outset that this is a typical application of filtering, but rarely do the results look this good in practice. The reader should not be discouraged when the results do not look as good with real data.

Suppose you are looking for electromagnetic pulses emitted by a distant source known to be sending in a band between 100 and 200 MHz. A sensitive antenna is aimed at the source, and the receiver output is digitized at 10^9 samples per second. One microsecond of data (i.e., 1000 samples) of the receiver output are plotted in the upper part of Figure 5.12.

Suppose also that most of the noise power is known to be concentrated at frequencies above 300 MHz. Having this information, you put the receiver output through a lowpass FIR filter cutting off at 250 MHz. The filter you choose is similar to the upper left filter in Figure 5.9. It uses the Hamming window and has $N = 127$ weights. The filter output is plotted in the lower half of Figure 5.12 and gives evidence of a single pulse at about 650 ns after the beginning of the 1-μs interval.

There are several things to note about the lower plot. First, the amplitude range is 1/10 of the upper plot range. The waveform in the vicinity of 650 ns is in exactly the same position in the upper plot but is obscured by the high-frequency noise. Second, to align the two signals, the lower plot has been shifted left 63 ns to account for the filter delay of $L = (N - 1)/2 = 63$ ns, that is, to account for the linear phase shift of the filter. [See the discussion with Equation (5.1) at the beginning of this chapter.] And third, the filter output component at 0 ns looks similar to the signal at 650 ns and might be mistaken for another pulse from the source, but it is only the filter startup transient, that is, the transient response of the filter with zero initial conditions to the onset of the upper waveform.

There are other things to consider as well, such as the choice of the number of filter weights and the cutoff frequencies, and so on, in addition to assuring that the filter output is being correctly interpreted.

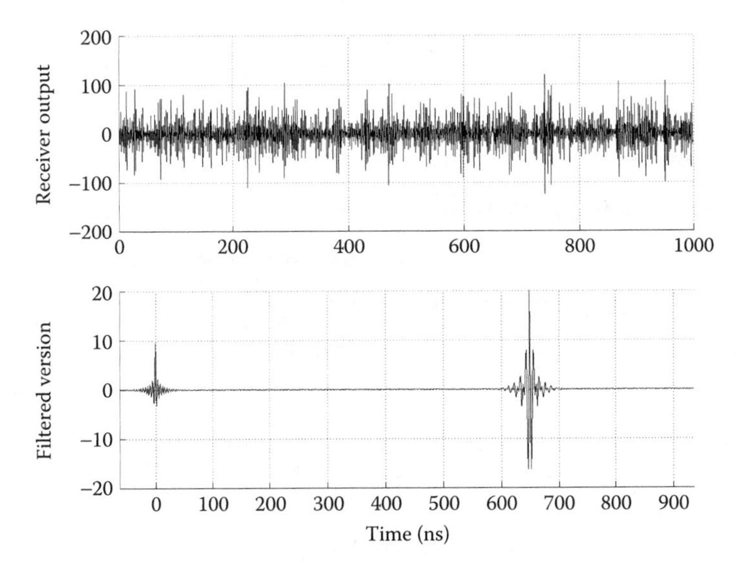

FIGURE 5.12
Receiver output consisting of a transient signal near 650 nanoseconds plus high-frequency random noise (upper plot), and filtered version (lower plot).

5.7 Other Types of FIR Filters

Besides the four principal types of FIR filters (lowpass, highpass, bandpass, and bandstop), other FIR filter types have found use in various special applications. It is easy to generalize the procedure we have been using to obtain the design of an FIR filter with any specified gain characteristic. We simply begin with the specified continuous gain function, find the ideal impulse response as in (5.2), translate to a realizable version as we did in (5.4), and apply a window to the realizable version. In other words,

To find the weights of an FIR filter:

1. Specify a real continuous gain, $H(j\omega)$, in the range $-\pi/T \le \omega \le \pi/T$ and make $H'(-j\omega) = H(j\omega)$.

2. Ideal impulse response:

$$h_{ideal}(kT) = \frac{T}{2\pi}\int_{-\infty}^{\infty} H(j\omega)e^{j\omega kT}d\omega; \quad -\infty < k < \infty \tag{5.17}$$

3. Realizable, causal version:

$$h_k = h_{ideal}((k-L)T); \quad 0 \le k \le N-1; \quad N = 2L+1$$

4. FIR filter weights:

$$b_k = w_k h_k; \quad 0 \le k \le N-1$$

5.8 Digital Differentiation

As an example of this procedure, we design an FIR *differentiator*, that is, a filter that differentiates the sampled input signal. The Fourier transform of the derivative of the continuous signal, $x(t)$, with respect to t is $j\omega X(j\omega)$, which may be proved by taking the inverse transform as follows:

$$\frac{1}{2\pi}\int_{-\infty}^{\infty} j\omega X(j\omega)e^{j\omega t}d\omega = \frac{d}{dt}\left[\frac{1}{2\pi}\int_{-\infty}^{\infty} X(j\omega)e^{j\omega t}d\omega\right] = \frac{d}{dt}(x(t)) \tag{5.18}$$

Therefore, in step 1 in Equation 5.17, we specify

$$H(j\omega) = j\omega; \quad |\omega| \le \pi/T \tag{5.19}$$

In step 2, we determine the ideal impulse response (after some effort):

$$h_{ideal}(kT) = \frac{T}{2\pi}\int_{-\pi/T}^{\pi/T} j\omega e^{jk\omega T}\,d\omega = \begin{cases} 0; & k=0 \\ \dfrac{(-1)^k}{kT}; & |k|>0 \end{cases} \tag{5.20}$$

In step 3, we choose a filter length, $N = 2L + 1$, and determine the realizable impulse response:

$$h_k = \begin{cases} 0; & k=L \\ \dfrac{(-1)^{k-L}}{(k-L)T}; & k=0,1,\ldots,L-1,L+1,\ldots,2L \end{cases} \tag{5.21}$$

Finally, in step 4, we apply a window function to obtain the weight vector, which in MATLAB notation is $b = w.*h$. The vector length (number of weights) is the overall range of k in (5.21), that is, $N = 2L + 1$.

An example of a differentiator with $N = 63$ weights using the Hamming window is presented in Figure 5.13. The amplitude gain, $|H(j\omega)|$, is plotted versus frequency in Hz on the left. The input signal on the upper right is a unit sine wave at 1 Hz, sampled at 20 samples/s. The output, which has been shifted left $L = 31$ samples to account for linear phase shift, is the derivative of the input with respect to time, that is, a cosine wave with amplitude equal

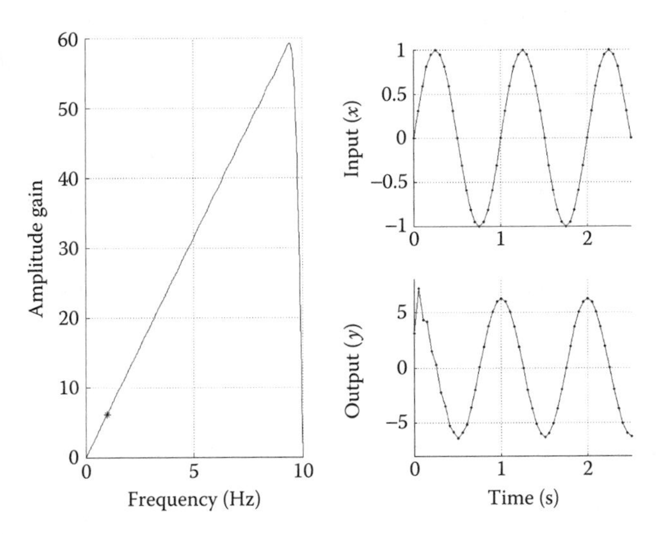

FIGURE 5.13
Amplitude gain of FIR differentiator with $N = 63$ weights and Hamming window (left). The "*" on the gain plot indicates the frequency (1 Hz) of an input unit sine wave (upper right). The differentiator output (lower right), after startup, is a cosine wave with amplitude $\omega = 2\pi$. The time step in this example is $T = 0.05$ seconds.

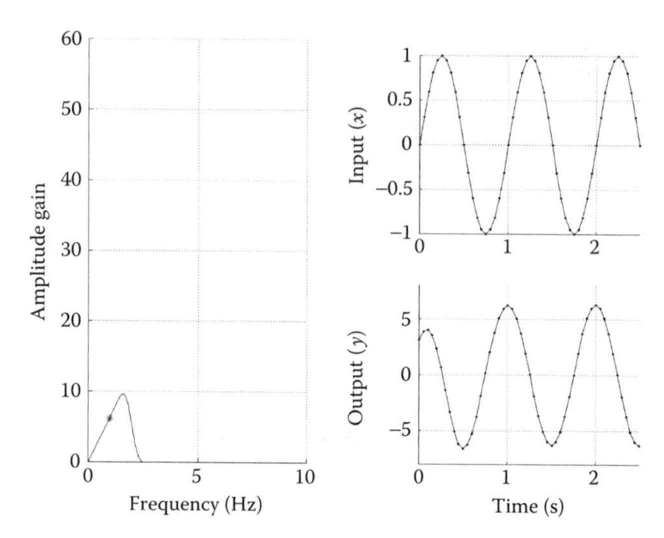

FIGURE 5.14
Repetition of Figure 5.13 with differentiator cutoff reduced from π/T to $f_c = 2$ Hz.

to ω, or 2π in this case, as shown by the marker (*) on the amplitude gain plot. Notice also the startup transient on the output (y) plot, with a duration of $L = 31$ samples.

Differentiators inherently amplify high frequencies. If a signal is known to have no components above, say, ω_c, the differentiator may be designed to cut off at ω_c and thus eliminate the amplification of high-frequency noise in the output. This would be done simply by changing the limits of integration in (5.20) from $\pm\pi/T$ to $\pm\omega_c$. The result, in place of (5.21), is as follows:

$$h_k = \begin{cases} 0; & k = L \\ \dfrac{\omega_c \cos((k-L)\omega_c T)}{(k-L)\pi} - \dfrac{\sin((k-L)\omega_c T)}{(k-L)^2 \pi T}; & 1 \le |k-1| \le L \end{cases} \tag{5.22}$$

An example of this with $\omega_c = 4\pi$ rad/s is shown in Figure 5.14. Because the input signal at $\omega = 2\pi$ rad/s is entirely below ω_c, the results after startup are the same as in Figure 5.13.

5.9 A Hilbert Transformer

Another example of the design procedure in (5.17), and another example of a filter with special use in signal processing, is the FIR *Hilbert transformer*, which is a digital filter designed ideally to provide a 90° phase shift at all frequencies below half the sampling rate. Its ideal transfer function is given in (5.23).

$$H(j\omega) = \begin{cases} j; & -\pi/T < \omega < 0 \\ -j; & 0 \le \omega < \pi/T \end{cases} \tag{5.23}$$

The Hilbert transformer is used typically with *complex* signals with samples that have real and imaginary parts. The output of a radar detector, for example, consists of real and imaginary or *quadrature* components on separate channels, and taken together, these become a complex signal with real and imaginary parts.

The operation of a Hilbert transformer on complex signals brings up an interesting spectral property of these signals. Suppose we have a complex signal vector, $s = x + jy$. Because the DFT is a linear operation, its DFT components must be $S_m = X_m + jY_m$. We recall from (3.11) and Table 3.3, Property 2, that, at negative frequencies, $X_{-m} = X_m'$ when x is real. The result is $S_{-m} = X_m' + jY_m'$ for the spectrum of a complex signal at negative frequencies. Now suppose the imaginary component, y, is the Hilbert transform of the real component, x. This implies $Y_m = -jX_m$ for $m \ge 0$ and $Y_m = jX_m$ for $m < 0$, and therefore, $S_m = X_m + j(-jX_m) = 2X_m$ and $S_{-m} = X_m' + j(jX_m') = 0$ for $m \ge 0$. Thus, we have the following result:

If s is the complex signal $s = x + jy$, and if y is the Hilbert transform of x, then the spectrum of s is one sided, that is,

$$S_m = \begin{cases} 2X_m; & m \ge 0 \\ 0; & m < 0 \end{cases} \tag{5.24}$$

The FIR Hilbert transformer is easy to design using the procedure in (5.17). The ideal impulse response in step 2 of the procedure is

$$h_{ideal}(kT) = \begin{cases} 0; & k \text{ even} \\ \dfrac{2}{k\pi}; & k \text{ odd} \end{cases} \tag{5.25}$$

The rest of the procedure, as well as the application of the Hilbert transformer to complex signals in communications and other related areas, is described in references Oppenheim and Schafer,[2] Orfanidis,[3] and Parks and Burrus.[8]

Exercises

General instructions: (1) When you make a plot in any of the exercises, be sure to label both axes so we can tell exactly what the units mean. (2) Make continuous plots unless otherwise specified. "Discrete plot" means to plot discrete symbols, unconnected unless otherwise specified. (3) In plots, and especially in subplots, set the axis limits for best use of the plotting area.

5.1 Plot the unweighted, linear-phase weights given by (5.4) in two subplots, one above the other. Set $L = 200$ for both plots. Set the cut-off frequency (ν_c) at 0.03 Hz-s in the upper plot and 0.3 Hz-s in the lower plot.

5.2 Complete the following:

a. Derive an expression for the "upper envelope" of (5.4), that is, an expression for a smooth curve passing through all the positive peaks beyond the second peak at $k - L = 5/4\nu_c$.

b. For $\nu_c = 0.03$, how far from the peak impulse response (b_L) does this envelope drop below 5% of the peak?

c. Verify your answers to (a) and (b) by plotting b_k/b_L as well as the upper envelope of b_k/b_L for $240 \le k \le 400$. On your plot, use abscissa limits $[L, 2L]$ and ordinate limits [0, 0.5].

5.3 Make a plot similar to Figure 5.3 of the power gains of two unwindowed FIR lowpass filters with weights given by (5.4). Use the following data:

Sampling frequency: 20 samples/s

Cutoff frequency: 1 Hz

Number of weights: 33 and 129

Power gain from −60 to 5 dB

Frequency from 0 to 10 Hz

5.4 Plot the power gains of the two filters described in Exercise 5.1 in a single figure with no subplots. Plot dB in the range [−80, 5] versus frequency in Hz-s in the range from zero to half the sampling rate.

5.5 Complete the following:

a. Show that the amplitude gain of any linear-phase filter with weights in the form of (5.3) may be expressed in the following form:

$$|H(\omega T)| = \left| b_0 + 2\sum_{k=1}^{L} b_k \cos(\omega kT) \right|$$

[Hint: Use the weight vector in (5.3) centered at zero, noting that the time shift has no effect on the amplitude spectrum.] Compare this result with (2.32), and note that here we have the magnitude of a continuous Fourier series for a function of *frequency* (not time).

b. Express the amplitude gain at half the sampling rate as a simple, nontrigonometric sum. Use this expression to compute the power gains of the two filters in Exercise 5.1 at half the sampling rate. If you have worked Exercise 5.4, use this result to verify the ends of the two dB plots.

5.6 Write a MATLAB function b = *my_fir_lowpass_weights(N,nuc)* that will compute unwindowed weights of any linear-phase lowpass FIR filter, where N is the number of weights, and *nuc* is ν_c, the cutoff frequency in Hz-s. Write the function so that if N is even, it will be increased by one with a warning.

a. Test the function with $N = 7$ and $N = 8$ to see that it prints the correct number of weights and issues a warning when $N = 8$.

b. Use the function to plot the unweighted impulse response of a lowpass filter with $\nu_c = 0.3$ and $N = 401$. Compare the plot with your answer to Exercise 5.1, or verify it by computing (5.2) and plotting the result. If the plots are the same, you may use the function in subsequent exercises where lowpass weights are required.

5.7 Write a MATLAB function called w = *my_window(N, type, beta)* to create w, a row vector of N window samples. Let *type* designate any of the three windows in Table 5.2 plus the boxcar window, and let *beta* designate the parameter β for the Kaiser window. (Note: You may use the MATLAB function *nargin* to determine whether there are two or three input arguments.) Plot all four windows in a single figure for $N = 15$, using $\beta = 7$ for the Kaiser window. Make connected discrete plots using a different symbol for each window, and print a legend on your plot.

5.8 Make connected discrete plots of the Hamming, Blackman, and Kaiser windows with $N = 63$ weights in three vertically stacked subplots. Make the Kaiser window look like the Blackman window. Label the plots to show which is which.

5.9 Design a lowpass FIR filter to cut off at 12.5 MHz, assuming a sampling frequency of 100 MHz. Use the Hamming window, and use $L = 12$. Plot the power gain in terms of dB versus frequency in MHz.

5.10 The specifications of an FIR lowpass filter are as follows:

Time step: 100 μs; Cutoff: 1.0 KHz;

Power gain ≤–40 dB for $f \geq 2$ KHz; window: Blackman

Design the filter by making 11 plots in the same graph of the power gain in terms of dB versus frequency in KHz using L = [5:15]. Indicate the smallest value of L that meets the specifications.

5.11 In a single graph, plot the power gains of two lowpass FIR filters with T and ν_c the same as in Exercise 5.10. Use the Blackman window for one filter and the Kaiser window for the other, with β set to match the Blackman window. Use $N = 31$ weights in both filters. Plot dB in the range [–100, 0] versus frequency in KHz, and identify each plot. Briefly compare the two filters.

5.12 Design a highpass FIR filter with 41 weights to operate with 10^6 samples/s, with passband beginning at 100 KHz. Use the Kaiser window with $\beta = 6$. Plot power gain in dB in the range [–100, 0] versus frequency in KHz. What is the *amplitude gain* at cutoff?

5.13 Design a bandpass FIR filter with $L = 30$ to operate with time step $T = 0.1$ s and a passband from 3 to 4 Hz. Plot power gain in dB in the range [–100, 0] versus frequency in Hz.

5.14 Complete the following:

a. Create a Gaussian noise vector using the following expressions:

```
N = 200;
randn('state',0);
x = 10*randn(1,N);
```

Treat this vector as a waveform sampled at 100 samples/s and beginning at $t = 0$. Next, create vector y by passing x through a lowpass FIR filter designed with 61 weights, the Hamming window, and cutoff at 30 Hz. Finally, add to y a unit 40-Hz sine wave beginning at $t = 1.0$ s and extending to the end of y. Plot y versus t in the upper of two subplots.

b. Create vector u by passing y through a highpass FIR filter with 61 weights, the Hamming window, and cutoff at 35 Hz. Plot u versus t in the lower subplot, and explain the result.

5.15 Complete the following:

a. Create vector x consisting of 200 samples of the sum of five unit sine waves with frequencies 10, 15, 20, 25, and 30 Hz, using time step $T = 10$ ms. (For full credit, create x with a single MATLAB expression.) Plot x versus t in seconds in the upper of two subplots.

b. Design a bandpass filter with 63 weights to select the component of x at 15 Hz. Use one of the windows in Table 5.2. Plot the filtered version of x in the lower subplot, and comment on the effectiveness of the filter.

5.16 Complete the following:

a. Create the vector x in Exercise 5.15 with $N = 300$ samples. Plot $x(t)$ in the range $t = [0, 2]$ s in the first of three subplots.

b. Filter x as in Exercise 5.15, but this time with a bandstop filter to remove the component at 15 Hz. Plot the filter output in the center subplot.

c. Shift the filter output to account for the delay caused by the filter. Then subtract it from x. Plot the result in the lower subplot over the same range of t. What is the frequency of the final waveform?

5.17 Derive the impulse response of the ideal differentiator, beginning with $H(j\omega) = j\omega$ and using integration by parts.

5.18 Complete the following:

a. Create the vector p = [0:2*T:1, 1–T:–T:0]. Let x consist of three cycles of p. Make a continuous plot of $x(t)$ versus t using $T = 0.001$ s in the upper of three subplots.

b. In the center subplot, make a connected discrete plot of the impulse response of a differentiator using $L = 10$ and the Hamming window. Plot the response versus time t.

c. In the lower subplot, show the differentiator output when the input is x. Shift the output so it is correctly aligned with x in the upper subplot.

5.19 Make two subplots. On the left, plot $N = 101$ samples of the sum of the impulse responses of an ideal allpass filter with gain equal to 2.0 and an ideal lowpass filter with gain equal to 1.0 and $\nu_c = 0.2$ Hz-s. On the right, plot the amplitude gain of this combined filter after applying the Hamming window.

References

1. Harris, F. J. 1978. On the use of windows for harmonic analysis with the discrete Fourier transform. *Proc IEEE* 66(1):51–83.
2. Oppenheim, A. V., and R. W. Schafer. 1989. *Discrete-Time Signal Processing*. Chap. 7. Englewood Cliffs, NJ: Prentice Hall.
3. Orfanidis, S. J. 1996. *Introduction to Digital Signal Processing*. Chap. 10. Upper Saddle River, NJ: Prentice Hall.
4. Kaiser, J. F. 1963. Design methods for sampled data filters. *Proceedings of First Allerton Conference on Circuit and System Theory,* 221.
5. Kaiser, J. F. 1966. Digital filters. In *System Analysis by Digital Computer,* ed. F. F. Kuo and J. F. Kaiser, Chap. 7. New York: John Wiley & Sons.
6. Blackman, R. B., and J. W. Tukey. 1958. *The Measurement of Power Spectra*. New York: Dover Publications.
7. Helms, H. D. 1968. Nonrecursive digital filters: Design methods for achieving specifications on frequency response. *IEEE Trans Audio Electroacoust* 16(3):336–42.
8. Parks, T. W., and C. S. Burrus. 1987. *Digital Filter Design*. New York: Wiley.

Further Reading

Antoniou, A. 1993. *Digital Filters: Analysis and Design.* 2nd ed. New York: McGraw-Hill.

Elliott, D. F., ed. 1987. *Handbook of Digital Signal Processing.* New York: Academic Press.

Jackson, L. B. 1989. *Digital Filters and Signal Processing.* Norwell, MA: Kluwer Academic Publishers.

Lutovac, M. D., D. V. Tosic, and B. L. Evans. 2001. *Filter Design for Signal Processing Using MATLAB and Mathematica.* Upper Saddle River, NJ: Prentice Hall.

Mitra, S. K., and J. F. Kaiser eds. 1993. *Handbook of Digital Signal Processing.* New York: Wiley.

Rabiner, L. R., and C. M. Rader, eds. 1972. *Digital Signal Processing.* New York: IEEE Press.

Roberts, R. A., and C. T. Mullis. 1987. *Digital Signal Processing.* Reading, MA: Addison-Wesley.

6

IIR Filter Design

6.1 Introduction

In Chapter 4, we saw that the transfer function of a linear recursive or infinite impulse response (IIR) filter is a ratio of polynomials in z, rather than just a single polynomial that describes a finite impulse response (FIR) filter. The direct-form algorithm, transfer function, and gain of an IIR filter were given in (4.10) through (4.12):

$$\begin{aligned}
&\text{Algorithm:} && y_k = \sum_{n=0}^{N-1} b_n x_{k-n} - \sum_{m=1}^{M-1} a_m y_{k-m} \\
&\text{Transfer function:} && H(z) = \frac{Y(z)}{X(z)} = \frac{b_0 + b_1 z^{-1} + \cdots + b_{N-1} z^{-(N-1)}}{1 + a_1 z^{-1} + \cdots + a_{M-1} z^{-(M-1)}} \\
&\text{Filter gain:} && H(e^{j\omega T}) = \frac{b_0 + b_1 e^{j\omega T} + \cdots + b_{N-1} e^{-j(N-1)\omega T}}{1 + a_1 e^{-j\omega T} + \cdots + a_{M-1} e^{-j(M-1)\omega T}}
\end{aligned} \tag{6.1}$$

Thus, the IIR transfer function has poles as well as (usually) zeros on the z-plane, and, as stated in (4.39), the poles must be inside the unit circle for stability. As we shall see in Section 6.2, this restriction on the location of the poles results in the impossibility of a stable, linear-phase IIR filter operating in real time.

On the other hand, as noted in Section 5.1, the presence of poles in $H(z)$ away from the origin can also result in a sharper transition from the passband to the stopband of an IIR filter, compared with an FIR filter with the same total number of weights. With this advantage, IIR filters are still used in situations where at least a small amount of phase distortion is allowable.

As in Chapter 5 for FIR filters, the primary purpose of this chapter is to develop weight-vector formulas for the standard designs of lowpass, highpass, bandpass, and bandstop IIR filters. Our approach will also be similar to the development in Chapter 5, that is, beginning with an "ideal" lowpass design, creating from this a realizable IIR lowpass filter, and finally transforming the latter into the other three filter types.

6.2 Linear Phase

As we saw in Chapter 5, the key to understanding linear phase is discrete Fourier transform Property 8 in Table 3.3, that is, a uniform shift in time transforms to a phase angle which is a linear function of frequency. This, in turn, led to the conclusion in (5.1), that is, if $H(e^{j\omega T})$ is the gain function of a linear-phase filter, the following condition must hold:

$$H(e^{j\omega T}) = |H(e^{j\omega T})|e^{-jn\omega T} = R(e^{j\omega T})e^{-jn\omega T} \tag{6.2}$$

where $R(e^{j\omega T})$ is a real function of ω and therefore equal to its own conjugate, $R(e^{-j\omega T})$. In terms of the z-transform, (6.2) becomes

$$H(z) = z^{-n}R(z); \quad R(z) = R(z^{-1}) \tag{6.3}$$

Now suppose $R(z)$ is a ratio of polynomials in z, as it must be for an IIR filter. Then, if $R(z)$ has a pole at z_0, the condition in (6.3) implies $R(z)$ must also have a pole at $z^{-1} = z_0$, that is, a pole at $z = 1/z_0$. Thus, $R(z)$ and therefore $H(z)$ cannot be stable, because if z_0 is inside the unit circle, $1/z_0$ must be outside the unit circle. Thus, any stable IIR filter in the form of (6.1) cannot have a linear phase characteristic.

Even though linear phase is not possible in (6.1), there are two points worth noting. First, the filter designs introduced in this chapter have, as we shall see, phase characteristics that are nearly linear through most of the passband. The stopband phase is usually not of concern, and so the nearly linear phase in portions of the passband results in many useful applications of IIR digital filters.

Second, if a digital filter (IIR or FIR) is not operating in real time, there is a way to operate the filter without phase shift, that is, a way to modify the operation of the filter so that the resulting gain is real. The method involves reversing the input or the output signal vectors.

In (4.6), we defined $X(z)$ as the z-transform of a signal vector x, that is, the z-transform of a sequence of samples beginning at $t = 0$ and ending at $t = (N-1)T$. Now, suppose we create a vector x^r by reversing this sequence so it begins at $t = 0$ and ends at $t = -(N-1)T$. Obviously, we could not reverse x in real time, but if x were recorded, we could reverse the elements of the signal vector. Thus,

$$x = [x_0, \ldots, x_{N-1}]; \quad x^r = [x_{-(N-1)}, \ldots, x_0] \tag{6.4}$$

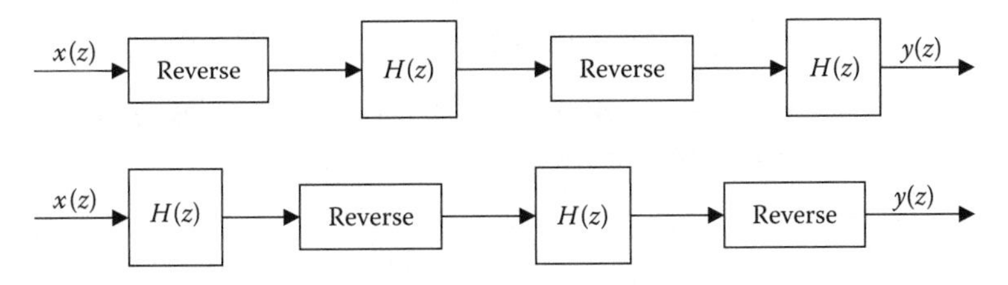

FIGURE 6.1
Filtering without phase shift via time-series reversal. In both cases, the overall gain is real and equal to $|H(e^{j\omega T})|^2$.

To transform x^r, we must extend the transform definition in (4.6) to include negative index values, that is,

$$X^r(z) = \sum_{n=-(N-1)}^{0} x_n z^{-n} = \sum_{m=0}^{N-1} x_{-m} z^m = X(z^{-1}) \tag{6.5}$$

Thus, reversing a signal vector is equivalent to substituting z for z^{-1} in its z-transform. An obvious corollary to this result is that reversal is its own inverse, that is, $\{X^r(z)\}^r = X(z)$. Furthermore, reversal is distributive, that is, $\{X(z)H(z)\}^r = X^r(z)H^r(z)$.

Now, consider the two equivalent operations in Figure 6.1. In each case, the "Reverse" labels mean that the output of the box is the reversed input vector. When we apply (6.5) and its corollaries to either of the two operations, the result is the same:

$$\begin{aligned} Y(z) &= \{X^r(z)H(z)\}^r H(z) = X(z)H(z^{-1})H(z) = X(z)\,|H(z)|^2 \\ Y(z) &= \{\{X(z)H(z)\}^r H(z)\}^r = X(z)H(z)H(z^{-1}) = X(z)\,|H(z)|^2 \end{aligned} \tag{6.6}$$

Thus, the overall gain in Figure 6.1, $|H(e^{j\omega T})|^2$, is real and equal to the squared magnitude of the gain of the filter used in each operation. In this limited sense, linear and zero phase shift are realizable with all digital filters, including IIR filters.

6.3 Butterworth Filters

In this section and Section 6.4, we introduce two of the most commonly used IIR filters, the Butterworth and Chebyshev filters. These filters were invented in the form of continuous filters long before the advent of digital signal processing (DSP). Each is optimal in a certain sense. Our purpose here is first to describe the lowpass version of each type of filter and then to show how to convert these to digital lowpass IIR filters that are optimal in the same sense.

We begin with the Butterworth filter. The power gain of the ideal, continuous, lowpass version of this filter is given by

$$|H(j\omega)|^2 = \frac{1}{1+\left(\dfrac{j\omega}{j\omega_c}\right)^{2L}} \tag{6.7}$$

where ω is frequency in rad/s, and ω_c is the cutoff frequency at the end of the passband, also in rad/s. The maximum power gain of this ideal filter is $|H(j0)|^2 = 1$, and the power gain at cutoff is by definition $|H(j\omega_c)|^2 = 1/2$ (unlike the FIR filters, with *amplitude gain* = 1/2 at cutoff).

The filter is named after Stephen Butterworth.[14] Butterworth was able to show that, of all possible rational functions of ω of order L, (6.7) has "maximum flatness" in the passband, that is, a maximum number of vanishing derivatives of the power gain function at $\omega = 0$, and is thus optimal in this sense. Using $\omega_c = 1$, examples of (6.7) for the first three even values of L, each illustrating the maximum flatness property, are shown in Figure 6.2.

When we use s in place of $j\omega$ to get the Laplace transfer function, (6.7) takes one of two forms depending on whether L, the order of the filter, is odd or

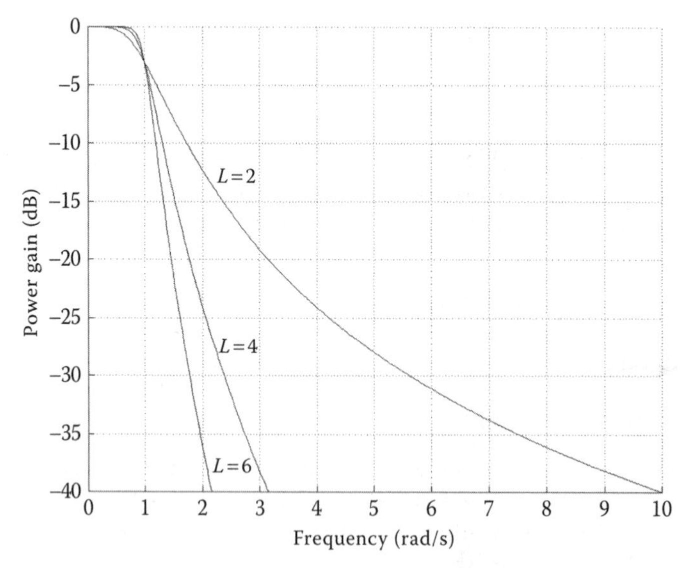

FIGURE 6.2
Butterworth power gain curves with $\omega_c = 1$ for the first three even values of L.

even. To simplify (6.7) as well as the rest of our discussion, we will assume L is even at this point. With this assumption, we have

$$|H(s)|^2 = \frac{1}{1+\left(\frac{s}{j\omega_c}\right)^{2L}} = \frac{1}{1+\left(\frac{s}{\omega_c}\right)^{2L}}; \quad L \text{ even} \tag{6.8}$$

A unique set of poles of this expression on the s-plane is given by

$$s = \omega_c(-1)^{\frac{1}{2L}} = \omega_c\left(e^{j(2n-1)\pi}\right)^{\frac{1}{2L}} = \omega_c e^{j\frac{(2n-1)\pi}{2L}}; \quad n = 1,2,\ldots,2L \tag{6.9}$$

These are spaced evenly around the circle with radius ω_c on the s-plane. The placement of poles for the first three even values of L, corresponding with the gain curves in Figure 6.2, is shown in Figure 6.3.

For continuous filters, the stability requirement for continuous systems implies that the s-plane poles must all be to the left of the imaginary axis, that is, on the left half plane. Because $|H(s)|^2$ in (6.8) is equal to $H(s)H(-s)$, we may therefore specify the stable Butterworth transfer function by using only the left half-plane poles in (6.9), that is, poles with angles that lie in the range:

$$\frac{\pi}{2} < \frac{(2n-1)\pi}{2L} < \frac{3\pi}{2}, \quad \text{or} \quad L < 2n-1 < 3L \tag{6.10}$$

Because we have assumed L is even, this means that the L values of n used in (6.9) to give the left-plane poles of $H(s)$ must be given by

$$n = \frac{L}{2}+1, \frac{L}{2}+2, \cdots, \frac{3L}{2}, \quad \text{or} \quad m = n - \frac{L}{2} = 1,2,\ldots,L \tag{6.11}$$

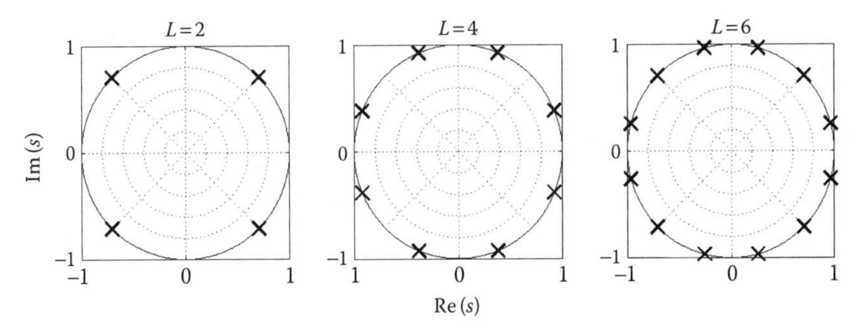

FIGURE 6.3
Butterworth power gain poles on the s-plane for the first three even values of L; $\omega_c = 1$.

Thus, using m in place of n in (6.9), the poles of the stable lowpass Butterworth filter are at

$$s_m = \omega_c e^{j\frac{(2m+L-1)\pi}{2L}}; \quad m = 1, 2, \ldots, L \tag{6.12}$$

For $L = 2$, for example, the pole angles of $H(s)$ are $3\pi/4$ and $5\pi/4$, in agreement with Figure 6.3.

When L is even, as we have been assuming, we can see that the poles of $H(s)$ are in conjugate pairs, that is, poles s_m and s_{L+1-m} are conjugates. Therefore, we can express $H(s)$ in cascade form as follows:

$$H(s) = H_1(s) \cdot H_2(s) \cdot \ldots \cdot H_{L/2}(s) \tag{6.13}$$

in which the transfer function of each section, normalized so that $H(0) = 1$, is given by

$$\begin{aligned} H_m(s) &= \frac{s_m}{s - s_m} \cdot \frac{s'_m}{s - s'_m} = \frac{|s_m|^2}{s^2 - (s_m + s'_m)s + s_m s'_m} \\ &= \frac{\omega_c^2}{s^2 - 2\omega_c\left(\cos\left(\frac{(2m+L-1)\pi}{2L}\right)\right)s + \omega_c^2}; \quad m = 1, 2, \ldots, \frac{L}{2} \end{aligned} \tag{6.14}$$

Thus, in (6.13) and (6.14), we have the transfer function of a continuous lowpass Butterworth filter of order L with cutoff at ω_c rad/s. We will see that this form with $\omega_c = 1$ is convenient when we transform $H(s)$ to construct an equivalent digital Butterworth filter.

As we noticed in Figure 6.2, the Butterworth power gain curve becomes more rectangular as L increases. It is not difficult, in fact, to determine L based on this sort of criterion. The traditional parameters described in the literature[1–11] specify two power gain limits. A simplified method that assumes the power gain is 0.5 (–3 dB) at the cutoff frequency, ω_c, is based on one's choice of the parameter λ, which is illustrated in Figure 6.4. Just as the passband is defined to end at ω_c, the *stopband* is defined to begin at ω_s, where the power gain of the lowpass filter is specified to be at or below $1/(1+\lambda^2)$ at all frequencies above ω_s. The steepness (and hence the rectangularity) of the gain curve then depends on the interval from ω_c to ω_s, and, when this interval is specified, on one's choice of λ. Larger values of λ result in greater steepness.

If we now substitute ω_s for ω in (6.7) and use the specifications just described for the stopband, the result is

$$\frac{1}{1+\lambda^2} = \frac{1}{1+(j\omega_s/j\omega_c)^{2L}}, \quad \text{or} \quad L = \frac{\log_{10}\lambda}{\log_{10}(\omega_s/\omega_c)} \tag{6.15}$$

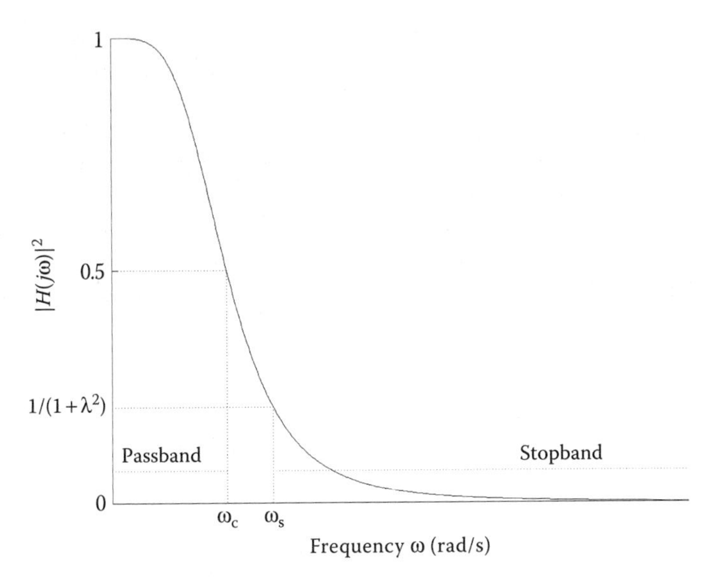

FIGURE 6.4
Passband and stopband specifications in terms of power gain.

Furthermore, practical values of λ are usually 10 or more, so we may approximate the maximum stopband gain with $1/(1+\lambda^2) \sim 1/\lambda^2$, that is, $-20\log_{10}\lambda$ dB. Using this approximation in (6.15), we can specify the filter order (L), that is, the number of poles, needed to meet given passband and stopband specifications:

Necessary number of lowpass analog Butterworth poles:

$$L \geq \frac{\text{(maximum stopband gain in dB)}}{-20\log_{10}(\omega_s/\omega_c)} \tag{6.16}$$

6.4 Chebyshev Filters

Chebyshev filters are named for P. L. Chebyshev, who discovered a unique set of polynomials that may be described in either of the following two ways:[13]

$$\begin{aligned} &V_0(x) = 1 \\ &V_1(x) = x \\ &\vdots \\ &V_L(x) = 2xV_{L-1}(x) - V_{L-2}(x);\ L \geq 2 \end{aligned} \qquad \begin{aligned} V_L(x) &= \cos(L\cos^{-1}x) \\ &= \cosh(L\cosh^{-1}x) \end{aligned} \tag{6.17}$$

The equivalence of these two forms of $V_L(x)$ may be shown[13] using the series expansions for $\cos(L\theta)$ and $\cosh(L\theta)$ in the right-hand formulas.

Chebyshev was able to show that, of all polynomials of degree L with leading coefficient equal to one, the polynomial $V_L(x)$ has the smallest maximum magnitude in the interval $|x| \le 1$. This magnitude is, in fact, one. It happens that there is a simple way to map the interval $|x| \le 1$ to either the passband or the stopband of the Butterworth power gain function, and thereby create a power gain curve with this maximum-magnitude or *equal-ripple* property in either of these two bands. Without going into detail, the results for the two types of lowpass Chebyshev filters, each of which is a simple modification of (6.7), is

$$\text{Type 1:} \quad |H(j\omega)|^2 = \frac{1}{1+\varepsilon^2 V_L^2(j\omega/j\omega_c)} \tag{6.18}$$

$$\text{Type 2:} \quad |H(j\omega)|^2 = \frac{1}{1+V_L^2(j\omega_s/j\omega_c)/V_L^2(j\omega_s/j\omega)} \tag{6.19}$$

The equal-ripple property of the Chebyshev polynomial is translated to the *passband* in the type 1 filter and into the *stopband* in the type 2 filter. In type 1, ε determines the amplitude of the passband ripple. In type 2, as for the Butterworth filter, the power gain is by definition equal to 1/2 at ω_c rad/s, and there are no parameters other than ω_c and ω_s.

The varying passband gain of the type 1 filter is undesirable in most applications, and thus, the type 2 filter is usually chosen. We will therefore focus on the design of the type 2 filter, calling it simply the *Chebyshev filter* from here on, and beginning with the derivation of its poles. (The derivation of the type 1 poles is similar and actually easier.)

The s-plane poles of the Chebyshev filter are similar to the Butterworth poles, but they are located on an ellipse rather than a circle. To find the poles, it is convenient to substitute $s = -\omega_s\omega_c/\sigma$ into (6.19), thus transforming from the s-plane to the σ-plane, and also to define the constant $\zeta = 1/V_L(\omega_s/\omega_c)$. With these definitions, (6.19) may be written

$$|H(j\omega)|^2 = \frac{\zeta^2 V_L^2(\sigma/j\omega_c)}{1+\zeta^2 V_L^2(\sigma/j\omega_c)} \tag{6.20}$$

To derive the pole locations on the σ-plane, we set the denominator in (6.20) equal to zero and substitute the "cosine" definition in (6.17). The result is

$$V_L(\sigma/j\omega_c) = \cos(L\cos^{-1}(\sigma/j\omega_c)) = \pm j/\zeta \tag{6.21}$$

Now the inner term $\cos^{-1}(\sigma/j\omega_c)$ is, in general, complex. Let us represent this term with $\beta + j\alpha$ and then solve for σ:

$$\begin{aligned} \cos^{-1}(\sigma/j\omega_c) &= \beta + j\alpha; \\ \sigma &= j\omega_c \cos(\beta + j\alpha) \\ &= \omega_c(\sinh\alpha\sin\beta + j\cosh\alpha\cos\beta) \end{aligned} \tag{6.22}$$

If we substitute the first line of (6.22) into (6.21), we will be able to solve for α and β, and thus for the pole locations in terms of σ:

$$\begin{aligned} V_L(\sigma/j\omega_c) &= \cos(L(\beta + j\alpha)) \\ &= \cos L\alpha\cosh L\alpha - j\sin L\beta\sinh L\alpha = \pm j/\zeta \end{aligned} \tag{6.23}$$

Because the real part of this expression is zero, we must have $\cos L\beta = 0$, and therefore,

$$\beta_n = \frac{(2n-1)\pi}{2L}; \quad n = 1, 2, \ldots, 2L \tag{6.24}$$

Note that these values of β are the same as the Butterworth pole angles in (6.9). Next, using these values of β and equating the imaginary components in (6.23), we have $\sin L\beta = \pm 1$, and therefore,

$$\sinh L\alpha = \pm 1/\zeta, \quad \text{or} \quad \alpha = \pm\frac{1}{L}\sinh^{-1}(1/\zeta) \tag{6.25}$$

Using (6.24) and (6.25) in (6.22), and switching to index m in a manner similar to (6.11), we can express the *right*-half σ-plane poles of the lowpass Chebyshev filter as follows:

$$\begin{aligned} &\sigma_m = \omega_c(\sinh\alpha\cos\beta_n + j\cosh\alpha\sin\beta_n); \\ &\alpha = \frac{1}{L}\sinh^{-1}\frac{1}{\zeta}; \quad \beta_m = \frac{(2m-L-1)\pi}{2L}; \quad m = 1, 2, \ldots, L \end{aligned} \tag{6.26}$$

Now, we may return to the s-plane, again using $s = -\omega_s\omega_c/\sigma$ and noting that the minus sign will map the right half of the σ-plane to the left half of the s-plane. Thus, we have a complete definition of the Chebyshev poles. The poles of the power gain function, that is, with $m = 1$ through $2L$, are plotted in Figure 6.5 with $\omega_c = 1$, $\omega_s = 2$, and $L = 2$, 4, and 6 for comparison with the Butterworth poles in Figure 6.3. In this case, the poles are on an ellipse with a shape that depends on ω_c and ω_s.

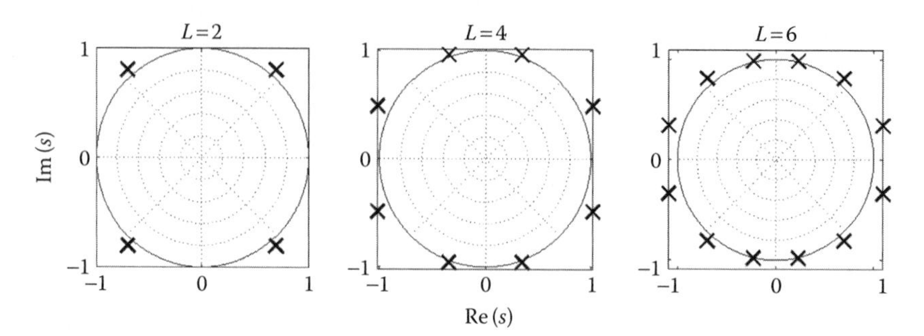

FIGURE 6.5
Chebyshev power gain poles on the s-plane with $\omega_c = 1$ and $\omega_s = 2$ for the first three even values of L.

The power gain function (6.20) also has zeros on the σ-plane that must be mapped to the s-plane. These are easier to find than the poles, because the inverse cosine is not complex. Setting the numerator of (6.20) equal to zero, we have

$$V_L(\sigma/j\omega_c) = \cos(L\cos^{-1}(\sigma/j\omega_c)) = 0 \tag{6.27}$$

In this case, the cos argument may be any odd multiple of $\pi/2$. The result is the following expression for the roots of (6.27), all of which are on the imaginary axis of the σ-plane:

$$\sigma_m = j\omega_c \cos\left(\frac{(2n-1)\pi}{2L}\right); \quad n = 1, 2, \ldots, L \tag{6.28}$$

As before, we transform these using $s = -\omega_s\omega_c/\sigma$ and thus obtain the zeros of the lowpass Chebyshev filter, which, we note, are all on the imaginary axis of the s-plane.

Thus, beginning with (6.19) and ending with (6.28), we have derived the poles and zeros of the type 2 Chebyshev analog lowpass filter. Although the derivation is complicated compared with the Butterworth derivation, the result is easily summarized in MATLAB® language:

```
% Poles.
zeta=1/cosh(L*acosh(ws/wc));
alpha=(1/L)*asinh(1/zeta);
beta=(2*(1:L)-L-1)*pi/(2*L);
sigma=wc*(sinh(alpha)*cos(beta)+j*cosh(alpha)*sin(beta));
poles=-wc*ws./sigma;
% Zeros.
sigma=j*wc*cos((2*(1:L)-1)*pi/(2*L));
zeros=-wc*ws./sigma;
```
(6.29)

Using these expressions, which locate poles and zeros in the left half plane, all poles for $\omega_c = 1$, $\omega_s = 2$, and $L = 2, 4$, and 6 may be calculated for the plot in Figure 6.5. The zeros, which are on the imaginary axis above and below the unit circle, are not plotted in Figure 6.5.

When L is even, as we assumed for the Butterworth filter, the poles and zeros are in complex pairs, and the lowpass Chebyshev transfer function may be expressed in cascade form as follows, in a manner similar to (6.13) and (6.14), in which p_m and z_m are the poles and zeros just described.

$$\begin{aligned} H(s) &= H_1(s)\cdot H_2(s)\cdot\ldots\cdot H_{L/2}(s) \\ H_m(s) &= \frac{p_m(s-z_m)}{z_m(s-p_m)}\cdot\frac{p'_m(s-z'_m)}{z'_m(s-p'_m)} \\ &= \frac{|p_m|^2\,(s^2+|z_m|^2)}{|z_m|^2\,(s^2-2sRe\{p_m\}+|p_m|^2)};\quad m=1,2,\ldots,\frac{L}{2} \end{aligned} \tag{6.30}$$

Note that the transfer function is normalized in this case so that $H_m(0)$, and therefore $H(0)$, is equal to one, that is, the filter gain is one at $\omega = 0$.

On account of the stopband ripple, it is important to have a formula like (6.16) for the Chebyshev filter so the maximum stopband gain may be specified. Using the same parameter, λ, illustrated in Figure 6.4, we can write a Chebyshev power gain expression similar to (6.15):

$$\frac{1}{1+\lambda^2} = \frac{1}{1+V_L^2(j\omega_s/j\omega_c)/V_L^2(1)} \tag{6.31}$$

Noting from (6.17) that $V_L^2(1)=1$ and using the "cosh" representation of $V_L(j\omega_s/j\omega_c)$, we conclude

$$\begin{aligned} \lambda &= V_L(\omega_s/\omega_c) = \cosh(L\cosh^{-1}(\omega_s/\omega_c)), \\ \text{or}\quad L &= \frac{\cosh^{-1}\lambda}{\cosh^{-1}(\omega_s/\omega_c)} \end{aligned} \tag{6.32}$$

To manipulate this into a rule like (6.16), we may use the identity $\cosh^{-1}\lambda = \log_e(\lambda + \sqrt{\lambda^2-1})$ with stopband gain dB = 10 $\log_{10}(1/(1+\lambda^2))$. If we again assume λ is large (say 10 or more), these become $\cosh^{-1}\lambda \approx \log_e(2\lambda)$ and dB $\approx$ $-20\log_{10}\lambda$, that is,

$$\cosh^{-1}\lambda = \frac{\log_{10}2+\log_{10}\lambda}{\log_{10}e} \approx \frac{\log_{10}2-\text{dB}/20}{\log_{10}e} \approx \frac{6.02-\text{dB}}{8.69} \tag{6.33}$$

With this approximation in (6.32), we approximate the required number of Chebyshev poles:

Necessary number of lowpass analog Chebyshev poles:

$$L \geq \frac{6.02 - (\text{maximum stopband gain in dB})}{8.69 \cosh^{-1}(\omega_s/\omega_c)} \tag{6.34}$$

There are times when one may need to solve this expression for ω_s—for example, when the poles and zeros are found using (6.29) for a fixed value of L—in order to have a specified stopband gain. This is easily done and the result is

$$\omega_s = \omega_c \cosh\left(\frac{6.02 - (\text{maximum stopband gain in dB})}{8.69L}\right) \tag{6.35}$$

This result was used in Figure 6.6 to plot lowpass Chebyshev power gain curves for comparison with the Butterworth curves in Figure 6.2. In each case, the maximum stopband gain was set to −40 dB, and ω_s was then determined with (6.35) and used in (6.19) to find the power gain.

Another comparison of Chebyshev and Butterworth gains is shown in Figure 6.7. Here, the maximum stopband gain was set to the bottom of the plot, and in each case, we can note that the Chebyshev gain is more rectangular.

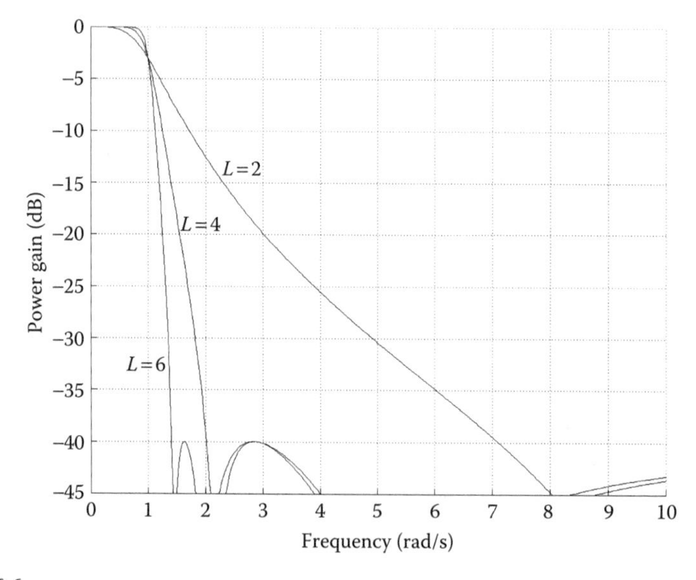

FIGURE 6.6
Chebyshev power gain curves with $\omega_c = 1$ for the first three even values of L. In each case, the maximum stopband gain was set to −40 dB and (6.35) was used to find ω_s.

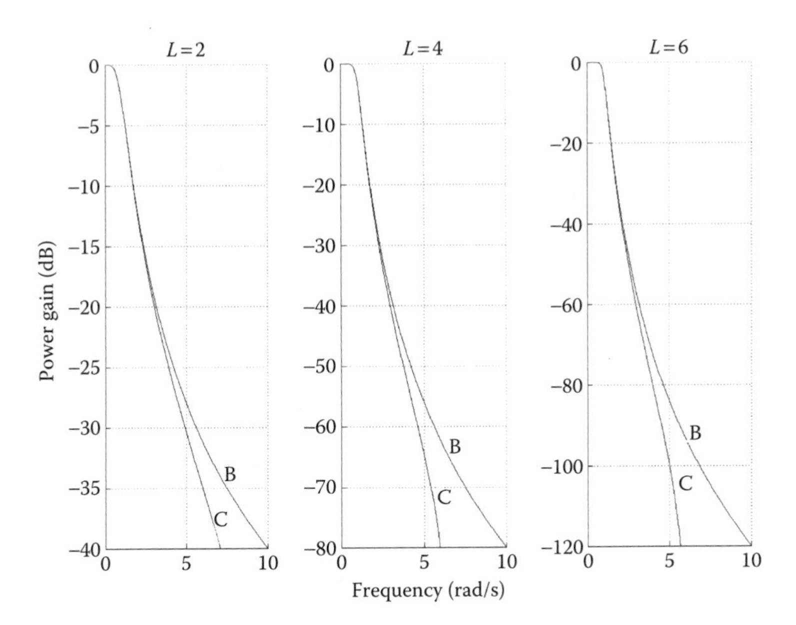

FIGURE 6.7
Comparison of Butterworth (B) and Chebyshev (C) power gain characteristics for the first three even values of L.

With the analog lowpass Butterworth and Chebyshev filters described, our next steps are to describe the translation from lowpass to other analog passbands and then the conversion to digital filters with optimal characteristics.

6.5 Frequency Translations

In the previous chapter, we transformed lowpass FIR filters into highpass, bandpass, and bandstop filters by adding and subtracting lowpass impulse response vectors to obtain the appropriate weight vectors. To transform the analog lowpass filters in this chapter, we use a different technique that accomplishes the transformation in the frequency domain.

The three frequency translations from lowpass to highpass, bandpass, and bandstop are accomplished by substituting three different functions of s for s in the lowpass Butterworth or Chebyshev transfer function, $H(s)$. The three substitutions for s, along with the corresponding substitutions for ω (found by setting $s = j\omega$), are as follows:

Lowpass to highpass:

$$s \leftarrow \frac{\omega_c^2}{s}; \quad \omega \leftarrow \frac{-\omega_c^2}{\omega} \tag{6.36}$$

Lowpass to bandpass:

$$s \leftarrow \frac{s^2 + \omega_1\omega_2}{s}; \quad \omega \leftarrow \frac{\omega^2 - \omega_1\omega_2}{\omega} \quad \text{with} \quad \omega_c = \omega_2 - \omega_1 \tag{6.37}$$

Lowpass to bandstop:

$$s \leftarrow \frac{s\omega_c^2}{s^2 + \omega_1\omega_2}; \quad \omega \leftarrow \frac{\omega\omega_c^2}{\omega_1\omega_2 - \omega^2} \quad \text{with} \quad \omega_c = \omega_2 - \omega_1 \tag{6.38}$$

Each of these is illustrated in Figure 6.8 using ideal amplitude gain functions. In each case, without much effort, you can see how the substitutions (6.36) through (6.38) translate the lowpass passband from $-\omega_c$ to ω_c into the passbands of the transformed gain functions. For example, the lowpass frequency sequence from $-\omega_c$ to 0 to ω_c is translated to the highpass range from

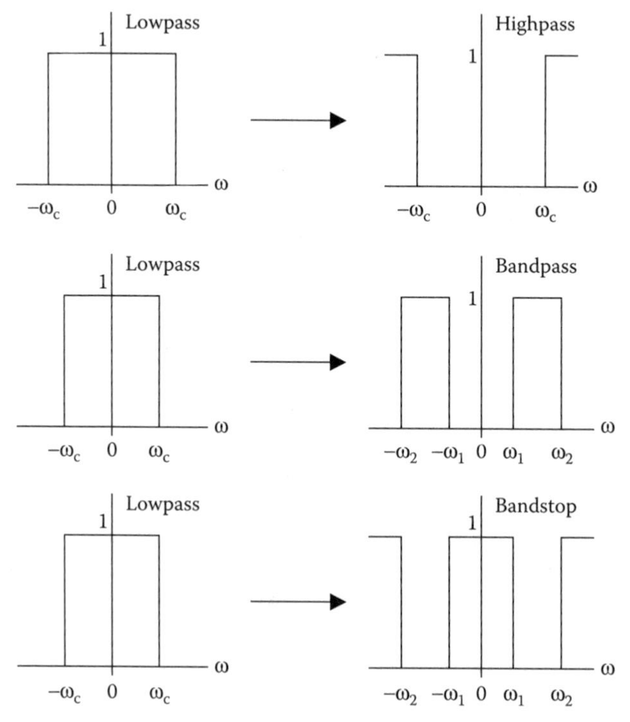

FIGURE 6.8
Illustration of the analog frequency translations in (6.36) through (6.38).

ω_c to ∞ to $-\infty$ to $-\omega_c$. Notice especially how the lowpass cutoff frequency, ω_c, becomes the *bandwidth* of the bandpass and bandstop filters.

Now suppose a lowpass transfer function, $H(s)$, is in the *cascade* form of either (6.14) for the Butterworth filter or (6.30) for the Chebyshev filter. This will turn out to be the most useful form to transform from analog to digital. Then, if we apply the substitutions (6.36) through (6.38) to one of the factors of $H(s)$, we get the terms shown in Table 6.1 for each of the other three filter types. In Table 6.1, we omit the index (m) in (6.14) and (6.30) for brevity.

In each of these translations, there is an inherent conversion of lowpass zeros and poles into highpass, bandpass, or bandstop poles and zeros. The translated zeros and poles may be derived by finding the roots of the numerators and denominators, respectively, in Table 6.1. This task is accomplished by two MATLAB functions to be discussed in Section 6.7. The function outputs are the weights of the target analog filter. We may note here, however, that the quadratic forms in the translations, where they appear in the bandpass and bandstop translations, cause the numbers of poles and zeros to double in the target filter. This and other translation features are illustrated in Figures 6.9 and 6.10, using specific examples of all six transformations in Table 6.1.

At this point, we have complete descriptions of analog Butterworth and Chebyshev filters with all four passbands. Our next step is to convert these to comparable digital filters. For this operation, we need one last translation. This is the *bilinear transformation* discussed in Section 6.6. After that, we will be able to form a complete picture of IIR digital filter design.

TABLE 6.1

Frequency Translation of One Factor of $H(s)$ in Cascade Form

Filter	Butterworth	Chebyshev
Lowpass	$\frac{p}{s-p}$	$\frac{p(s-z)}{z(s-p)}$
Highpass	$\frac{s}{s-\omega_c^2/p}$	$\frac{s-\omega_c^2/z}{s-\omega_c^2/p}$
Bandpass	$\frac{ps}{s^2-ps+\omega_1\omega_2}$	$\frac{p(s^2-zs+\omega_1\omega_2)}{z(s^2-ps+\omega_1\omega_2)}$
Bandstop	$\frac{s^2+\omega_1\omega_2}{s^2-\omega_c^2 s/p+\omega_1\omega_2}$	$\frac{s^2-\omega_c^2 s/z+\omega_1\omega_2}{s^2-\omega_c^2 s/p+\omega_1\omega_2}$

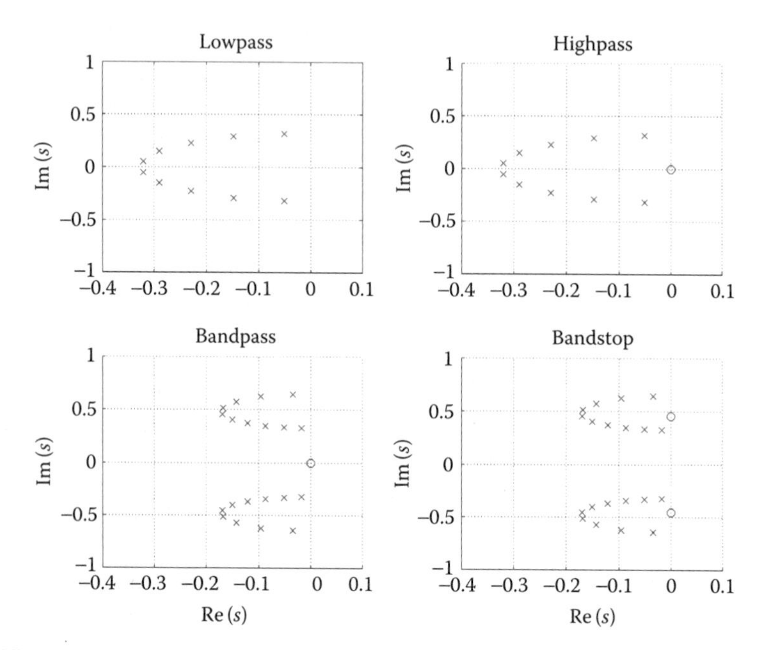

FIGURE 6.9
Butterworth poles on the s-plane translated from lowpass to highpass, bandpass, and bandstop configurations using (6.36) through (6.38). Parameters for the lowpass poles at upper left are $L = 10$ and $\omega_c = 0.32$. Also, $\omega_1 = 0.32$ (and hence, $\omega_2 = 0.64$) for the bandpass and bandstop translations.

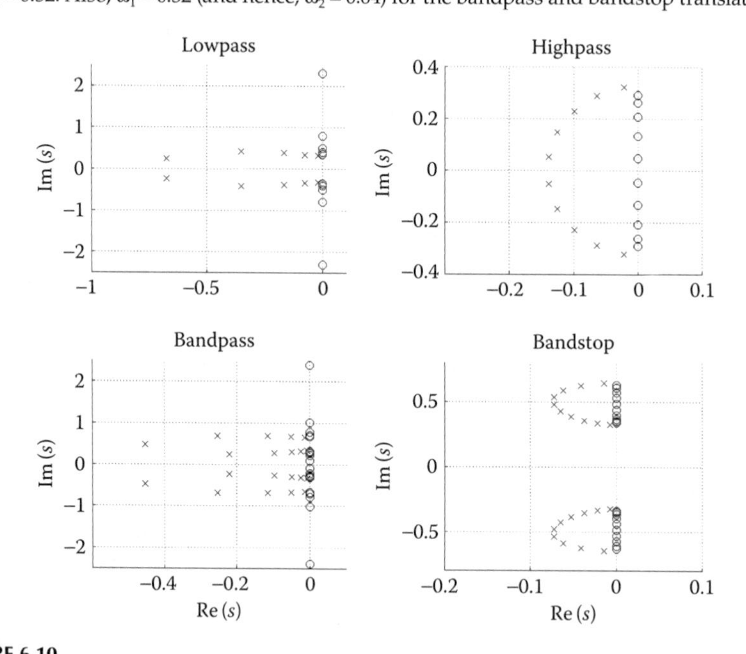

FIGURE 6.10
Chebyshev poles and zeros on the s-plane translated from lowpass to highpass, bandpass, and bandstop configurations using (6.36) through (6.38). Parameters are the same as for Figure 6.9.

6.6 The Bilinear Transformation

The bilinear transformation, described by Kaiser and colleagues,[11,12] is used to transform a continuous analog transfer function, $H(s)$, into a discrete digital transfer function, $H(z)$, and thus transform an analog filter into a digital filter. We will see that the power gain of the resulting digital filter approximates, and in one sense actually improves, the power gain of its analog counterpart, that is, any of the filters discussed in Section 6.5.

The bilinear transformation is accomplished simply by substituting a function of the discrete transform variable, z, for the continuous transform variable, s, to map points on the z-plane to points on the s-plane, and vice versa, that is, substituting a function of s for z to map from the s-plane to the z-plane.

The mappings are as follows:

z-plane to s-plane:	s-plane to z-plane:
$s \leftarrow \dfrac{z-1}{z+1}$	$z \leftarrow \dfrac{1+s}{1-s}$

(6.39)

The geometry of this transformation is easily understood if we take the squared magnitude of the right-hand expression with $s = \alpha + j\omega$:

$$|z|^2 = \frac{|1+\alpha+j\omega|^2}{|1-\alpha-j\omega|^2} = \frac{(1+\alpha)^2+\omega^2}{(1-\alpha)^2+\omega^2} \tag{6.40}$$

Here, we can see that if α, the real part of s, is negative, then $|z|$ is less than one, and also if α is zero, then $|z| = 1$. In other words, in the bilinear transformation from the s-plane to the z-plane,

1. The left half of the s-plane is mapped to the interior of the unit circle on the z-plane.
2. The $j\omega$ axis on the s-plane is mapped to the unit circle on the z-plane.

These are the essential properties of the bilinear transformation, which is also the simplest mapping function that produces these results.

Now suppose we convert a lowpass analog filter with transfer function $H_A(s)$ to a lowpass digital filter with transfer function $H(z)$ using the bilinear transformation with the properties just described. What happens to the filter gain in the transformation? The answer is found simply by expressing the gain function. First, from the definition (6.39), we have

$$H(z) = H_A\left(\frac{z-1}{z+1}\right) \tag{6.41}$$

From this, we can write the frequency response (gain) by substituting $z = e^{j\omega T} = e^{j2\pi\nu}$ into (6.41). We will use the latter form, $e^{j2\pi\nu}$, in order to emphasize that $H(e^{j2\pi\nu})$ is the *digital* gain:

$$H(e^{j2\pi\nu}) = H_A\left(\frac{e^{j2\pi\nu}-1}{e^{j2\pi\nu}+1}\right) = H_A\left(\frac{e^{j\pi\nu}-e^{j2\pi\nu}}{e^{j\pi\nu}+e^{j\pi\nu}}\right) = H_A(j\tan(\pi\nu)) \qquad (6.42)$$

Thus, the gain of the digital filter is the same as the gain of the analog filter, but at different frequencies. We also have from (6.42) a more explicit statement of rule 2 on the previous page, that is,

> Under the bilinear transformation, points given by $e^{j2\pi\nu}$ in the range $0 \le \nu < 0.5$ on the unit circle on the z-plane map to corresponding points given by $\omega = \tan(\pi\nu)$ in the range $0 \le \omega < \infty$ on the $j\omega$-axis on the s-plane.

We recall from Chapter 4 that with digital filters, the usual frequency range is from zero to half the sampling rate, that is $0 \le \nu \le 0.5$. But now when the frequency goes through this range in the digital filter, we see in (6.42) that the frequency in the analog filter, that is, $\tan(\pi\nu)$, goes through the range $[0, \infty]$.

The effect of this is illustrated in Figure 6.11, in which power gain rather than dB is plotted in order to illustrate the effect clearly. In the case of a lowpass filter, the *critical frequency* is the cutoff frequency, ν_c. [The power gain at cutoff for the Butterworth and type 2 Chebyshev filters is $|H(j\omega_c)|^2 = 1/2$, as shown in (6.7) and (6.19), respectively.] In Figure 6.11, we see that if the analog lowpass filter is designed to cut off at $\omega_c = \tan(\pi\nu_c)$ rad/s, and if this analog filter is converted to a digital lowpass filter using the bilinear transformation, then the digital filter will cut off at ν_c Hz-s. Figure 6.11 simply illustrates

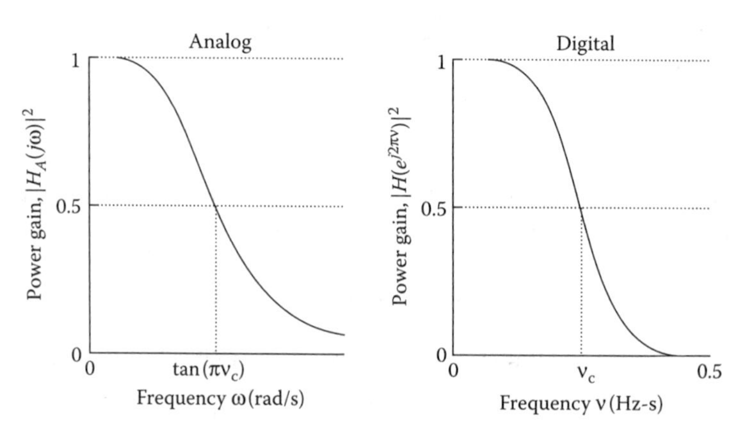

FIGURE 6.11
Illustrating the effect of the bilinear transformation. The analog filter on the left is transformed to the digital filter on the right.

the property of the bilinear transformation shown in (6.42), that is, the gain of the analog filter at $\tan(\pi\nu_c)$ rad/s is the same as the gain of the digital filter at ν_c Hz-s, where ν_c is any critical frequency in the range [0, 0.5] Hz-s.

Another important effect of the bilinear transformation, which is also illustrated in Figure 6.11, is the mapping of analog filter gain in the range $\tan(\pi\nu_c) < \omega < \infty$ to digital filter gain in the range $\nu_c < \nu < 0.5$. As we can see, this part of the mapping warps the analog filter gain in a way that improves the digital filter gain, because the digital filter gain goes to zero at $\nu = 0.5$.

Mapping frequencies in this way also affects the rules in (6.16) and (6.34) for the necessary number of lowpass poles. Each of the critical frequencies must be mapped as just described, and so the equivalent rules for lowpass digital filters are

Necessary number of lowpass digital Butterworth poles:

$$L \geq \frac{(\text{maximum stopband gain in dB})}{-20\log_{10}(\tan(\pi\nu_s)/\tan(\pi\nu_c))} \tag{6.43}$$

Necessary number of lowpass digital Chebyshev poles:

$$L \geq \frac{6.02 - (\text{maximum stopband gain in dB})}{8.69\cosh^{-1}(\tan(\pi\nu_s)/\tan(\pi\nu_c))} \tag{6.44}$$

In summary, we can see that the design of an IIR lowpass digital filter with cutoff at ω_c rad/s may be accomplished in three simple steps:

1. Determine the number of poles using (6.43) or (6.44).
2. Design the analog lowpass filter, $H_A(s)$, to cut off at $\tan(\pi\nu_c)$ rad/s.
3. Transform $H_A(s)$ into $H(z)$ with the bilinear transformation, that is,

$$H(z) = H_A\left(\frac{z-1}{z+1}\right) \tag{6.45}$$

Lowpass analog filters may be transformed into highpass, bandpass, or bandstop filters as described in Section 6.5. Therefore, these same rules are applicable to the design of all four types of IIR digital filters, provided all critical frequencies of the digital filter are transformed in accordance with (6.42), that is,

$$\begin{aligned} \omega_c &= \tan(\pi\nu_c) \\ \omega_1 &= \tan(\pi\nu_1) \\ \omega_2 &= \tan(\pi\nu_2) \end{aligned} \tag{6.46}$$

Thus, the rules in (6.45) may be expanded to include the design of all four types of filters:

To design an IIR digital filter:

1. Determine ω_c, ω_1, and ω_2, as needed using (6.46).
2. Determine the number of lowpass poles using (6.43) or (6.44).
3. Design the analog lowpass filter, $H_A(s)$, to cut off at ω_c rad/s.
4. Translate $H_A(s)$ if necessary, as shown in Table 6.1.
5. Transform $H_A(s)$ into $H(z)$ with the bilinear transformation, that is,

$$H(z) = H_A\left(\frac{z-1}{z+1}\right) \tag{6.47}$$

It is also possible, and just as easy, to exchange steps 5 and 4 and perform the frequency translations in the z-plane.[15] In Section 6.7, we describe the process (6.47) in more detail.

6.7 IIR Digital Filters

The ability described in Chapter 1, Section 1.7, to write functions that perform routine low-level tasks and to use these functions in higher-level functions, is a feature that gives MATLAB and similar high-level programming languages a great advantage over older programming languages, where defining and using functions were more difficult and restricted. We can take advantage of this capability to develop modules for the design of IIR digital filters. For example, expressions (6.43) and (6.44) for the necessary number of poles are easy enough to calculate without writing a program, but on the other hand, a MATLAB function also is easy to implement and use in other programs and has the advantage of one-time debugging. For example, the function may easily be written to make L even, so poles will be in conjugate pairs, and disallow obvious errors in the parameters such as specifying a critical frequency greater than half the sampling rate, and so on. The function

```
L=bw_lowpass_size(vc,vs,dB)
```

solves (6.43) with these constraints. All functions in this text are available in the *functions* folder on the publisher's website for this text.

The process of designing a lowpass IIR digital filter typically begins with the selection of the filter type, which for now we assume is either Butterworth or Chebyshev type 2. Then, the process follows the course outlined in (6.47), which is summarized in Table 6.2.

There are many different ways to proceed from a lowpass analog filter to a desired digital filter, and Table 6.2 represents only one. The details of each algorithm (how it implements the corresponding equations) may be seen by examining the expressions in its complete listing. The reader may prefer to find or invent another set of algorithms. But, if the procedure in the table and the accompanying MATLAB functions are used, some remarks will be helpful. First, the frequency arguments for the first set of algorithms are in Hz-s. The frequency arguments for the second set, which are meant primarily for internal use (by other functions), are in radians, and for our purposes, are usually bilinear translations using $\tan(\pi\nu)$ as in (6.42). Also, for the second set of functions, the analog weights are coefficients of *single-pole* analog cascade sections and are, in general, complex. Only $L/2$ of these are generated, because the other $L/2$ are complex conjugates. It is simpler to use single-pole sections this way and pass the weights on to the final operation, which applies the bilinear translation and generates the weights of the digital filter in cascade form with quadratic sections and real coefficients. Thus, b and a are real arrays dimensioned $L/2 \times 3$, each row containing the coefficients of the corresponding quadratic filter section given in (6.1) as $b(z)/a(z)$ with $M = N = 3$.

In the procedure in Table 6.2, and others like it, with bandpass and bandstop filters, the frequency translations and bilinear transformations combine to warp the skirts of the filter gain function, so that only the gain at the critical frequencies is accurate. Thus, it is a good idea to plot the power gain and decide whether to accept the design or change one or more of the parameters.

TABLE 6.2

Designing an Infinite Impulse Response Digital Filter

What Is Computed	MATLAB Function	Equation
Number of poles	`L=bw_lowpass_size(vc,vs,dB)`	(6.43)
	`L=ch_lowpass_size(vc,vs,dB)`	(6.44)
Analog weights (single-pole sections)	`[d,c]=bw_analog_weights(band,L,w1,w2)`	(6.12), (6.14)
	`[d,c]=ch_analog_weights(band,L,dB,w1,w2)`	(6.29) through (6.30), (6.36) through (6.38)
Digital weights (quadratic sections)	`[b,a]=bilin(d,c)`	(6.41)

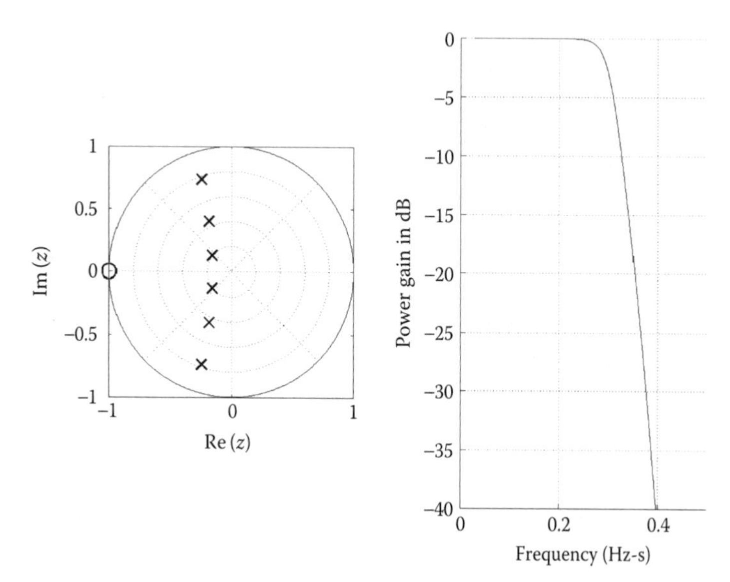

FIGURE 6.12
Lowpass Butterworth digital filter designed with the procedure in (6.47). Poles and zeros are plotted on the left and power gain on the right.

A couple of examples will illustrate this. In each case, the results of the MATLAB expressions are the weights of the digital filter in cascade form with $L/2$ quadratic sections, each section having two zeros and two poles. Also, for each example, we assume that the first algorithm in Table 6.2 has provided $L = 6$ for the number of lowpass poles.

In our first example, we design a lowpass Butterworth digital filter to cut off at $\nu_c = 0.3$ Hz-s using the following expressions (see Table 6.2):

```
[d,c]=bw_analog_weights(1,6,tan(pi*0.3));
[b,a]=bilin(d,c);
```
(6.48)

The results are shown in Figure 6.12 in terms of a pole-zero plot and a power gain plot. Note that the bilinear transformation creates a zero of order L at $z = -1$, even though the lowpass analog filter has only poles. Notice also how the distribution of poles causes the gain to be uniformly high until the frequency ν approaches 0.3 Hz-s, that is, $e^{j2\pi\nu}$ approaches $e^{j0.6\pi}$, where the polar angle is 108°. Above this frequency, the operating point moves toward the six zeros at $z = -1$, and the gain decreases rapidly toward zero.

In the second example, with results shown in Figure 6.13, we produce the weights of a digital Butterworth bandpass filter with passband from $\nu_1 = 0.3$ to $\nu_2 = 0.4$ using the following expressions:

```
[d,c]=bw_analog_weights(3,6,tan(pi*0.3),tan(pi*0.4));
[b,a]=bilin(d,c);
```
(6.49)

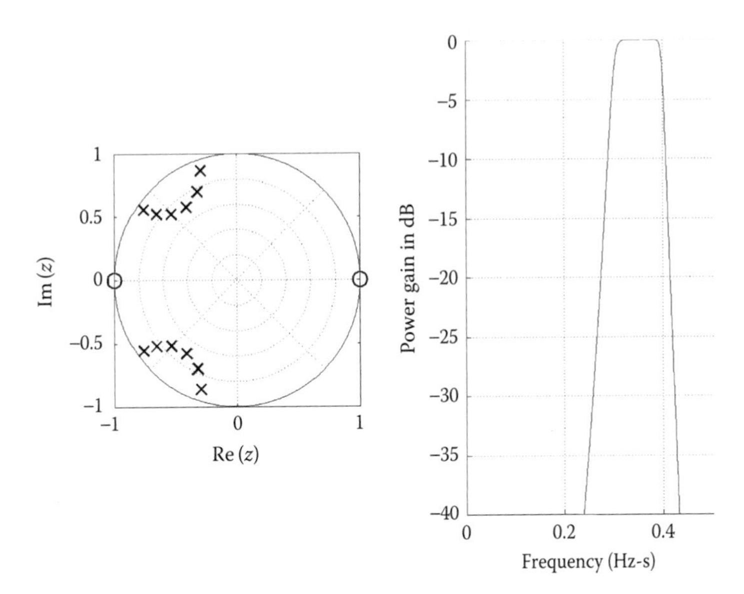

FIGURE 6.13
Bandpass Butterworth digital filter designed with the procedures in (6.47). Poles and zeros are plotted on the left and power gain on the right.

On account of the quadratic substitution for s in (6.37), which causes L to increase from 6 to 12, the bandpass filter has six quadratic sections—twice as many as the lowpass filter. On the pole-zero plot in Figure 6.13, notice how the poles cover the passband from $\Omega_1 = 0.6\pi$ to $\Omega_2 = 0.8\pi$ rad, and the −3 dB points on the power gain curve are at $\nu_1 = 0.3$ and $\nu_2 = 0.4$. In this case, there are also 12 zeros, six at $z = 1$ and six at $z = -1$.

Phase characteristics for these two examples are illustrated in Figure 6.14, with the passband phases signified by heavy lines. The passband phases are seen to be nearly linear and would be acceptable in many filtering applications. But if absolutely linear phase is required, we must revert either to an FIR filter or the time-reversal technique described in Section 6.2.

Instead of the two expressions in (6.48) and (6.49), one may wish to write a single function that computes analog weights and transforms these to digital weights using the bilinear transformation. Two such functions are provided along with the other functions in Table 6.2. These are

```
[b,a]=bw_weights(band,L,v1,v2)
[b,a]=ch_weights(band,L,dB,v1,v2)
```
(6.50)

If you examine the listings of either of these functions, you will see that most of the effort in composing the function is in the comments. But if your memory is not perfect or if the function is for use by others as well as yourself, it is justified.

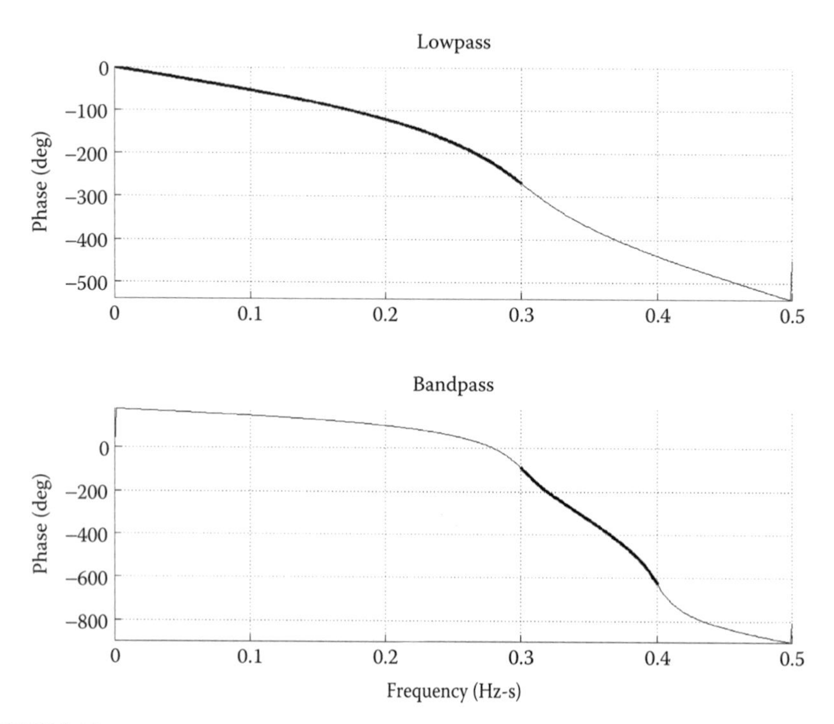

FIGURE 6.14
Phase characteristics of the lowpass and bandpass filters used for Figures 6.12 and 6.13. Phase in the passband of each filter is shown as a heavier line.

Techniques for IIR filtering, that is, using the IIR digital filters in this chapter to filter sample vectors, have been described in Chapter 4. Any of the algorithms discussed in Chapter 4, including the MATLAB *filter* function, may be applied using the filters in this chapter. The MATLAB *filter* function is the best choice if MATLAB is being used; however, the current version of *filter* requires the direct form of $H(z)$ rather than the cascade form preferred in this chapter. To implement the cascade form, the function

```
y = filters(b, a, x)
```

may be used. This function allows b and a to be arrays, with each row containing the weights of one section of the cascade structure. In other words, filters accept weights as computed by the functions in (6.50). Internally, the *filters* function calls the *filter* function to implement each quadratic section of the cascade structure.

The exercises contain additional examples of designing and applying IIR filters and filtering using the techniques described in this section.

6.8 Digital Resonators and the Spectrogram

In addition to the four standard types of IIR filters, there is an unlimited variety of different transfer functions that one may wish to invent for special purposes. We offer a few examples in this section and in Sections 6.9 and 6.10, but in general, a useful way to think about obtaining a given amplitude gain or phase characteristic is to imagine how poles and zeros might be placed on the z-plane to achieve the desired results. The *digital resonator* illustrates this process.

Suppose we need a digital filter that "resonates," that is, has high gain at a specified frequency and low gain at all other frequencies. We saw in Chapter 4, Section 4.5, how gain is affected by distances from the operating point, $e^{j\omega T}$ on the unit circle, to poles and zeros on the z-plane. That is, the gain is high when the operating point is close to a pole and low when the operating point is near a zero. Now suppose we want a resonator that resonates at $\omega = \omega_0$. Then, it is reasonable to place a pole at angle $\theta = \omega_0 T$ just inside the unit circle (for stability), say at radius $r = 1 - \varepsilon$. This pole, along with its conjugate, would give us the following transfer function:

$$H_r(z) = \frac{z^2}{(z - re^{j\omega_0 T})(z - re^{-j\omega_0 T})} = \frac{z^2}{z^2 - 2rz\cos\omega_0 T + r^2} \tag{6.51}$$

Now suppose we wish to place zeros all around on the unit circle to suppress the gain at frequencies away from $\omega_0 T$. We could do this with

$$H_2(z) = 1 - z^{-N} \tag{6.52}$$

which has zeros at $1^{1/N} = e^{j2\pi/N}$ in a manner similar to (6.9):

$$\text{zeros at } z = e^{j2\pi n/N}; \quad n = 0, 1, \ldots, N-1 \tag{6.53}$$

We then construct the digital resonator as the cascade combination $H_z(z)H_r(z)$ shown in Figure 6.15, with a constant gain A to make the gain equal to one at resonance, that is,

$$A = \frac{1}{|H_z(e^{j\omega_0 T})H_r(e^{j\omega_0 T})|} \tag{6.54}$$

Thus, the digital resonator is a simple structure consisting of the delay section (6.52) followed by the two-pole IIR filter section (6.51). An example with $N = 16$, $r = 0.99$, and $\omega_0 T = 2.5(2\pi/N)$ is given in Figure 6.16, which shows

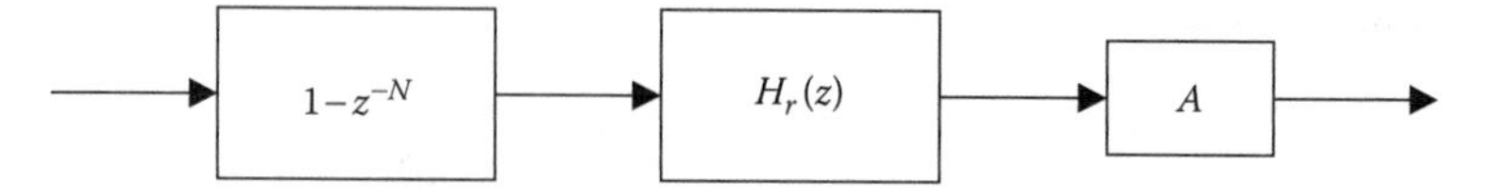

FIGURE 6.15
Digital resonator: cascade combination of $H_z(z)$ in (6.52), $H_r(z)$ in (6.51), and constant gain A in (6.54).

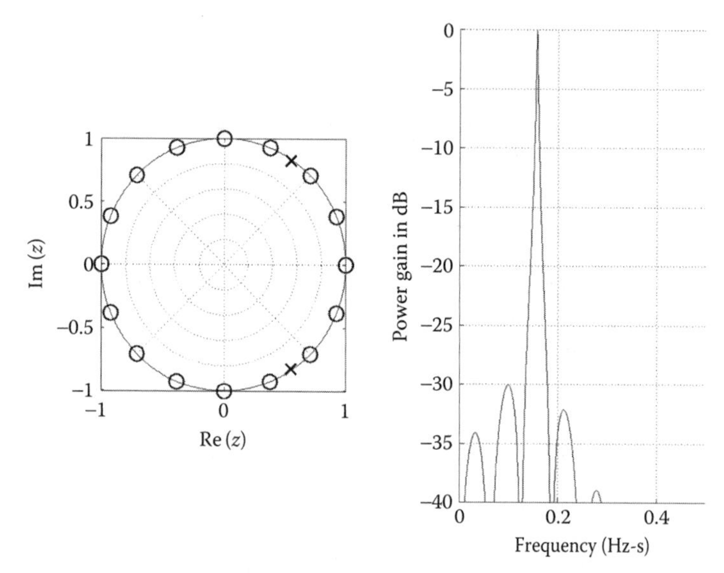

FIGURE 6.16
Pole-zero and power gain plots for the configuration in Figure 6.15 with $N = 16$ zeros, pole radius $r = 0.99$, and resonant frequency $2.5(2\pi/N)$ rad.

pole-zero and power gain plots for the resonator. Note that the pole is placed halfway between two zeros and just inside the unit circle at $r = 0.99$, so the main lobe of the gain is nearly symmetric around ω_0.

The *comb filter* shown in Figure 6.17 is another type of IIR resonator, in which several units in the form of $H_r(z)$ in (6.51) and (6.54) are connected to a single FIR delay:

$$H_z(z) = 1 - r^N z^{-N} \tag{6.55}$$

The factor r^N causes all the zeros to move in from the unit circle to a circle of radius r. Each pole of the comb filter is placed directly on a zero, and therefore, cancels its effect on gain, thus producing a resonant peak at the corresponding frequency. The result is shown in Figure 6.18—a "comb" filter with outputs in adjacent frequency bands.

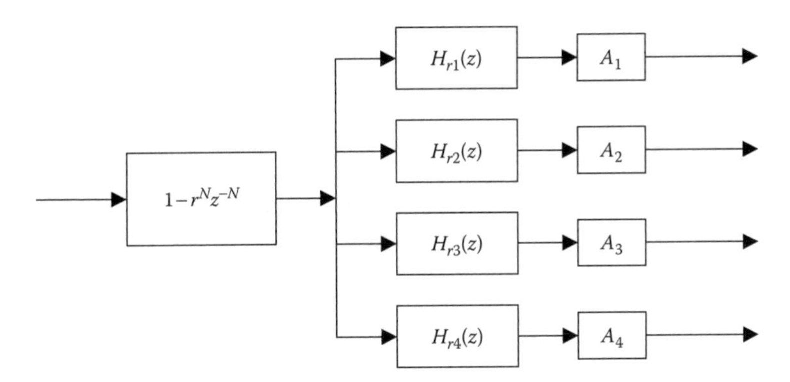

FIGURE 6.17
Comb filter configuration described by (6.51) through (6.54).

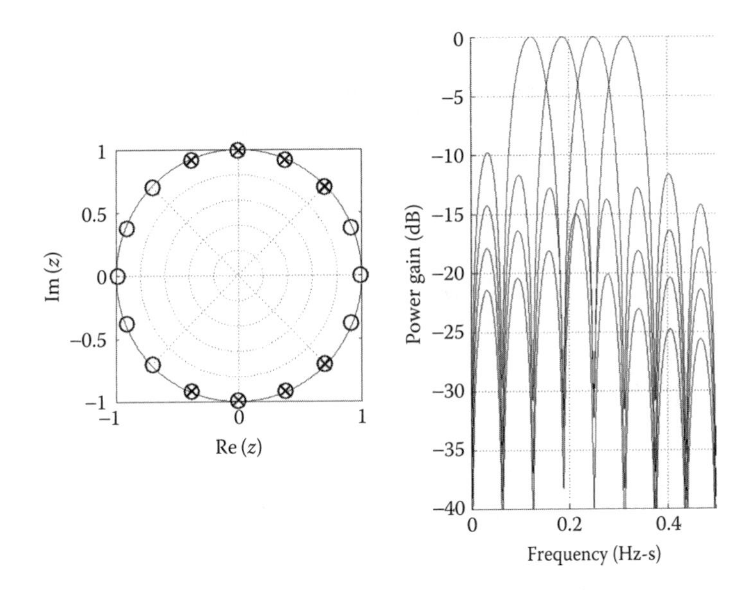

FIGURE 6.18
Pole-zero and power gain plots for the configuration in Figure 6.17 with $N = 16$ zeros, pole and zero radii $r = 0.99$, and resonant frequencies at $[2:5]*(2*\pi/N)$ radians.

The *spectrogram* is an interesting application of the comb filter or any similar set of bandpass filters. It is useful with *nonstationary* signals (signals with spectral properties that change with time), which are discussed in Chapter 7. An example of a nonstationary signal, in this case, a "chirping" signal that increases in both amplitude and frequency is shown in Figure 6.19.

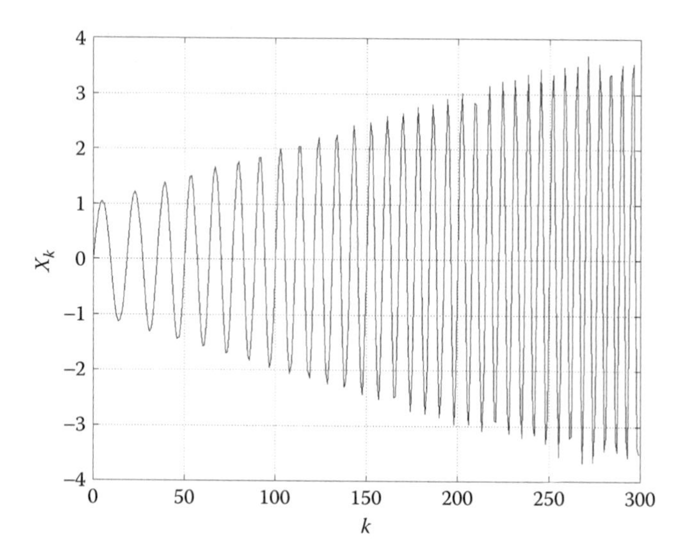

FIGURE 6.19
Plot of 300 samples of a nonstationary signal increasing in both frequency and amplitude.

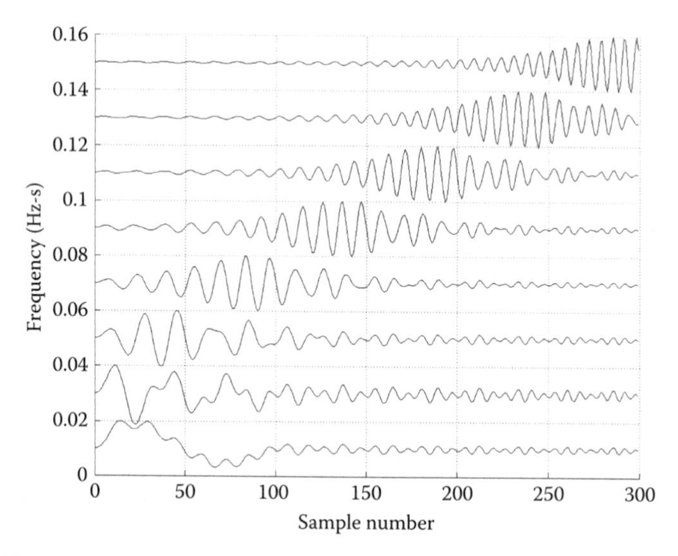

FIGURE 6.20
Spectrogram of the signal in Figure 6.19 produced using a comb filter with $N = 100$, $r = 0.99$, and resonances at $[1:2:16]*(2*\pi/N)$.

A spectrogram of the signal, consisting of the outputs of a comb filter with eight bands, is shown in Figure 6.20. With a properly constructed spectrogram like this, we can observe the arrival of different frequency components of a signal at different times, which is often useful with certain types of data, such as acoustic or seismic waves.

6.9 The All-Pass Filter

Our second example of a different type of IIR filter is the *all-pass filter*, which has constant gain at all frequencies; that is, has the effect of altering only the *phase* of the incoming signal. Suppose we place poles at p and p' (the conjugate of p) on the z-plane, and then, to compensate for the contributions of these to the amplitude gain, we place zeros at $1/p$ and $1/p'$ and construct the following filter:

$$H(z) = \frac{|p|^2 (z - 1/p)(z - 1/p')}{(z-p)(z-p')} = \frac{(pz-1)(p'z-1)}{(z-p)(z-p')} \tag{6.56}$$

We can see that the amplitude response is one at all frequencies:

$$\begin{aligned} |H(e^{j\omega T})| &= \frac{|pe^{j\omega T} - 1|\,|p'e^{j\omega T} - 1|}{|e^{j\omega T} - p|\,|e^{j\omega T} - p'|} = \frac{|pe^{j\omega T} - 1|\,|p'e^{j\omega T} - 1|}{|1 - pe^{-j\omega T}|\,|1 - p'e^{-j\omega T}|} \\ &= \frac{|pe^{j\omega T} - 1|\,|p'e^{j\omega T} - 1|}{|(1 - p'e^{j\omega T})'|\,|(1 - pe^{j\omega T})'|} = 1 \end{aligned} \tag{6.57}$$

The steps in (6.57) were first to multiply each denominator term by $|e^{-j\omega T}| = 1$ and then use the distributive property of the conjugate plus the property $|\pm\xi'| = |\xi|$ for any complex variable or function ξ. Thus, the all-pass filter is a filter with unit gain and phase shift depending on the value of p. All-pass filter sections, each with its own value of p, may be connected in cascade to produce different phase characteristics. For further analysis of all-pass phase and group-delay characteristics, the reader is referred to more advanced DSP texts, such as Oppenheim and Schafer.[8]

6.10 Digital Integration and Averaging

Ordinary numerical operations like averaging and integration have existed since the dawn of civilization. Many of these are *linear* in the sense that they involve multiplying samples of a variable, x, by constants and summing the result, either recursively or nonrecursively. *Integration* is an example of the former; simple *averaging* is an example of the latter.

One of the first to recognize the importance of viewing the operations in the frequency domain was R. W. Hamming.[16] His text, *Numerical Methods for Scientists and Engineers*, has been a classic on this subject since it was first published. Integration and averaging, which we discuss in this section, are used frequently in DSP operations, and it is interesting to consider the effects of these in the frequency domain.

Integration of sampled data is closely associated with *interpolation*, or *resampling*, which was the subject of Chapter 3, Section 3.12, as we shall see. The simplest approaches to computing the integral of a continuous waveform from its samples amount to constructing steps or straight lines (or polynomials of degree greater than one) from one sample to the next and adding incremental areas. An example is shown in Figure 6.21. The waveform is "accurately sampled" in the sense that the samples themselves are without quantizing errors, and the sampling theorem is satisfied. But, the time step is nevertheless large enough to cause a noticeable error in the integral estimates, which are given below the figure.

Suppose we now apply the procedure in Section 3.12, which, when the sampling theorem is satisfied, preserves the frequency content of the continuous waveform, $x(t)$, at and between the sample points. Recall that we first subtracted a line from x_0 to x_{N-1} to obtain from x the vector u with $u_0 = u_{N-1} = 0$. Then, we appended u to itself in the following way to form the vector v in (3.60):

$$v = \left[u_0 \; u_1 \cdots u_{N-2} \; u_{N-1} - u_{N-2} \cdots - u_2 - u_1\right] \tag{6.58}$$

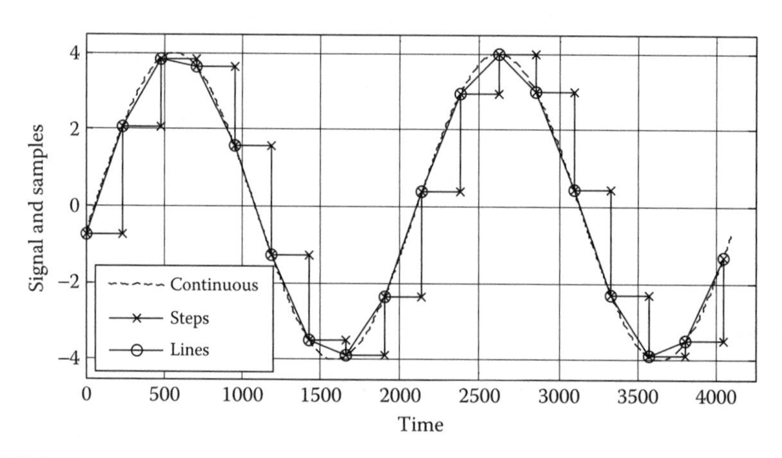

FIGURE 6.21
Accurately sampled sinusoidal waveform with steps (x) and lines (o) from one sample to the next. In terms of the units shown on the axes, the true integral from the first sample to the last is about 51.5. The result of rectangular integration (using steps) is about 119.7, and the result of trapezoidal integration (using lines) is about 49.2.

Finally, we expressed the periodic extension of v as a Fourier sine series in continuous time, noting that the period of v is $2(N-1)$ samples.:

$$v(t) = \sum_{m=1}^{N-1} b_m \sin\frac{\pi m t}{(N-1)T}; \quad 0 \le t \le (N-1)T$$
$$b_m = \frac{1}{N-1}\sum_{n=0}^{2N-3} v_n \sin\frac{\pi m n}{N-1} \tag{6.59}$$

Since $v(t)$ has no frequencies above half the original sampling rate, we should be able to obtain an accurate estimate of the integral:

$$I_v(t) = \int_0^t v(\tau)\,d\tau = \sum_{m=1}^{N-1} b_m\left(-\frac{(N-1)T}{m\pi}\right)\cos\frac{\pi m t}{(N-1)T}$$
$$= -\frac{(N-1)T}{\pi}\sum_{m=1}^{N-1}\frac{b_m}{m}\cos\frac{\pi m t}{(N-1)T} \tag{6.60}$$

and so the integral of u in (6.58) is

$$I_u(t) = I_v(t); \quad 0 \le t \le (N-1)T \tag{6.61}$$

Now we can estimate the continuous integral of $x(t)$ by adding the integral of the straight line originally subtracted from x to obtain u; that is,

$$I_x(t) = I_u(t) + x_0 t + \frac{x_{N-1} - x_0}{(N-1)T}\int_0^t \tau\,d\tau$$
$$= I_u(t) + x_0 t + \frac{t^2\left(x_{N-1} - x_0\right)}{2(N-1)T}$$
$$= x_0 t + \frac{t^2\left(x_{N-1} - x_0\right)}{2(N-1)T} - \frac{(N-1)T}{\pi}\sum_{m=1}^{N-1}\frac{b_m}{m}\cos\frac{\pi m t}{(N-1)T} \tag{6.62}$$

The MATLAB function *integral.m* calculates (6.62) at the original time steps of the input vector, x. An example of the use of *integral* is in Figure 6.22. The signal in this case is

$$x(t) = c + e^{a\omega t}\sin\omega t, \quad \text{with } c = 0.1, a = -0.1, \text{and } \omega = 2\pi/9 \tag{6.63}$$

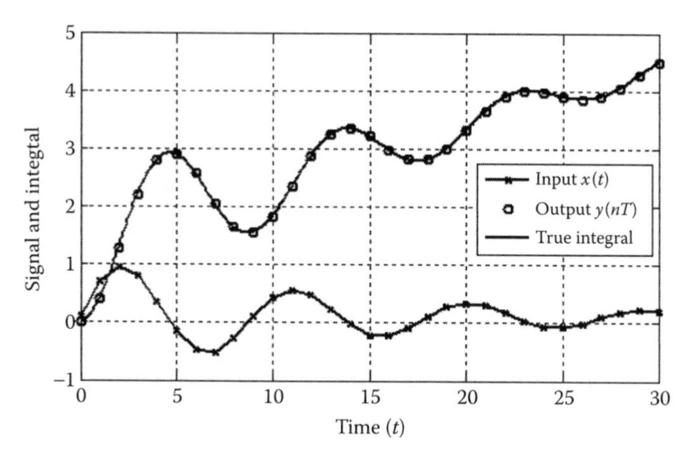

FIGURE 6.22
Frequency-domain integration of a band-limited signal using the *integral* function. The integral is accurate if the original signal is sampled at a rate greater than twice the highest signal frequency.

The true integral is

$$y(t)=\int_0^t x(\tau)\,d\tau = ct+\frac{1}{\omega\left(a^2\omega^2+1\right)}\left(1+e^{a\omega t}\left(a\omega\sin\omega t-\cos\omega t\right)\right) \tag{6.64}$$

Since in this case there is no significant signal power beyond half the sampling rate, the output samples produced by *integral.m*, $y(nT)$ agree almost exactly with the true integral.

Averaging is another common numerical method. Averaging and integration are related in the sense that the integral of a signal, $x(t)$, over an interval of t, divided by the length of the interval, is called the *average value* or *mean value* of x over that interval.

Suppose we process a sampled signal, x, by computing a "running average", y, as the average of the most recent N samples of x:

$$y_k=\frac{1}{N}\sum_{n=0}^{N-1}x_{k-n};\quad k\ge N-1 \tag{6.65}$$

This expresses averaging in the form of an FIR filter with amplitude gain

$$\left|H(e^{j\omega T})\right|=\left|\frac{Y(e^{j\omega T})}{X(e^{j\omega T})}\right|=\left|\frac{1}{N}\sum_{n=0}^{N-1}e^{-jn\omega T}\right| \tag{6.66}$$

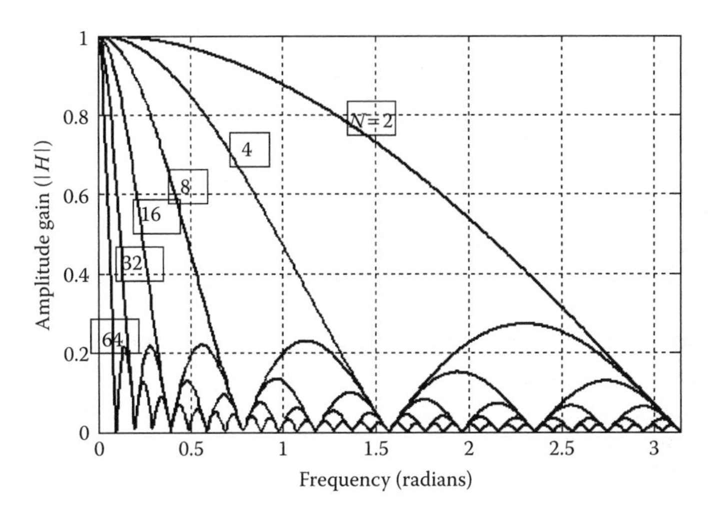

FIGURE 6.23
Amplitude gain produced by taking a running average of N samples, with values of N ranging from 2 to 64.

Using the geometric series formula (1.19), this becomes

$$|H(e^{j\omega T})| = \frac{1}{N}\left|\frac{1-e^{-jN\omega T}}{1-e^{-j\omega T}}\right| = \frac{1}{N}\left|\frac{e^{jN\omega T/2}-e^{-jN\omega T/2}}{e^{j\omega T/2}-e^{-j\omega T/2}}\right| = \frac{1}{N}\left|\frac{\sin(N\omega T/2)}{\sin(\omega T/2)}\right| \tag{6.67}$$

(To get this result we have used the fact that individual terms of the form $e^{j\theta}$, as well as j itself, all have unit magnitude.) This amplitude gain produced by averaging $N = 2, 4, 8, 16, 32$, and 64 is illustrated in Figure 6.23. Averaging is easy to implement, has the same smoothing effect as a lowpass filter, and is sometimes used to reduce high-frequency variations in a long data vector. However, the "filters" illustrated in Figure 6.23 do not reject high frequencies nearly as well as any of the FIR or IIR lowpass filters we have already discussed in Chapter 5 or in this chapter.

Exercises

General instructions: (1) Whenever you make a plot in any of the exercises, be sure to label both axes, so we can tell exactly what the units mean. (2) Make continuous plots unless otherwise specified. "Discrete plot" means to plot discrete symbols, unconnected unless otherwise specified. (3) In plots, and especially in subplots, use the *axis* function for best use of the plotting area.

6.1 Make two subplots. On the left subplot, plot the s-plane poles of an analog lowpass Butterworth filter having $L = 12$ poles and cutoff frequency equal to 2.0 rad/s. On the right, plot the power gain from –200 to 0 dB versus rad/s from 0 to 10.

6.2 Do Exercise 6.1 using a Chebyshev filter. The stopband should begin at 5 rad/s. Plot zeros as well as poles on the s-plane.

6.3 Make two subplots. On the left, plot the power gains of four digital lowpass Butterworth filters with cutoff at 0.3 Hz-s and $L = 2, 4, 8$, and 16. Plot dB in the range [–120, 0] versus Hz-s in the range [0, 0.5]. On the right, plot the same curves for the corresponding Chebyshev filters with stopband gain set to –100 dB.

6.4 Make two subplots. On the left, plot the poles and zeros of a digital Butterworth filter with $L = 10$ poles and passband from 80 to 100 KHz, assuming a time step of 0.005 ms. On the right, make the same plot for a Chebyshev filter, assuming stopband gain = –80 dB.

6.5 Complete the following:

   a. Make two subplots. On the left, plot the amplitude gain versus frequency in KHz for a digital Butterworth filter with $L = 10$ poles and passband from 80 to 100 KHz, assuming a time step of 0.005 ms. On the right, make the same plot for a Chebyshev filter, assuming stopband gain = –80 dB.

   b. Explain the forms of the plots in terms of the pole and zero locations in Exercise 6.4.

   c. Comment on the differences between the two amplitude gain plots.

6.6 Make two subplots. On the left, plot the amplitude gain versus frequency in MHz for a digital Butterworth filter with 16 poles and passband from 70 to 90 MHz, assuming a sampling frequency of 300 MHz. On the right, make the same plot for a Chebyshev filter, assuming stopband gain = –80 dB.

6.7 Repeat Exercise 6.6, plotting power gain in the range [–100, 0] dB instead of amplitude gain.

6.8 Make two subplots. Plot the phases of the two filters in Exercise 6.6 over just the passband from 70 to 90 MHz. In both plots, plot phase shift in the range [–10, 20] rad.

6.9 An analog filter has s-plane poles at –4 and $-2 \pm j4$, zeros at $\pm j$, and a maximum gain of 1. Make two subplots. On the left, plot the amplitude gain versus frequency in the range [0, 50] rad/s. On the right, plot amplitude gain versus frequency in the range [0, 0.5] Hz-s for the digital filter obtained via the bilinear transformation. Explain how the analog gain peak at 4.0 rad/s is mapped to the digital gain peak at 0.42 Hz-s.

6.10 Generate a signal vector x with 200 samples of $x(t)$, which consists of a unit sine wave at 5 KHz plus another sine wave with amplitude 10 at 15 KHz, assuming a time step of 0.025 ms. Design a Butterworth digital filter with cutoff at 10 KHz and power gain down at least 60 dB at 15 KHz.

a. Find L, the necessary number of poles.

b. Design the filter and plot power gain in the range [–100, 0] dB versus frequency in KHz. State your estimation of the effect of the filter on x.

c. In a separate figure, make two subplots. In the upper subplot, plot x versus time in ms. In the lower subplot, plot the filtered version of x.

6.11 Test a digital highpass Chebyshev filter by filtering a sinusoid with frequency equal to the cutoff frequency. Use $L = 8$, dB = –100, and $\nu_c = 0.05$ in your test, and filter 200 samples of a unit sine wave. Plot the input and output signals together.

a. Comment on whether or not the filter gain is correct at the cutoff frequency.

b. Why are the two signals out of phase?

c. How would you estimate the approximate duration of the filter's impulse response?

6.12 A signal $x(t)$ has unit sinusoidal components at 10, 20, 30, 40, and 50 Hz. A 2-second segment of $x(t)$ is sampled at 400 samples/s. Make three subplots, each with frequency range [0, 80] Hz.

a. On the left, plot the amplitude spectrum, $|X|$.

b. In the middle, plot the amplitude gain of a digital Chebyshev bandstop filter designed to eliminate just the component at 40 Hz.

c. On the right, plot the amplitude spectrum of the filtered version of x, indicating the effect of the filter.

6.13 Make three subplots. On the upper left, plot the power gain in the range [–80, 0] dB of a highpass digital Butterworth filter with cutoff at 0.3 Hz-s. On the upper right, plot the poles and zeros of this filter. Construct a sample vector with 200 samples of the sum of two sine waves: $c = 10\sin(2\pi(0.1)k) + \sin(2\pi(0.35)k);\ 0 \le k \le 199$. Filter x to produce y. In the lower subplot, plot x and y together.

6.14 Construct a comb filter in the configuration of Figure 6.17. Use $N = 100$ zeros and pole/zero radius $r = 0.99$. Set the poles for resonance at frequencies [1:2:16]/N Hz-s. Make a plot similar to Figure 6.18, with a pole-zero plot on the left and a set of power gain plots on the right. Plot power gain in the range [–40, 0] dB versus frequency in the range [0, 0.5] Hz-s.

6.15 Generate 300 samples of the chirping signal in Figure 6.19 with the following expressions:

```
k=0:299;
x=(1+3*k/300).*sin(2*pi*k.*(.05+(k/300)*.06));
```

a. Plot the signal and observe that it is the same as Figure 6.19.
b. In a separate plot, reproduce the spectrogram in Figure 6.20 using the comb filter in Exercise 6.14. Plot each waveform with an offset corresponding with the resonant frequency of the comb filter section.

6.16 Design a Butterworth lowpass digital filter with $L = 10$ poles and cutoff at 0.4 Hz-s. Plot the power gain in dB versus frequency in Hz-s. Set the axis ranges to [–80, 0] dB and [0, 0.5] Hz-s, respectively. Then, on the same plot, overlay the power gain of an FIR lowpass filter using the Kaiser window with $\beta = 8.0$ that matches the Butterworth gain as closely as possible. Make the FIR plot with 50 connected discrete points. Choose the FIR cutoff frequency (ν_c) and filter length (N) to produce the best match. Print your choices in the plot title.

References

1. Orfanidis, S. J. 1996. *Introduction to Digital Signal Processing*. Chap. 9. Upper Saddle River, NJ: Prentice Hall.
2. Mitra, S. K., and J. F. Kaiser, eds. 1993. *Handbook of Digital Signal Processing*. New York: John Wiley & Sons.
3. Antoniou, A. 1993. *Digital Filters: Analysis and Design*. 2nd ed. New York: McGraw-Hill.
4. Jackson, L. B. 1989. *Digital Filters and Signal Processing*. Norwell, MA: Kluwer Academic Publishers.
5. Oppenheim, A. V., and R. W. Schafer. *Discrete-Time Signal Processing*. 2nd ed. Chap. 5. Upper Saddle River, NJ: Prentice Hall.
6. Stearns, S. D., and D. R. Hush. 1989. *Digital Signal Analysis*. Chap. 12. Englewood Cliffs, NJ: Prentice Hall.
7. Parks, T. W., and C. S. Burrus. 1987. *Digital Filter Design*. New York: John Wiley & Sons.
8. Roberts, R. A., and C. T. Mullis. 1987. *Digital Signal Processing*. Reading, MA: Addison-Wesley.
9. Elliott, D. F., ed. 1987. *Handbook of Digital Signal Processing*. New York: Academic Press.
10. Rabiner, L. R., and C. M. Rader, eds. 1972. *Digital Signal Processing*. New York: IEEE Press.
11. Kaiser, J. F. 1966. Digital filters. In *System Analysis by Digital Computer*, ed. F. F. Kuo and J. F. Kaiser. Chap. 7. New York: John Wiley & Sons.

12. Kaiser, J. F. 1965. Some practical considerations in the realization of linear digital filters. In *Proceedings of the 3rd Annual Conference Circuits System Theory*, p. 621.
13. Storer, J. E. 1957. *Passive Network Synthesis*. Chap. 30. New York: McGraw-Hill.
14. Butterworth, S. 1930. On the theory of filter amplifiers. *Wirel Engr* 1:536–41.
15. Proakis, J., and D. Manolakis. 1996. *Digital Signal Processing: Principles, Algorithms, and Applications*. 3rd ed. Chap. 8. Upper Saddle River, NJ: Prentice Hall.
16. Hamming, R. W. 1962. *Numerical Methods for Scientists and Engineers*. Chaps. 21 and 25. New York: McGraw-Hill.

Further Reading

Kohlenberg, A. 1953. Exact interpolation of band-limited functions. *J Appl Phys* 24:1432–6.

Lin, K. 1976. *A Digital Signal Processing Approach to Interpolation and Interpolation Filters*. Dissertation, University of New Mexico, Albuquerque, New Mexico.

Lutovac, M. D., D. V. Tosic, and B. L. Evans. 2001. *Filter Design for Signal Processing Using MATLAB and Mathematica*. Upper Saddle River, NJ: Prentice Hall.

McDonald, T. S. 1977. *Frequency-Domain Considerations of Numerical Integration Techniques*. Dissertation, University of New Mexico, Albuquerque, New Mexico.

Rosko, J. S. 1972. *Digital Simulation of Physical Systems*. Chap. 6. Addison-Wesley.

Schafer, R. W., and Rabiner, L. R. 1973. A digital signal processing approach to interpolation. *IEEE Proc* 61(6):692–702.

Vich, R. 1971. Two methods for the construction of transfer functions of digital integrators. *Electron Lett* 7(15):422–5.

7

Random Signals and Spectral Estimation

7.1 Introduction

Random signals are signals with elements that cannot be predicted or derived exactly from other elements. In our previous examples of discrete signals, we used finite data vectors or analytic signals that are either periodic or transient; that is, we used signals that can be described completely in simple terms.

When a signal has random components and cannot be described analytically, we may be able to describe it in terms of its *statistical properties*, which we discuss in this chapter. There are two basic sources of the most commonly used statistical properties: (1) the *amplitude distribution* of the random signal and (2) the autocorrelation function or, equivalently, the *power density spectrum*. The amplitude distribution is something we have not yet considered. The autocorrelation function is an application of the correlation function in Section 3.2. The power density spectrum is like the power spectra we have been using to analyze filters, but not quite the same thing, as we shall see.

Before we discuss these representations of the signal, we should consider the assumption of *stationarity* that is made or implied when statistical signal processing techniques are applied in digital signal processing (DSP). A *stationary* signal is a signal with local statistics that are invariant over the entire duration of the signal. Thus, a periodic signal is a stationary signal, but a transient signal that occurs locally in a long time domain is not stationary. A random signal is stationary if average local estimates of the amplitude distribution and the autocorrelation function do not change over the duration of the signal. Figure 7.1 shows an example of a stationary signal with two nonstationary signals. The upper signal is nonstationary because its average power is steadily increasing. The middle signal is nonstationary because one of its components is steadily increasing in frequency; thus, we could say that its power spectrum is changing with time. Only the lower signal is stationary, with unchanging distributions of amplitude and power.

One could argue that no process in the natural world is truly stationary. But, there are many situations in engineering where the *assumption* of

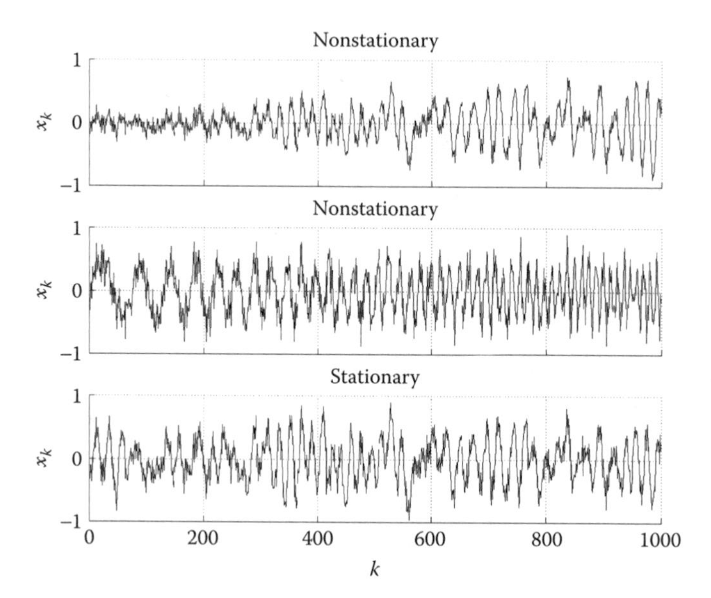

FIGURE 7.1
Vectors taken from nonstationary and stationary signals. The upper signal is nonstationary, because its average power is increasing steadily. The center signal is nonstationary, because its frequency is increasing steadily. The lower signal is stationary.

stationarity can be very useful. For example, if a long recording of the signal from a noise source is obtained, the estimated *power density spectrum*, which we describe in this chapter, is usually the best way we have to characterize the signal.

In this chapter, we begin by discussing the amplitude distributions of random signals and then go on to discuss power spectral estimates.

7.2 Amplitude Distributions

Suppose we have a long recording of a signal, such as a measurement of temperature, pressure, radiation from a particular source, or pulse rate (there are many examples like these), and suppose the recorded signal is stationary and includes random components. If the signal is digitized and processed using DSP techniques, we may consider two kinds of *amplitude distributions*,[1–5] that is, functions that give the relative likelihood, or *frequency*, of different signal amplitudes. (This is a different use than we have had for "frequency," but also a common use of the word in statistical analysis.) To analyze the continuous signal before it is digitized, we use the *continuous amplitude distribution*. To analyze the signal after it has been digitized and can only have discrete values, we use the *discrete amplitude distribution*.

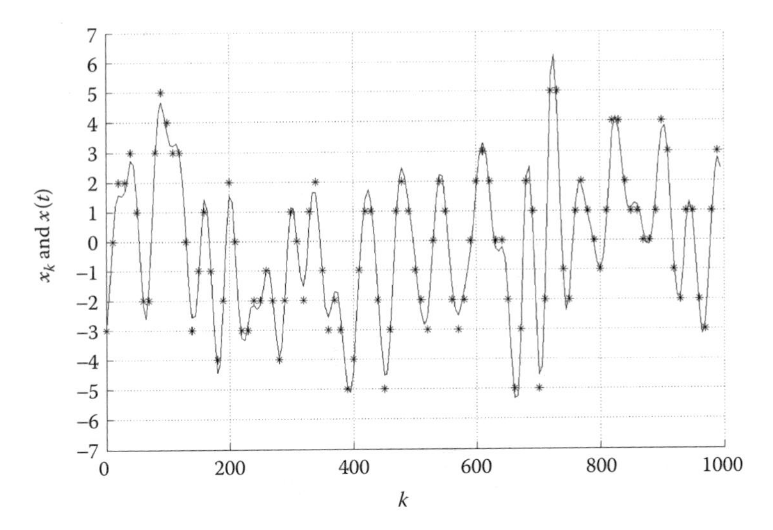

FIGURE 7.2
Stationary random waveform with discrete samples. Each sample, x_k, is the integer nearest to the corresponding waveform value, $x(kT)$.

These two cases are shown in Figure 7.2. For the continuous signal, all values of $x(t)$ in the range [–7, 7] are theoretically possible. But, the digitized signal can only have integer values in this range, that is, each x_k is the integer nearest to $x(kT)$. The continuous and discrete amplitudes are distributed in accordance with the following functions:

$$\text{Continuous: } p(x); \quad \int_{-\infty}^{\infty} p(x)\,dx = 1$$

$$\text{Discrete: } P_n = Pr\{x = n\}; \quad \sum_{n=-\infty}^{\infty} P_n = 1 \tag{7.1}$$

Thus, $p(x)dx$ is the *probability* that $x(t)$ is in the range $[x, x + dx]$ and P_n is the probability that x takes the integer value n. The function $p(x)$ is called the *probability density* function of x. Thus, assuming the analog-to-digital conversion is unbiased, we may say that

$$P_n = \int_{n-1/2}^{n+1/2} p(x)\,dx \tag{7.2}$$

In signal processing applications, the discrete probability function is often estimated as the *frequency function*, f_n, which is simply the relative frequency of occurrence of integer n in a digitized sample vector, that is, for a digitized sample vector of length N,

$$P_n \approx f_n = \frac{\#\text{ samples equal to } n}{N} \tag{7.3}$$

Before examining important amplitude distributions, that is, examples of $p(x)$ and P_n, a few fundamental properties may be defined in general terms. First, the *mean value* (*expected value, average value*) of any function of x, say $y(x)$, is defined as follows:

$$\begin{aligned} &\text{Continuous: } E[y] = \int_{-\infty}^{\infty} y(x)p(x)dx \\ &\text{Discrete: } E[y] = \sum_{n=-\infty}^{\infty} y(x_n)P_n \end{aligned} \tag{7.4}$$

Thus, $E[y]$ is essentially the sum of all values of $y(x)$, each value being weighted by $p(x)$. If the integral or sum in (7.4) does not converge, then $y(x)$ has no mean value. Two particular mean-value functions are so important that they are given special symbols. First is the mean value of x, given the symbol μ_x. Applying the continuous form of (7.4) with $y(x) = x$, the mean value of x is as follows:

$$\mu_x \equiv E[x] = \int_{-\infty}^{\infty} xp(x)dx \tag{7.5}$$

Thus, μ_x is essentially the sum of weighted values of x, that is, the "center of mass" coordinate of the distribution given by $p(x)$.

The second important expected value function is the *variance* of x, which is given the symbol σ_x^2. The variance is the most common measure of the variability of $x(t)$ about its mean value, μ_x. It is defined as the expected squared deviation of x from its mean value, that is, σ_x^2 is the expected value of $y(x) = (x - \mu_x)^2$. Accordingly,

$$\begin{aligned} \sigma_x^2 = E[(x-\mu_x)^2] &= \int_{-\infty}^{\infty} (x-\mu_x)^2\, p(x)dx \\ &= \int_{-\infty}^{\infty} x^2 p(x)dx - 2\mu_x \int_{-\infty}^{\infty} xp(x)dx + \mu_x^2 \int_{-\infty}^{\infty} p(x)dx \\ &= E[x^2] - 2\mu_x\mu_x + \mu_x^2 = E[x^2] - \mu_x^2 \end{aligned} \tag{7.6}$$

Thus, the variance turns out to be the difference between the mean squared value of x and the square of μ_x. The square root of σ_x^2, σ_x, called the *standard*

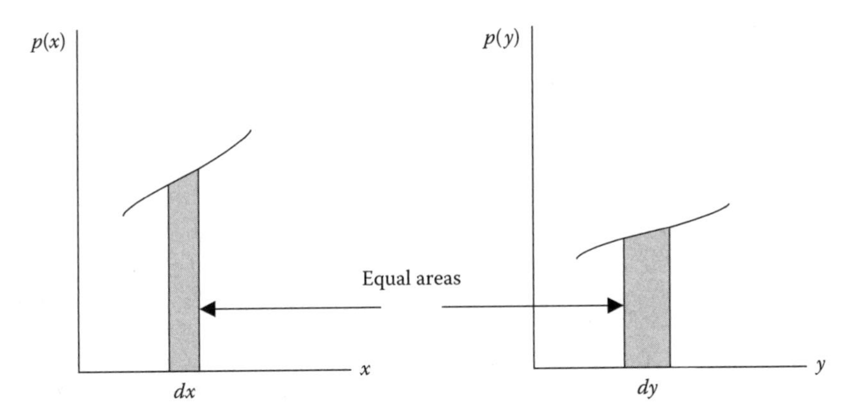

FIGURE 7.3
Equal probabilities shown as equal areas.

deviation of x, is the standard measure of the deviation of x from its mean value, the measure having the same units as x.

An important related problem arising often in signal analysis is that of determining the probability density of a function $y(x)$, given $p(x)$. This problem is solved in general[1] by partitioning x into segments over which $y(x)$ is monotonic and viewing the product $p(x)dx$ defined in (7.1) as the probability that $x(t)$ lies between x and $x + dx$, that is, viewing $p(x)dx$ as a vanishingly small increment of area above the sample space of x. In the simplest case, which is illustrated in Figure 7.3, $y(x)$ is a unique, one-to-one mapping between the x and y sample spaces, so that $p(x)$ and $p(y)$ are different functions, but $p(x)dx$ and $p(y)dy$ are the same probabilities. That is, assuming x and y are differentiable,

$$\text{Equal areas: } p(y)\,|\,dy\,| = p(x)\,|\,dx\,| \quad \text{or} \quad p(y) = \frac{p(x)}{|\,dy/dx\,|} \tag{7.7}$$

where $|dy/dx|$ is the absolute value of the derivative of $y(x)$. Exercises 7.1 and 7.2 provide illustrations of this relationship.

The case where y is a linear function of x, say $y = ax + b$, is another useful example. In this case, $|dy/dx| = |a|$, and so $p(y) = p(x)/|a|$. Two useful corollaries follow from this result:

$$\begin{aligned} \mu_{ax+b} &= a\mu_x + b \\ \sigma^2_{ax+b} &= a^2\sigma^2_x \end{aligned} \tag{7.8}$$

These both follow from (7.7) and the definitions of mean and variance in (7.5) and (7.6), and they are the subject of Exercise 7.1 at the end of this chapter.

7.3 Uniform, Gaussian, and Other Distributions

Turning now to specific examples of amplitude distributions, two distributions are of particular interest in digital signal analysis. First is the *uniform* distribution illustrated in Figure 7.4. The uniform probability density function expresses the assumption that all values of x in the interval $[a,b]$ are "equally likely." (This notion has real meaning only in the discrete case; nevertheless, it has heuristic appeal in the continuous case.) In the continuous case, the sample space is the horizontal (x) axis, and the integral of $p(x)$ satisfies (7.1). In the discrete case, the sample space consists of N integer values of x in the range $[a,b]$ and the sum of P_n over n again satisfies (7.1). In either case, if the amplitudes of $x(t)$ and its samples are uniformly distributed, $x(t)$ is called a *uniform variate.*

Using the continuous probability function, the *mean value* of the uniform variate in Figure 7.4 is found as follows:

$$\mu_x = \int_{-\infty}^{\infty} xp(x)dx = \frac{1}{b-a}\int_a^b xdx = \frac{b+a}{2} \tag{7.9}$$

Using the result in (7.6), the *variance* of the uniform variate is

$$\sigma_x^2 = E[x^2] - \mu_x^2 = \frac{1}{b-a}\int_a^b x^2dx - \frac{(b+a)^2}{4} = \frac{(b-a)^2}{12} \tag{7.10}$$

Thus, the *standard deviation*, σ_x, is $1/\sqrt{12}$ times the range of values of x, that is, $b-a$.

The second important example of an amplitude distribution is the *Gaussian* or *normal* probability function, illustrated in Figure 7.5. This function is of interest in DSP primarily, because, with many types of data (electromagnetic, acoustic, seismic, vibration, etc.), the amplitude distributions of random

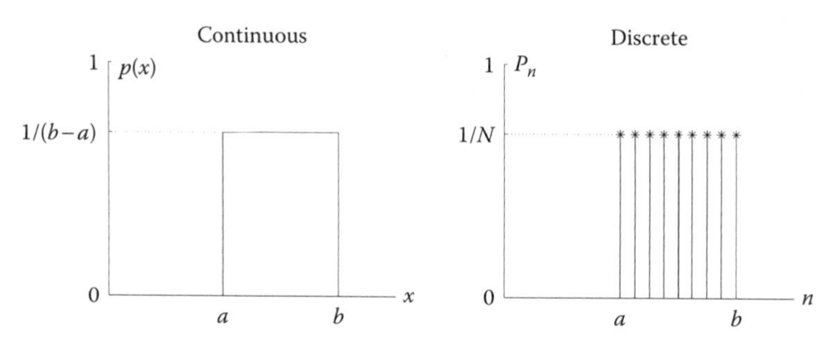

FIGURE 7.4
Continuous and discrete uniform amplitude distribution functions.

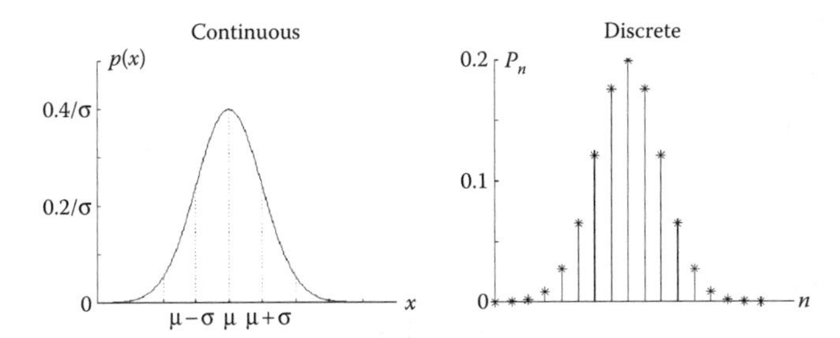

FIGURE 7.5
Continuous and discrete Gaussian amplitude distribution functions.

noise are Gaussian, or approximately so. The general form of the Gaussian probability density function is

$$p(x) = N(\mu,\sigma) \equiv \frac{1}{\sigma\sqrt{2\pi}} e^{\frac{(x-\mu)^2}{2\sigma^2}} \tag{7.11}$$

In this expression, μ is the mean value, μ_x, and σ is the standard deviation, σ_x. The discrete form, shown on the right in Figure 7.5, is a discrete distribution function, P_n, having an envelope given by (7.11) and satisfying the condition (7.1).

To see that $N(\mu,\sigma)$ has a unit integral as in (7.1), we may substitute $y = (x-\mu)/(\sigma\sqrt{2})$, so the integral of $p(x)$ becomes

$$\int_{-\infty}^{\infty} p(x)dx = \frac{1}{\sigma\sqrt{2\pi}} \int_{-\infty}^{\infty} e^{-(x-\mu)^2/2\sigma^2}\, dx = \frac{1}{\sqrt{\pi}} \int_{-\infty}^{\infty} e^{-y^2} dy \tag{7.12}$$

In this form, the integral can be evaluated simply by evaluating its square with the use of polar coordinates:

$$\left[\int_{-\infty}^{\infty} p(x)dx\right]^2 = \frac{1}{\pi}\int_{-\infty}^{\infty}\int_{-\infty}^{\infty} e^{-y^2}e^{-v^2}dy\,dv = \frac{1}{\pi}\int_{0}^{2\pi}\int_{0}^{\infty} e^{-r^2} r\,dr\,d\theta = 2\int_{0}^{\infty} re^{-r^2}dr = 1 \tag{7.13}$$

The demonstration that μ is the mean value of x is made by placing the Gaussian function into the definition (7.5) and again making the substitution in (7.12) for x, that is, $x = \sigma y\sqrt{2} + \mu$

$$\begin{aligned}
\mu_x &= \int_{-\infty}^{\infty} \frac{x}{\sigma\sqrt{2\pi}} e^{\frac{(x-\mu)^2}{2\sigma^2}}\, dx = \int_{-\infty}^{\infty} \frac{\sigma y\sqrt{2}}{\sqrt{\pi}} e^{-y^2} dy + \frac{\mu}{\sqrt{\pi}} \int_{-\infty}^{\infty} e^{-y^2} dy \\
&= \frac{\sigma\sqrt{2}}{\sqrt{\pi}} \int_{-\infty}^{\infty} ye^{-y^2} dy + \frac{\mu}{\sqrt{\pi}} \int_{-\infty}^{\infty} e^{-y^2} dy \\
&= 0 + \mu = \mu
\end{aligned} \tag{7.14}$$

Thus, the use of μ in (7.11) is justified. Similarly, for the variance as given in (7.6), we can let $\mu = 0$ and obtain σ_x^2 as the expected value of x^2 as follows, doing the integration by parts (Table 1.3):

$$\sigma_x^2 = E[x^2] = \int_{-\infty}^{\infty} \frac{x^2}{\sigma\sqrt{2\pi}} e^{\frac{x^2}{2\sigma^2}} dx = \sigma^2 \tag{7.15}$$

Thus, because $\mu = \mu_x$, σ^2 is the variance of the Gaussian distribution.

Unfortunately, when the distribution of $x(t)$ is given by $N(\mu, \sigma)$ as in (7.11), the probability that x lies between some pair of values [as in (7.2)] cannot be easily expressed, as it obviously can be, for example, in the case of the uniform distribution. Therefore, normalized versions of this probability, which is called the *error function* (*erf*), are approximated in various ways. In MATLAB®, the computation is done by *erf(v)*, which computes the *normalized error function* of each element of a vector v as follows:

$$\text{erf}(v) = \frac{2}{\sqrt{\pi}} \int_0^v e^{-y^2} dy \tag{7.16}$$

To apply *erf(v)* to the form of $p(x)$ shown in (7.11) and in Figure 7.5, we again use the substitution used to obtain (7.12), in this case, $y = (x - \mu)/\sigma\sqrt{2}$:

$$\text{erf}(v) = \frac{2}{\sqrt{\pi}} \int_{\mu}^{\mu+v\sigma\sqrt{2}} e^{-\frac{(x-\mu)^2}{2\sigma^2}} \frac{dx}{\sigma\sqrt{2}} = 2 \int_{\mu}^{\mu+v\sigma\sqrt{2}} N(\mu,\sigma)dx \quad \text{or}$$
$$\int_{\mu}^{\mu+x} N(\mu,\sigma)dx = \frac{1}{2}\text{erf}\left(\frac{x}{\sigma\sqrt{2}}\right) \tag{7.17}$$

Assuming $x \ge 0$, the second result gives us the probability that a normal variate lies in the range $(\mu, \mu + x)$ in terms of the MATLAB function, *erf.*

An example is shown in Figure 7.6, which illustrates the approximately Gaussian amplitude distribution $N(2000, 600)$ of a signal (x) from a 12-bit analog-to-digital converter. The distribution of x is discrete, but with 2^{12} = 4096 possible values, it may be viewed as continuous. In this case, because the values of x are separated by unit distance, the integral of $p(x)$ and the sum of discrete probabilities (P_n) must be the same. To compute the dark area shown in the figure, we would use (in MATLAB notation) `darkarea = .5*erf((2500-mu)/(sigma*sqrt(2)))` with *mu* = 2000 and *sigma* = 600.

Two MATLAB functions are used to produce random sequences and arrays. (Other computing languages have similar functions.) *rand(M,N)* produces an M by N array of uniform random variates with each element in the range (0,1). Similarly, *randn(M,N)* produces an array of Gaussian variates from the distribution $N(0,1)$. These functions actually generate *pseudorandom* sequences and arrays using number-generating algorithms with very long cycles. To use the same random sequence, say in a test or simulation, the expressions *rand('seed', K)* and *randn('seed', K)* may be used to reset the starting point.

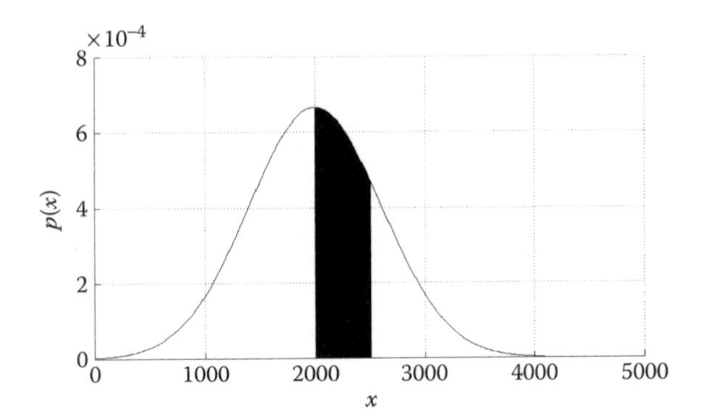

FIGURE 7.6
Amplitude distribution of signal sampled with a 12-bit A-D converter. The dark area, which is $Pr\{2000 < x < 2500\}$, is computed as shown in (7.17) and the text.

Given (7.8), which allows us to make a linear transformation of any random variate, we can deduce the following rules for generating zero-mean random sequences with variance σ^2:

$$\begin{array}{|l|}\hline \text{Random vectors: length} = N\text{, mean} = 0\text{, and standard deviation} = \sigma \\ \text{Uniform: } x = \sigma * \sqrt{12} * (rand(1,N) - 0.5) \\ \text{Gaussian: } x = \sigma * randn(1,N) \\ \hline \end{array} \quad (7.18)$$

Two examples are shown in Figure 7.7. In each case, $N = 10{,}000$ samples, and each element of x is rounded to the nearest integer to simulate the output of a digital device such as an analog-to-digital converter. On the left is the frequency function of a uniform sequence with $\sigma = 4.33$, which translates via (7.18) to equally likely integer values in the range [–7, 7]. On the right is the frequency function of a Gaussian sequence with $\sigma = 2$. In both cases, we note that, due to the finite value of N, the distributions are only approximately the same as the theoretical distributions.

Obviously, a sequence does not need to be random or even have random components in order to have an amplitude distribution. Essentially, all sequences and arrays of samples have amplitude distributions. And although the amplitude distribution is used mainly in the analysis of random functions, it is also used with all types of functions in the analysis of coding and compression techniques, which are important areas of DSP discussed in Chapter 10. So to conclude our discussion here, we offer also Figure 7.8, which consists of two nonrandom examples, both with $N = 2^{16}$ samples. On the left is a sine wave with amplitude 7.5, with samples rounded to the nearest integer before compiling the amplitude distribution, $[f_n]$. Note how the amplitude distribution is symmetric and indicates that values of $f(t)$ near the minimum and maximum are more likely than values near zero.

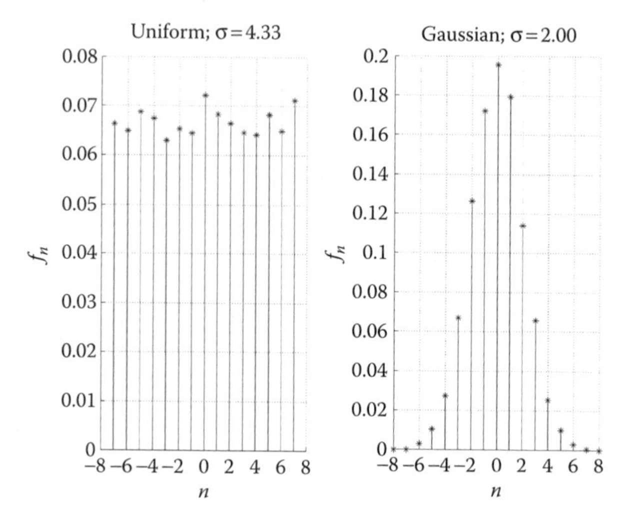

FIGURE 7.7
Amplitude distributions of digitized uniform and Gaussian random sequences. In each case, N = 10,000 samples, μ = 0, and σ is shown in the figure.

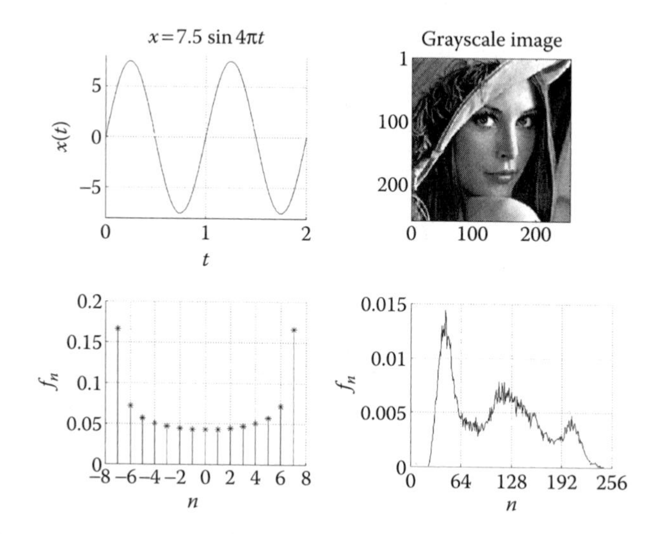

FIGURE 7.8
Amplitude distributions that are neither uniform nor Gaussian. The amplitude distribution of a digitized sine wave is on the left, and the distribution of grayscale image pixel values is on the right.

The amplitude distribution of a continuous sine wave $y(x)$ may be found using (7.7), beginning with a uniform distribution of x and transforming $p(x)$ into $p(y)$. See Exercise 7.2 at the end of this chapter.

On the right in Figure 7.8 is the image known as "Lena," which has been a standard gray-scale image in DSP literature for decades at the time of this publication. (We pause here to thank the subject for her noteworthy contribution to image processing and hope the years have been good to her.) In this version, the image is 256 × 256 pixels, and each pixel value is on a black-to-white scale from 0 to 255 (although not all pixel values are present in the image). The amplitude distribution in this case shows the relative frequencies of different pixel values occurring in the image. In both cases, the frequency functions sum to one in accordance with (7.1) and (7.3).

7.4 Power and Power Density Spectra

In Chapter 4, Section 4.13, we measured the power gain of a linear system in terms of the square of the transfer function magnitude, which in turn is measured by the discrete Fourier transform (DFT) of the impulse response. Thus, we could say that the square of the DFT magnitude of any function $x(t)$ is a measure of the distribution of the power in $x(t)$ over the frequency domain. Now, we wish to pursue this idea and arrive at precise definitions of *power* and *power density*.

The *instantaneous power* of $x(t)$ at any time t is the squared signal magnitude, $|x(t)|^2$. We assume the signal is real so that $x^2(t)$ may be used, but the general definition holds even for complex signals. It follows that the *average* or *expected power* of a stationary function $x(t)$ is

$$\text{Average power} = E[x^2(t)] = \lim_{T\to\infty}\frac{1}{T}\int_{-T/2}^{T/2} x^2(t)dt \tag{7.19}$$

If x is a vector consisting of N samples of $x(t)$, we say the average power in x is

$$\text{Average power} = \frac{1}{N}\sum_{n=0}^{N-1} x_n^2 \approx \frac{1}{NT}\int_0^{NT} x^2(t)dt \tag{7.20}$$

The average power in this sense is an *estimate* of the true average power in (7.19).

To express the average power in terms of the DFT, we substitute the inverse DFT formula in (3.23) for x_n in (7.20) and write the result as follows:

$$\begin{aligned}\text{Average power} &= \frac{1}{N}\sum_{n=0}^{N-1}\left[\frac{1}{N}\sum_{m=0}^{N-1} X_m e^{j2\pi mn/N}\right]^2 \\ &= \frac{1}{N^3}\sum_{m=0}^{N-1}\sum_{k=0}^{N-1} X_m X_k \sum_{n=0}^{N-1} e^{j2\pi(m+k)n/N}\end{aligned} \tag{7.21}$$

In this result, we may substitute $k = N - i$, use the redundancy in (3.10) to note that $X_0 = X_N$, and use the redundancy in (3.11) to let $X_{N-i} = X_i'$. Then, the average power expression becomes

$$\text{Average power} = \frac{1}{N^3}\sum_{m=0}^{N-1}\sum_{i=0}^{N-1} X_m X_i' \sum_{n=0}^{N-1} e^{j2\pi(m-i)n/N} = \frac{1}{N^2}\sum_{m=0}^{N-1}|X_m|^2 \quad (7.22)$$

The result on the right follows, because the inner sum in the center equals N when $m = i$ and is zero otherwise. Thus, in combination with (7.20), we have the result

$$\boxed{\frac{1}{N}\sum_{n=0}^{N-1} x_n^2 = \frac{1}{N^2}\sum_{m=0}^{N-1} | X_m |^2} \quad (7.23)$$

This result presents average signal power in terms of the power spectrum. Known as *Parseval's theorem*, it provides an important insight and link between the time and frequency domains.

Our final step in the development of the notion of a *power density spectrum* is to define the *periodogram*. The periodogram has N components given by

$$\boxed{\text{Periodogram:} \quad P_{xx}(m) = \frac{1}{N} | X_m |^2; \quad m = 0, 1, \ldots, N-1} \quad (7.24)$$

The periodogram is thus a real function of m, periodic in accordance with (3.10) and with $P_{xx}(N - m) = P_{xx}(m)$ in accordance with (3.11). Furthermore, we may combine (7.23) and (7.24) and write

$$\text{Average power} = \frac{1}{N}\sum_{n=0}^{N-1} x_n^2 = \frac{1}{N}\sum_{m=0}^{N-1} P_{xx}(m) \quad (7.25)$$

Thus, in the same way that x_n^2 gives a measure of signal power at a point in the time domain, $P_{xx}(m)$ gives a measure of signal power at a point in the frequency domain. Therefore, the vector $P_{xx} = [P_{xx}(m)]$ is said to be a measure of the *power density spectrum* of $x(t)$, because its average value in (7.25) is another expression of the average power estimate.

Power density in this form is best applied to stationary signals and images from which segments are extracted for analysis, as described in Section 7.5. If a signal component is considered to have a definite beginning and end and if the sampling process covers the entire range, then we use *energy* measures instead of power. Energy is the integral of power over time or space. In terms of samples, the measure of signal energy is

$$\text{Energy} = T\sum_{m=0}^{N-1} P_{xx}(m) = T\sum_{n=0}^{N-1} x_n^2 \approx \int_0^{NT} x^2(t)dt \quad (7.26)$$

Thus, the periodogram *average* is our measure of signal power, and the *total* periodogram is our measure of signal energy.

Examples of power and energy spectra in terms of the periodogram are shown in Figures 7.9 and 7.10. Figure 7.9 shows one cycle of a periodic function and its periodogram. The latter represents power density in the sense that (7.25) holds, with the average power being about 0.09 in this example. Figure 7.10 shows a transient signal and its periodogram. The periodogram in this case represents energy density in the sense that (7.26) holds, with the energy being approximately 11.3 in the example, assuming $T = 1$. The two periodograms are computed in the same manner, but there is a difference in how they are used to represent power or energy. To make the discussion easier, from here on, we will discuss the periodogram as representing *power density*, because this is how it is used with random signals.

When $x(t)$ represents a time-varying physical quantity, the power spectrum is sometimes plotted versus the frequency index as it is in Figure 7.9, but more often, it is plotted in terms of power per unit of frequency—usually *power per Hz* or *power per Hz-s*. In this case, a *bar graph* is used in place of the discrete plot in Figure 7.9. The bar graph is made such that, in the power spectrum,

$$\begin{aligned}\text{Power in range } (|m| \pm 0.5)/N \text{ Hz-s} &= (\text{area of bar at } m) + (\text{area of bar at } -m)\\ &= 2(\text{area of bar at } m)\end{aligned} \qquad (7.27)$$

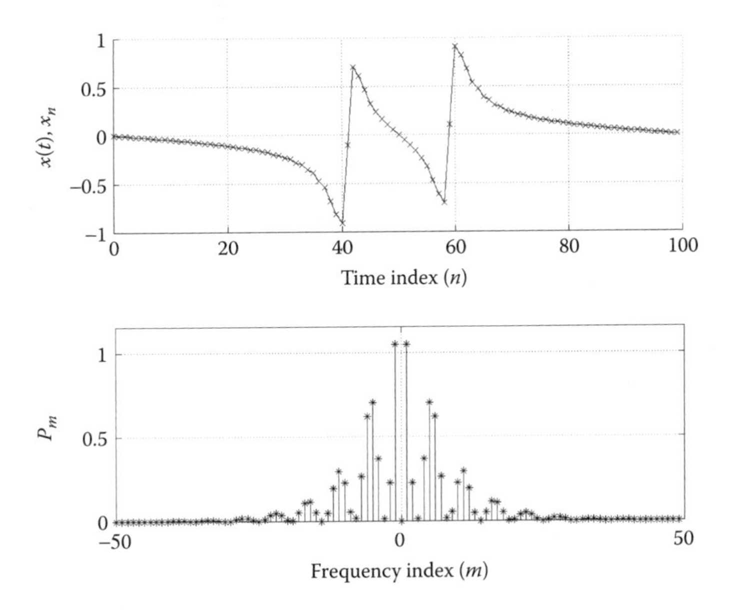

FIGURE 7.9
A sampled periodic signal and its periodogram. One hundred samples of a single cycle of $x(t)$ are shown in the upper plot, and the periodogram is shown in the lower plot.

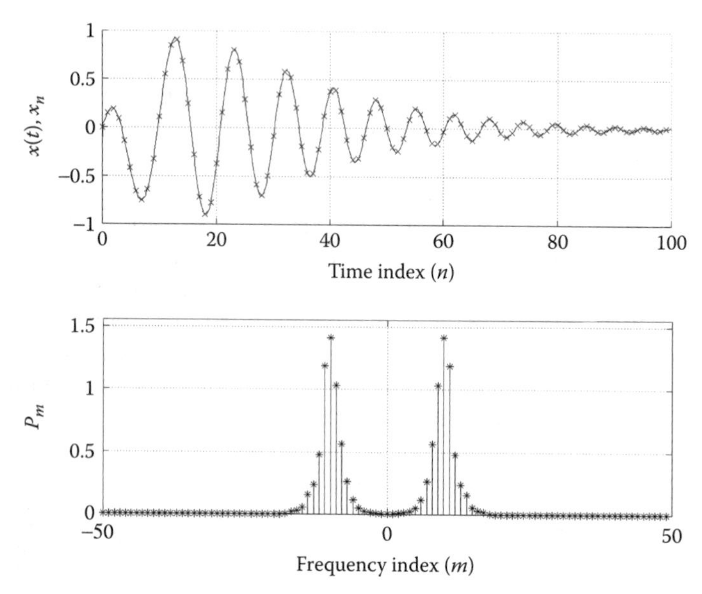

FIGURE 7.10
A transient signal and its periodogram. One hundred samples of $x(t)$ are shown in the upper plot, and the periodogram is shown in the lower plot.

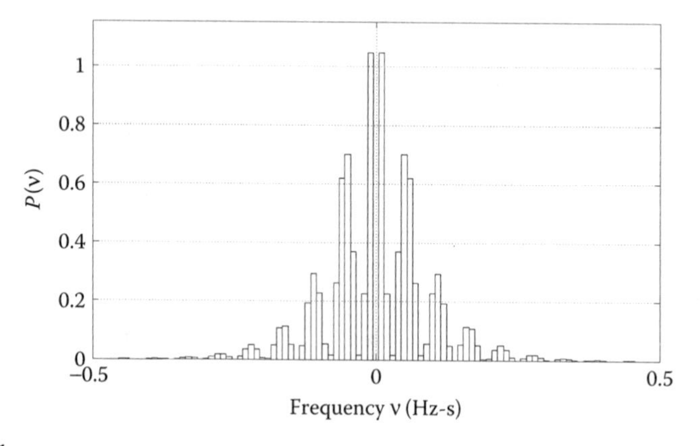

FIGURE 7.11
Periodogram in Figure 7.9 redrawn as a power density spectrum. A bar centered over $\nu = m/N$ represents power in the range $(m \pm 0.5)/N$, and the integral of the entire spectrum is equal to the average squared signal value.

(and similarly for the energy spectrum). Figure 7.11 shows the periodogram in Figure 7.9 modified in this manner. That is, the power represented by each line in Figure 7.9 is represented by the area of the corresponding bar in Figure 7.11. Thus, the total power, or average power, is the integral of the power spectrum, and (7.25) holds in either case, because the bar width in Figure 7.11 is $1/N$. Note, however, that if the frequency units are changed from Hz-s to Hz, the bar width is now $1/NT$, and so the power density must be scaled by T for the integral to agree with (7.25). The general version of (7.25) accommodating different familiar frequency notations is

$$\text{Average power} = \frac{1}{N}\sum_{n=0}^{N-1} x_n^2 = \frac{1}{N}\sum_{m=0}^{N-1} P_{xx}(m) = \int_{-0.5}^{0.5} P_{xx}(\nu)\,d\nu = \int_{-1/2T}^{1/2T} P_{xx}(f)\,df = \int_{-\pi/T}^{\pi/T} P_{xx}(\omega)\,d\omega \tag{7.28}$$

To see how each spectral measure must be scaled, assume $P_{xx}(m)$ is constant over m and equal to P. Then, we can see that $P_{xx}(\nu) = P$, $P_{xx}(f) = TP$, and $P_{xx}(\omega) = TP/2\pi$ will all produce the same average power, that is, P.

7.5 Properties of the Power Spectrum

There are many examples in DSP where the source of a signal, $x(t)$, is such that $x(t)$ is stationary over a long period and contains random components. The power spectrum, estimated from recorded segments of $x(t)$, often reveals facts about the source that we would not notice in the raw data. The power spectrum and the amplitude distribution together contain the basic signal statistics that are normally required for analysis, filtering, signal compression, and other types of processing. In Section 7.6, we discuss a method for estimating the power spectrum of a stationary random sequence. Here, we review its properties, the most important being its relationship to the autocorrelation function.

Suppose a signal vector x is acquired in accordance with Figure 7.12. In practice, aliasing may be prevented by analog prefiltering, or simply by sampling at a rate greater than twice the maximum frequency response of the sensor. The frequency-limited signal, $x(t)$, is sampled to produce the vector x consisting of integer samples. For the following analysis, we assume a periodic extension of x, which is reasonable when the signal is stationary.

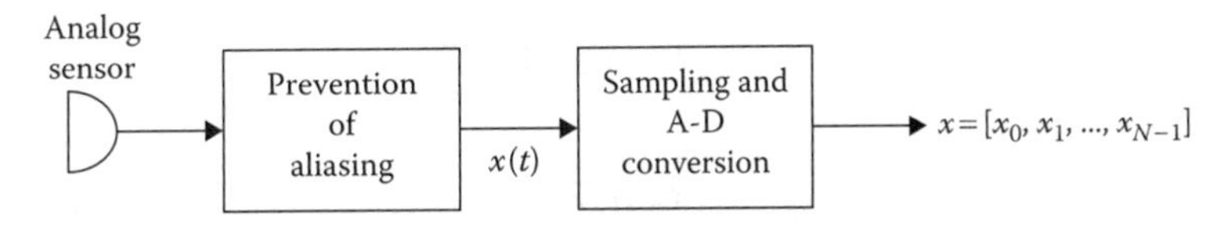

FIGURE 7.12
Acquiring the sample vector, x, from the measurement of a continuous signal.

Then, in accordance with (3.2) in Chapter 3, the autocorrelation function of x (extended periodically) is

$$\varphi_{xx}(k) = \frac{1}{N}\sum_{n=0}^{N-1} x_n x_{n+k}; \quad k = 0, 1, \ldots, N-1 \tag{7.29}$$

From this definition and (7.25), we notice that $\varphi_{xx}(k)$ is also periodic, and furthermore,

$$\text{Average power} = \frac{1}{N}\sum_{n=0}^{N-1} x_n^2 = \varphi_{xx}(0) \tag{7.30}$$

We can show by the following that the periodogram is the DFT of $\varphi_{xx}(k)$. First, using the definition (3.7) of the DFT,

$$\text{DFT}\{\varphi_{xx}(k)\} = \frac{1}{N}\sum_{k=0}^{N-1}\sum_{n=0}^{N-1} x_n x_{n+k} e^{-j2\pi km/N}; \quad m = 0, 1, \ldots, N-1 \tag{7.31}$$

Now, substitute the inverse DFT (3.23) for x_n and x_{n+k}, using the linear phase shift property (Table 3.3, Property 8) on the latter (the range of m remains the same throughout):

$$\begin{aligned}\text{DFT}\{\varphi_{xx}(k)\} &= \frac{1}{N^3}\sum_{k=0}^{N-1}\sum_{n=0}^{N-1}\sum_{\alpha=0}^{N-1} X_\alpha e^{j2\pi\alpha n/N}\sum_{\beta=0}^{N-1} X_\beta e^{j2\pi\beta(n+k)N} e^{-j2\pi km/N} \\ &= \frac{1}{N^3}\sum_{\alpha=0}^{N-1}\sum_{\beta=0}^{N-1} X_\alpha X_\beta \sum_{k=0}^{N-1}\sum_{n=0}^{N-1} e^{j2\pi[(\alpha+\beta)n+(\beta-m)k]/N}\end{aligned} \tag{7.32}$$

Into this equation, we substitute $\alpha = N - \gamma$. The reduction is then similar to the reduction of (7.21), and we obtain the final result:

$$\begin{aligned}\text{DFT}\{\varphi_{xx}(k)\} &= \frac{1}{N^3}\sum_{\beta=0}^{N-1}\sum_{\gamma=0}^{N-1} X_\beta X'_\gamma \sum_{k=0}^{N-1} e^{j2\pi(\beta-m)k/N}\sum_{n=0}^{N-1} e^{j2\pi(\beta-\gamma)n/N} \\ &= \frac{1}{N}|X_m|^2 = P_{xx}(m); \quad m = 0, 1, \ldots, N-1\end{aligned} \tag{7.33}$$

The final result follows because, similar to (7.22), the sums over k and n are nonzero and equal to N only when $\alpha = \gamma = m$. Thus, we have the following important result:

$$\boxed{\begin{array}{c}\text{The periodogram of a vector } x \text{ with periodic extension is the}\\ \text{DFT of the autocorrelation function of } x\text{, that is,}\\ \text{DFT}\{\varphi_{xx}(k)\} = P_{xx}(m); \quad m = 0,1,\ldots,N-1\end{array}} \tag{7.34}$$

Our usual view of the autocorrelation function is that φ_{xx} gives us information on the dependence of a given sample, x_n, on nearby samples, $x_{n\pm1}$, $x_{n\pm2}$, and so on. The periodogram, as we have seen, tells us the distribution of signal power, x_n^2, over frequency. Thus, these two kinds of information about the signal are equivalent via (7.34). Notice also that neither φ_{xx} nor P_{xx} contain any phase information about x, that is, the functions are not affected by shifting x in the time domain.

Two special examples of (7.34) are worth mentioning. First is the case where x is any sinusoidal component of a waveform, sampled over K cycles. It is easy to show in this case that $\varphi_{xx}(k)$ is a cosine function, regardless of the phase of x, and $P_{xx}(m)$ is zero everywhere except at $m = \pm K$. In the second case, x is a segment of *white noise*, that is, a random signal with the same power density at all frequencies. In this case, $\varphi_{xx}(k)$ is an impulse at $k = 0$, showing that the signal elements are statistically independent. Both of these cases are illustrated in Figure 7.13.

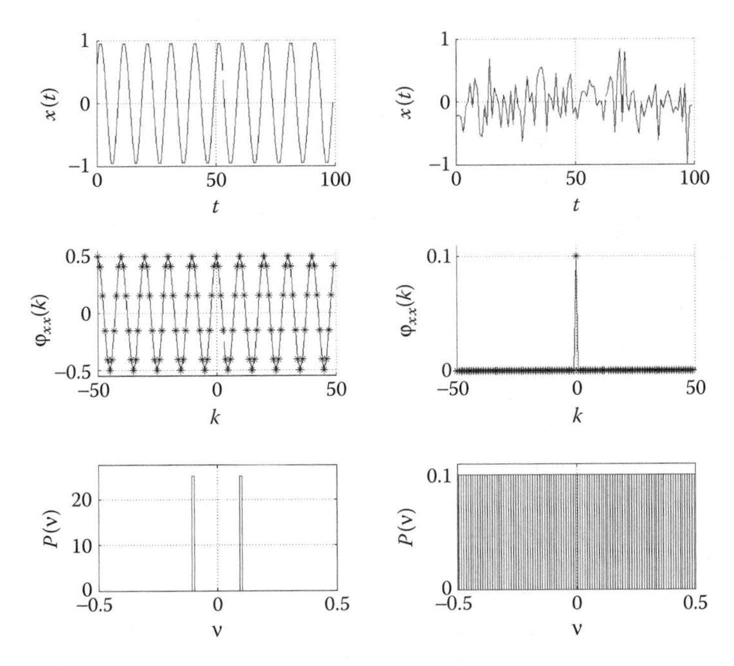

FIGURE 7.13
Signals, autocorrelation functions, and periodograms for a unit sinusoid (left) and white Gaussian noise (right). The average power, $\varphi_{xx}(0)$, is 0.5 for the sinusoid and 0.1 for the noise.

On the left in Figure 7.13, $x(t)$ is sampled with $T = 1$. The autocorrelation function $\varphi_{xx}(k)$, plotted below $x(t)$, is a cosine function. Note that $\varphi_{xx}(0) = 0.5$, which is the average squared value of $x(t)$. The power spectrum, plotted at the bottom, shows that all the power is at $\nu = 0.1$, which is the frequency of $x(t)$ in Hz-s. The integral of the power spectrum, in accordance with (7.28), is the average power or 0.5.

On the right in Figure 7.13, the random signal $x(t)$ is again sampled with $T = 1$. The functions $\varphi_{xx}(k)$ and $P_{xx}(\nu)$ are the statistical properties of the stationary signal from which the segment $x(t)$ was taken, not the measured properties of $x(t)$, in order to better illustrate the properties of white noise.

7.6 Power Spectral Estimation

As in Section 7.5, let us assume that $x(t)$ is stationary and contains random components. Let us also assume that we have a recording of $x(t)$ in terms of a long sample vector obtained as in Figure 7.12. From this vector, we wish to estimate the power spectrum of $x(t)$. There are many applications like this in statistical analysis, where the characteristics of a population are estimated using samples taken from the population.

Having the sample vector, intuition would suggest at first that we use its periodogram as the estimate of the power spectrum, but this approach turns out to be not useful. The reason has to do with the number of degrees of freedom (and thus the variance) of the periodogram [as if we had a large collection of periodograms of $x(t)$ and could measure the variance of each component, $P_{xx}(m)$]. Because, by definition, $P_{xx}(m)$ is found from the DFT magnitude, the periodogram must always have $N/2$ degrees of freedom. Therefore, the degrees of freedom of the periodogram increases linearly with the segment length, and the periodogram of a long sample vector taken from a random function, $x(t)$, is essentially no more reliable than the periodogram of a short sample vector as an estimate of the true power spectrum.

This concept is illustrated in the two upper power spectra in Figure 7.14. (Here, we plot only the positive half of the spectra, using connected points instead of bars on account of the large values of N.) Referring to Figure 7.12, for this illustration, we simulated the sampling, 12-bit conversion, and scaling to units of volts (v) of a white random signal with average power equal to 100 volts.[2] The upper periodogram is based on 2^{16} samples and the center periodogram on 2^9 samples of the signal. Both periodograms are essentially useless estimators of the true power spectrum.

On the other hand, the periodogram at the bottom of Figure 7.14 gives an estimate of the true power spectrum that is quite accurate. It was constructed by partitioning the same segment with 2^{16} samples into 2^7 successive,

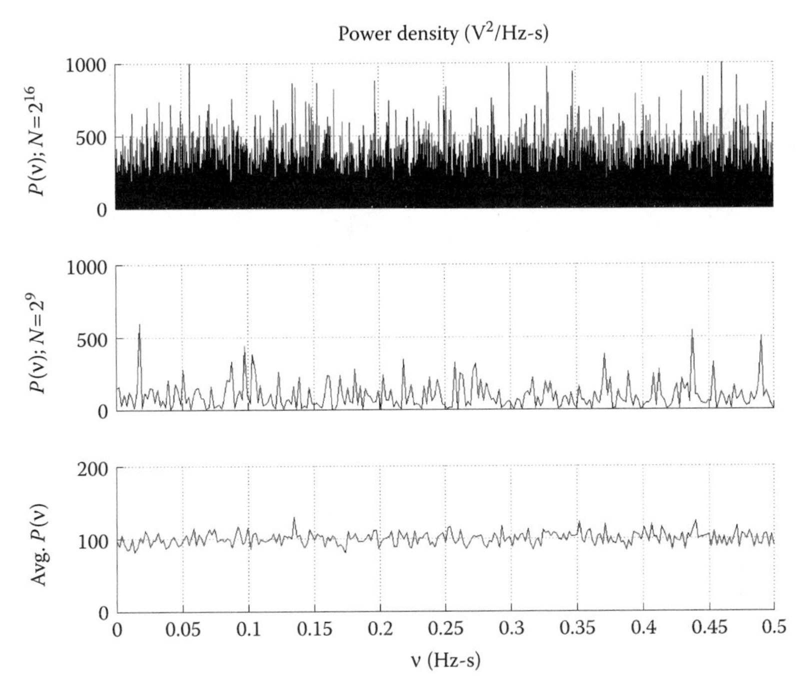

FIGURE 7.14
Power spectral estimates. The two upper periodograms, based on single periodograms, are not good estimates. The lower is the average of 128 periodograms and is a good estimate of the true power density, which is 100 at all frequencies.

nonoverlapping segments, each of length 2^9, and by then averaging the segment periodograms to produce an *average periodogram*. We can see in this case that the power density is approximately 100 volts^2/Hz-s at all frequencies.

Thus, if we have a recorded segment of a stationary function with random components, our method for obtaining a useful estimate of the power spectrum consists of partitioning the recorded segment into smaller segments and averaging the periodograms of the smaller segments. It even helps to *overlap* the smaller segments. At first, it may appear that overlapping does not improve the estimate, because samples are reused. But although the samples are reused, overlapping improves the estimate, because it allows more segments in a given recording, and even though the segments overlap, each segment is different.

This partitioning concept is illustrated in Figure 7.15. At the top, $x(kT)$ is a plot of $N = 1024$ samples of a stationary random signal. The segments x_1 and x_2 in the middle of the figure, each with 512 samples, are obtained simply by partitioning x into two halves. The segments x_1 through x_5 at the bottom, each again with 512 samples, are the result of applying a rectangular, 512-sample window to x, with the window shifting 128 samples for each successive segment. Thus, the figure represents three choices for

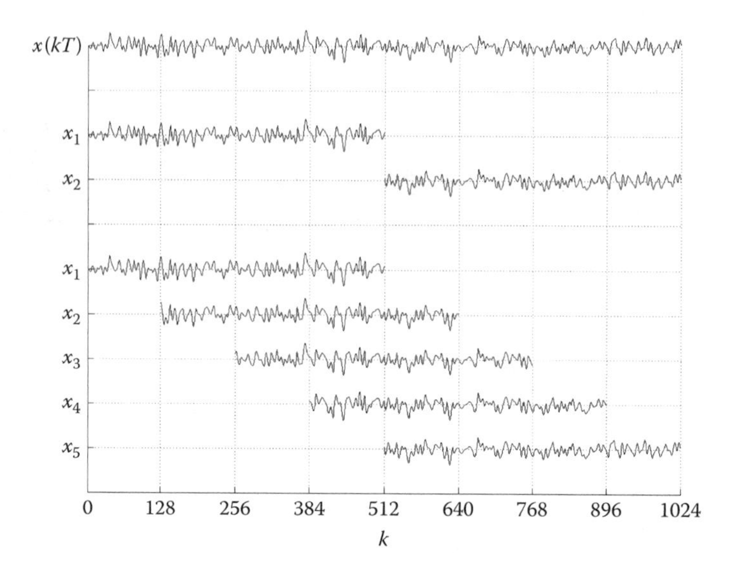

FIGURE 7.15
Partitioning $N = 1024$ samples of a recorded random waveform into segments of length $N/2 = 512$. Segments x_1–x_2 illustrate two nonoverlapping segments, and segments x_1–x_5 illustrate five segments with 75% overlap.

power spectral estimation using the same recorded data, and these choices may be rated as follows:

Worst:	Periodogram of x
Better:	Average periodogram of two independent segments
Best:	Average periodogram of five overlapping segments

Even though the overlapping segments contain redundant data, no two segments are alike, and a larger averaged number of periodograms means a better estimate.

The *average periodogram method* for estimating the power spectrum of a stationary waveform has been described by Welch.[11] Welch suggested using half-overlapping segments. One reason for this choice is that more overlap, although it helps, does not produce proportionately less variance in the spectral estimate, and half-overlapping is a good compromise. A second reason is that an FFT is required for each segment, so more overlap translates into more computing. However, the second reason is not as important as it was in the "old days" when computers were orders of magnitude slower than they are now. More overlap, even to the point where the window moves only one sample at a time, generally, produces a better spectral estimate. The accuracy of spectral estimation is discussed further in the literature.[6–16]

This principle is illustrated using $N = 1024$ samples of filtered white Gaussian noise, generated as shown in Figure 7.16. The input signal vector, x, is a record of white Gaussian noise with zero mean and unit standard deviation, in other words, a recording from a signal having unit power and unit power density in accordance with (7.6). The true power spectrum of the bandpass filter output, y, is therefore the same as the power gain of the filter.

Three power spectral computations are shown in Figure 7.17. The true bandpass power spectrum is plotted as the heavy line in each case. Each spectral computation was made with the same record (y) having length $N = 1024$, using 64-sample segments and a rectangular data window. Only the overlap is changed, and we can see that the accuracy seems to improve with the degree of overlap. But we also note that the number of segments, and thus the number of FFTs, only approximately doubles as we go to 50%

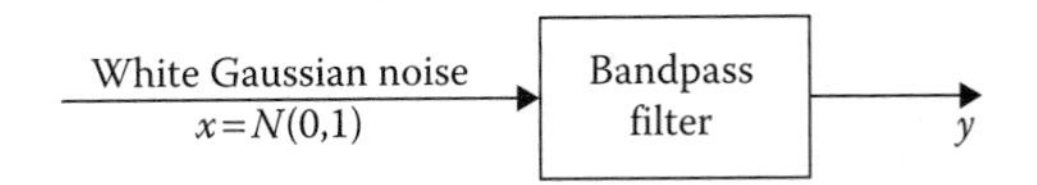

FIGURE 7.16
Illustration showing how the signal (y) is produced for the examples of power spectra in Figure 7.17. The input (x) is white Gaussian noise with $\mu = 0$ and $\sigma = 1$, that is, unit power density. The power spectrum of y is therefore the squared amplitude gain of the bandpass filter.

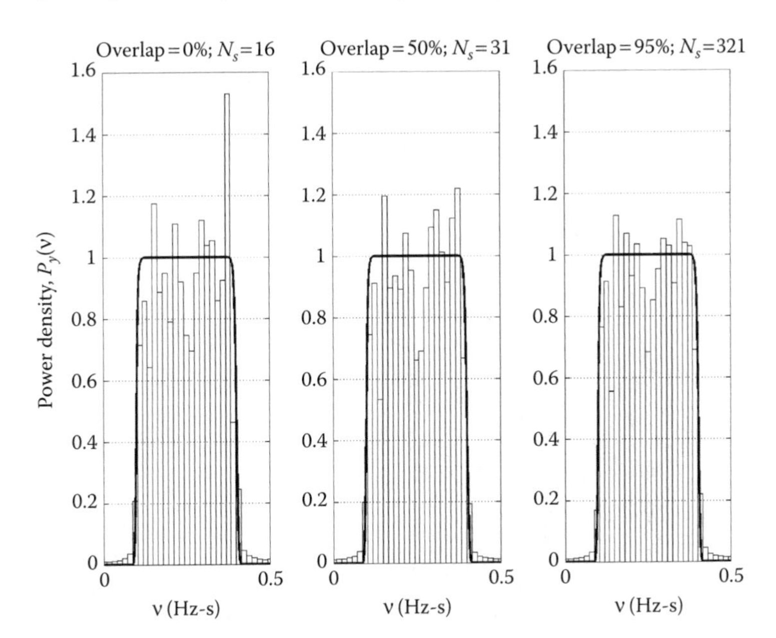

FIGURE 7.17
Power density, $P_y(\nu)$, computed from 1024 samples of the signal y in Figure 7.16. In each plot, the segment size is 64 samples, and the window is rectangular. The segment overlap, and hence, the number of segments (N_s), is the only change from one spectrum to the next. The heavy line in each case shows the true power spectrum.

overlap, whereas this number increases by another order of magnitude as we go to 95% overlap. Thus, the choice of overlap becomes a matter of judgment, depending on the accuracy we wish to attain.

We should also note that the foregoing is based on the idea of getting the most we can out of a fixed signal record of fixed length, such as the recording from an instrument, or telemetry, of a source that has finite duration. If an unlimited length of the stationary random signal is available, then it is always preferable to use independent, nonoverlapping segments.

Another matter of judgment concerns the selection of the segment size, again assuming a recording of fixed length. Obviously, a shorter segment length means more segments and a more accurate spectral estimate. On the other hand, if the segment length is K, then the frequency resolution is $1/K$ Hz-s, and so a shorter segment length also means less resolution. On account of this trade-off, the value of K becomes a matter of choice. The effect of reducing K in order to get a better spectral estimate is illustrated in Figure 7.18, which again is produced using a signal y as produced in Figure 7.16. In this example, the overlap was 95% in all three cases, K was reduced from 64 to 32 to 16, and we can see the improvement in the spectral estimate.

The illustrations in Figures 7.17 and 7.18 are random in the sense that examples like these will be quite different for different random sequences due to the short record length of only $N = 2^{10}$ samples (compared with 2^{16}

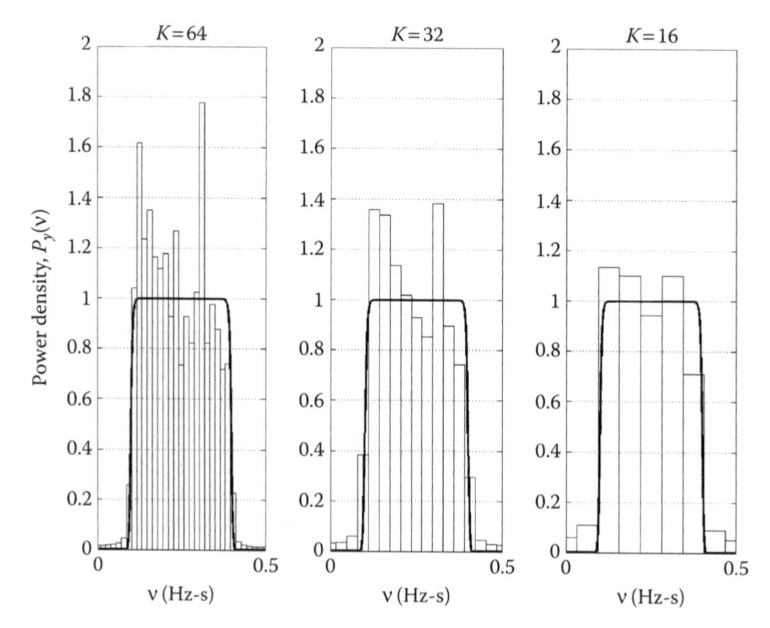

FIGURE 7.18
Power density, $P_y(\nu)$, computed from 1024 samples of the signal y in Figure 7.16. In each plot, the segment overlap is 95%, and the window is rectangular. The segment size (K) is the only change from one spectrum to the next. The heavy line in each case shows the true power spectrum.

samples used for the lower plot in Figure 7.14). In other words, the examples in Figures 7.17 and 7.18 illustrate the concepts of improving spectral estimates, but they are not "typical" in any other sense.

7.7 Data Windows in Spectral Estimation

In Chapter 5, Section 5.4, we used window functions to modify the impulse response of an ideal FIR filter in order to obtain a smoothed version of the filter's power gain function. In the same way, the application of a window function to a random signal segment should, in general, result in a smoothed periodogram. Windows are often used for this purpose in spectral estimation.

The effect of the data window is governed by the relationship (5.7), which states in effect that when any signal vector is multiplied by a window vector, the spectral result is the periodic convolution of the DFTs of the two vectors. That is,

$$\mathrm{DFT}\{w_k x_k\} = \frac{1}{N}\sum_{n=0}^{N-1} W_n X_{m-n}; \quad m = 0, 1, \ldots, N-1 \tag{7.35}$$

where w and x are the window and signal vectors of length N. The convolution of the window spectrum with the signal spectrum has the effect discussed in Chapter 5, Section 5.4. In the present case, it improves the spectral estimate by smoothing it but also smears abrupt changes in the estimated power density, making the estimated spectrum somewhat less accurate in this respect.

When we use a window vector (w) to modify a data segment (x) in power spectral estimation, it is also important to remember that the window alters the total squared value of the data. That is, unless we are using the boxcar window, (w .* x) .^2 is not equal to x .^2. Therefore, to preserve (7.28) and have the integral of the periodogram equal to the average power, we must remove this effect of the window by dividing each periodogram component by $w * w'$, that is, we must divide each periodogram by the total squared window value.

An illustration of the use of data windows (scaled as just described) is given in Figure 7.19, which again shows power spectral estimates of the signal y generated as in Figure 7.16. In this illustration, the record size was increased to 2^{12} samples to give a better estimate for segment length 64. The overlap was again set at 95%. We can note that the Hanning and Blackman windows produce smoothing along the top of the spectrum, but (on close inspection) they have the adverse effect of causing the edges at 0.1 and 0.4 Hz-s to be less defined.

The use of windows in power spectral estimation is generally recommended with stationary random signals, especially signals with broad spectra. The windows generated by the function *window*, described in Chapter 5, (5.13), are illustrated in Figure 7.20. Any of these may be used to smooth the periodograms that are averaged to obtain an estimated power spectrum.

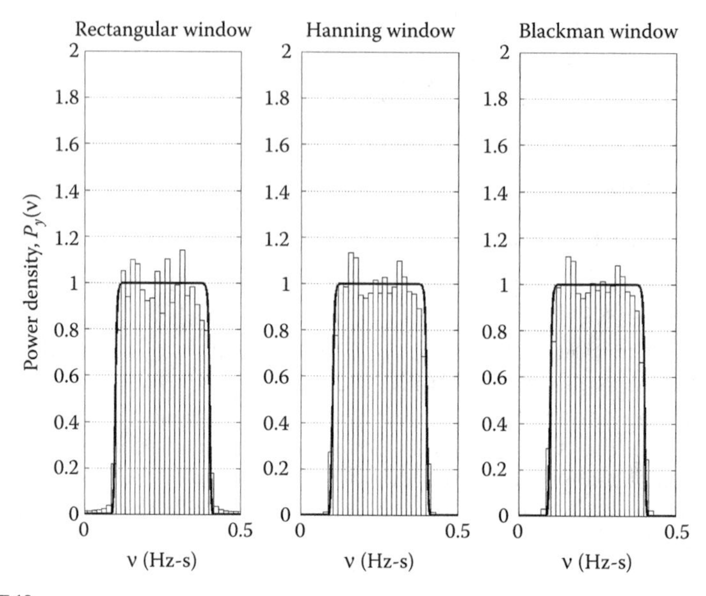

FIGURE 7.19
Power density, $P_y(\nu)$, computed from 4096 samples of the signal y in Figure 7.16. In each plot, the segment overlap is 95%, and the segment size is 64. The plots show the effect of using different data windows. The heavy line in each case shows the true power spectrum.

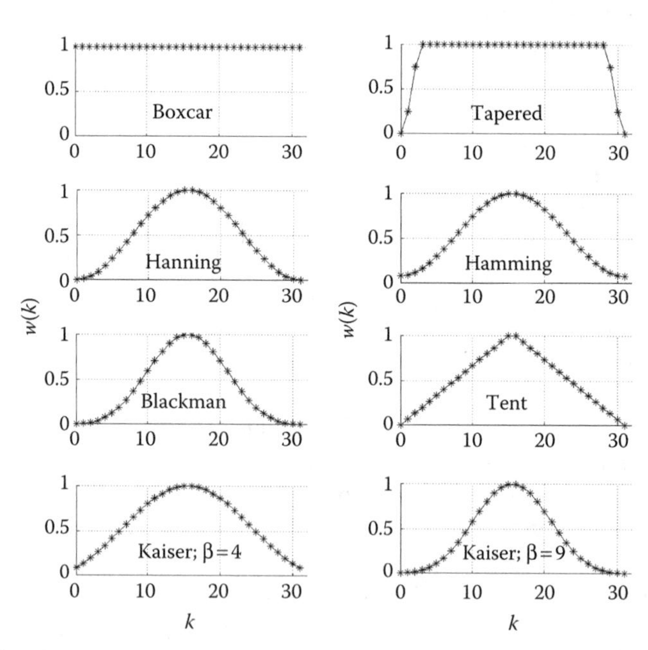

FIGURE 7.20
Data windows generated by the *windows* function; $N = 32$.

If a signal is not stationary or has a true power spectrum with sharp discontinuities, data windows must be used with caution. In this case, the *boxcar* and the *tapered rectangular* window, which are among the choices illustrated in Figure 7.20, are generally preferable. The tapered rectangular window is almost rectangular, having a cosine taper over 10% of the length at each end.

7.8 The Cross-Power Spectrum

The *cross-power spectrum*, or *cross spectrum* for short, is like the power spectrum we have been discussing, except it involves two time series instead of just one. Suppose x and y are vectors of length N taken from two real waveforms, $x(t)$ and $y(t)$. As before, we assume for this analysis that both x and y extend periodically. Then, analogous to the periodogram in (7.24), we have the *cross-periodogram*:

$$\boxed{\text{Cross-periodogram:}\quad P_{xy}(m) = \frac{1}{N} X'_m Y_m; \quad m = 0, 1, \ldots, N-1} \tag{7.36}$$

Just as $P_{xx}(m)$ was a measure of power density, so also $P_{xy}(m)$ is a measure of *cross-power density*, in the following sense. Similar to (7.27) and (7.28), a single component $P_{xy}(m)$ gives the cross-power in range $(m \pm 0.5)/N$ Hz-s. And as illustrated in Figure 7.11, we add the components at positive and negative frequencies, that is, $P_{xy}(m)$ and $P_{xy}(-m) = P_{xy}(N - m)$, to get the total power in this frequency range. But in this case, these components are complex conjugates in accordance with (3.11). The development that led from (7.20) to (7.22) in this case leads to the following:

$$\begin{aligned} \text{Average cross-power} &= \frac{1}{N}\sum_{n=0}^{N-1} x_n y_n \approx \frac{1}{NT}\int_0^{NT} x(t)y(t)dt \\ &= \varphi_{xy}(0) = \frac{1}{N^2}\sum_{m=0}^{N-1} X'_m Y_m = \frac{1}{N}\sum_{m=0}^{N-1} P_{xy}(m) \end{aligned} \tag{7.37}$$

The sum on the right is real, because $P_{xy}(0)$ and $P_{xy}(N/2)$ are real, and $P_{xy}(m)$ and $P_{xy}(N - m)$ are conjugates for the other values of m, and also, of course, because $\varphi_{xy}(0)$ is real.

Thus, only the real part of the cross-periodogram contributes to the cross-power. In physical terms, if $v(t)$ and $i(t)$ are voltage and current waveforms, the average cross-power given by (7.37) is the actual power in watts. For example, if $v(t)$ and $i(t)$ are both unit sine waves, the average power is the average product or 0.5 watt; if $v(t)$ is a sine wave and $i(t)$ is a cosine wave, the average power is again the average product or zero.

Another important use of the cross-power spectrum comes from the following relationship, which may be derived just as (7.34) was derived. The cross-periodogram of vectors x and y with periodic extension is the DFT of the correlation function $\varphi_{xy}(k)$, that is,

$$\text{DFT}\{\varphi_{xy}(k)\} = P_{xy}(m); \quad m = 0, 1, \ldots, N-1 \tag{7.38}$$

Thus, $P_{xy}(m)$ tells us how the cross-correlation function is distributed over frequency. In this sense, the magnitude of $P_{xy}(m)$, that is, $|P_{xy}(m)|$, is a measure of the *coherence* of signals x and y at different frequencies. A function called the *Magnitude-Squared Coherence function* (MSC)[14,17,18] is defined as

$$\text{Magnitude-squared coherence:} \quad \Gamma_{xy}^2(\nu) = \frac{|P_{xy}(\nu)|^2}{P_{xx}(\nu)P_{yy}(\nu)} \tag{7.39}$$

The MSC is a normalized measure of the coherence of signals x and y at ν Hz-s, in the sense that it indicates whether the cross-correlation function, $\varphi_{xy}(k)$, has a component at that frequency.

The MSC is useful in situations where the source of a signal is weak and inaccessible, and only noisy versions of the signal can be recorded by placing sensors away from the source. Medical situations where invasive procedures are ruled out, or geophysical applications where the source cannot be reached, are examples. Then, one may use multiple sensors and look for coherence between pairs of sensor outputs using the MSC. (The concept also extends to more than two sensors using multidimensional spectra.)

The MSC concept is illustrated in the situation shown in Figure 7.21, where two signals, x and y, are recorded by placing sensors where signals from an unknown source (s) are able to be received. In this example, white Gaussian noise (n_1) is added to s to produce x, and *independent* white Gaussian noise

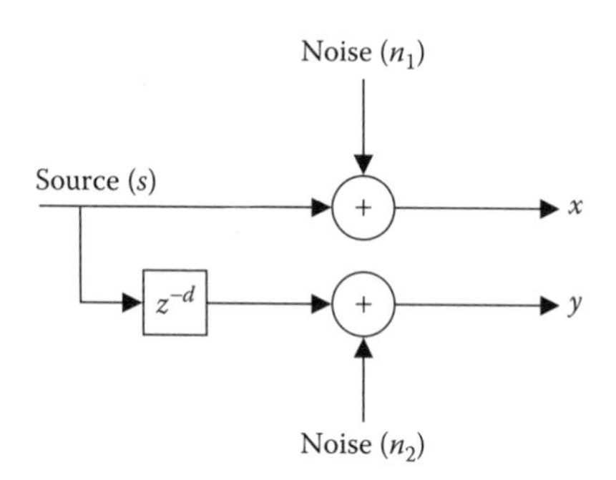

FIGURE 7.21
Generation of signals x and y, which are used to produce the power spectra in Figure 7.22.

(n_2) is added to a delayed version of s to produce y. The power spectrum of s, $P_{ss}(\nu)$, is shown on the left in Figure 7.22, with $P_{xx}(\nu)$ and $P_{yy}(\nu)$ in the center of Figure 7.22. In situations like this, the signal quality is expressed in terms of the *signal-to-noise ratio* (SNR), which is

$$\text{SNR} = \frac{\text{signal power}}{\text{noise power}} \tag{7.40}$$

In this example, s is limited to the frequency range [0.1, 0.3] Hz-s, and in this range, the SNR is 0.03, which is representative of a very noisy signal.

The magnitude of the cross-spectrum, $|P_{xy}(\nu)|$, is shown on the right in Figure 7.22. All four spectra are based on a record length of 10^5 samples and periodograms of length 20 with 50% segment overlap. The segment window in each case was a tapered boxcar. The cross-spectrum magnitude, which is the nonnormalized version of the MSC in (7.39), illustrates the "recovery" of the signal spectrum, $P_{ss}(\nu)$, in this situation. Note that all four spectra are plotted on the same scale, and $|P_{xy}(\nu)|$ is (theoretically) the same as $P_{ss}(\nu)$, because n_1 and n_2 are statistically independent and, therefore, uncorrelated. Other illustrations of MSC may be found at the end of this chapter in Exercises 7.14 and 7.15.

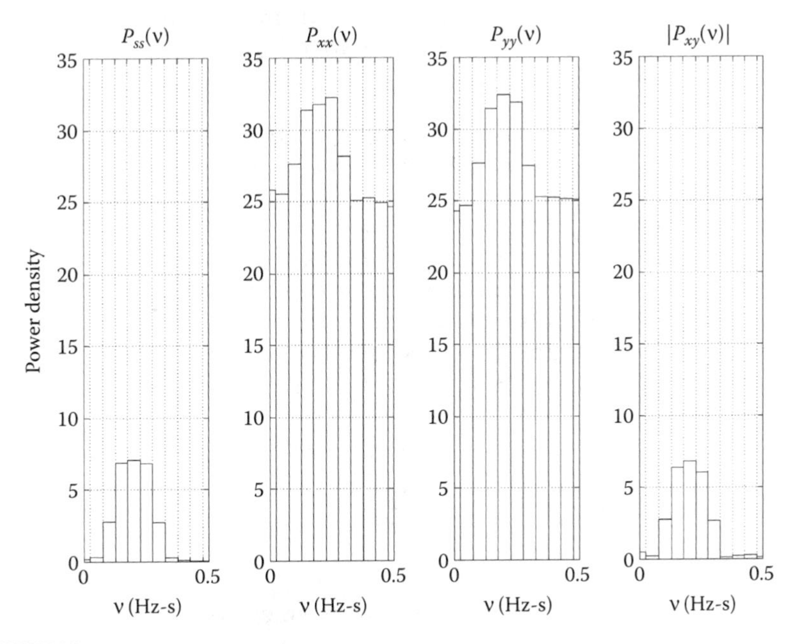

FIGURE 7.22
Power density spectra for the signals in Figure 7.21. Each spectrum is based on a recording of 10^5 samples of the signal(s). The SNR is 0.03.

Thus, we have seen that the cross-power spectrum has meaning in terms of electrical power (and similarly in terms of mechanical power) and that it is also useful in various ways as a measure of the distribution over frequency of the cross-correlation of two signal vectors.

7.9 Algorithms

Besides the various MATLAB functions that form the basis for statistical signal analysis, three additional functions included in the *functions* folder on the publisher's website were used to help produce examples in this chapter. First is the function

```
[f,xmin,xmax]=freq(x)                    (7.41)
```

This function computes the frequency function (f_n) given by (7.3), as well as the minimum and maximum elements, of any integer array or vector, x. (If the elements of x are not integers, they are rounded.)

The second function is

```
bar2(v,p,color)                          (7.42)
```

This is a simple modification of the MATLAB function *bar,* which may be used to plot continuous amplitude distributions or power spectra.

The third set of functions is

```
[P,nsgmts,v]=pds1(x,N,windo,overlap)
[P,nsgmts,v]=pds2(x,y,N,windo,overlap)   (7.43)
```

The first function computes the power spectrum of a signal and the second computes the cross-power spectrum of two signals. The first output P(1:N/2) is the power spectrum at positive frequencies. The second output (nsgmts) is the number of segments the function was able to use, given N and the amount of overlap, the latter being specified within the range (0, 1). The third output (v) is the frequency vector in Hz-s, the elements of which correspond with the elements of P.

Exercises

General instructions: (1) Whenever you make a plot in any of the exercises, be sure to label both axes, so we can tell exactly what the units mean. (2) Make continuous plots unless otherwise specified. "Discrete plot" means to plot discrete symbols, unconnected unless otherwise specified. (3) In plots, and especially in subplots, use the *axis* function for best use of the plotting area. (4) In exercises that require random numbers, initialize the seed to "123."

7.1 Complete the following:

a. Given $y(t) = ax(t) + b$, begin with the definition of the mean in (7.4) and prove the formula for μ_y in (7.8).

b. Begin with the definition of variance in (7.6) and prove the formula for σ_y^2 in (7.8).

7.2 Let $y = \sin x$. Assuming x is uniformly distributed in the range $(0, 2\pi)$, use (7.7) over a range of x, where y increases monotonically, say $x = (0, \pi/2)$, to derive the probability density function of y. Make a plot of $p(y)$ versus y over the range $y = (-1, 1)$.

7.3 Generate 5000 samples of a continuous uniform random variate with zero mean and average power equal to 27. Make a single plot of the theoretical continuous probability density, and on the same graph make a histogram plot of the sample amplitude distribution, using nine bars of equal width that exactly span the range of the variate.

7.4 Generate 5000 samples of a continuous Gaussian random variate with mean equal to 6 and variance equal to 4. Make a single plot of the theoretical continuous probability density, and on the same graph, make a histogram plot of the sample amplitude distribution, using bars centered at integers ranging from $\mu_x - 3\sigma_x$ to $\mu_x + 3\sigma_x$. Comment on the integral of the histogram plot in this case.

7.5 Is the amplitude distribution of $x + y$ equal to $p(x) + p(y)$? Prove your answer.

7.6 Suppose $x(t)$ is a Gaussian signal with mean equal to zero and power equal to 10. At any time t, determine the following:

a. What is the probability that $x(t)$ lies in the range [0, 5]?

b. What is the probability that $|x(t)|$ lies in the range [1, 2]?

7.7 For the signal vector (x), use the 5000-sample sequence in Exercise 7.3. Make three subplots. On the left, make a *continuous* plot of the periodogram of x over the range [0, 0.5] Hz-s. In the center, make a bar plot similar to Figure 7.11, but only over [0, 0.5] Hz-s, using half-overlapping 100-sample segments. On the right, make a similar bar plot using 40-sample segments. Use [0, 100] for the periodogram range on all three plots.

7.8 Generate 1200 samples of uniform white noise with zero mean and average power equal to 3.

a. Plot the amplitude distribution as a histogram with each bar width equal to 0.5.

b. Make two subplots, each with a power density spectrum like Figure 7.11. For the left-hand plot, use nonoverlapping segments of length 80. For the right-hand plot, use segments of length 80 that overlap 50%. Use the tapered boxcar window in both cases.

c. How does the integral of each plot compare with the theoretical average power? How do the two spectra compare with each other and with the theoretical power spectrum?

7.9 Generate 5000 samples of

$$x_k = \sin(2\pi k/50) + \sqrt{2}\cos(2\pi k/20)$$

Test the operation of function *pds1* by making two subplots. On the left, plot power density versus Hz-s using segment size 100, a boxcar window, and no overlap. On the right, make a similar plot using the Hamming window and 50% overlap. Make the plots over the range [0, 0.5] Hz-s as in Figure 7.17.

a. Explain the differences between the two plots.

b. Equate the average power in x to the integral of each plot.

7.10 Generate 8192 samples of uniform white noise with mean zero and average power equal to 3 in vector x. Filter x to produce vector y. Use an FIR lowpass filter with Hamming window, 31 weights, and cutoff at 12.5 Hz. The time step between samples is 0.02 s.

a. Explain how you would compute the theoretical power density spectrum of y.

b. Estimate the power density spectrum using 64-sample segments, 50% overlap, and the Hamming window. Make a bar plot of the spectrum over the range [0, 0.5] Hz-s, and overlay a continuous plot of the theoretical spectrum.

7.11 Describe how a window is *normalized*. In a 2×2 array of subplots, plot nonnormalized and normalized versions of the tapered rectangular, Hamming, Tent, and Blackman windows. Begin on the upper left with normalized and nonnormalized plots of the tapered rectangular window.

7.12 Generate $N = 8000$ samples of uniform white noise with mean $= 0$ and average power $= 3$.

a. Plot the amplitude distribution as a histogram with each bar width of 0.5.

b. Make three subplots. On the left, make a continuous plot of the periodogram of all N samples versus frequency from -0.5 to 0.5 Hz-s. In the middle, make a bar plot of the power density using nonoverlapping segments of length 80. On the right, make a similar plot of 75% overlapping segments of length 80. Use the Hamming window in the latter two plots.

7.13 Generate vector x having 8192 uniform white noise samples with average power equal to one. Generate y as a similar vector of Gaussian white noise. Make four subplots. On the upper left, plot the amplitude distribution of x using a histogram with 15 bars. On the lower left, plot the power density of x over [0, 0.5] Hz-s using half-overlapping segments of length 32 and the Hamming window. Do the same on the right for y, using amplitude range $\pm 3\sigma$.

7.14 Let $k = [0{:}3999]$. Generate vector $x = \sin(0.1\pi k)$. Then, generate vector $y = \cos(0.1\pi k) + n$, where n is a Gaussian white noise vector with zero mean and average power 100.

a. Assuming the "signal" is the cosine wave, what is the SNR of y in dB?

b. Make two subplots. On the left, make a bar plot of the power density spectrum of y over [0, 0.5] Hz-s. Use 75% overlapping segments of length 40 and the boxcar window. On the right, make a similar plot of $|P_{xy}(\nu)|$, the magnitude of the cross-power density of x and y. Observe the differences in the two plots.

7.15 Generate $N = 8000$ samples of signals x and y in Figure 7.21, using $s = \sin[0.1\pi(0{:}N-1)]$, $d = 2$, and independent white Gaussian noise with power 50 for n_1 and n_2. Now let x and y simulate signals from two sensors, in which the signal from the unknown source, s, may or may not be present.

a. Plot x and y, one above the other, in two subplots. Is there any sign of the source in these plots?

b. Make three subplots. Use segment length 20, the boxcar window, and 75% overlap. From left to right, plot $P_{xx}(\nu)$, $P_{yy}(\nu)$, and the mean-squared coherence, $\Gamma^2_{xy}(\nu)$, versus ν over range [0, 0.5]. Is there any sign of the source in these plots? If so, explain.

References

1. Shammugan, K. S. 1988. *Random Signals: Detection, Estimation, and Data Analysis*. Chap. 2. New York: John Wiley & Sons.
2. Douglas, C. M., and G. C. Runger. 1999. *Applied Statistics and Probability for Engineers*. 2nd ed. New York: John Wiley & Sons.
3. Derman, C., L. J. Gleser, and I. Olkin. 1973. *A Guide to Probability Theory and Application*. New York: Holt, Rinehart and Winston.
4. Dwass, M. 1970. *Probability Theory and Applications*. New York: W. A. Benjamin.
5. Feller, W. 1957. *An Introduction to Probability Theory and Its Applications*. 2nd ed. Vol. 1. New York: John Wiley & Sons.
6. Kay, S. 1987. *Modern Spectral Estimation: Theory and Application*. Englewood Cliffs, NJ: Prentice Hall.

7. Marple, S. L. 1987. *Digital Spectral Analysis with Applications*. Englewood Cliffs, NJ: Prentice Hall.
8. Oppenheim, A. V., and R. W. Schafer. 1989. *Discrete-Time Signal Processing*. Chaps. 8 and 11. Englewood Cliffs, NJ: Prentice Hall.
9. Rabiner, L. R., and B. Gold. 1975. *Theory and Application of Digital Signal Processing*. Chap. 6. Englewood Cliffs, NJ: Prentice Hall.
10. Richards, P. I. 1967. Computing reliable power spectra. *IEEE Spectr* 4(1):83–90.
11. Welch, P. D. 1967. The use of the fast Fourier transform for the estimation of power spectra. *IEEE Trans* 15(2):70–3.
12. Bingham, C., M. D. Godfrey, and J. W. Tukey. 1967. Modern techniques of power spectrum estimation. *IEEE Trans* 15(2):56–66.
13. Yuen, C. K. 1977. A comparison of five methods for computing the power spectrum of a random process using data segmentation. *Proc IEEE* 65(6):984–6.
14. Hinich, M. J., and C. S. Clay. 1968. The application of the discrete Fourier transform in the estimation of power spectra, coherence, and bispectra of geophysical data. *Rev Geophys* 6(3):347–63.
15. Carter, G. C., and A. H. Nuttall. 1980. On the weighted overlapped segment averaging method for power spectral estimation. *Proc IEEE* 68(10):1352–4.
16. Nuttall, A. H., and G. C. Carter. 1980. A generalized framework for power spectral estimation. *IEEE Trans* 28(3):334–5.
17. Carter, G. C., C. H. Knapp, and A. H. Nuttall. 1973. Estimation of the magnitude-squared coherence function via overlapped FFT processing. *IEEE Trans* 21(2):337–44.
18. Benignus, V. A. 1969. Estimation of the coherence spectrum and its confidence interval using the fast Fourier transform. *IEEE Trans* 17(2):145–50.

Further Reading

Bendat, J. S., and A. G. Piersol. 1971. *Random Data: Analysis and Measurement Procedure*. Chap. 9. New York: John Wiley & Sons.

Blanc-LaPierre, A., and R. Fortet. 1965. *Theory of Random Functions*. Vol. 1. New York: Gordon and Breach.

Jenkins, G. M., and D. G. Watts. 1968. *Spectral Analysis and Its Applications*. San Francisco, CA: Holden-Day.

Koopmans, L. H. 1974. *The Spectral Analysis of Time Series*. New York: Academic Press.

Otnes, R. K., and L. Enochson. 1972. *Digital Time Series Analysis*. New York: John Wiley & Sons.

Therrien, C., and T. Murali. 2004. *Probability for Electrical and Computer Engineers*. Boca Raton, FL: CRC Press.

8

Least-Squares System Design

8.1 Introduction

The least-squares principle is widely applicable to the design of digital signal processing (DSP) systems. In Chapter 2, we saw how to fit a general linear combination of functions to a desired function or sequence of samples, and how the finite Fourier series, as an example, provides a least-squares fit to a sequence of data. In Chapter 5, Exercise 5.5, we saw that the finite impulse response (FIR) filter gain may be expressed as a Fourier series in frequency and is therefore a least-squares approximation to the ideal rectangular gain function. In this chapter, we wish to extend the least-squares concept to the design of other kinds of signal processing systems. We will see how several kinds of DSP systems may be realized using a common linear least-squares approach.

First, we describe briefly a variety of tasks involving *prediction, modeling, equalization,* and *interference canceling,* where least-squares design is useful. We show that all these tasks lead to the same least-squares design problem, which is to match a particular signal to a desired signal so that the difference between the two signals is minimal in the least-squares sense.

Next, we discuss the solution to the least-squares system design problem, which, for nonrecursive systems, amounts to the inversion of a matrix of correlation coefficients, similar to the symmetric coefficient matrix in Chapter 2, (2.10). We show that finding this solution is equivalent to finding the minimum point on a quadratic mean-squared error (MSE) performance surface (a concept that will become useful when we discuss *adaptive signal processing* in Chapter 9).

We also include various examples of least-squares design. The examples illustrate a variety of system configurations used in different applications, and also different ways to estimate correlation coefficients. However, the examples do not even begin to cover all of the many applications of this topic. A number of additional applications are included in the exercises, and therefore, the exercises are an important part of this particular chapter. Furthermore, this chapter forms the basis for adaptive signal processing, which is the subject of Chapter 9.

8.2 Applications of Least-Squares Design

In this section, we describe system configurations where the least-squares design is applicable. The first configuration, illustrated in Figure 8.1, is the *linear predictor*. The prediction concept is illustrated in its simplest form in the upper diagram. In the least-squares design, the coefficients, or *weights*, of a causal linear system, $H(z)$, are adjusted to minimize the MSE, $E[e_k^2]$, thereby making the system output, g_k, the least-squares approximation to a desired signal, d_k. In this case, d_k is the same as the input, s_k, and must be "predicted" using the past history of s_k. That is, d_k must be predicted in terms of s_k delayed by m samples, and processed through $H(z)$. Again, the least-squares design process consists of adjusting the weights of $H(z)$ to make this processed version of the history of the input the best "prediction" of the present sample value, s_k.

The upper diagram in Figure 8.1 with a unit delay, that is, with $m = 1$, is the most common form of the linear predictor, even though it does not actually predict a future value of the input. In most signal processing applications, the error, e_k, is needed rather than a future value of s_k. If an actual prediction of a signal is needed, the augmented form in the lower diagram of Figure 8.1

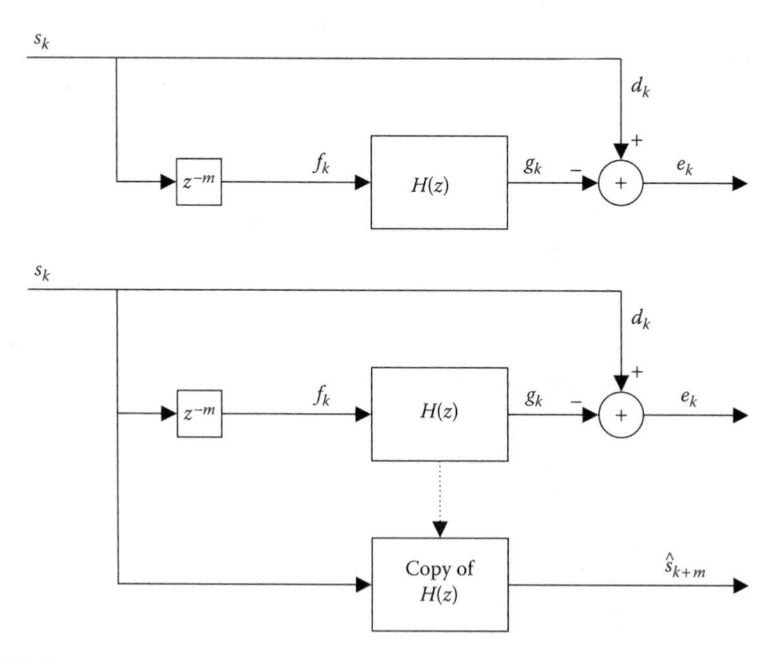

FIGURE 8.1
Least-squares linear prediction. In the upper diagram, g_k is the prediction of the current input, s_k, given the past history of s. In the lower diagram, $\hat{s}_k$ is the prediction of a future input, s_{k+m}, based on the history of s.

may be implemented. Here, s_k is processed through a copy of the predictor, $H(z)$, which produces the predicted future value, $\hat{S}_{k+m}$.

The linear predictor is useful in waveform encoding and data compression,[7,14] spectral estimation,[5,6,10,11] spectral line enhancement,[9] event detection,[8] and other areas. Its effect in all these applications, as one might guess from the previous discussion, is to produce a signal e_k, which is a *decorrelated* or *whitened* version of s_k. The input is a signal that may be predictable in terms of its past values, and the output, one might say, is everything that is left after the predictable part is taken out.

The second configuration in which least-squares design is applicable is illustrated in Figure 8.2 and is called *modeling* or *system identification*. Here, a linear system, $H(z)$, models or identifies an unknown "plant" (see Section 8.8) consisting of an unknown system with internal noise. The least-squares design forces the linear system output, g_k, to be a least-squares approximation to the desired plant output, d_k, for a particular input signal, f_k. When f_k has spectral content at all frequencies and when the plant noise contributes at most a small part of the power in d_k, we expect $H(z)$ to be similar to the transfer function of the plant's unknown system. Note, however, that $H(z)$ is not necessarily a least-squares approximation, as it was, for example, in Chapter 5. Thus, the modeling concept is applicable in cases where the best approximation to a signal, rather than to a transfer function, is the objective. The type of modeling illustrated in Figure 8.2 has a wide range of applications, including modeling in the biological, social, and economic sciences,[17] in adaptive control systems,[2,4] in digital filter design,[18] and in geophysics.[2]

Another application of least-squares system design, known as *inverse modeling* or *equalization*, is illustrated in Figure 8.3. Here, the desired output of

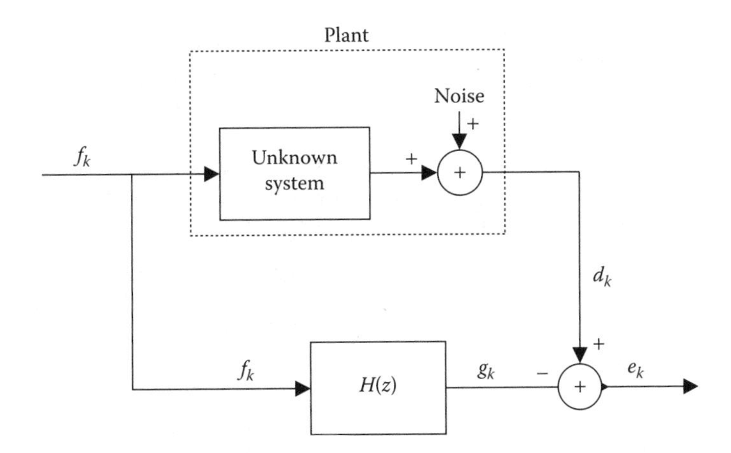

FIGURE 8.2
Least-squares modeling or system identification. When the noise is independent of the input, f, $H(z)$ becomes a "model" of the unknown system in the sense that its output, g_k, is a least-squares approximation to the output of the unknown system.

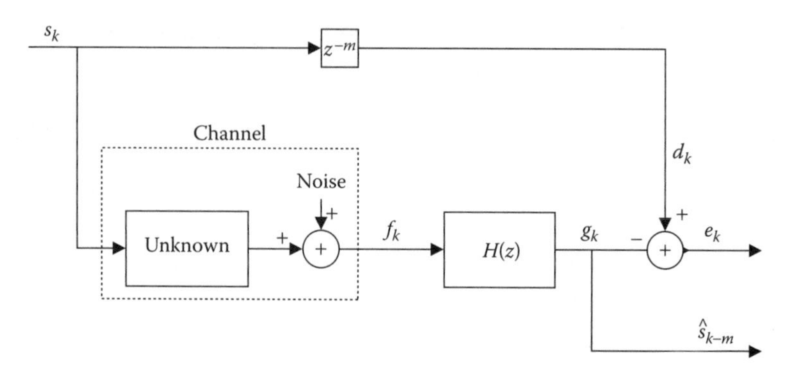

FIGURE 8.3
Least-squares inverse modeling or equalization. When the noise is independent of the input, $H(z)$ becomes the inverse of the channel and thus "equalizes" the effect of the channel on the input, producing a delayed least-squares approximation to the input.

$H(z)$, again labeled d_k, is a delayed version of the input signal. To cause the output g_k to approximate d_k, $H(z)$ is adjusted to model the inverse of, or *equalize*, the unknown system (or "channel") with internal noise. When the unknown system and the linear system are both causal, the delay, z^{-m}, is used to compensate for the propagation delay through the two systems in cascade. When the internal noise of the unknown system is only a small part of f_k, the output of the equalizer, g_k, is a good approximation to the delayed input, s_{k-m}. Hence, the linear system is used here to invert, or equalize, the effect of the unknown system on the input signal.

Equalization is used in communication systems to remove distortion by compensating for nonuniform channel gain and multipath effects and to improve the signal-to-noise ratio if the channel introduces band-limited noise.[2,12,13] Equalization and inverse modeling are also used in adaptive control,[2] speech analysis,[3,10,11] deconvolution,[15] digital filter design,[2,18] and other areas.

Our final example of a configuration where least-squares design is used, *interference canceling*, is illustrated in Figure 8.4. The interference-canceling principle[1,2,16] is applicable in cases in which there is a signal, s_k, with additive noise, n_k, and also a source of correlated noise, n'_k. Ideally, n_k and n'_k are correlated with each other but not with s_k, although the principle is applicable even if the signal and noise are correlated. The least-squares design objective is to adjust $H(z)$ so that its output, g_k, is a least-squares approximation to n_k, thereby canceling, by subtraction, the noise from the incoming waveform. When the signal and noise are independent, this is equivalent to minimizing the mean-squared value of the error, $E[e_k^2]$, because the independent noise cannot be made to cancel the signal, s_k. The delay is placed in the configuration to compensate for propagation through the causal linear system and allow the noise sequences, n_k and n'_k, to be aligned in time.

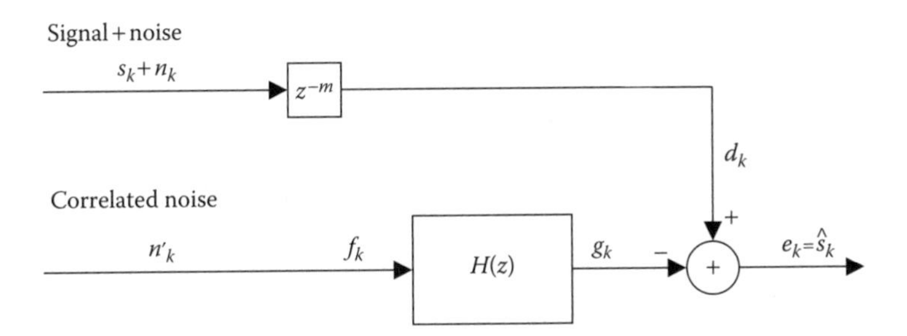

FIGURE 8.4
Least-squares noise cancellation. When the noise (n_k) is independent of the signal, and correlated noise (n'_k) can be measured independently, $H(z)$ produces a least-squares approximation to n_k, and thus improves the SNR, even though s_k and n_k may have overlapping spectra.

Interference canceling is an interesting and sometimes preferable alternative to bandpass filtering for improving the signal-to-noise ratio. For example, suppose we have an underground seismic sensor that receives, in addition to a seismic signal component, s_k, an acoustic noise component, n_k, coupled into the ground from the atmosphere above, and the two components have similar spectra so that n_k cannot be removed from $(s_k + n_k)$ using bandpass filtering. We could then add an aboveground microphone to the system to receive n'_k, an acoustic component correlated with (but not exactly the same as) n_k, and process the signals as in Figure 8.4 to reduce the acoustic noise and increase the signal-to-noise ratio. The number of examples of this type is limited only by one's imagination.

Although the applications illustrated in Figures 8.1 through 8.4 are distinctly different, they all involve the same least-squares design problem. The important features may be summarized as follows. There is a linear system, $H(z)$, with adjustable coefficients. These coefficients are adjusted to cause the output, g_k, of the linear system to be a least-squares approximation to the desired signal, d_k, and thereby minimize the MSE, $E[e_k^2]$. We now proceed to examine the nature of the resulting least-squares design problem, using a geometrical interpretation.

8.3 System Design via the Mean-Squared Error

If we compare Figures 8.1 through 8.4, we can see a common least-squares design problem, which is illustrated in Figure 8.5. The parameters of a causal linear system, $H(z)$, are to be adjusted or selected to minimize a MSE, $E[e_k^2]$. [Actually, there is no need in what follows to assume $H(z)$ is causal, but we assume causality for convenience, and to include real-time applications.] For example, if $H(z)$ is a linear system of the form $B(z)/A(z)$, the parameters to be selected are b and a, the coefficient vectors with transforms $B(z)$ and $A(z)$.

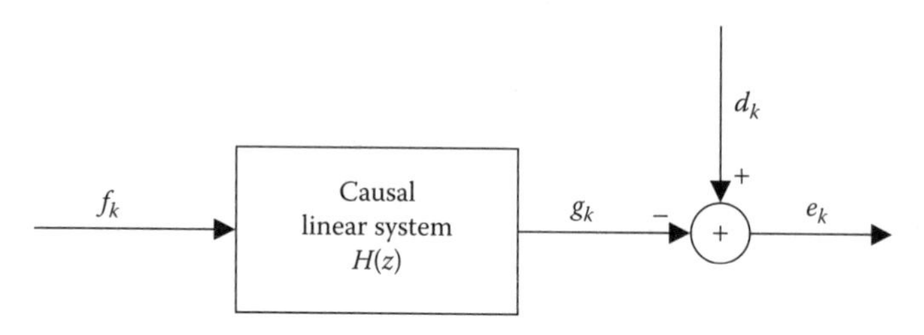

FIGURE 8.5
Essential ingredients common to Figures 8.1 through 8.4. The weights of a causal linear system, $H(z)$, are adjusted to minimize the MSE, that is, the mean value of e_k^2, and thereby make g_k a least-squares approximation to d_k.

The error, e_k, is the difference between the desired signal, d_k, and the linear system output, g_k:

$$e_k = d_k - g_k \tag{8.1}$$

Let us now assume that the signals in Figure 8.5 are stationary so that expected values and correlation functions are defined. Then, the mean-squared error (MSE) is

$$\begin{aligned} \text{MSE} \equiv E[e_k^2] &= E[(d_k - g_k)^2] \\ &= E[d_k^2 + g_k^2 - 2d_k g_k] \\ &= \varphi_{dd}(0) + E[g_k^2] - 2E[d_k g_k] \end{aligned} \tag{8.2}$$

The last result follows from the definition of the correlation function in Chapter 3, (3.2). For the causal linear system in Figure 8.5, we have

$$g_k = \sum_{n=0}^{N-1} h_n f_{k-n}; \quad 0 \le k < \infty \tag{8.3}$$

where vector h is the impulse response of $H(z)$. (We assume for now that only the first N samples of h affect the MSE.) Using this in (8.2), we obtain

$$\begin{aligned} \text{MSE} &= \varphi_{dd}(0) + E\left[\sum_{n=0}^{N-1}\sum_{m=0}^{N-1} h_n h_m f_{k-n} f_{k-m}\right] - 2E\left[d_k \sum_{n=0}^{N-1} h_n f_{k-n}\right] \\ &= \varphi_{dd}(0) + \sum_{n=0}^{N-1}\sum_{m=0}^{N-1} h_n h_m \varphi_{ff}(n-m) - 2\sum_{n=0}^{N-1} h_n \varphi_{df}(-n) \end{aligned} \tag{8.4}$$

In the center term, f_{k-n} and f_{k-m} are displaced $n - m$ time steps from each other; hence, the expected product is $\varphi_{ff}(n - m)$. Similarly, the expected value of $d_k f_{k-n}$ is $\varphi_{df}(-n)$.

Our next objective is to formulate the minimization of (8.4) as a linear least-squares problem so we can apply the methods in Chapter 2, Section 2.2 to obtain its solution for the optimal impulse response. Because the MSE in (8.4) is a quadratic function of the N samples in h, we can see that the gradient of the MSE with respect to these samples is a linear function of the samples; that is,

$$\frac{\partial \text{MSE}}{\partial h_k} = 2\sum_{n=0}^{N-1} \varphi_{ff}(n-k)h_n - 2\varphi_{fd}(k); \quad k = 0, 1, \ldots, N-1 \tag{8.5}$$

[To obtain this, we used the fact that $\varphi_{ff}(n)$ is an even function of n and also substituted $\varphi_{fd}(n)$ in place of $\varphi_{df}(-n)$.] Setting this gradient vector equal to zero, we have the N equations we need to solve for the impulse response of the "optimal" $H(z)$ that minimizes the MSE in a stationary signal environment:

$$\begin{bmatrix} \varphi_{ff}(0) & \varphi_{ff}(1) & \varphi_{ff}(2) & \cdots & \varphi_{ff}(N-1) \\ \varphi_{ff}(1) & \varphi_{ff}(0) & \varphi_{ff}(1) & \cdots & \varphi_{ff}(N-2) \\ \varphi_{ff}(2) & \varphi_{ff}(1) & \varphi_{ff}(0) & \cdots & \varphi_{ff}(N-3) \\ \vdots & \vdots & \vdots & \vdots & \vdots \\ \varphi_{ff}(N-1) & \varphi_{ff}(N-2) & \varphi_{ff}(N-3) & \cdots & \varphi_{ff}(0) \end{bmatrix} * \begin{bmatrix} h_0 \\ h_1 \\ h_2 \\ \vdots \\ h_{N-1} \end{bmatrix} = \begin{bmatrix} \varphi_{fd}(0) \\ \varphi_{fd}(1) \\ \varphi_{fd}(2) \\ \vdots \\ \varphi_{fd}(N-1) \end{bmatrix} \tag{8.6}$$

Thus, we have formulated the problem of minimizing the MSE as a linear least-squares problem. The coefficient matrix, called the *autocorrelation matrix* of the signal f, is known as a *Toeplitz* matrix on account of its special form. Note that each column (or row) is a rotation of the *autocorrelation vector*, φ_{ff}, which is $N \times 1$. Let Φ_{ff} denote the coefficient matrix. In MATLAB®, to create Φ_{ff} from the elements of φ_{ff}, we create an $N \times N$ *index array*. MATLAB indices must begin at one, so the elements of the index array are one plus the indices of φ_{ff} in (8.6), that is, 1 through N on the first row, 2 followed by 1 through $N-1$ in the second row, and so on, and finally N, $N-1$, …, 1 on the last row. The autocorrelation matrix, Φ_{ff}, is then created using the index array. For example, the following expressions, in which *index_mat* is the index matrix, will produce Φ_{ff} from the autocorrelation vector, *phi*, as follows:

```
N=length(phi);
q=[phi(N:-1:2);phi];
ix1=[0:N-1]'*ones[1,N];
ix2=ones(N,1)*[N:-1:1];
index_mat=ix1+ix2
Phi_ff=q(index_mat)
```

(8.7)

By running this algorithm with "phi" equal to a column vector of length $N = 3$ or 4 and allowing q and *index_mat* to be printed, one can observe how the index matrix is used to create Φ_{ff} from the correlation vector, *phi*. For example, with *phi* equal [90, 80, 70]′ and therefore *q* equal [70, 80, 90, 80, 70]′,

```
index_mat=
           3   2   1
           4   3   2
           5   4   3
Phi_ff     =
           90   80   70
           80   90   80
           70   80   90
```
(8.8)

The solution of (8.6) is the impulse response vector, *h*, of the optimal causal linear system that minimizes the MSE. If the system is nonrecursive, then $h = b$, that is, the elements of *h* are the weights of the optimal FIR filter. If the system is recursive, then *h* is the inverse transform of $B(z)/A(z)$, and the solution for *a* and *b* is, in general, a nonlinear problem. Therefore, we will restrict our discussion to the causal FIR configuration in Figure 8.6 with impulse response $h = b$, and *N* in (8.6) becomes the length of the *b vector.*

We can write (8.4) and (8.6) using vector and array notation, again using the notation where $\varphi_{xy}(n)$ stands for the *n*th of *N* elements of column vector φ_{xy}, and assuming $b = h$ is also a column vector of length *N*. The two vector expressions are, respectively, as follows:

$$\begin{aligned} (8.4):\quad & \text{MSE} = \varphi_{dd}(0) + b'\Phi_{ff}b - 2b'\varphi_{fd} \\ (8.6):\quad & \Phi_{ff}b = \varphi_{fd} \end{aligned} \tag{8.9}$$

An algorithm due to Levinson[19,20] uses the special properties of Φ_{ff} to solve for *b* in (8.6). If Levinson's algorithm is not available, the methods of Chapter 2 may be used. The solution without the use of Levinson's algorithm and the resulting minimum MSE are given in Table 8.1, where we use b_{opt} to represent the solution to (8.6), that is, the optimal weight vector.

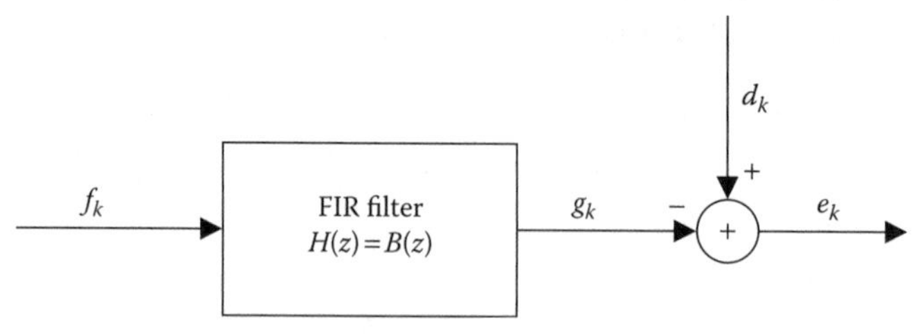

FIGURE 8.6
The same essential ingredients as in Figure 8.5, but now with $H(z)$ in the form of an FIR filter with weights $b = [b_0, b_1, \ldots, b_{N-1}]'$.

TABLE 8.1

Least-Squares Equations Using Correlation: Algebraic and MATLAB®

Mean-squared error (MSE)	$\text{MSE} = \varphi_{dd}(0) + b'\Phi_{ff}b - 2b'\varphi_{fd}$ `MSE=phi_dd(1)+b'*Phi_ff*b-2*b'*phi_fd`
Optimal weights	$b_{\text{opt}} = \Phi_{ff}^{-1}\varphi_{fd}$ `bopt=Phi_ff\phi_fd`
Minimum MSE	$\text{MSE}_{\min} = \varphi_{dd}(0) - \varphi'_{fd}b_{opt}$ `MSEmin=phi_dd(1)-phi_fd'*bopt`

The result for the minimum MSE is proved by substituting the optimal weight expression into the MSE and noting that $\Phi'_{ff} = \Phi_{ff}$ and $[\Phi_{ff}^{-1}\varphi_{fd}]' = \varphi'_{fd}\Phi_{ff}^{-1}$.

We note in these equations that the MSE is, by its nature, always positive, and therefore, because it is a quadratic function of the weights, it describes a bowl-shaped surface in $(N + 1)$ dimensional Cartesian space. Thus, the MSE must have the single global minimum given by the optimal weight vector. This minimum might be distributed due to a nonsingular correlation matrix, that is, the bowl might have a flat bottom, but local minima cannot exist.

For some system design problems, the correlation vectors are known or assumed exactly. In other applications, they must be estimated. In still other applications, they may drift slowly with time. This leads to the design of *adaptive* signal-processing systems, which is the subject of Chapter 9. Adaptive systems are systems that adjust the weight vector, b, continually as they continuously seek the solution to (8.6).

8.4 A Design Example

We now consider a simple example illustrating system design via the minimum MSE. Perhaps the easiest type of system to understand and to solve is the *least-squares predictor* with a simple periodic input, an example of which is shown in Figure 8.7. The predictor is an example of the upper diagram in Figure 8.1, with the delay (m) set to one time step. It is called a *one-step predictor.* For the signal s_k, we use a sine wave:

$$s_k = \sqrt{2}\sin(2\pi k/12); \quad -\infty < k < \infty \tag{8.10}$$

The correlation functions in (8.6) are found by averaging over one cycle (12 samples) of s_k. Using the fourth line of Table 1.2 in Chapter 1, we have

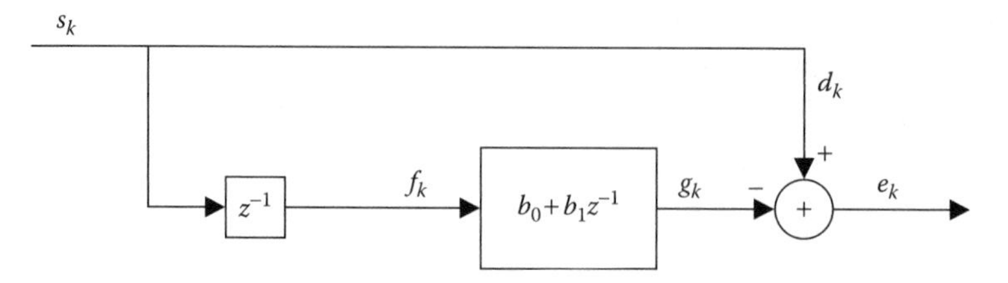

FIGURE 8.7
A simple example of the linear predictor in Figure 8.1, known as a "one-step predictor with two weights."

$$\varphi_{ff}(n)=\frac{2}{12}\sum_{k=0}^{11}\sin\left(\frac{2\pi k}{12}\right)\sin\left(\frac{2\pi(k+n)}{12}\right)=\cos\left(\frac{2\pi n}{12}\right)$$
$$\varphi_{fd}(n)=\frac{2}{12}\sum_{k=0}^{11}\sin\left(\frac{2\pi(k-m)}{12}\right)\sin\left(\frac{2\pi(k+n)}{12}\right)=\cos\left(\frac{2\pi(m+n)}{12}\right) \tag{8.11}$$

With the delay (m) set at one, these computations give us the following results for the present example:

$$\Phi_{ff}=\begin{bmatrix}1.0000 & 0.8660\\ 0.8660 & 1.0000\end{bmatrix};\quad \varphi_{fd}=\begin{bmatrix}0.8660\\ 0.5000\end{bmatrix} \tag{8.12}$$

We also note that, in this example, $\varphi_{dd}=\varphi_{ff}$, which is the first row or column of Φ_{ff}. Thus, the MSE in Table 8.1 for this example is

$$\text{MSE}=1+\begin{bmatrix}b_0 & b_1\end{bmatrix}\cdot\begin{bmatrix}1.0000 & 0.8660\\ 0.8660 & 1.0000\end{bmatrix}\cdot\begin{bmatrix}b_0\\ b_1\end{bmatrix}-2\begin{bmatrix}b_0 & b_1\end{bmatrix}\cdot\begin{bmatrix}0.8660\\ 0.5000\end{bmatrix} \tag{8.13}$$

The MSE, a quadratic function of b, is plotted in Figure 8.8. The optimal weight values and the minimum MSE are, in accordance with Table 8.1, as follows:

$$b_{\text{opt}}=\Phi_{ff}^{-1}\varphi_{fd}=\begin{bmatrix}\sqrt{3}\\ -1\end{bmatrix};\quad \text{MSE}_{\min}=\varphi_{dd}(0)-\varphi_{fd}'b=0 \tag{8.14}$$

The surface in Figure 8.8 reaches zero at the point where the heavy lines cross, that is, where the weights are at their optimal values. Thus, the one-step predictor with optimal weights exactly cancels the signal, s_k, and the error, e_k, is zero.

The MSE is illustrated in Figure 8.8 as a three-dimensional bowl-shaped surface. A *contour plot* of the MSE is illustrated in Figure 8.9. Most of the time, the contour plot is preferable for analysis. It shows clearly how the MSE

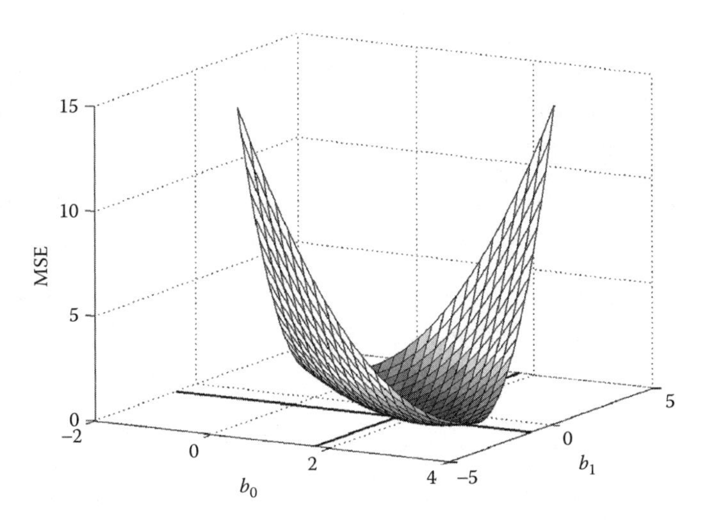

FIGURE 8.8
Mean-squared error plot for the example in Figure 8.7 with $s_k = \sin(2\pi k/12)$.

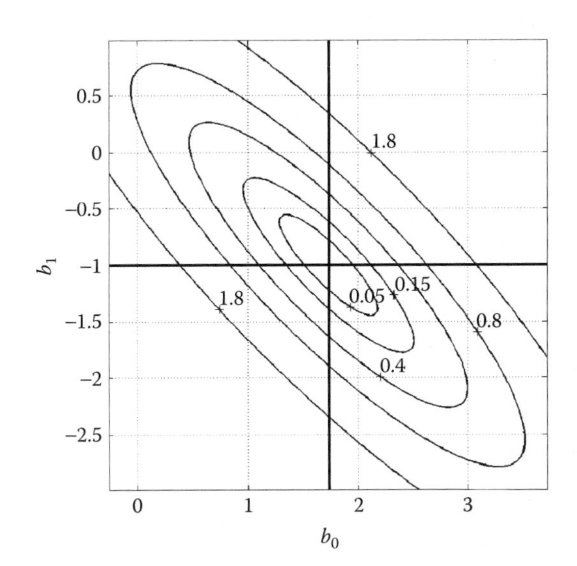

FIGURE 8.9
Contour plots, as if viewing a portion of the MSE in Figure 8.8 from directly above.

decreases to a minimum at the optimal value of b and gives a better picture of the gradient of the MSE surface. The four contours in Figure 8.9 were plotted with the following MATLAB instructions:

```
v=[.05 .15 .4 .8 1.8];
[c,h]=contour(b0,b1,MSE,v,'k');grid;
clabel(c);
```

(8.15)

For the plot, the elements of vectors b_0 and b_1 were spaced evenly over the ranges shown, and MSE was a matrix of MSE values computed (in this case) on a grid of 150^2 evenly spaced pairs of elements of b_0 and b_1. The *clabel* function was used to label the contours.

This example is artificial, because the signal is simple and is known exactly, and also because the filter has only two weights. It is meant to illustrate the quadratic error surface in three dimensions and to show a simple derivation of the optimal weights.

As a final step in this simple example, let us observe the effect of adding a third weight, b_2, to the filter in Figure 8.7. With the third weight, the arrays in (8.12) are now

$$\Phi_{ff} = \begin{bmatrix} 1.0000 & 0.8660 & 0.5000 \\ 0.8660 & 1.0000 & 0.8660 \\ 0.5000 & 0.8660 & 1.0000 \end{bmatrix}; \quad \varphi_{fd} = \begin{bmatrix} 0.8660 \\ 0.5000 \\ 0.0000 \end{bmatrix} \tag{8.16}$$

In this case, Φ_{ff} is singular and has no inverse. That is, (8.16) imposes constraints on the weight vector b but does not provide one specific vector. The constraints may be expressed as

$$\begin{aligned} b_0 - b_2 &= \sqrt{3} \\ b_0 + \sqrt{3}b_1 + b_2 &= 0 \end{aligned} \tag{8.17}$$

They imply a distributed minimum of the error surface, that is, a subspace in which the MSE is everywhere equal to zero. The solution for the optimal vector with $b_2 = 0$ is seen to agree with the two-weight solution in (8.14).

8.5 Least-Squares Design with Finite Signal Vectors

In Sections 8.3 and 8.4, we dealt with stationary signals described in terms of correlation functions and derived least-squares design equations from this basis. These we summarized in Table 8.1. Now, we turn our attention to the design of FIR systems that are optimal with respect to finite signal vectors. There are many ways in which these systems may be applicable to the tasks of prediction, noise cancellation, and so on, as described in Section 8.2. Even in real-time applications, the high processing rates of DSP chips may be used in a *block processing* mode, in which the system is optimized for each block of time and processes the blocks of signals in order as they occur in real time.

When the signals in Figure 8.5 are finite vectors, each with K elements, the least-squares equations are quite similar to the equations in Table 8.1. Instead of the MSE, it is easier to use the *total squared error* (TSE) and avoid having to divide by K; that is, similar to (8.4),

$$\begin{aligned} \text{TSE} &= \sum_{k=0}^{K-1} d_k^2 + \sum_{k=0}^{K-1}\sum_{n=0}^{N-1}\sum_{m=0}^{N-1} b_n b_m f_{k-n} f_{k-m} - 2\sum_{k=0}^{K-1}\sum_{n=0}^{N-1} b_n d_k f_{k-n} \\ &= r_{dd}(0,0) + \sum_{n=0}^{N-1}\sum_{m=0}^{N-1} b_n b_m r_{ff}(m,n) - 2\sum_{n=0}^{N-1} b_n r_{df}(-n) \end{aligned} \tag{8.18}$$

Here, in place of the correlation functions of the stationary signals in (8.4), we substituted the *covariance functions* of the finite signal vectors, and these now completely determine the geometry of the least-squares design problem, just as the correlation functions did in (8.4). The covariance functions in (8.18) are as follows:

$$\begin{aligned} \text{Autocovariance: } r_{xx}(m,n) &= \sum_{k=0}^{K-1} x_{k-m} x_{k-n} = r_{xx}(n,m); \quad 0 \le (m,n) \le N-1 \\ \text{Cross-covariance: } r_{xy}(n) &= \sum_{k=0}^{K-1} x_k y_{k+n} = r_{yx}(-n); \quad 0 \le n \le N-1 \end{aligned} \tag{8.19}$$

We note that covariance is defined in different ways. These definitions are convenient here, because they apply directly to the TSE formula (8.18).

Notice that the covariance functions, as defined in (8.19), require knowledge of the signal sequences beyond the range $0 \le k \le K-1$. Specifically, from the way the functions in (8.19) are used in (8.18), the following sequences are required:

$$\begin{aligned} f &= [f_{-N+1} \quad f_{-N+2} \quad \cdots \quad f_0 \quad \cdots \quad f_{K-1}]' \\ d &= [d_0 \quad d_1 \quad \cdots \quad d_{K-1}]' \end{aligned} \tag{8.20}$$

That is, these sequences are required to produce the error vector, e, on which the TSE is based. Depending on the application, it may be appropriate to set elements f_{-K+1} through f_{-1} to zeros.

With the TSE expression (8.18), the principal results for the error surface and the least-squares weights in Table 8.1 are applicable, provided that the covariance functions in (8.19) are substituted for the correlation functions used previously. Table 8.2 is the covariance equivalent of Table 8.1, with b_{opt} again representing the optimal weight vector.

In Table 8.2, as in Table 8.1, notice that the algebraic indices begin at zero and the MATLAB indices at one, and also that r_{df} in (8.18) is replaced with r_{fd} here in order to make the range of the index (n) positive. Also, the *autocovariance matrix*, R_{ff}, replaces the autocorrelation matrix in Table 8.1. Notice that

TABLE 8.2

Least-Squares Equations Using Covariance: Algebraic and MATLAB®

Total-squared error (TSE)	$\text{TSE} = r_{dd}(0,0) + b'R_{ff}b - 2b'r_{fd}$ `TSE=d'*d + b'*R_ff*b - 2*b'*r_fd`
Optimal weights	$b_{\text{opt}} = R_{ff}^{-1}r_{fd}$ `bopt=R_ff\r_fd`
Minimum TSE	$\text{TSE}_{\min} = r_{dd}(0,0) - r'_{fd}b_{\text{opt}}$ `TSEmin=d'*d - r_fd'*bopt`

the autocovariance matrix is symmetric, that is, $r_{ff}(m, n) = r_{ff}(n, m)$, but is not Toeplitz, that is, the rows are, generally, not permutations of each other.

In summary, least-squares design using covariance is almost the same as least-squares design using correlation, the only essential difference being in the way we extend the ends of the signal vectors. Before considering more applications of these methods, we describe methods for computing the correlation and covariance functions for given signal vectors.

8.6 Correlation and Covariance Computation

In this section, we consider two simple methods for estimating correlation or computing covariance that are useful in a variety of applications but do not cover all cases. Here, we assume the functions are estimated from samples of the signals; however, especially in the case of correlation, this may not be true. The correlation functions may be derived theoretically from assumed properties of the signals, or they may be computed using the inverse discrete Fourier transform (DFT) of the power spectrum as described in Chapter 7.

Suppose the cross-correlation function $\varphi_{xy}(n)$ is to be estimated in terms of two signal vectors, x and y. Suppose we allow these vectors to be extended periodically. Then, we may compute $\varphi_{xy}(n)$ as follows:

$$\varphi_{xy}(n) = \frac{1}{K}\sum_{k=0}^{K-1} x_k y_{k+n(\pm K)} = \varphi_{yx}(-n); \quad 0 \le |n| \le K-1 \tag{8.21}$$

The computation of $\varphi_{xy}(n)$ in this form requires K^2 products and would be lengthy if (1) all K products were required and (2) K were large. Let us assume that both (1) and (2) are true and examine the computation in terms of a DFT product. From (8.21), we may conclude

$$\varphi_{xy}(n) = \varphi_{yx}(-n) = \frac{1}{K}\sum_{k=0}^{K-1} y_k x_{k-n(\pm K)} = \frac{1}{K}\sum_{k=0}^{K-1} y_k x^r_{n-k(\pm K)}; \quad 0 \le |n| \le K-1 \tag{8.22}$$

where r is the operation in (6.4) that reverses the order of the elements of a vector. This result gives $\varphi_{xy}(n)$ in the form of a convolution, and thus, as the inverse DFT of a DFT product in accordance with (4.8). Furthermore, using $z = e^{j\omega T}$ in (6.5), we can see that the DFT of a reversed vector x^r is X', that is, the conjugate of the DFT of x. Therefore, when (4.8) is applied to (8.22), the result is

$$\boxed{\text{Correlation vector } \varphi_{xy} = \frac{1}{K}\text{DFT}^{-1}\{X'Y\}} \tag{8.23}$$

Although this result, which is analogous to (4.8), but for correlation instead of convolution, is of general interest, it is only useful if the *entire* autocorrelation vector is required. If K is the length of x and y, then a computation of N elements of φ_{xy} requires NK products. In Chapter 4, (4.81), we argued that the computation of (8.23) requires $12K\log_2(2K)$ real products. Therefore, the computation of $\varphi_{xy}(1{:}N)$ becomes more efficient using (8.23) only when

$$NK > 12K\log_2(2K), \quad \text{or} \quad N > 12\log_2(2K) \tag{8.24}$$

In least-squares applications, this condition is not very likely when K is large. In typical applications, K is at least an order of magnitude greater than N. Therefore, we usually prefer to compute correlation and covariance computations in the form of (8.21) rather than (8.23).

Three correlation functions are coded in the form of (8.21). The first is *autocorr*, which is illustrated with MATLAB expressions in (8.25). The arguments are x, the signal vector, *type*, which is set to zero to extend x with zeros and one to extend x periodically, and N, the number of values of φ_{xx} to compute.

Autocorrelation example:

`x=[x1 x2 x3 x4]; phi=autocorr(x,type,3)`

	type=0	type=1
phi(1)	$(x_1x_1+x_2x_2+x_3x_3x+x_4x_4)/4$	$(x_1x_1+x_2x_2+x_3x_3+x_4x_4)/4$
phi(2)	$(x_1x_2+x_2x_3+x_3x_4)/4$	$(x_1x_2+x_2x_3+x_3x_4+x_4x_1)/4$
phi(3)	$(x_1x_3+x_2x_4)/4$	$(x_1x_3+x_2x_4+x_3x_1+x_4x_2)/4$

(8.25)

The second correlation function is *crosscorr*, which has similar arguments and is illustrated with MATLAB expressions in (8.26).

Cross-correlation example: $x=[x_1\ x_2\ x_3\ x_4]$; $y=[y_1\ y_2\ y_3\ y_4]$; phi=crosscorr(x,y,type,3)		
	type=0	type=1
phi(1)	$(x_1y_1+x_2y_2+x_3y_3+x_4y_4)/4$	$(x_1x_1+x_2x_2+x_3x_3+x_4x_4)/4$
phi(2)	$(x_1y_2+x_2y_3+x_3y_4)/4$	$(x_1y_2+x_2y_3+x_3y_4+x_4y_1)/4$
phi(3)	$(x_1y_3+x_2y_4)/4$	$(x_1y_3+x_2y_4+x_3y_1+x_4x_2)/4$

(8.26)

The third correlation function is `Phi=autocorr_mat(x,type,N)`, which computes the N-element autocorrelation vector φ_{xx} using *autocorr* and then produces the $N \times N$ autocorrelation matrix $\boldsymbol{\Phi}_{xx}$ using code similar to (8.7).

In the case of covariance as it is used in least-squares design, there is not much reason to compute an autocovariance vector. Therefore, there are just two MATLAB functions, the first being `R=autocovar_mat(x,N)`, which implements the computation of the elements $r_{xx}(m,n)$ in (8.19). For covariance computation, the signals (x in this case) are always extended with zeros. If nonzero values of the startup elements in (8.20) are necessary in the computation, then all signal vectors may be extended to the *left* with $N-1$ elements, and K increased to $K+N-1$. That is, in (8.20), we would append $N-1$ startup values to the *beginning* of f, and $N-1$ zeros to the beginning of d. With this understanding, the autocovariance equation in (8.19) becomes

$$r_{xx}(m,n) = \sum_{k=0}^{K-1} x_{k-m}x_{k-n} = \sum_{i=0}^{K-m-1} x_i x_{i+(m-n)}; \quad m \geq n \tag{8.27}$$

Comparing (8.21) with the second form of $r_{xx}(m, n)$, we note that as the signal vector becomes long enough to neglect end effects, the covariance function approaches K times the autocorrelation function. [The same is true for the cross-covariance function in (8.19).]

Because R is symmetric, we only need to compute elements for $m \geq n$. An example of the covariance computation using MATLAB expressions with $N=2$ is given in (8.28). As in the previous examples, the indices begin at one instead of zero.

Autocovariance example: `x=[x1 x2 x3 x4];` `R=autocovar_mat(x,N)`	
`r(1,1)`	$x_1x_1+x_2x_2+x_3x_3+x_4x_4$
`r(2,1)`	$x_1x_2+x_2x_3+x_3x_4$
`r(2,2)`	$x_1x_1+x_2x_2+x_3x_3$

(8.28)

The second cross-covariance function is `r=crosscovar(x,y,N)`. The computation is made as in the cross-covariance formula in (8.19). An example is shown in (8.29).

Cross-covariance example: `x=[x1 x2 x3 x4];` `y=[y1 y2 y3 y4];` `r=crosscovar(x,y,3)`	
`r(1)`	$x_1y_1+x_2y_2+x_3y_3+x_4y_4$
`r(2)`	$x_1y_2+x_2y_3+x_3y_4$
`r(3)`	$x_1y_3+x_2y_4$

(8.29)

With the functions described in this section and with the power of MATLAB in expressions such as those in Tables 8.1 and 8.2, it is easy to design least-squares systems for the kinds of applications described in Section 8.2. Examples are described next in Sections 8.7 through 8.10.

8.7 Channel Equalization

In this section, we consider another example of nonrecursive least-squares design in order to illustrate the use of the functions just described, and also to design a system different from the predictor designed previously. In this example, illustrated in Figure 8.10, we are to design a least-squares equalizer for an "unknown channel," which, in this case, is represented by an all-pole transfer function given by

$$U(z)=\frac{e^{-0.1}\sin(0.1\pi)z^{-1}}{1-2e^{-0.1}\cos(0.1\pi)z^{-1}+e^{-0.2}z^{-2}} \tag{8.30}$$

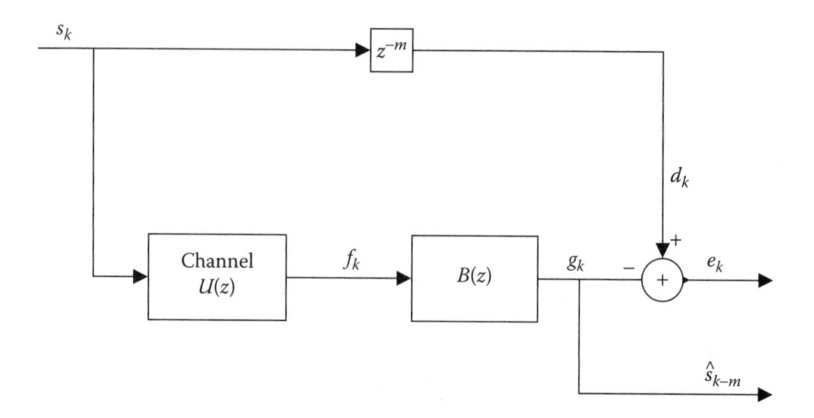

FIGURE 8.10
Equalizer design example. The "unknown" channel is specified by $U(z)$ in (8.30). The problem is to adjust the weights of $B(z)$ so that the squared difference between $\hat{s}_{k-m}$ and s_k is minimal.

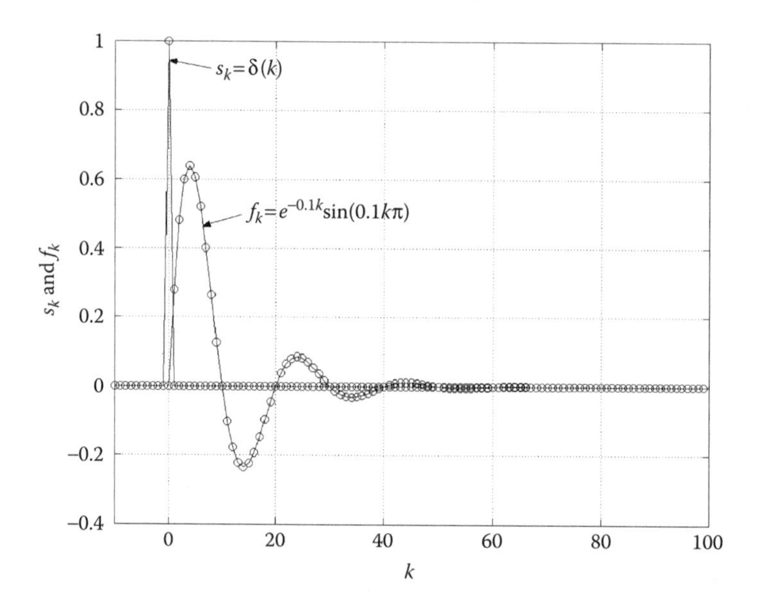

FIGURE 8.11
Signals in the equalizer example shown in Figure 8.10. The input signal, s_k, is a unit impulse, and f_k is the response of the unknown channel to s_k.

Unlike Figure 8.3, which is more realistic, we are not including noise in Figure 8.10. This description of the unknown channel assures us a simple design of the all-zero equalizer, $B(z)$, with zeros that must cancel the poles of $U(z)$. To obtain the equalizer, we assume $U(z)$ is unknown, but that we have sequences of the signals s_k and f_k, the latter being the response of $U(z)$ to the input signal, s_k. These are shown in Figure 8.11, where s_k is seen to be an impulse at $k = 0$, and f_k is the impulse response, that is, the inverse transform of $U(z)$, which is a decaying sinusoid. (See Table 4.2, lines D and 4.) In MATLAB, we would express the signal vectors as follows:

```
s=[1, zeros(1,K-1)];
f=exp(-.1*k).*sin(.1*k*pi);
d=[zeros(1,m), s(1:K-m)];
```
(8.31)

Notice that d is created by appending m zeros to the beginning of s, affecting the delay of m samples in Figure 8.10.

Thus, we are designing an equalizer that corrects the impulse response of the unknown channel so that its output, $\hat{s}_{k-m}$, is a delayed least-squares approximation to the signal, s_k. Using the functions described in Section 8.6, the optimal weights and minimum TSE for given values of N and m are computed as follows (see Table 8.2):

```
Rff=autocovar_mat(f,N);
rfd=crosscovar(f,d,N);
b=Rff\rfd;
TSEmin(N+1)=d*d' + b'*Rff*b - 2*b'*rfd;
```
(8.32)

It is interesting to consider the effect of the delay (m) on the required number of weights, which is illustrated in Figure 8.12. The latter is a plot of the minimum TSE versus the number of FIR weights for different values of the delay, m. It shows first that if the delay (m) is at least 1, an equalizer that

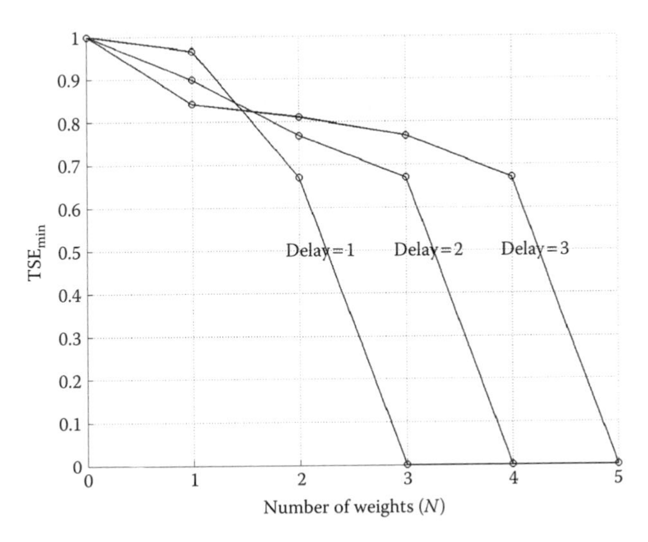

FIGURE 8.12
Minimum total squared error (TSE_{min}) versus number of optimal weights (N) in the equalizer example in Figure 8.10 for delay values 1, 2, and 3, illustrating the effect of the delay value.

exactly compensates for the channel and drives the TSE to zero is realizable. Second, the plot shows that there is also an optimal delay in systems like this, in the sense that the number of equalizer weights is minimized.

8.8 System Identification

The next example in Figure 8.13 is one in which we are trying to identify (model) an unknown system or "plant." In this case, the plant has poles, so we cannot exactly match the desired signal with an FIR filter, that is, we cannot drive the TSE to zero using $B(z)$ to model the plant. In this example, as in the previous example, the transfer functions $U(z)$ and $B(z)$ in Figure 8.13 are

$$U(z) = \frac{e^{-0.1}\sin(0.1\pi)z^{-1}}{1 - 2e^{-0.1}\cos(0.1\pi)z^{-1} + e^{-0.2}z^{-2}}$$
$$B(z) = \sum_{n=0}^{N-1} b_n z^{-n} \tag{8.33}$$

The input signal, f_k, in Figure 8.13, is a random white signal with unit power. The input sequence is given by

```
f=randn(1, K)
```
(8.34)

That is, f is a white Gaussian noise sequence with K elements and unit power.

As noted in Section 8.6, the covariance elements in the matrix R should approach scaled values of the correlation coefficients, $\varphi_{ff}(n)$, as K increases.

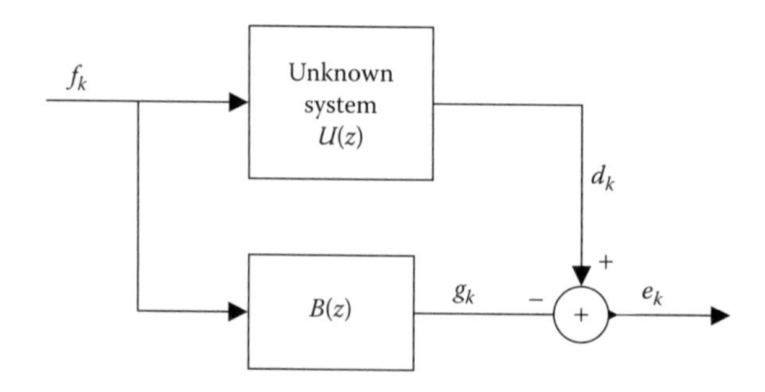

FIGURE 8.13
System identification example, in which $B(z)$ becomes a model of the unknown system when the total squared error is minimized.

Because f_k is a white random sequence with independent samples and unit power, we have

$$\varphi_{ff}(n) = \begin{cases} 0; & n \neq 0 \\ 1; & n = 0 \end{cases} \tag{8.35}$$

Thus, the theoretical autocorrelation matrix is diagonal, that is, $\Phi_{ff} = I$ theoretically, and from Table 8.1, the optimal weights are given by the cross-correlation vector; that is,

$$b = \varphi_{fd} \tag{8.36}$$

Now let us multiply the transfer function relationship for $U(z)$, in Figure 8.13, by the conjugate of the transform of f to obtain

$$F'(z)D(z) = F'(z)F(z)U(z) \tag{8.37}$$

Using $\Phi_{fd}(z)$ and $\Phi_{ff}(z)$ to represent the z-transforms of the corresponding correlation functions, we can apply (8.23) through (8.37), and, noting that φ_{ff} is a unit impulse, and therefore, $\Phi_{ff}(z) = 1$, obtain

$$\Phi_{fd}(z) = \Phi_{ff}(z)U(z) = U(z) \tag{8.38}$$

Therefore, φ_{fd}, and thus the theoretical value of b, is the impulse response of $U(z)$, which, as we saw in (8.31), is

$$b_{n(\text{theoretical})} = e^{-0.1n}\sin(0.1n\pi) \tag{8.39}$$

Thus, with a broadband input signal, the least-squares weight vector b tends to match a finite portion of the impulse response of the unknown system. One might suppose that a very large length (K) of the random input sequence would be required for the covariance solution to match the correlation solution, but in fact, a short sequence suffices. As an example, the sequences of length $K = 64$ illustrated in Figure 8.14 were used to derive a least-squares model of size $N = 50$. These were produced with the following expressions:

```
K=64;
randn('seed',0);
f=randn(1,K);
c=exp(-0.1);
d=filter([0 c*sin(0.1*pi)],[1 -2*c*cos(0.1*pi) c^2], f);
```

(8.40)

The least-squares weight vector, b, is computed using the first three expressions in (8.32), with $N = 50$ and compared in Figure 8.15 with samples of the continuous plant impulse response, that is, the theoretical weights in (8.39).

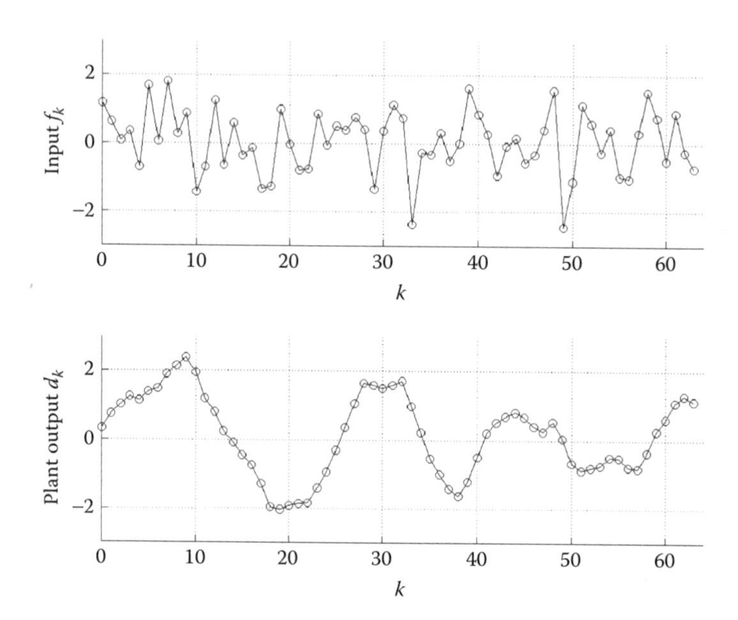

FIGURE 8.14
Signal vectors f and d in the system identification example.

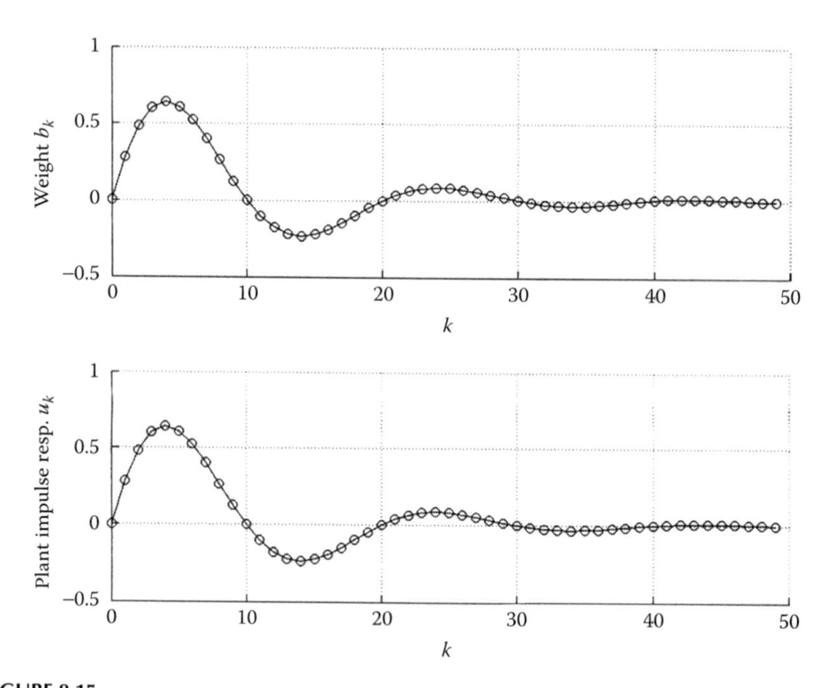

FIGURE 8.15
Impulse responses of the model, $B(z)$, and the unknown system. In this noiseless example, the model is nearly perfect in the sense that the TSE is nearly zero.

The two plots are nearly identical, and the minimum TSE in this example is on the order of 10^{-5}, indicating a nearly perfect model. Notice that the FIR filter length $N = 50$ covers essentially all of the nonzero portion of the impulse response of the unknown system.

8.9 Interference Canceling

Our next example illustrates the interference-canceling concept in Figure 8.4 in the communication situation shown in Figure 8.16. A caller is speaking in a construction environment, where the acoustic noise is so loud that when it is picked up by his telephone and added to his speech, the combined signal becomes unintelligible to the listener. Luckily, a DSP engineer is on hand to design the least-squares noise canceller, which operates by processing correlated noise picked up on a microphone in the same environment. The noise canceller works by *subtracting* the processed correlated noise, g_k, from the combined signal, $s_k + n_k$.

The reason for the necessity to subtract the noise instead of using a noise-canceling filter is shown in Figure 8.17. A 1-second sample of the speech signal, $s(t)$, is shown at the left along with the power density spectrum of the entire signal. The construction noise, $n(t)$, and its spectrum are on the right.

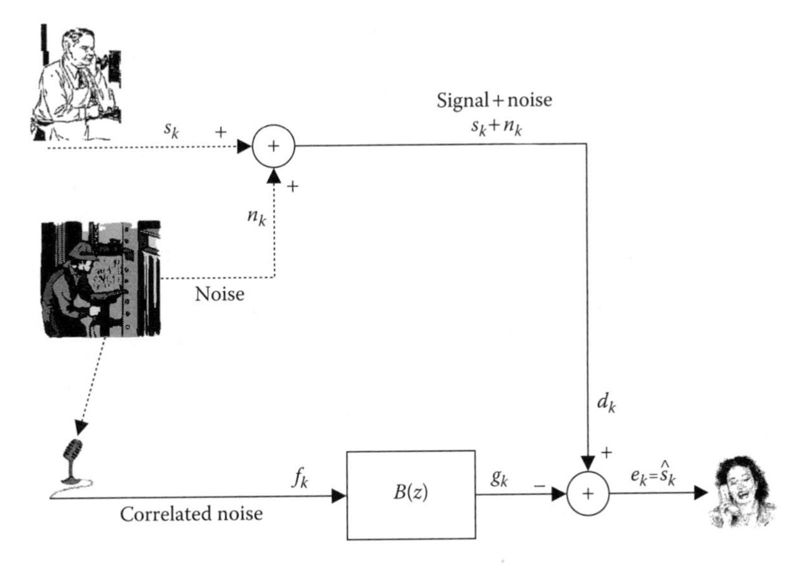

FIGURE 8.16
Least-squares interference canceling. Loud shop noise is added to the speaker's voice. Using a correlated version of the noise without the speech, the interference-canceling filter, $B(z)$, produces an output that cancels most of the noise. The filter weights are chosen to minimize the total squared value of the output, e_k.

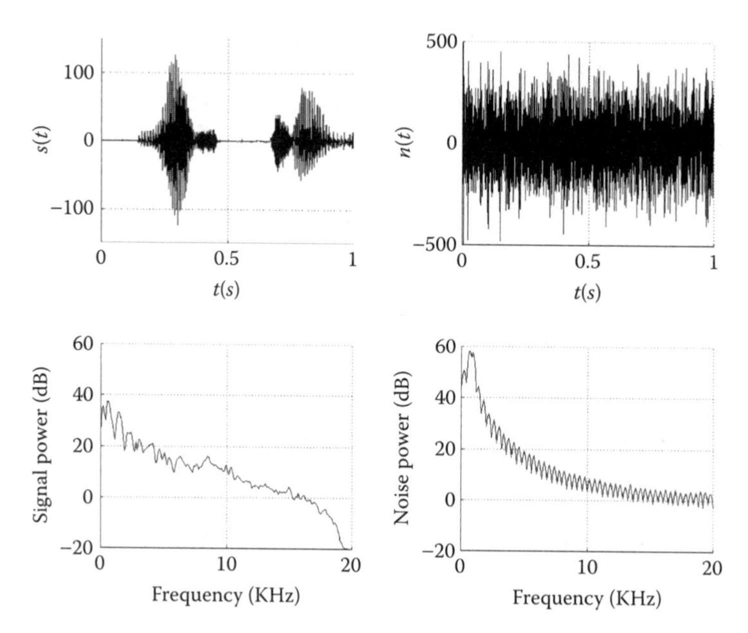

FIGURE 8.17
Speech signal, $s(t)$, and noise, $n(t)$, with power density spectra, for the interference-canceling example in Fig. 8.16. The signal in this case is around 20 dB below the noise level.

The sampling rate in both cases is 40 KHz. We can see that filtering the combined signal, $s(t) + n(t)$, to improve the SNR might be possible to some degree but would also result in the loss of some of the signal. If $s(t)$ and $n(t)$ are independent, the configuration in Figure 8.16 should be a better choice for improving the SNR.

The combined signal is shown in Figure 8.18, in which we can observe that the low SNR in this case tends to obscure the speech component almost completely to the eye. The human ear is more adept than the eye at separating the two components, but even a good listener would lose most of the speech when this much noise is added.

Short (0.1-second) segments of the noise, $n(t)$, and correlated noise, $f(t)$, are shown in Figure 8.19. In this idealized example, the correlated noise is a linear transformation of the noise. In reality, we must usually accept at least a small degree of independence between the measured noise and the actual noise added to the signal.

As in previous examples, the optimal weight vector computation is described in Table 8.2. The required MATLAB expressions, in which N is the number of weights, are

```
Rff=autocovar_mat(f,N);
Rfd=crosscovar(f,s+n,N);
b=Rff\Rfd;
```
(8.41)

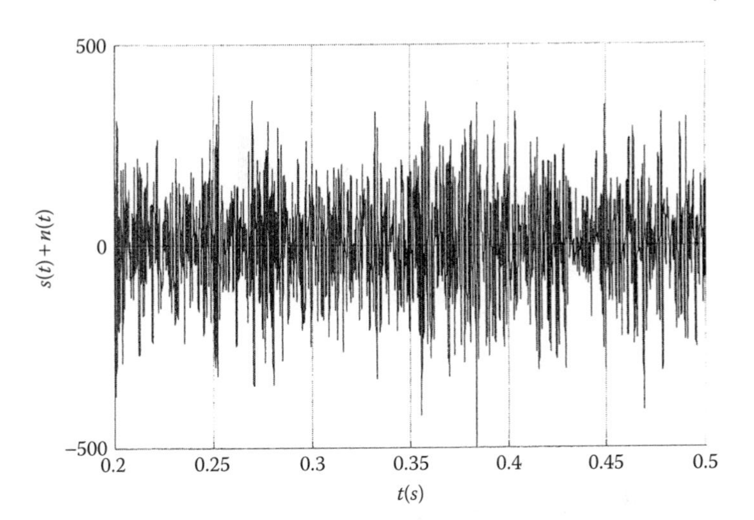

FIGURE 8.18
A segment of signal plus noise, illustrating a poor signal-to-noise ratio.

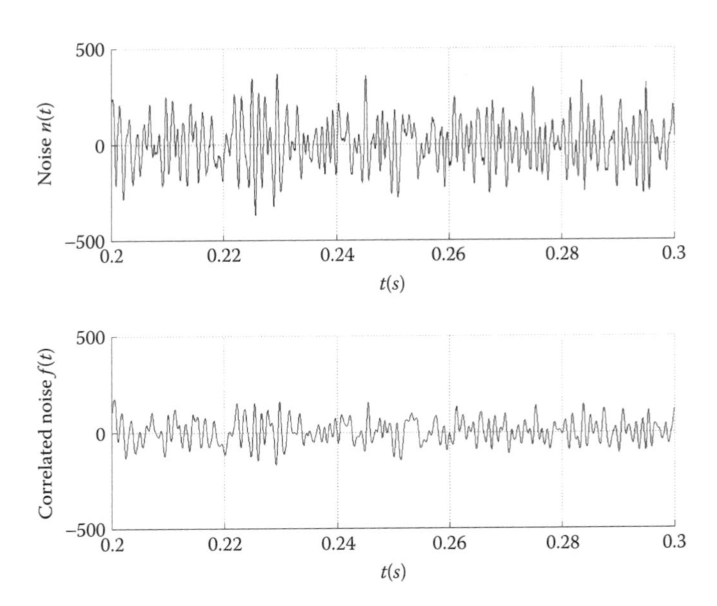

FIGURE 8.19
Segments of noise and correlated noise.

The TSE, computed as in Table 8.2 for increasing values of N, is plotted in dB in Figure 8.20. The TSE is seen to decrease slowly after $N = 4$ or 5. In practice, because quantities such as speech quality or image quality require subjective judgments, the choice of N may best be made on the basis of "best listening" or "best viewing," rather than on the basis of the TSE.

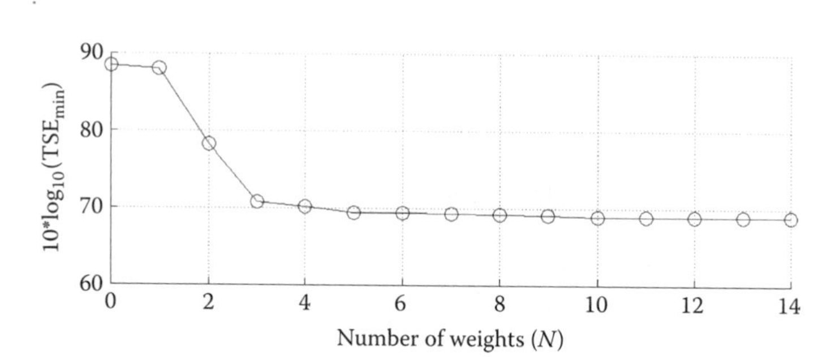

FIGURE 8.20
Minimum TSE versus number of weights (N) in the interference-canceling example, indicating that the SNR increases only slightly as N increases above 4 or 5.

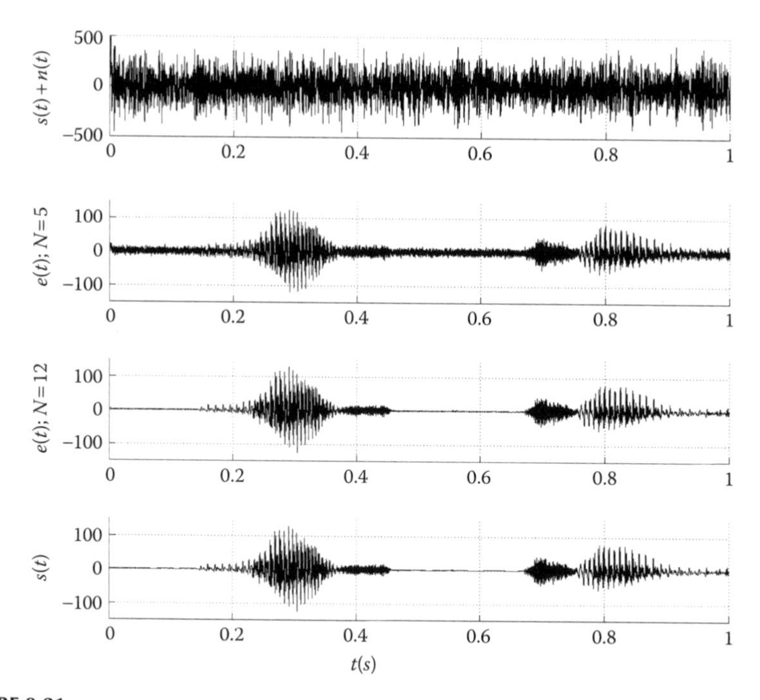

FIGURE 8.21
Final plot for the interference-canceling example, showing, from top to bottom, (1) the received signal without canceling, (2, 3) the received signal with canceling using $N = 5$ and 12 weights, and (4) the original speech signal without noise.

In any case, we can see that least-squares interference canceling is potentially capable of producing a significant increase in the SNR. The improvement is illustrated in the final figure, Figure 8.21, which shows first the noisy signal, $s(t) + n(t)$, then the system output, $e(t) = \hat{s}(t)$, for $N = 5$ and $N = 12$, and finally, the original speech signal, $s(t)$. Thanks to DSP, the listener in Figure 8.16 is able to hear the message.

8.10 Linear Prediction and Recovery

In Chapter 4 (Figure 4.8), we discussed how *inverse filtering* could be used to recover a filtered signal. In Figure 8.22, we illustrate an application of inverse filtering together with linear prediction in Figure 8.1. In this application, when $B(z)$ is chosen to minimize the total squared error, the predictor removes predictable (correlated) components from the input d, leaving a smaller residue, e. The residue becomes the compressed or *encoded* version of d, and the operation is called *linear predictive coding*.

The recovery of the original signal, d, from e is shown on the right in Figure 8.22. The recovery follows the description of inverse filtering in Chapter 4, and (4.44) in particular. Recovery requires only knowledge of the weight vector, b, and the prediction error vector, e. With the exception of roundoff errors, $\hat{d}$ and d should be identical.

An example of predictive compression and recovery is given in Figures 8.23 through 8.25. The upper part of Figure 8.23 is a plot of 40,000 samples of a seismic event recorded by the USGS Seismological Laboratory at Albuquerque, New Mexico. The sampling rate was 20 samples/s, so the plot spans 2000 s. The lower part of Figure 8.23 is a plot of the prediction error using a predictor with $N = 6$ weights. Notice the difference in vertical scales on the two plots. The prediction error is seen to be much smaller and gives the appearance of being more random than the upper plot.

It is remarkable, but not unusual, that a such a small linear predictor is capable of subtracting so much content in this way from a recorded signal. The reason for prediction working so well on many instrumentation and other waveforms is not hard to find. In Section 8.4, we saw that two weights were required to cancel a sinusoidal component at a given frequency with arbitrary amplitude and phase. Thus, it is reasonable that, when the input signal is nonwhite and is concentrated in one or a few narrow frequency bands, a small linear predictor can remove a large part of the signal and leave a small, white residue.

This *whitening* effect is seen in the power spectra of d and e in Figure 8.24. The spectrum of d is seen to be concentrated at low frequencies and to have a large total power. The spectrum of e, on the other hand, has a relatively small

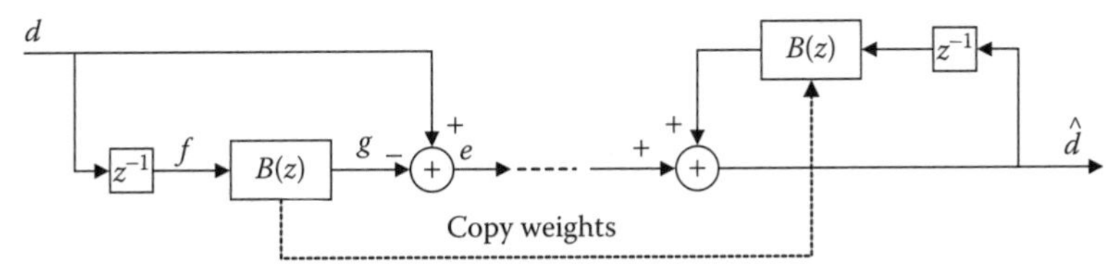

FIGURE 8.22
Waveform compression and recovery using least-squares prediction and inverse prediction. In this application, the weight vector, b, and the prediction error, e, contain all the information needed to reconstruct the original signal, d, except possibly for roundoff errors.

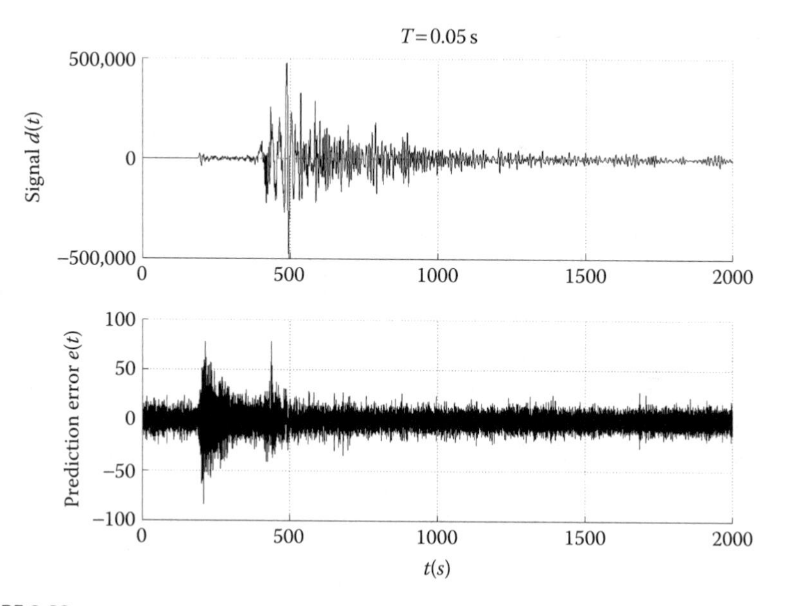

FIGURE 8.23
Short-period seismic event recorded by the USGS Albuquerque Seismological Laboratory, New Mexico, and prediction error using a least-squares predictor with $N = 6$ weights. The entire prediction error, e, is shown with the exception of $e(1{:}N)$. Note the different vertical scales on the two plots.

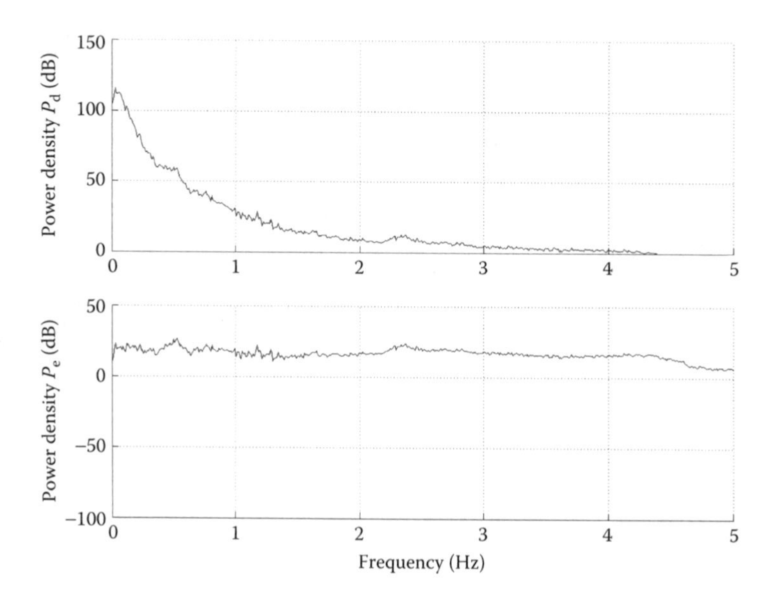

FIGURE 8.24
Power density spectra of the seismic waveform (upper plot) and the prediction error without startup values (lower plot). The vertical range is the same in both plots to emphasize the whitening effect of least-squares linear prediction.

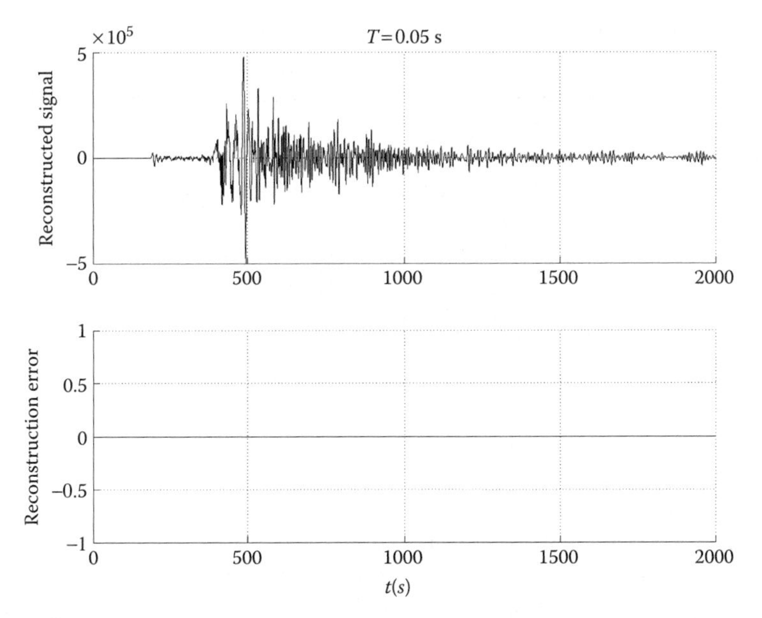

FIGURE 8.25
Reconstructed signal, $\hat{d}$, and reconstruction error, $d-\hat{d}$. In this example, even though computations are made with a floating-point processor, the accuracy is sufficient to produce a perfect reconstruction.

total power and, when plotted over the same range of 150 dB, looks almost flat (white).

As indicated in Figure 8.22, recovery requires knowledge of the weight vector, b, and the prediction error, e. The recovery process is an application of inverse filtering described in Chapter 4, (4.44). The reconstructed signal, $\hat{d}$, and the reconstruction error, $d-\hat{d}$, are plotted in Figure 8.25. In this case, the reconstruction is exact, and the error is zero.

The recovery operation and the predictive coding operation are both accomplished easily using MATLAB functions we have already described, as follows, in this case with $K = 40{,}000$ and $N = 6$:

Predictive Coding	**Recovery**
`f=[0; d(1:K-1)];`	`dhat=filter(1,[1 -b'],e)`
`Rff=autocovar_mat(f,N);`	
`rfd=crosscovar(f,d,N);`	
`b=Rff\rfd;`	
`e=d-filter(b,1,f);`	

Without going into too much detail, we can easily see the application to signal compression in this example. If integer samples of the original waveform, d, are stored or transmitted without the use of any other coding scheme, we can see in Figure 8.23 that a range of 10^6 is required, that is, 20 bits/sample. Because d consists of 40,000 samples, 800,000 bits are required. If the prediction data is used instead, then we need to store or transmit the weight vector, b, and the prediction error, e. The first N values of e (where N is the number of weights) require 20 bits each, but afterwards, as seen in Figure 8.23, a range of only 200, or 8 bits/sample, is required. Thus, the total number of compressed data bits with $N = 6$ in this example is

$$\text{Compressed bits} = 6(20) + 6(20) + 39994(8) = 320192 \tag{8.42}$$

The *compression ratio* in this case is about 2.5, meaning that 2.5 times as many bits are required to describe the signal in its original form, compared with the compressed version.

Predictive coding has been applied in this manner to many kinds of signals and data, including speech and image data, telemetry, and so on. Predictive coding can also be made *adaptive*, that is, made to adapt to changing signal statistics, as discussed later in Chapters 9 and 10.

Predictive coding is also applied in many areas besides compression. Encryption was mentioned in Chapter 4, for example, where the weight vector becomes the *encryption key*, and the prediction error is sent in the clear.

Another application, used in spectral estimation, is called *prewhitening*. If a power spectrum is estimated using the methods described in Chapter 7, and if the signal contains one or more large narrowband components, these may bias the power density estimates at other frequencies. If so, the prewhitening operation may be used to reduce the bias. Suppose x is the signal vector and y is the prediction error after passing x through a least-squares linear predictor. Then, the transform relationship is

$$Y(z) = (1 - B(z))X(z) \tag{8.43}$$

Taking the squared magnitude of both sides of (8.43) and then applying (8.23), we have

$$Y'(z)Y(z) = |1 - B(z)|^2\, X'(z)X(z), \quad \text{or} \quad \Phi_{yy}(z) = |1 - B(z)|^2\, \Phi_{xx}(z) \tag{8.44}$$

From this, we conclude

$$\Phi_{xx}(z) = \frac{\Phi_{yy}(z)}{|1 - B(z)|^2} \tag{8.45}$$

Thus, the power spectrum of x may be estimated from the spectrum of the whitened signal, y, and the power gain of the least-squares predictor.

8.11 Effects of Independent Broadband Noise

In many practical design cases, noise is added to the signals with properties that are used to design least-squares systems, that is, to the vectors d and f in Figure 8.6. The noise could be *plant noise* added to d as shown in Figure 8.2, or independent *channel noise* associated with the unknown channel and added to f as shown in Figure 8.3, or *uncorrelated noise* added to either s or f in Figures 8.4 and 8.16, and so on. Our purpose, in this section, is to derive the general effects of noise in least-squares systems, assuming the noise is independent, has a broad spectrum, and is *additive*, that is, adds to the signals. These assumptions are somewhat restrictive, but nevertheless, cover many cases of interest, and the approach used here is applicable with modifications, if the noise is not independent or does not have a flat power spectrum.

The noisy least-squares design configuration is illustrated in Figure 8.26. It is the same as Figure 8.6 with noise added to the signals f and d. Either of the two noise sequences, n or m, or both, may be present depending on the application of Figure 8.26. If the noises are independent of the signals, and of each other, the cross-correlation and cross-spectral terms are zero. That is, for example,

$$\begin{aligned}\varphi_{f+n,f+n}(i) &= E[(f_k+n_k)(f_{k+i}+n_{k+i})] \\ &= E[f_k f_{k+i}] + E[n_k n_{k+i}] = \varphi_{ff}(i) + \varphi_{nn}(i)\end{aligned} \tag{8.46}$$

The second line in (8.46) follows, because if f and n are independent, the expected cross-products in the first line are zero. Furthermore, when the noise is white, as we will assume here, $\varphi_{nn}(i)$ is zero except at $i = 0$, hence, the corresponding autocorrelation matrix, Φ_{nn}, is diagonal.

Therefore, if we use the noisy version in Figure 8.26 in place of Figure 8.6, we replace the autocorrelation functions with noisy versions as in (8.46), leaving the cross-correlation functions unchanged, because n and m are independent.

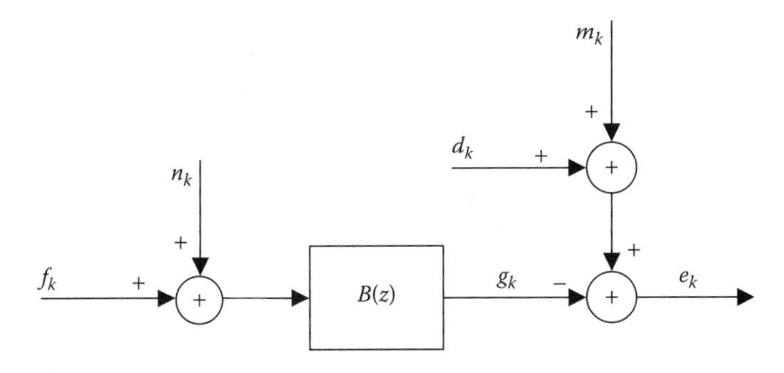

FIGURE 8.26
The essential components of the least-squares system with independent broadband noise, m_k and n_k, added to the signals d_k and f_k, respectively.

The result is the following modification of the formula (8.4) for the MSE in terms of the weight vector b:

$$\text{MSE} = \varphi_{dd}(0) + \varphi_{mm}(0) + \sum_{i=0}^{N-1}\sum_{j=0}^{N-1} b_i b_j \varphi_{ff}(i-j) + \varphi_{nn}(0)\sum_{i=0}^{N-1} b_i^2 - 2\sum_{i=0}^{N-1} b_i \varphi_{df}(-i) \quad (8.47)$$

Note that the third term in this expression is the same as the second term in (8.4), and the fourth term results from diagonal autocorrelation matrix, Φ_{nn}, of the white noise.

The result in (8.47), and all that follows here, is essentially the same using either correlation or covariance functions. In terms of covariance, Table 8.3 shows how the covariance equations in Table 8.2 are modified to include additive white noise.

From the formulas in Table 8.3, we reach the following conclusions:

- Adding independent white noise to the input signal (f) is equivalent to adding a constant to the diagonal of R_{ff}, the autocovariance matrix.
- Adding independent noise to the desired signal (d) simply adds a constant to the total MSE and to its minimum value, $\text{TSE}_{\min}$.

If R_{ff}, the autocovariance matrix, is a diagonal matrix, then cross terms of the form $b_i b_j$ with $i \neq j$ do not appear in the MSE and TSE formulas, as seen, for example, in (8.47) with $\varphi_{ff}(n-m) = 0$ for $n \neq m$, and hence, the error contours (in Figure 8.9, for example) are circular. Furthermore, the $\varphi_{nn}(0)$ term in (8.47) increases the b^2 terms in the MSE formula, thereby causing the sides of the bowl-shaped error surface to become steeper. Therefore, we also conclude the following:

- Adding independent white noise to the input signal tends to steepen the sides of the performance function and to make its contours more circular.

TABLE 8.3

Least-Squares Covariance Equations under Noisy Conditions (Figure 8.26)

Total-squared error (TSE)	$\text{TSE} = r_{dd}(0,0) + r_{mm}(0,0) + b'(R_{ff} + r_{nn}(0,0)I)b - 2b'r_{fd}$ `TSE=d'*d+m'*m+b'*(R_ff+n'*n*I)*b-2*b'*r_fd`
Optimal weights	$b = (R_{ff} + r_{nn}(0,0)I)^{-1} r_{fd}$ `b=(R_ff+n'*n*I)\r_fd`
Minimum TSE	$\text{TSE}_{\min} = r_{dd}(0,0) + r_{mm}(0,0) - r'_{fd}\, b$ `TSEmin=d'*d+m*m'-r_fd'*b`

Note: m and n are the noise sequences in Figure 8.26.

In Table 8.3, we also note that the optimal coefficients with input noise are modified by having the increased diagonal elements in the autocovariance matrix. We note that if the input is dominated by the noise sequence, n, then the optimal coefficients are just scaled values of the cross-covariance terms, that is,

$$b = r_{fd}/\varphi_{nn}(0) \tag{8.48}$$

Furthermore, in the limiting case where the input is entirely independent noise, the cross-correlation terms are zero, and therefore, the weight vector, b, is zero. We also observe in Table 8.3 that the solution for optimal coefficients is not affected by the noise m added to the desired signal, which only affects the TSE by adding a constant, $r_{mm}(0, 0)$.

Thus, in these results, we have expressed the effects of independent broadband noise on least-squares system design. The results in Table 8.3 for least-squares design using covariance functions, and similar results with correlation functions, indicate that least-squares system design is a valid and useful process, even if random noise is added to the signals involved.

Exercises

8.1 In a least-squares interference-canceling system, suppose the filter $B(z)$ has two weights, b_0 and b_1, that the total power in the desired signal power is σ_d^2, and that the correlation functions are $\varphi_{ff}(n)$ and $\varphi_{df}(n)$.

a. Express the MSE as a function of b_0 and b_1.

b. Express the optimal weights, b_0 and b_1, in terms of the correlation functions.

c. Express the minimum MSE in terms of the signal statistics.

8.2 In Exercise 8.1, suppose $E[f_k f_{k+1}] = \varphi_{ff}(1) = 0$.

a. What are the optimal weights?

b. What relationship between the signal statistics must hold in order for the minimum MSE to equal zero?

8.3 Consider a general sinusoid, $x_k = A\sin(\omega kT + \alpha)$, sampled at K samples per cycle.

a. What is the average power in x_k?

b. Express the autocorrelation function of x_k, and show that it is independent of the phase, α.

c. Show that the autocorrelation function of x_k has a period equal to that of x_k.

8.4 A long sequence, x_k, with a period equal to eight samples is

$$x_k = [\ldots,1,1,1,1,-1,-1,-1,-1,1,1,\ldots]$$

Plot x_k and the autocorrelation function, $\varphi_{xx}(n)$, for $-15 \le n \le 15$.

8.5 If the samples of two sequences x_k and y_k are *complex*, the correlation function $\varphi_{xy}(n)$ is defined to be $\varphi_{xy}(n) \equiv E[x_k' y_{k+n}]$, that is, $\varphi_{xy}(n)$ is the average product of the conjugate of x_k times y_{k+n}.

a. Show that $\varphi_{xy}(n)$ agrees with the previous definition when x_k is a real sequence.

b. Show that $\varphi_{xy}'(n) = \varphi_{yx}(-n)$.

c. Express $\varphi_{xx}(n)$ when x_k is a complex sinusoidal function, $x_k = Ae^{j(2\pi k/N+\alpha)}$. Show that $\varphi_{xx}(n)$ does not depend on the phase of x_k and has a period equal to that of x_k.

8.6 A sequence x has samples given by $x_k = A\cos(\omega k + \alpha) + r_k$, where r_k is a uniform white random variate with variance σ_r^2. Express the autocorrelation function, $\varphi_{xx}(n)$.

8.7 Suppose x_k and y_k are sinusoidal signals, either real or complex, with frequencies ω_x and ω_y. What condition must hold in order for $\varphi_{xy}(n)$ to be zero for all values of n?

8.8 The predictor in Figure 8.27 is a specific example of Figure 8.1 with delay $m = 2$.

a. Let $s_k = \sqrt{2}\cos(2\pi k/15 + \pi/4)$. Determine matrix Φ_{ff} and vector φ_{fs} for this example.

b. Express the MSE as a quadratic function of b_0 and b_1.

c. What are the optimal values of b_0 and b_1?

d. What is the minimum MSE, $E[e_k^2]$?

8.9 Do Exercise 8.8 with $s_k = \sqrt{2}\sin(2\pi k/15 + r_k)$ where r_k is a uniform white random sequence in the range [–1, 1]. Then, replace r_k with αr_k and plot the minimum MSE, $\text{MSE}_{\min}$, versus α in the range [0, 10].

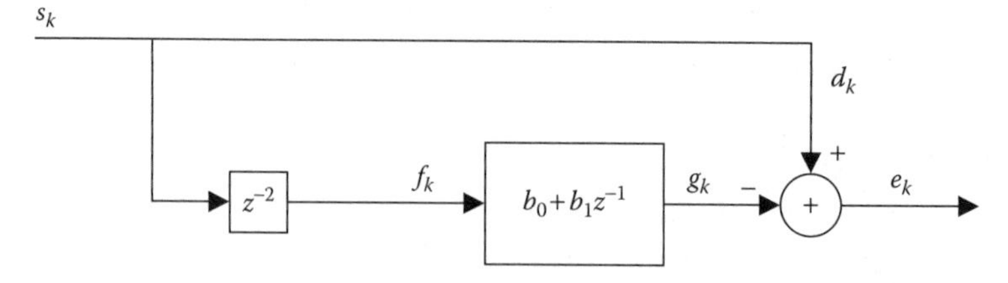

FIGURE 8.27
Two-step predictor.

8.10 In Exercise 8.8, how are matrix Φ_{ff} and vector φ_{fs} altered if we replace $(b_0 + b_1 z^{-1})$ with $(b_0 + b_1 z^{-i})$, where i is a positive integer greater than one?

8.11 Give a simple expression for the MSE in (8.4) when the input signal, f, is a white random sequence not necessarily independent of d.

8.12 A least-squares system has an adjustable "bias weight," which causes the constant c to be added to the filter output, as shown in Figure 8.28. The bias weight is used in cases where one of the signals has a nonzero mean value and the other does not. Suppose $E[f_k] = \mu_f$, $E[d_k] = 0$, $B(z)$ is a highpass filter, and the weight vector, b, is optimized. What is the optimal value of c in terms of μ_f?

8.13 The noise-canceling system in Figure 8.29 is used to cancel sinusoidal interference from a broadband signal. The signal is represented as shown by a filtered version of r_k, where r_k is a uniform white random sequence with $\sigma_r^2 = 2.0$. The noise is $n_k = \sin(2\pi k/31)$ and reference signal (correlated noise) is $f_k = \cos(2\pi k/31)$.

   a. Find the power density spectrum of s_k and the total signal power, σ_s^2.

   b. Write an expression for the MSE, $E[e_k^2]$, assuming s_k and n_k are independent.

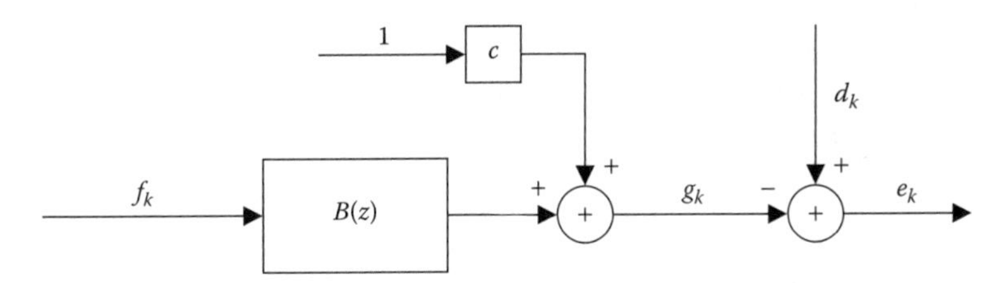

FIGURE 8.28
Least-squares system with bias weight.

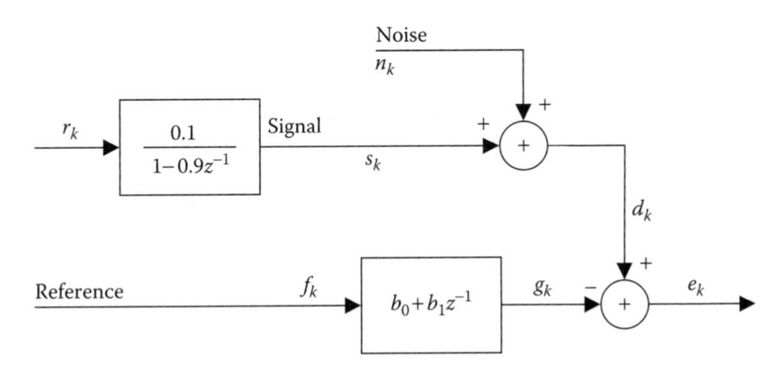

FIGURE 8.29
Noise-canceling system.

c. Find the optimal weights, b_0 and b_1.

d. Show that the minimum MSE equals σ_s^2.

8.14 For the situation described in Exercise 8.8, draw contours for MSE = 1, 0.8, 0.6, 0.4, and 0.2 on the $b_0 - b_1$ plane. (Note: the MATLAB *contour* function is useful for this purpose.)

8.15 A least-squares filter is to be used to adjust the amplitude and phase of f_k to match the amplitude and phase of d_k as shown in Figure 8.30, where f_k and d_k are described as follows:

$$f_k = 5\sin(2\pi k/32 + \pi/4); \quad d_k = \sin(2\pi k/32)$$

a. Find b_0 and b_1 assuming the signals are unbounded.

b. Using correlation functions with sequence lengths $K = 8$, 16, 32, and 64 and periodic extension, print b_0, b_1, and $\text{TSE}_{\min}$ for each value of K. Compare the results with part (a) and explain.

8.16 Do Exercise 8.13 using covariance functions. Find b_0, b_1, and $\text{TSE}_{\min}$ for K = 10, 100, 1000, and 10,000. Compare these results with the answers to Exercise 8.13c and d, and explain.

8.17 Solve the following sets of linear equations:

a. $b_0 + 2b_1 + 3b_2 = 8; \quad 4b_0 + b_1 + 6b_2 = 17; \quad 9b_0 + b_1 + b_2 = 19.$

b. $4b_0 + 3b_1 + 2b_2 + b_3 = 2; \quad 3b_0 + 4b_1 + 3b_2 + 2b_3 = 0; \quad 2b_0 + 3b_1 + 4b_2 + 3b_3 = 0; \quad b_0 + 2b_1 + 3b_2 + 4b_3 = -2.$

c. Five equations of the form $B = \Phi_{ff}^{-1}\varphi_{fd}$, in which the correlation functions are $\varphi_{ff}(n) = \cos(2\pi n/15)$ and $\varphi_{fd}(n) = 2\sin(2\pi n/15)$.

d. $b_0 + b_1 = 1; \quad b_0 + b_2 = 4; \quad b_0 + b_3 = 0; \quad b_0 + b_4 = 2; \quad b_1 + b_2 = 3.$

8.18 In this exercise, we use the modeling configuration in Figure 8.13 to design a lowpass FIR filter. The concept is illustrated in Figure 8.31. The input signal, f_k, is composed of N sine waves at frequencies evenly distributed *between* zero to half the sampling rate, that is,

$$f_k = \sum_{i=1}^{N} \sin\left(\frac{2\pi kn}{2N-1}\right); \quad k = 0, 1, \ldots, 2N$$

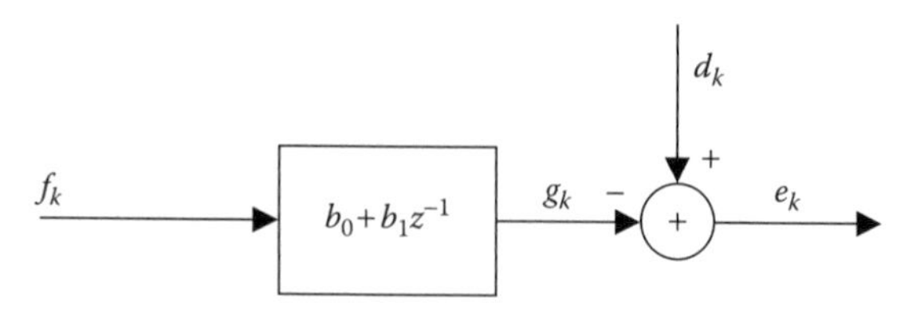

FIGURE 8.30
System with two weights.

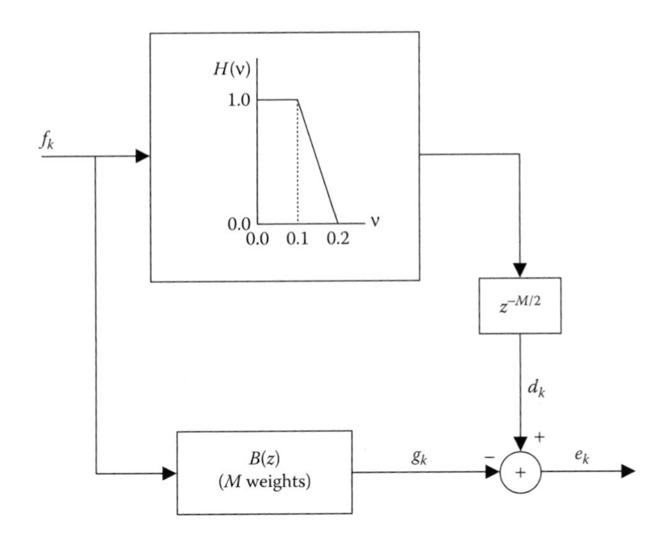

FIGURE 8.31
Least-squares finite impulse response filter design.

The desired (real) gain characteristic, $H(\nu)$, is applied to each of these components, and the filtered version of f_k is then delayed to produce d_k.

a. Write an expression for d_k similar to the expression for f_k above.
b. Write an expression for $\varphi_{ff}(n)$.
c. With $M = 20$ and $N = 199$, find the optimal weight vector, b, the minimum TSE, and the ratio of TSE_{min} to the maximum TSE.
d. In the upper of two subplots, plot the first halves of the DFTs of f and d versus ν in Hz-s.
e. In the lower subplot, make a continuous plot of the filter amplitude gain, $|B(\nu)|$, versus ν in Hz-s.

8.19 Do Exercise 8.18 using the covariance method with signal vectors just long enough to cover one fundamental period. Comment on any differences in the results, especially in the minimum TSE.

8.20 Suppose the least-squares FIR filter in Figure 8.6 is a *linear-phase* filter of the following form:

$$B(z) = b_0 + \sum_{n=1}^{N-1} b_n (z^n + z^{-n})$$

a. Express the error signal, e_k.
b. Express the MSE, $E[e_k^2]$, in terms of correlation functions. Is the MSE a quadratic function of the weights?
c. Derive a set of linear equations to solve for the optimal weights.

8.21 Design the least-squares equalizer in Figure 8.10 with $U(z)$ given by

$$U(z) = \frac{1}{1 - 1.82z^{-1} + 0.81z^{-2}}$$

For s_k, use the impulse function shown in Figure 8.11, and use $K = 50$ samples. Make a plot of the minimum TSE versus the number of weights similar to Figure 8.12, and explain the results.

8.22 Design the equalizer in Exercise 8.21 with $N = 3$ weights and delay $m = 0$. For the input signal vector $s(1{:}K)$, use a white Gaussian sequence with $\sigma_s^2 = 1$ and $K = 50$. Compare your solutions for b_{opt} and TSE_{min} with the corresponding solutions in Exercise 8.21.

8.23 The *equation error* method for least-squares infinite impulse response (IIR) system identification is illustrated in Figure 8.32. Note that the method produces a combined least-squares system consisting of $B(z)$, a forward model of $P(z)$, and $1 - A(z)$, an inverse model of $P(z)$. The method could be applied to model a plant with poles and zeros, to design an IIR filter using the method described in Exercise 8.18, to equalize a channel with zeros as well as poles, and so on.

a. Assume f is a finite signal vector. In terms of z-transforms, show that the overall transfer function from f to e_q is $H_q(z) = P(z)(1 - A(z)) - B(z)$.

b. Now suppose the same plant, $P(z)$, is used in Figure 8.13, and the model in that figure, $B(z)$, is replaced with a recursive model given by

$$H(z) = \frac{B(z)}{1 - A(z)}$$

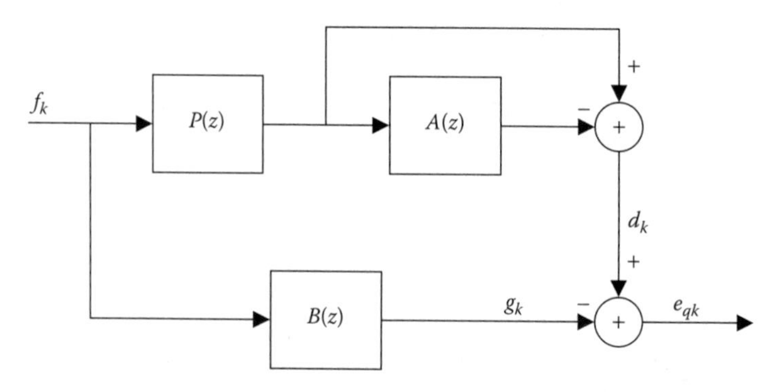

FIGURE 8.32
Infinite impulse response least-squares system identification.

Express $E_q(z)$ in Figure 8.32 in terms of $E(z)$ in Figure 8.13, and prove the following:

$$E_q(z) = (1 - A(z))E(z)$$

Thus, if the model in Figure 8.32 is perfect in the sense that $e_{qk} = 0$, then $e_k = 0$ in Figure 8.13 as well. Therefore, as the error approaches zero, $H(z)$ above becomes a valid least-squares IIR model of the plant in Figure 8.32.

c. Show how to minimize the MSE, first with a fixed and then with b fixed.

d. Let $P(z) = H(z)$ in Chapter 4, Exercise 4.5. Iterate, solving for $[b_0, b_1, b_2]$ with a fixed and then $[a_1, a_2]$ with b fixed, until $B(z)/(1 - A(z))$ converges to $P(z)$. Plot all five weights versus the iteration number as they converge to the weights in $P(z)$.

8.24 A simple direction-finding system is shown in Figure 8.33. A plane wave traveling at velocity c m/s arrives with direction angle θ at two receptors spaced x_0 meters apart, such that

$$s_{0k} = A\cos\left(\frac{2\pi k}{36}\right); \quad s_{1k} = A\cos\left(\frac{2\pi(k-\Delta)}{36}\right)$$

where $\Delta = (x_0 \sin\theta)/c$ seconds, c being the speed of light in m/s. The second signal, s_{1k}, is sent through a *quadrature filter*, in which there is a 90-degree (nine-sample) phase delay from b_0 to b_1.

a. Express the optimal weights in terms of the signal direction, θ.

b. Describe how the system could be used as a direction finder, that is, to acquire a signal and determine the direction θ.

c. Comment on the effects of adding mutually independent white noise at the two inputs.

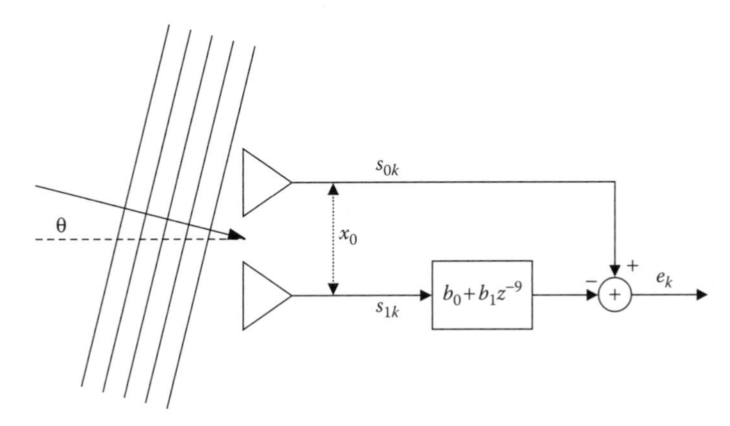

FIGURE 8.33
Direction finder.

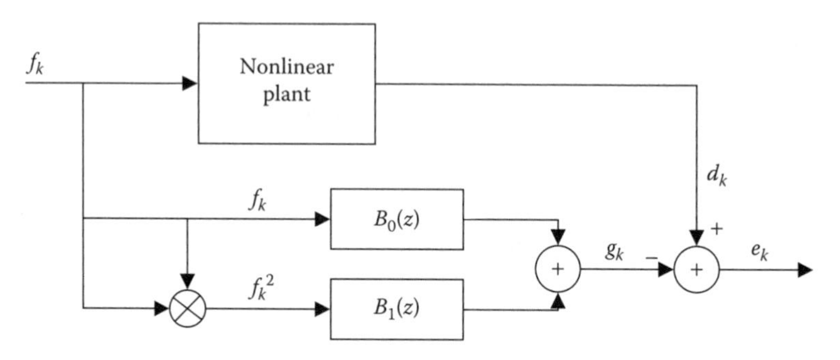

FIGURE 8.34
Least-squares nonlinear modeling.

8.25 A *nonlinear* modeling system is shown in Figure 8.34. The plant being modeled is nonlinear, and the system attempts to construct a least-squares *polynomial model* of the plant. In this case, the degree is 2, but it is easy to see how higher degrees of f_k could be appended to the bottom of the figure and added to g_k. A bias weight could also be added as in Exercise 8.12 to complete the polynomial.

   a. Derive an expression for the TSE similar to (8.18) for Figure 8.34, and show that the error surface is a quadratic function of the weights having a single global minimum.

   b. Prove that the solution for optimal weights amounts to solving a set of linear equations.

8.26 In Exercise 8.13, suppose uniform white noise with power equal to 3 is added to the reference input.

   a. Write the revised expression for the MSE.

   b. Find the optimal weights.

   c. Compute the minimum MSE and compare the result with σ_s^2.

References

1. Haykin, S. S. 1991. *Adaptive Filter Theory*. 2nd ed. Englewood Cliffs, NJ: Prentice Hall.
2. Widrow, B., and Stearns, S. D. 1985. *Adaptive Signal Processing*. Englewood Cliffs, NJ: Prentice Hall.
3. Rabiner, L. R., and Schafer, R. W. 1978. *Digital Processing of Speech Signals*. Chap. 8. Englewood Cliffs, NJ: Prentice Hall.
4. Franklin, G. F., and Powell, J. D. 1980. *Digital Control of Dynamic Systems*. Reading, MA: Addison-Wesley.
5. Marple, S. L. 1987. *Digital Spectral Analysis with Applications*. Englewood Cliffs, NJ: Prentice Hall.

6. Kay, S. 1987. *Modern Spectral Estimation: Theory and Application*. Englewood Cliffs, NJ: Prentice Hall.
7. Bordley, T. E. 1983. Linear predictive coding of marine seismic data. *IEEE Trans Acoust Speech Signal Process* 31(4):828–35.
8. Clark, G. A., and Rodgers, P. W. 1981. Adaptive prediction applied to seismic event detection. *Proc IEEE* 69(9):1166–8.
9. Zeidler, J. R., et al. 1978. Adaptive enhancement of multiple sinusoids in uncorrelated noise. *IEEE Trans Acoust Speech Signal Process* 26(3):240–54.
10. Itakura, F., and Saito, S. 1971. Digital filtering techniques for speech analysis and synthesis. *Proceedings of the 7th International Congress on Acoustics,* p. 261. Budapest.
11. Makhoul, J. 1973. Spectral analysis of speech by linear prediction. *IEEE Trans Audio Electroacoust* 21(3):140–8.
12. Lucky, R. W. 1966. Techniques for adaptive equalization of digital communication systems. *Bell System Tech J* 45:255.
13. Gersho, A. 1969. Adaptive equalization of highly dispersive channels for data transmission. *Bell System Tech J* 48:55.
14. Ives, R. 2001. Near lossless compression of SAR imagery using gradient adaptive lattice filters. *Proc 35th Asilomar Conf Signals Syst Comput* 1:663–6.
15. Griffiths, L. J., Smolka, F. R., and Trembly, L. D. 1977. Adaptive deconvolution: A new technique for processing time-varying seismic data. *Geophysics* 42(4):742.
16. Widrow, B., et al. 1975. Adaptive noise canceling: Principles and applications. *Proc IEEE* 63(12):1692–1716.
17. Kailath, T., ed. 1974. Special issue on system identification and time series analysis. *IEEE Trans Automat Contr* 19(6):638–40.
18. Widrow, B., Titchener, P. P., and Gooch, R. P. 1981. Adaptive design of digital filters. *Proc IEEE Int Conf Acoust Speech Signal Process'* 6:243–6.
19. Levinson, N. 1946. The Wiener RMS error criterion in filter design and prediction. *J Math Phys* 25:261–78.
20. Zohar, S. 1969. Toeplitz matrix inversion: the algorithm of W. F. Trench. *J Assoc Comput Mach* 16(4):592–601.

Further Reading

Atal, B. S., and Hanauer, S. L. 1971. Speech analysis and synthesis by linear prediction of the speech wave. *J Acoust Soc Amer* 50(2B):637–55.

Blahut, R. 1985. *Fast Algorithms for Digital Signal Processing*. Reading, MA: Addison-Wesley.

Jain, A. 1989. *Fundamentals of Image Processing*. Englewood Cliffs, NJ: Prentice Hall.

Orfanidis, S. 1985. *Optimum Signal Processing: An Introduction*. New York: Macmillan.

Saywood, K. 1996. *Introduction to Data Compression*. San Francisco, CA: Morgan Kaufman.

Wiener, N. 1949. *Extrapolation, Interpolation and Smoothing of Stationary Time Series with Engineering Applications*. New York: John Wiley & Sons.

9

Adaptive Signal Processing

9.1 Introduction

In this chapter, which introduces a type of system we have not yet considered, we will be relying on the discussions of linear least-squares systems in Chapter 8. Adaptive signal processing systems, as introduced here, are time-varying versions of the systems in Chapter 8, in which the weights are varied to allow the systems to adapt to trends or variations in the statistical properties of *nonstationary* signals.

Adaptive signal processing systems appear in many different forms, including all of the different applications mentioned in Chapter 8. The diversity has been promoted by developments in microelectronics that have greatly increased the amount of computing possible in real-time digital signal processing (DSP) systems. Seismic signals at 10^2 Hz and below, speech and acoustic signals at 10^2–10^5 Hz, electromagnetic signals at 10^7 Hz and beyond, and signals in control systems at all frequencies, as well as other similar signals appearing in time and space, are all reasonable candidates for adaptive signal processing.

The *adaptive* notion derives from the human desire to emulate living systems found in nature, which adapt to their environments using various remarkable methods that we can copy only in very limited ways. In the development of adaptive systems, the journey from concept to reality has taken us away from direct emulation, somewhat like the development of manned and robotic aircraft. An adaptive signal processor usually resembles its natural counterpart about as much as, or if anything less than, an airplane resembles a bird.

Adaptive signal processing has its roots in adaptive control and in the mathematics of iterative processes, where the first attempts were made to design systems that adapt to their environments. In this chapter, we focus on the basic theory and applications of modern adaptive signal processing. Much of what we describe here is based on the work of Bernard Widrow and his colleagues,[1–3,5,6,7,8,10,14,16,19,20,23] which began around 1960. We use

Widrow's development of the geometry of adaptation,[5] and we introduce the *least-mean-square* (LMS) algorithm as the simplest and most widely applicable adaptive processing scheme.

Adaptation involves a process known as *performance feedback*,[5] in which the performance of a system, measured in terms of the same squared error that we used in Chapter 8, is used to adjust the adaptive weights. The weights may be adjusted at each step in the sampling process, or after N steps in what is known as *block adaptive processing*.

In any adaptive signal processing-system, we are usually able to identify the basic structure and signals shown in Chapter 8, Figures 8.5 or 8.6, depending on whether the *adaptive filter* is recursive or not. Thus, we begin with the structure and signals in Figure 9.1. The variable weights are adjusted repeatedly, at regular intervals, in accordance with an *adaptive algorithm*. The adaptive algorithm usually uses, either explicitly or implicitly, the signals shown in Figure 9.1, that is, the input signal, f_k, the desired signal, d_k, and the error signal, e_k, which is the difference between d_k and g_k, the filtered version of f_k. The details of the adaptive algorithms are given in Sections 9.4 through 9.6. The main point here is that the weights are adjusted continually during adaptation to reduce the mean-squared error (MSE), $E[e_k^2]$, toward its minimum value. Thus, the adaptive system seeks continually to reduce the difference between the desired response, d_k, and its own response, g_k. In this way, as we shall see, the system "adapts" to its prescribed signal environment.

In this chapter, we can only hope to introduce some of the basic ideas and applications of adaptive signal processing, because the subject is so broad. Texts devoted to this subject, some of which are listed in the References section, are recommended for better coverage. The text by Haykin[4] has an especially comprehensive and complete treatment of adaptive signal processing and filter theory.

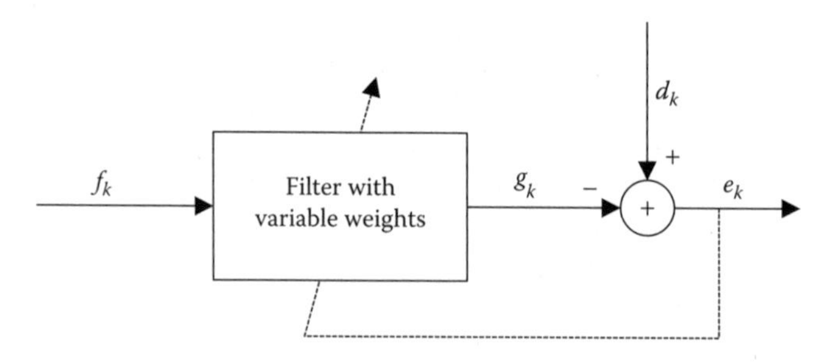

FIGURE 9.1
Essential elements in adaptive signal processing: These are the same as the essential elements in least-squares design, except that here the filter weights are adjusted continuously to minimize the squared error, e_k^2.

9.2 The Mean-Squared Error Performance Surface

In Chapter 8, Section 8.3, we saw that the MSE was a quadratic function of the samples, $[h_k]$, of the impulse response of the least-squares filter. The function was derived in Equation 8.4, which is repeated as follows:

$$\text{MSE} = \varphi_{dd}(0) + \sum_{n=0}^{N-1}\sum_{m=0}^{N-1} b_n b_m \varphi_{ff}(n-m) - 2\sum_{n=0}^{N-1} b_n \varphi_{df}(-n) \tag{9.1}$$

To keep the quadratic form and have a finite filter, in Chapter 8, we set the impulse response equal to b, the weight vector of a finite impulse response (FIR) filter, as we have done here in (9.1). Later in this chapter, we will consider the consequences of using an adaptive infinite impulse response (IIR) filter.

In adaptive systems, (9.1) describes what is called a quadratic *performance surface*,[5] meaning that the performance of an adaptive system may be described in terms of the geometry of this surface, which turns out to be a helpful vehicle for visualizing the adaptive process. Examples of two ways to view the performance surface with two weights were given in Chapter 8, Figures 8.8 and 8.9.

The key equations describing the performance surface were given in Chapter 8, Table 8.2 in terms of correlation functions, that is,

$$\text{Mean-squared error: MSE} = \varphi_{dd}(0) + b'\Phi_{ff}b - 2b'\varphi_{fd} \tag{9.2}$$

$$\text{Optimal weights: } b = \Phi_{ff}^{-1}\varphi_{fd} \tag{9.3}$$

$$\text{Minimum MSE: MSE}_{\min} = \varphi_{dd}(0) - \varphi'_{fd}b \tag{9.4}$$

Under stationary conditions, the correlation functions are constant properties of the signals. If, as we assume in this chapter, the signals are nonstationary, then we may view the correlation functions as changing with time. In (9.2), we can see that this implies an MSE, that is, a quadratic performance surface that moves and changes shape as the correlation functions change. The goal of the adaptive signal processor is then to track these changes and continuously adjust the elements of the weight vector (b) so that the MSE is kept near its minimum value.

In the rest of this chapter, we will consider methods for searching the performance surface and tracking the movement of the minimum MSE. The most general adaptive algorithms for searching performance surfaces under unknown signaling conditions are *random search* algorithms. One type of random search algorithm[14] selects a random *direction* in the performance space in which to look at each iteration, and then moves in directions that happen to reduce the MSE. An even more general random search algorithm[15] selects random *locations* on the weight plane (or hyperplane), estimates the MSE at these points, and retains points with the lowest MSE from one iteration to the next.

In this chapter, we focus on the two most common methods for accomplishing such adjustments in real-time adaptive systems; namely, *steepest descent* and *sequential regression*. These methods work best with quadratic error surfaces, and therefore, are applicable to FIR adaptive filters, including the equation-error structure described in Chapter 8, Exercise 8.23. However, these are by no means the only methods that work, especially with other types of filters and nonquadratic error surfaces.

9.3 Searching the Performance Surface

With the types of systems and the quadratic performance surfaces we have been discussing, as well as with similar performance functions that are continuous and have continuous derivatives, we can implement powerful, deterministic (i.e., nonrandom) adaptive search algorithms. In this section, we provide a basis for these algorithms in terms of the parabolic MSE in (9.2), viewing the MSE as a function of the N elements of the weight vector, b, that is, viewing the MSE as a performance surface in (N + 1)-dimensional space. As discussed in Section 9.2, in adaptive processing, we also view the performance surface as moving and changing shape; so the algorithms must continually adjust b to track the minimum MSE.

We will see that to track the minimum MSE in this manner, deterministic search algorithms, in one way or another, must employ an estimate of the *gradient* of the MSE performance surface. The gradient of the MSE is a vector with elements that are derivatives with respect to the weights; that is, similar to Chapter 8, (8.5),

$$\nabla = \frac{\partial \text{MSE}}{\partial b} = \begin{bmatrix} \dfrac{\partial \text{MSE}}{\partial b_0} & \dfrac{\partial \text{MSE}}{\partial b_1} & \cdots & \dfrac{\partial \text{MSE}}{\partial b_{N-1}} \end{bmatrix} \tag{9.5}$$

We can write the MSE gradient as the vector form of (9.5) by differentiating the MSE in (9.2) with respect to b. To differentiate the second term in (9.2), let $\delta(n)$ represent a vector of length N, with the nth element equal to one and all other elements equal to zero. Then, we have

$$\begin{aligned} \frac{\partial}{\partial b_n}\{b'\Phi_{ff}b\} &= \delta(n)'\,\Phi_{ff}b + b'\Phi_{ff}\delta(n) \\ &= \Phi_{ff}(n,:)b + b'\Phi_{ff}(:,n) = 2\sum_{m=0}^{N-1}\varphi_{ff}(n-m)\,b_m \end{aligned} \tag{9.6}$$

The final result is due to the symmetry of Φ_{ff}, that is, $\Phi_{ff}(n - m) = \Phi_{ff}(m - n)$. Thus, the gradient vector, which is the derivative of (9.2) with respect to b, is given by

$$\nabla = 2\Phi_{ff}b - 2\varphi_{fd} \tag{9.7}$$

Setting $\nabla = 0$ produces the solution in (9.3) for optimal weights.

Now we consider a weight vector, $b(k)$, as a function of its location in the sample space. We assume sampling takes place in time, so $b(k)$ is the weight vector and $\nabla(k)$ the gradient vector at time kT in the time domain of the adaptive processor. We denote the *optimal* weight vector in (9.3) as b_{opt}. If we multiply (9.7) by $\Phi_{ff}^{-1}/2$ and add the result to (9.3), we may obtain the following "one-step" algorithm for moving from an arbitrary point on the performance surface to the minimum MSE in a single step:

$$b_{\text{opt}} = b(k) - \frac{1}{2}\Phi_{ff}^{-1}\nabla(k) \tag{9.8}$$

This result is essentially Newton's root-finding method[24] applied to find the zero of a linear function (∇ in this case). It can also be derived by writing the MSE in the form of a Taylor series.[25] Given any weight vector $b(k)$ along with Φ_{ff} and the corresponding gradient $\nabla(k)$, we can move from $b(k)$ to the optimum weight vector b_{opt} in a single step by adjusting b in accordance with (9.8).

If we could apply (9.8) in practical situations such as those illustrated in Figures 8.1 through 8.4 in Chapter 8, we would always be able to adapt in one iteration to the optimal weight vector. As with natural systems, however, practical adaptive signal processing-systems do not have enough information to adapt perfectly in just one iteration. There are two specific problems with (9.8) in practice, under nonstationary conditions:

1. The correlation functions change and can, at best, only be estimated.
2. The gradient must be estimated at each adaptive iteration, using local statistics.

In other words, we must work with estimated values of Φ_{ff}^{-1} and $\nabla(k)$ in (9.8).

To anticipate the use of noisy estimates of these values, we modify (9.8) to obtain an algorithm that adjusts b in smaller increments and converges to b_{opt} after many iterations. The small increments have the effect of smoothing the noise in the estimates of Φ_{ff}^{-1} and $\nabla(k)$, thus allowing the adaptive system to have stable behavior. The modified version of (9.8), which allows us to compute $b(k + 1)$ in terms of $b(k)$ and a convergence parameter, u, is

$$\boxed{b(k+1) = b(k) - u\,\Phi_{ff}^{-1}\nabla(k)} \tag{9.9}$$

We can examine the convergence of (9.9) by first subtracting $2u$ times (9.8) from (9.9) to obtain

$$b(k+1) = (1-2u)b(k) + 2ub_{\text{opt}} \tag{9.10}$$

Taking the z-transform of (9.10) and solving for $B(z)$, we have

$$\begin{aligned} zB(z) &= (1-2u)B(z) + \frac{2uzb_{\text{opt}}}{z-1} \\ B(z) &= \frac{2uzb_{\text{opt}}}{(z-1)(z-(1-2u))} \end{aligned} \tag{9.11}$$

Taking the inverse transform (Chapter 4, Table 4.2, line 6), we have

$$b(k) = [1-(1-2u)^k]b_{\text{opt}} \tag{9.12}$$

Here, we have a formula that begins at $b(0) = 0$ and converges (conditionally) to b_{opt} at $k = \infty$. Because the formula follows a linear path on the b-plane, we may move the starting point to $b(0)$ by substituting $b(k) - b(0)$ for $b(k)$ and $b_{\text{opt}} - b(0)$ for b_{opt}. Then (9.12) becomes

$$\boxed{b(k) = b_{\text{opt}} + (1-2u)^k(b(0) - b_{\text{opt}})} \tag{9.13}$$

Here, we have a description of the relaxation of the weight vector from $b(0)$ to b_{opt} under ideal conditions, where Φ_{ff} and $\nabla(k)$ are known at each step.

In (9.13), we can see that $b(k)$ will converge to b_{opt} only if the geometric ratio $1 - 2u$ is less than 1 in magnitude, that is,

$$\boxed{\text{For convergence under ideal conditions: } 0 < u < 1} \tag{9.14}$$

In (9.13), the one-step solution is found at $k = 1$ with $u = 1/2$. With $u > 1/2$ the convergence is oscillatory, with b jumping back and forth across the bowl and converging toward b_{opt}. In practical adaptive systems with noise, values of u well below $1/2$, typically on the order of 0.01, are used, giving a convergence *time constant* of many iterations. We define the convergence time constant to be τ iterations, where the relaxation toward b_{opt} goes in proportion to $(1 - e^{-k/\tau})$. Then, using the linear approximation $(1 - 1/\tau)$ for $e^{-1/\tau}$ when τ is large and comparing it with (9.13), we have

$$\begin{aligned} 1 - e^{-k/\tau} &= 1 - (1 - 1/\tau)^k \\ &= 1 - (1-2u)^k \end{aligned} \tag{9.15}$$

Thus, we conclude that $1/\tau \approx 2u$, and we have the following rule:

$$\boxed{\begin{array}{l} \text{Time constant for convergence of weights under ideal conditions:} \\ \tau_w \approx \dfrac{1}{2u} \text{ iterations;} \quad 0 < u << 1/2 \end{array}} \tag{9.16}$$

In addition to weight convergence in (9.13), the convergence of the MSE to its minimum value is also often used as a performance measure in adaptive systems. A plot of the MSE versus the iteration number k is called euphemistically a *learning curve*.[5] To derive a formula similar to (9.13) for the learning curve, it is convenient to use a translated weight vector, c, that allows the minimum MSE to occur at $c_{\text{opt}} = 0$:

$$c = b - b_{\text{opt}} \tag{9.17}$$

Then, noting that $b'\varphi_{fd} = \varphi'_{fd}b$, the MSE in (9.2) becomes

$$\begin{aligned} \text{MSE} &= \varphi_{dd}(0) + (c + b_{\text{opt}})'\Phi_{ff}(c + b_{\text{opt}}) - 2\varphi'_{fd}(c + b_{\text{opt}}) \\ &= \varphi_{dd}(0) + b'_{\text{opt}}\Phi_{ff}b_{\text{opt}} - 2\varphi'_{fd}b_{\text{opt}} + c'\Phi_{ff}c + c'\Phi_{ff}b_{\text{opt}} + b'_{\text{opt}}\Phi_{ff}c - 2\varphi'_{jd}c \end{aligned} \tag{9.18}$$

We can simplify this result in two steps. First, if we substitute (9.3) for the first occurrence in the second line of b'_{opt} in (9.18), the first three terms become equal to MSE_{min} in (9.4). Second, again using (9.3) in the last three terms, we can substitute $\Phi_{ff}b_{\text{opt}} = \varphi_{fd}$ and also, because Φ_{ff} is symmetric, $b'_{\text{opt}}\Phi_{ff} = \varphi'_{fd}$. Therefore, the last three terms sum to zero, and (9.18) reduces to

$$\boxed{\text{MSE} = \text{MSE}_{\text{min}} + c'\Phi_{ff}c} \tag{9.19}$$

This is the simplest expression for the quadratic performance surface, given in terms of coordinates normalized so that the optimal weight vector, c_{opt}, is zero.

If we substitute (9.17) into (9.13), we have a simpler formula for the relaxation of the weight vector from a starting value, c_0, to its optimum value, that is, to zero:

$$c(k) = (1 - 2u)^k c(0) \tag{9.20}$$

Substituting (9.20) for c in (9.19), we have a formula for the learning curve, that is, the relaxation of the MSE to its minimum, under ideal conditions:

$$\text{MSE}_k = \text{MSE}_{\text{min}} + (1 - 2u)^{2k} c'(0)\Phi_{ff}c(0) \tag{9.21}$$

Comparing this with (9.20) or (9.13), we see that the convergence ratio is now $(1 - 2u)^2$ instead of $(1 - 2u)$. Thus, similar to (9.16), the time constant for the learning curve is

Time constant for convergence of the MSE under ideal conditions:
$\tau_{\text{MSE}} = \dfrac{\tau_w}{2} \approx \dfrac{1}{4u}$ iterations; $0 < u << 1/2$

(9.22)

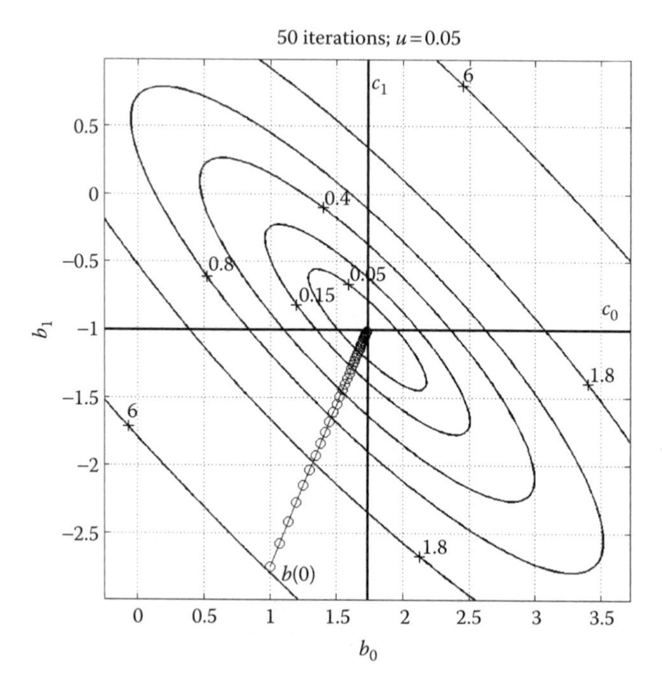

FIGURE 9.2
Weight convergence in the predictor described in Chapter 8, Section 8.4, with error contours illustrated in Figure 8.9: Convergence is in accordance with (9.13), with $u = 0.05$, beginning at $b(0) = [1.00, -2.75]$ and continuing to $b(49)$.

Examples of weight convergence and a learning curve are given in Figures 9.2 and 9.3. These examples illustrate ideal convergence of the weights as described by (9.9), beginning at $b(0) = [1.00; -2.75]$ and using $u = 0.05$, the latter being well below the one-step convergence value, $u = 0.5$. The examples are for the prediction example in Chapter 8, Section 8.4, with performance surface illustrated in Figures 8.8 and 8.9. In Figure 9.2, note the straight, noise-free track from $b(0)$ to $b(49)$, which is close to b_{opt}, and also the translated coordinates of c with origin at b_{opt}. With $u = 0.05$, the weight convergence time constant is approximately 10 iterations in accordance with (9.16), and we note in Figure 9.2 that there is convergence after 5 time constants, or 50 iterations.

In Figure 9.3, we observe the corresponding exponential decrease of the MSE from its starting value of 5.8 toward the minimum value, which is zero in this case. The MSE time constant, τ_{MSE}, is approximately five iterations in accordance with (9.22), and we observe in Figure 9.3 that the approximation is valid. After four time constants, or 20 iterations in this example, the MSE has relaxed to about 1.1% of its original value, which is close to $5.8e^{-4}$.

In this section, we discussed the process of searching the quadratic performance surface. The principal formulas (9.9) and (9.13) describe how,

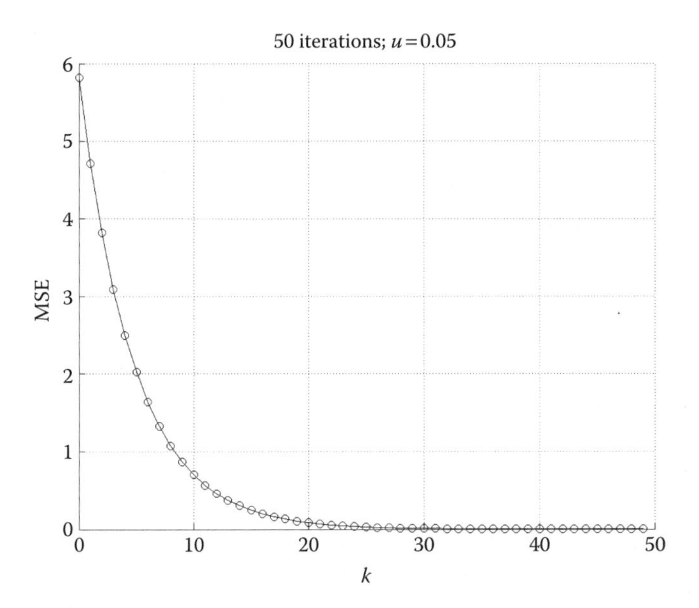

FIGURE 9.3
Convergence of the mean-squared error (learning curve) corresponding with weight convergence in Figure 9.2.

under ideal conditions, the weight vector changes at each step and converges finally to its optimum value.

We must now examine how adaptive systems reach convergence under realistic conditions, with imperfect knowledge of the gradient and correlation properties of the performance surface, which may change from one iteration to the next. First, we will assume that we have little knowledge of the correlation properties and can only estimate the input signal power, that is, $\varphi_{ff}(0)$. This assumption leads us to the steepest descent class of algorithms and, in turn, to the widely used Widrow-Hoff LMS algorithm, which is the simplest and most useful algorithm in adaptive signal processing.

9.4 Steepest Descent and the LMS Algorithm

A steepest-descent adaptive algorithm is a formula like (9.9) for step-by-step adjustment of the weight vector, but it is unlike (9.9) in that it does not move along a straight track in the weight plane (or hyperplane) as in Figure 9.2 but rather follows the path of steepest descent down the error surface, perpendicular to the MSE contours in Figure 9.2.

One of the goals in developing such an algorithm is to revise (9.9) by removing the correlation information in Φ'_{ff}, and yet be able to control convergence as we did in (9.9) by restricting u to the range (0, 1/2). To accomplish this, we recall that Φ_{ff} is a square matrix with rows and columns that are permutations of the autocorrelation vector, φ_{ff}. If Φ_{ff} is nonsingular so that Φ_{ff}^{-1} exists, which is the case in all our applications, the *eigenvalues* of Φ_{ff} are developed from the following equation,[5] in which λ_n is a scalar, I is the identity matrix, and q_n is a column *vector* (not a single element) with N elements, $q_{n0}, q_{n1}, \ldots, q_{n,N-1}$:

$$[\Phi_{ff} - \lambda_n I]q_n = [0 \ldots 0]' \tag{9.23}$$

This equation has nontrivial solutions for λ_n and q_n if the following determinant vanishes:

$$\det[\Phi_{ff} - \lambda_n I] = 0 \tag{9.24}$$

This is called the *characteristic equation* of Φ_{ff}. Its N solutions, designated λ_0, $\lambda_1, \ldots, \lambda_{N-1}$, are the *eigenvalues* of Φ_{ff}. Corresponding with each distinct eigenvalue, λ_n, there is an *eigenvector*, q_n, satisfying (9.23). In (9.23), we note that q_n may be normalized so that

$$q_n' q_n = 1; \quad n = 0, 1, \ldots, N-1 \tag{9.25}$$

We will assume the eigenvalues are distinct, and the eigenvectors are normalized in this manner.

For each eigenvalue λ_n, we may express the equality in (9.23) by writing

$$\Phi_{ff} q_n = \lambda_n q_n; \quad n = 0, 1, \ldots, N-1 \tag{9.26}$$

From this expression, we can show that the N eigenvectors are *mutually orthogonal*. We take two versions of (9.26), one for q_m and the other for q_n:

$$\Phi_{ff} q_m = \lambda_m q_m; \quad \Phi_{ff} q_n = \lambda_n q_n$$

We multiply the first expression by q_n' and the second by q_m':

$$q_n' \Phi_{ff} q_m = \lambda_m q_n' q_m; \quad q_m' \Phi_{ff} q_n = \lambda_n q_m' q_n$$

We transpose the first of these expressions, recalling that $\Phi_{ff}' = \Phi_{ff}$. The result is the same as the second expression, that is,

$$(q_n' \Phi_{ff} q_m)' = q_m' \Phi_{ff} q_n = \lambda_n q_m' q_n = \lambda_m q_n' q_m$$

Because the expressions are equal, and because λ_m and λ_n are different, we conclude that

$$q'_m q_n = q'_n q_m = 0; \quad n \neq m \tag{9.27}$$

Thus, the eigenvectors are mutually orthogonal.

Next, we form the *eigenvector matrix*, Q, as a square matrix with q_n as the nth column, and also Λ, a diagonal matrix with λ_n as the nth diagonal element:

$$Q = [q_0 \; q_1 \; \dots \; q_{N-1}]; \quad \Lambda = \text{diag}[\lambda_0 \; \lambda_1 \; \dots \lambda_{N-1}] \tag{9.28}$$

With these definitions, (9.26) may be applied N times, once for each column of Q, to give

$$\Phi_{ff} Q = Q\Lambda, \quad \text{or} \quad \Phi_{ff} = Q\Lambda Q^{-1} \tag{9.29}$$

This is called the *normal form* of the autocorrelation matrix Φ_{ff}, in which the eigenvalues appear explicitly. Furthermore, because (9.25) is true for each column of Q, and because the columns of Q are mutually orthogonal, it follows that

$$QQ' = I; \text{ that is, } Q' = Q^{-1} \tag{9.30}$$

And, therefore, another expression of the normal form of Φ_{ff} is

$$\Phi_{ff} = Q\Lambda Q', \quad \text{or} \quad \Lambda = Q'\Phi_{ff} Q \tag{9.31}$$

The inverse of the normal form is easy to find, because Λ is diagonal and so Λ^{-1} is also diagonal with elements; $1/\lambda_n$, that is,

$$\Lambda^{-1} = \text{diag}[\,1/\lambda_0 \; 1/\lambda_1 \; \dots \; 1/\lambda_{N-1}\,] \tag{9.32}$$

Furthermore, for any two matrices A and B with inverses, $(AB)^{-1} = B^{-1}A^{-1}$. Therefore, using (9.29), the inverse of (9.31) is

$$\Phi_{ff}^{-1} = (Q')^{-1}(Q\Lambda)^{-1} = Q\Lambda^{-1}Q' \tag{9.33}$$

With the normal form and its inverse, we may now discuss the convergence of a steepest descent algorithm, starting with a revised version of the ideal weight recursion formula (9.9), which contains no information about the input correlation function, φ_{ff}:

$$b(k+1) = b(k) - \mu\nabla(k) \tag{9.34}$$

First, we translate (9.34) using (9.17) so that the optimal weight vector is $c_{\text{opt}} = 0$:

$$c(k+1) = c(k) - \mu\nabla_c(k) \tag{9.35}$$

where ∇_c is the gradient vector translated so that the gradient is 0 at $c = 0$. Next, we substitute (9.7) for the *translated* gradient (noting that $\varphi_{fd} = \Phi_{ff} b_{\text{opt}}$):

$$c(k+1) = c(k) - 2\mu\,\Phi_{ff} c(k) = (I - 2\mu\,\Phi_{ff})c(k) \tag{9.36}$$

Our next step is to *rotate* the *c*-coordinates to new coordinates, *v*, using the eigenvector matrix, as follows:

$$c = Qv, \quad \text{or} \quad Q'c = v \tag{9.37}$$

The latter version follows from (9.30). Now, (9.36) becomes

$$Qv(k+1) = (I - 2\mu\Phi_{ff})Qv(k) = (Q - 2\mu\Phi_{ff}Q)v(k)$$

Multiplying by Q' and substituting (9.31), we have the normalized form of the ideal convergence formula:

$$\begin{aligned} v(k+1) &= (Q'Q - 2\mu Q'\Phi_{ff}Q)v(k) \\ &= (I - 2\mu\Lambda)v(k) \end{aligned} \tag{9.38}$$

In this normal form, because the off-diagonal elements of Λ are all zero, we now have *N independent* convergence formulas:

$$v_n(k+1) = (1 - 2\mu\lambda_n)v_n(k); \quad n = 0, 1, \ldots, N-1 \tag{9.39}$$

Here, similar to (9.10) or (9.13), the geometric convergence ratio is $(1 - 2\mu\lambda_n)$, and so we see that the convergence criterion is now

$$\boxed{\text{For steepest-descent convergence:} \quad 0 < \mu < 1/\lambda_{\max}} \tag{9.40}$$

where $\lambda_{\max}$ is the maximum eigenvalue (corresponding with fastest convergence). Similarly, the time constants for convergence under ideal conditions, τ_w in (9.16) and τ_{MSE} in (9.22), may be modified as follows to reflect convergence along the principal axes:

$$\boxed{\text{Steepest descent:} \quad \tau_{wn} = \frac{1}{2\mu\lambda_n}; \quad \tau_{\text{MSE}n} = \frac{1}{4\mu\lambda_n}; \quad 0 \le n < N} \tag{9.41}$$

Having used the rotated *c*-coordinates of (9.37) to obtain a formula (9.41) for the MSE convergence time constant, we can also use (9.37) to express the MSE itself in terms of eigenfunctions. Substituting (9.37) into the MSE formula (9.19), we have

$$\begin{aligned} \text{MSE} &= \text{MSE}_{\min} + c'\Phi_{ff}c \\ &= \text{MSE}_{\min} + (Qv)'\,\Phi_{ff}Qv = MSE_{\min} + v'Q'\Phi_{ff}Qv \end{aligned} \tag{9.42}$$

Now we can substitute (9.31) and express the MSE as follows:

$$\text{MSE} = \text{MSE}_{\min} + v'\Lambda v \tag{9.43}$$

This gives the MSE explicitly in terms of eigenfunctions. Furthermore, when we apply the definition of the diagonal matrix Λ in (9.28) and perform the product in (9.42), the MSE expression becomes

$$\text{MSE} = \text{MSE}_{\min} + \sum_{n=0}^{N-1} \lambda_n v_n^2 \tag{9.44}$$

In (9.41), note that the *slowest* descent (longest time constant) corresponds with the *smallest* eigenvalue, $\lambda_{\min}$, which tends to dominate the latter part of the convergence process, as will be seen in the examples that follow.

Concerning the range of μ in (9.40), we do not yet have a practical convergence criterion because $\lambda_{\max}$ in (9.40) is known only if φ_{ff} is known. Hence, our final step in the development of a steepest descent algorithm is to express the *sum* of the eigenvalues in terms of a local estimate of signal power and use this in place of $\lambda_{\max}$ in (9.40) to provide an upper limit on μ.

Now the sum of the diagonal elements of any square matrix A is called the *trace* of A, or Tr(A). Thus, the trace of Λ, which is a diagonal matrix, is the sum of the eigenvalues, and the trace of Φ_{ff} is the sum of the N diagonal elements, that is,

$$\text{Tr}(\Lambda) = \sum_{n=0}^{N-1} \lambda_n; \quad \text{Tr}(\Phi_{ff}) = \sum_{n=0}^{N-1} \varphi_{ff}(0) = N\varphi_{ff}(0) \tag{9.45}$$

These two traces are equal. The demonstration of equality can be made using an interesting property of the trace, namely, that the trace of the product of two square matrices is the trace of the reverse product; that is, if A and B are square matrices, then using the definition of the product AB in Chapter 1, (1.7) with the definition of the trace in (9.45),

$$\text{Tr}(AB) = \sum_{i=0}^{N-1} \sum_{n=0}^{N-1} a_{in} b_{ni} = \sum_{n=0}^{N-1} \sum_{i=0}^{N-1} b_{ni} a_{in} = \text{Tr}(BA) \tag{9.46}$$

The equality of Tr(Λ) and Tr(Φ_{ff}) follows when we use (9.31) in Tr(Λ) and apply (9.46) along with (9.30):

$$\text{Tr}(\Lambda) = \text{Tr}(Q'\Phi_{ff}Q) = \text{Tr}(\Phi_{ff}Q'Q) = \text{Tr}(\Phi_{ff}) \tag{9.47}$$

Thus, the trace of Λ, that is, the sum of the eigenvalues, is equal to the trace of Φ_{ff}, that is, the sum of the diagonal elements of the autocorrelation matrix, each of which is $\varphi_{ff}(0)$ as in (8.6). Therefore,

$$\mathrm{Tr}(\Lambda) = \sum_{n=0}^{N-1} \lambda_n = N\varphi_{ff}(0) = N\sigma_f^2 \tag{9.48}$$

Here, we use σ_f^2 to represent the variance, or average power, of the signal f, which is the same as $\varphi_{ff}(0)$ in Chapter 7, (7.30). If σ_f^2 changes during the adaptive process, then it must be adjusted along with the weights. It is easy to estimate the average power locally in time as the average squared signal amplitude (one estimation method is described in Section 9.6).

Thus, we now have a practical version of the convergence criterion in (9.40) for steepest descent algorithms. To be consistent with the ideal weight convergence described in (9.9), we set

$$\mu = \frac{u}{N\sigma_f^2(k)} \tag{9.49}$$

Then, the steepest-descent algorithm in (9.34) and the convergence criterion in (9.40) produce the following result:

$$\boxed{\text{Steepest descent:}\quad b(k+1) = b(k) - \frac{u}{N\sigma_f^2(k)}\nabla(k);\quad 0 < u < 1} \tag{9.50}$$

With this result, we are ready to derive the LMS algorithm, which was first published by Widrow and Hoff[1] and has since become the most widely applied algorithm in adaptive signal processing. All we need is a local estimate of the gradient vector, which the LMS algorithm supplies. We recall that the ideal gradient in (9.5) and (9.9) is the derivative of the MSE with respect to the weight vector, b. In the Widrow LMS algorithm, the local estimate of the MSE is the squared error itself, e_k^2. First, we define the *input signal vector* of length N, consisting of the most recent N samples of the input signal, f, in *reverse order*:

$$f(k) = [f_k \;\; f_{k-1} \;\; \cdots \;\; f_{k-N+1}]' \tag{9.51}$$

Using (8.1) and (8.3) in Chapter 8, with $b(k)$ as the current weight vector, our expression for e_k is

$$e_k = d_k - b(k)' f(k) \tag{9.52}$$

(Remember here that *subscript k* denotes the kth sample, and *argument k* denotes the kth vector.) Using (9.52), Widrow's estimate of the gradient vector for use in the LMS algorithm is

$$\hat{\nabla}(k) = \frac{\partial e_k^2}{\partial b} = 2e_k \frac{\partial e_k}{\partial b} = -2e_k f(k) \tag{9.53}$$

Using this gradient vector estimate in place of $\nabla(k)$ in (9.50) produces the LMS algorithm:

$$\boxed{\text{LMS algorithm:} \quad b(k+1) = b(k) + \frac{2u}{N\sigma_f^2(k)} e_k f(k); \quad 0 < u << 1} \tag{9.54}$$

The condition $u << 1$ is in (9.54) to emphasize that u is small in practice (sometimes on the order of 0.1 and often much less) depending on how fast the signal environment is changing. The LMS gradient estimate causes the weight vector to converge on a noisy version of the steepest descent path, as we will see in the first example in Section 9.5.

Convergence of the LMS weights and MSE may be approximated by the steepest descent time constants in (9.41), with μ specified in (9.49). In general, the convergence of the LMS algorithm is governed increasingly as the MSE approaches $\text{MSE}_{\min}$ by the minimum eigenvalue, $\lambda_{\min}$, which corresponds with the longest convergence time constant. For a conservative description of LMS convergence, we therefore substitute $\lambda_{\min}$ for λ_n and (9.49) for μ in (9.41) and obtain these LMS convergence time constants:

$$\boxed{\text{LMS convergence:} \quad \tau_w = \frac{N\sigma_f^2}{2u\lambda_{\min}}; \quad \tau_{\text{MSE}} = \frac{N\sigma_f^2}{4u\lambda_{\min}}; \quad 0 < u << 1} \tag{9.55}$$

Thus, convergence of the LMS algorithm is slower than ideal convergence when $\lambda_{\min}$ is less than the average eigenvalue. If the adaptive system is designed with $\lambda_{\min}$ unknown, which is usually true, one must plan for the worst case. For example, suppose the input to an adaptive filter with two weights has a sinusoidal component at ν Hz-s. Then, it is easy to show (Exercise 9.13) that the ratio of $\lambda_{\min}$ to the average eigenvalue is

$$\frac{\lambda_{\min}}{\lambda_{av}} = 1 - \cos(2\pi\nu)$$

Thus, other factors being equal, convergence times for the LMS algorithm are longer at lower frequencies.

9.5 LMS Examples

Our first example of adaptive weight convergence using the LMS algorithm is illustrated in Figure 9.4, which repeats Figure 9.2 for one-step prediction of a sine wave with two weights and ideal convergence, and also includes the weight track for LMS convergence. The adaptive gain is $u = 0.05$ for both weight tracks; however, convergence requires about 50 iterations on the ideal track and 350 iterations on the LMS track. On the LMS track, we can observe erratic direction and step size due to inaccuracy in the gradient estimate (9.50), as well as the steepest descent weight track, which is orthogonal to the MSE contours.

A plot of e_k^2 versus k in this example is shown in Figure 9.5. The plot is equivalent to the learning curve described in Section 9.3, which is a plot of $E[e_k^2]$ versus k. It covers 51 iterations of the LMS algorithm and is seen to converge toward the minimum MSE (zero) over this range. Convergence is comparable to the ideal convergence plotted in Figure 9.3. The latter is plotted as a dashed line for comparison with LMS convergence in this example.

The error signals used for the LMS track in Figure 9.4 and the learning curve in Figure 9.5 were generated using a MATLAB® function on the

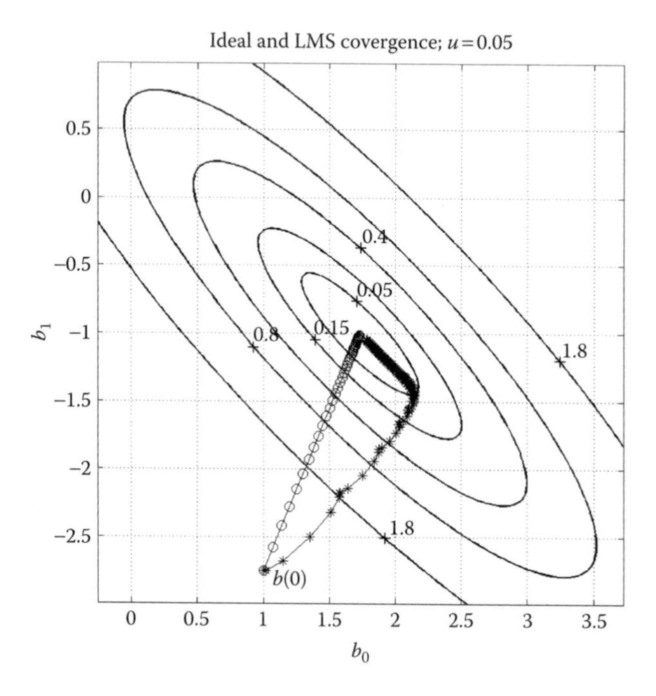

FIGURE 9.4
Comparison of ideal and least-mean-square (LMS) weight convergences for the predictor with two weights (Figure 9.2): The number of iterations is $K = 50$ for ideal convergence and $K = 350$ for LMS convergence.

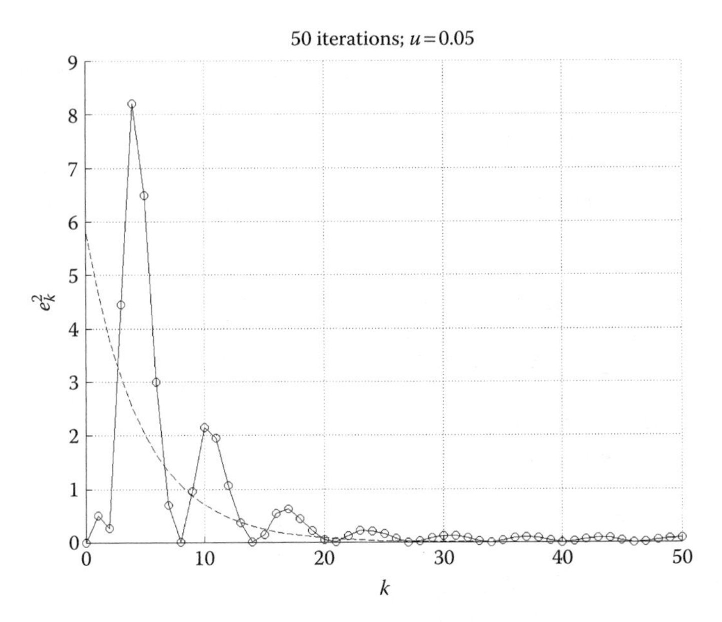

FIGURE 9.5
Comparison of convergence of e_k^2 (connected points) using the least-mean-square (LMS) algorithm with ideal convergence of the mean-squared error (dashed line) for the prediction example presented in Figure 9.3: The adaptive gain is $u = 0.05$ in both cases.

publisher's website called *lms_ filter*. Here is part of the result of entering "help lms_filter" in the command window:

```
>> help lms_filter
   [g,b]=lms_filter(b,u,d,f,ind)

   Inputs:
      b  = initial weight vector with length(b)=N.
      u  = adaptive gain constant; 0 <= u << 1.
      d  = desired signal vector.
      f  = input vector to FIR adaptive filter.
      ind= indicator (see output "b" below.)
   Output:
      g  = adaptive filter output vector; error=d-g.
      b  = Final weight vector if ind is omitted or zero, or
         = NxK array of successive weight vectors if ind=1.
```

With *ind* = 1, this function produces vector g in Figure 9.1 as the result of filtering f and adapting at each step using the error, $e_k = d_k - g_k$. The weight vector $b(k)$ is updated and stored at each step, so that the entire weight track is stored as an output. This is a simple implementation of the LMS algorithm in which σ_k^2 is held fixed. It is useful for producing results like those in Figures 9.4 and 9.5, but may need to be modified for other uses.

Our next example is an illustration of why adaptive signal processing may be preferred with nonstationary signals. We will keep the same, simple two-weight predictor described in Chapter 8, Section 8.4, and illustrated in Figure 8.7. To have a simple nonstationary signal, we use a single "frame" of signal d_k, which is shown in the upper plot in Figure 9.6. The frame consists of 1000 samples of a unit sine wave with frequency changing from $\nu = 0.05$ to $\nu = 0.10$ Hz-s at sample $k = 750$. We assume that prior to the time of this frame the frequency of d_k has been constant at $\nu = 0.05$ Hz-s. Now, suppose we are using a two-weight predictor to compress this signal and we wish, therefore, to produce a small residue, e_k. If fixed optimal weights are computed using the method in Chapter 8, the residue, shown in the center plot in Figure 9.6, cannot be reduced to zero due to the frequency change in d_k. On the other hand, if an LMS *adaptive predictor* has previously adapted to the initial sine wave and again adapts to the frequency change in this frame, the residue remains close to zero over most of the frame, as shown in the lower plot in Figure 9.6. The only noticeable residue is where the change occurs at $k = 750$. In this particular example, the nonadaptive total squared error (TSE) is 7.6 and the adaptive TSE is only 0.2, indicating a much smaller average residue and, hence, significant compression of the signal.

If an adaptive filter is used in signal compression, the question of recovery arises. Can the original signal be recovered from the residue? The answer is

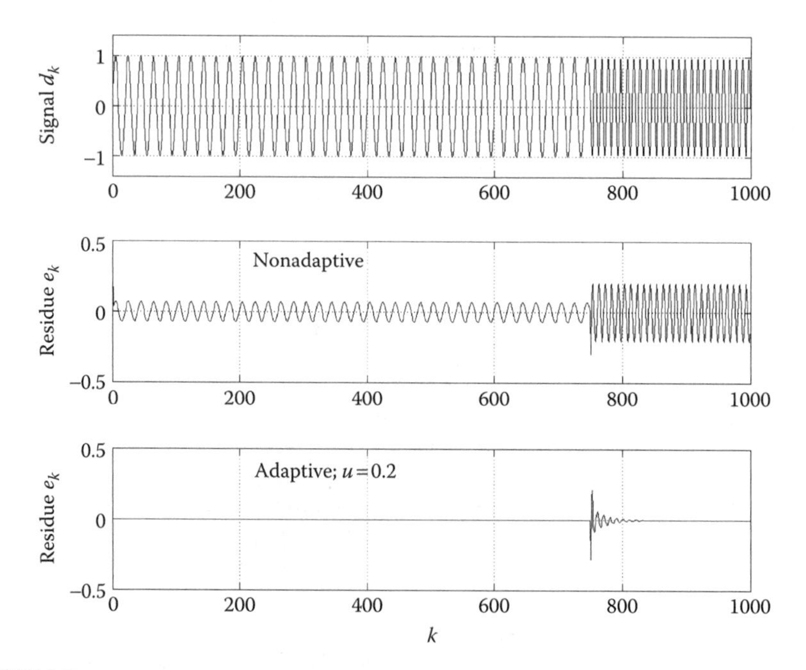

FIGURE 9.6
Nonadaptive and adaptive signal compressions using a one-step predictor with two weights. The frequency of the signal, d_k, doubles at $k = 750$. The fixed (nonadaptive) predictor has weights optimized for the entire frame. The adaptive predictor has adapted previously to the first frequency, so its average prediction error (residue) is much smaller.

yes, and with no additional information except the adaptive gain constant, u. Recovery proceeds as described in Chapter 4, Section 4.32, and as discussed in Chapter 8, Section 8.10. The only difference is that the LMS algorithm must be applied at each step to adjust the weight vector during recovery. It is easy to see in the formulation (9.54) of the LMS algorithm that the adjustment is realizable, because at each step, the weights for the *next* step are computed in terms of quantities that are available during recovery. A more detailed discussion of adaptive prediction and recovery is presented in Chapter 10, Section 10.4.

9.6 Direct Descent and the RLS Algorithm

Comparing the steepest descent formula (9.50) with the ideal formula (9.9), and comparing the weight tracks in Figure 9.4, we may note that the presence of Φ_{ff}^{-1} in (9.9) has the effect of steering the weights away from the direction of the gradient to a direct path leading to the minimum MSE. Thus, it is reasonable to look for a formula like (9.54) that, in addition to estimating the gradient at each step, also estimates Φ_{ff}^{-1} at each step. The *Recursive Least Squares* (RLS) algorithm is such a formula.

The RLS algorithm exists in different forms.[4,6,27] The version we use here is based on an approximation to (9.9) of the form

$$b(k+1) = b(k) - u\hat{\Phi}_{ff}^{-1}\hat{\nabla}(k)$$

The simplest local gradient estimate is the LMS gradient estimate (9.53). Substituting this estimate for $\hat{\nabla}(k)$ and using $\hat{\Phi}_{ff}^{-1}(k)$ for the current estimate of $\hat{\Phi}_{ff}^{-1}$, we have

$$b(k+1) = b(k) + 2u\hat{\Phi}_{ff}^{-1}(k)e_k f(k) \tag{9.56}$$

Let us now consider the problem of estimating the inverse autocorrelation matrix, $\hat{\Phi}_{ff}^{-1}$. The information in $\hat{\Phi}_{ff}$ is contained in N correlation elements, $\varphi_{ff}(0:N-1)$. With the exception of $\varphi_{ff}(0)$, which is the input power σ_f^2, a correlation value $\varphi_{ff}(n) = E[f_k f_{k-n}]$ cannot be estimated without using a history of at least n samples. Thus, the simplest local estimate of $\varphi_{ff}(n)$ is the sample product $f_k f_{k-n}$, and using the definition of the kth signal vector, $f(k)$, in (9.51), it follows that the simplest local estimate of $\hat{\Phi}_{ff}$ at the kth step is the square matrix given by

$$\hat{\Phi}_{ff}(k) = f(k)f'(k) \tag{9.57}$$

In Chapter 8, where stationary signals were assumed, we used the available history, that is, the entire signal vector f, to estimate $\varphi_{ff}(n)$ as the average (or total) product. If f is nonstationary such that φ_{ff} changes slowly, then it is reasonable to use only the recent history of f to estimate φ_{ff}. If the signal $f(t)$ is locally periodic, it is reasonable to average over one or more periods. If the signal statistics are changing slowly with time, we use the following modification of (9.57), which gradually "forgets" the past history of the estimate:

$$\hat{\Phi}_{ff}(k) = (1-\alpha)\hat{\Phi}_{ff}(k-1) + \alpha f(k)f'(k); \quad 0 < \alpha < 1 \tag{9.58}$$

When α is close to zero, current events in $f(t)$ have little effect on the estimate, and conversely, when α is close to one, current events in $f(t)$ have a large effect on the estimate.

The recursive estimate (9.58) produces the correlation estimate as essentially a lowpass-filtered version of the instantaneous estimate, $f(k)f'(k)$. The lowpass transfer function is seen to be

$$H(z) = \frac{\alpha}{1-(1-\alpha)z^{-1}} \tag{9.59}$$

The corresponding impulse response is

$$h_k = \alpha(1-\alpha)^k; \quad k \ge 0 \tag{9.60}$$

and the correlation estimate is the convolution of h_k with $f(k)f'(k)$, that is,

$$\begin{aligned}\hat{\Phi}_{ff}(k) &= \sum_{n=0}^{k} h_n f(k-n)f'(k-n)\\ &= \alpha\sum_{n=0}^{k}(1-\alpha)^n f(k-n)f'(k-n)\end{aligned} \tag{9.61}$$

Here again, as in (9.13), we can see that when α is small, there is an exponential decay in the memory of the past correlation estimate, and the "forgetting time constant" is

$$\boxed{\begin{array}{l}\text{Time constant for forgetting the correlation estimate:}\\ \tau_{\Phi_{ff}} \approx \dfrac{1}{\alpha} \text{ iterations}; \quad 0 < \alpha << 1\end{array}} \tag{9.62}$$

To eliminate the need for a matrix inversion at each step in (9.56), we now proceed to change (9.58) from a recursive estimate of $\hat{\Phi}_{ff}$ to a recursive estimate of $\hat{\Phi}_{ff}^{-1}$. First, we pre-multiply (9.58) by $\hat{\Phi}_{ff}^{-1}(k)$ and postmultiply by $\hat{\Phi}_{ff}^{-1}(k-1)$ to obtain

$$\hat{\Phi}_{ff}^{-1}(k-1) = (1-\alpha)\hat{\Phi}_{ff}^{-1}(k) + \alpha\hat{\Phi}_{ff}^{-1}(k)f(k)f'(k)\hat{\Phi}_{ff}^{-1}(k-1) \tag{9.63}$$

Next we postmultiply this result by $f(k)$:

$$\hat{\Phi}_{ff}^{-1}(k-1)f(k) = (1-\alpha)\hat{\Phi}_{ff}^{-1}(k)f(k) + \alpha\hat{\Phi}_{ff}^{-1}(k)f(k)f'(k)\hat{\Phi}_{ff}^{-1}(k-1)f(k)$$
$$= \hat{\Phi}_{ff}^{-1}(k)f(k)\left(1-\alpha+\alpha f'(k)\hat{\Phi}_{ff}^{-1}(k-1)f(k)\right) \tag{9.64}$$

To make this less complicated, we define

$$S(k) = \hat{\Phi}_{ff}^{-1}(k-1)f(k); \quad S'(k) = f'(k)\hat{\Phi}_{ff}^{-1}(k-1) \tag{9.65}$$

(The second form follows because $\hat{\Phi}_{ff}^{-1}$ is symmetric and equal to its transpose.) Now, (9.64) becomes

$$S(k) = \hat{\Phi}_{ff}^{-1}(k)f(k)(1-\alpha+\alpha f'(k)S(k)) \tag{9.66}$$

Note that the factor in parentheses here is a scalar. Dividing Equation 9.66 by this factor and postmultiplying by $S'(k)$ in (9.65), we have

$$\hat{\Phi}_{ff}^{-1}(k)f(k)f'(k)\hat{\Phi}_{ff}^{-1}(k-1) = \frac{S(k)S'(k)}{1-\alpha+\alpha f'(k)S(k)} \tag{9.67}$$

Our final step is to substitute α times this result for the last term in (9.63) and rearrange the terms to obtain

$$\hat{\Phi}_{ff}^{-1}(k) = \frac{1}{1-\alpha}\left(\hat{\Phi}_{ff}^{-1}(k-1) - \frac{\alpha S(k)S'(k)}{1-\alpha+\alpha f'(k)S(k)}\right) \tag{9.68}$$

In this result, with the definition of $S(k)$ in (9.65), we have an iterative algorithm for computing $\hat{\Phi}_{ff}^{-1}(k)$ at each step k. It is reasonable also to use the same smoothed estimate of total power, using either $\varphi_{ff}(0)$ or the instantaneous squared value, f_k^2, as the input to $H(z)$ in (9.59). In either case, the equivalent of (9.59) for the power estimate is

$$\sigma_f^2(k) = (1-\alpha)\sigma_f^2(k-1) + \alpha f_k^2 \tag{9.69}$$

Note that we could, if desired, replace f_k^2 with the average diagonal element of $f(k)f'(k)$ as a current estimate of $\varphi_{ff}(0)$.

Initial conditions must be specified for the functions in both these recursive estimation formulas. For $\sigma_k^2(0)$ in (9.69), we assign a value, say σ_{f0}^2, as an a priori estimate of the input signal power. Then σ_{f0}^2 is also the a priori estimate of $\varphi_{ff}(0)$ and, in the absence of any other correlation information, we may use $\sigma_{f0}^2 I$ as the initial estimate of $\hat{\Phi}_{ff}$. Thus, the initial conditions for (9.68) and (9.69) are as follows:

$$\text{Initial condtions:} \begin{cases} \sigma_f^2(0) = \sigma_{f0}^2 \\ \hat{\Phi}_{ff}^{-1}(0) = \dfrac{1}{\sigma_{f0}^2} I \end{cases} \tag{9.70}$$

With these initial conditions, we also require starting values $b(0)$ and $f(0)$ for the weight vector and signal vector, respectively, as we required with the LMS algorithm. After start-up, the RLS algorithm then consists of (9.65), (9.68), and (9.56) and may be summarized as follows:

RLS Algorithm:

$$S(k) = \hat{\Phi}_{ff}^{-1}(k-1)f(k)$$

$$\hat{\Phi}_{ff}^{-1}(k) = \frac{1}{1-\alpha}\left(\hat{\Phi}_{ff}^{-1}(k-1) - \frac{\alpha S(k)S'(k)}{1-\alpha+\alpha f'(k)S(k)}\right); \quad 0 < \alpha << 1 \tag{9.71}$$

$$e_k = d_k - b'(k)f(k)$$

$$b(k+1) = b(k) + 2u\hat{\Phi}_{ff}^{-1}(k)e_k f(k); \quad 0 < u << 1$$

Compared with the LMS algorithm (9.54), the RLS algorithm requires much more computation at each step and also requires the specification of two parameters, α and u. However, it has the potential, provided $\hat{\Phi}_{ff}^{-1}(k)$ is a good estimate of the inverse autocorrelation matrix, of producing faster convergence to b_{opt}, the optimal weight vector.

An example of RLS weight convergence is provided in Figure 9.7, which is like Figure 9.4 with the RLS weight track in place of the ideal weight track. Compared with the ideal and LMS tracks in Figure 9.4, the RLS track is more like the ideal track. It follows a more or less direct path to the optimal point, which, we recall, is at $[b_0, b_1] = [\sqrt{3}, -1]$ in this example. The LMS track contains twice as many iterations and, following the path of steepest descent, converges slowly as the gradient approaches zero.

A learning curve for the RLS algorithm is shown in Figure 9.8. The adaptive gain was set to $u = 0.05$ so the result could be compared with the LMS learning curve in Figure 9.5, and we can see that convergence is improved with the RLS algorithm in this example.

As in the case of Figures 9.4 and 9.5, the plots in Figures 9.7 and 9.8 were made using a MATLAB function on the publisher's website. The function is called *rls_filter* and has arguments as follows:

```
[g,b]=rls_filter(b,u,alpha,d,f)
```

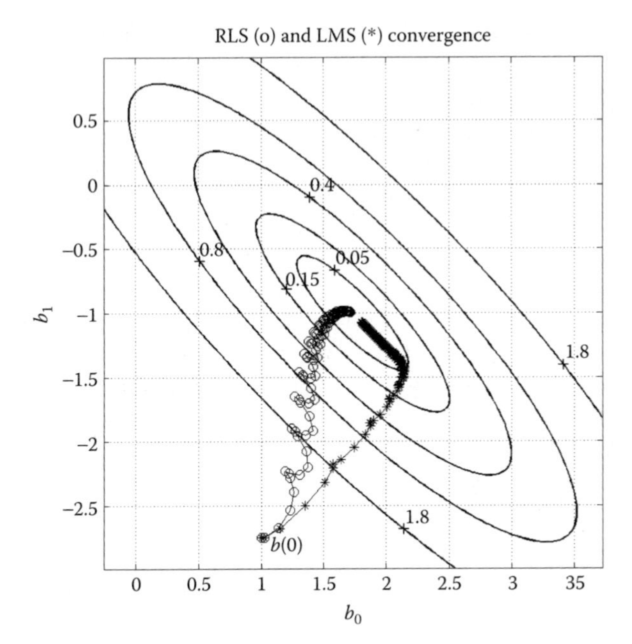

FIGURE 9.7
Comparison of recursive least-squares (RLS) and least-mean-square (LMS) weight convergences for the predictor with two weights (Figure 9.2): The number of iterations was $K = 150$ for RLS convergence and 300 for LMS convergence. The RLS parameters were $[u, \alpha] = [0.02, 0.2]$ and the LMS adaptive gain was $u = 0.05$.

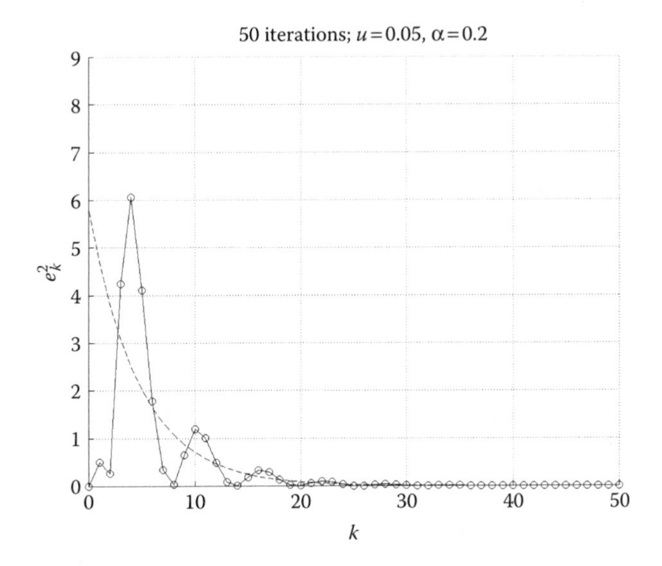

FIGURE 9.8
Comparison of convergence of e_k^2 (connected points) using the recursive least-squares algorithm with ideal convergence of the mean-squared error (dashed line) for the prediction example presented in Figure 9.3: The adaptive gain is $u = 0.05$ in both cases.

This function is the same as the *lms_filter* function described in Section 9.5, except that it implements the RLS algorithm and its arguments include the forgetting factor, α, used to adjust the correlation and power estimates.

Having seen these examples and having discussed and compared the LMS and RLS adaptive algorithms, the reader may wonder why, if one or the other of the algorithms does not converge as fast as desired, the adaptive gain cannot be increased until it does so. The answer is that the behavior of the system *after* convergence may suffer if u is too close to its stability limit (one). After convergence under stationary conditions, the weight vector, $b(k)$, continues to "adapt," and wanders around under the error surface in the vicinity of its optimum value, b_{opt}. That is, due to inaccuracies in local estimates, the MSE climbs up the bowl-shaped quadratic error surface and is then driven back down toward MSE_{min}. The continuation of this process results in a deviation of the MSE from its minimum value, MSE_{min}, which is undesirable. Thus, there is a trade-off between convergence speed and convergence accuracy, which we discuss in Section 9.7.

9.7 Measures of Adaptive System Performance

The two principal measures of adaptive system performance are those we discussed in the three preceding sections. First are the time constants for convergence of the weights and the MSE, which measure the speed of adaptation. Second is a measure called *misadjustment*, which is a normalized measure of the average deviation of the MSE from its minimum value.

For convergence time constants, we have the weight and MSE time constants, τ_w and τ_{MSE}, which we derived in Section 9.3, (9.16) and (9.22), and in Section 9.4, (9.55):

$$\begin{array}{|lll|}\hline \text{Convergence time constants:} & 0 < u << 1 & \\ \text{Ideal conditions:} & \tau_w \approx \dfrac{1}{2u}; \quad \tau_{MSE} \approx \dfrac{1}{4u} & \text{iterations} \\ \text{Steepest descent:} & \tau_w \approx \dfrac{N\sigma_f^2}{2u\lambda_{min}}; \quad \tau_{MSE} \approx \dfrac{N\sigma_f^2}{4u\lambda_{min}} & \text{iterations} \\ \hline \end{array} \tag{9.72}$$

Because these time constants are inversely proportional to u, why not increase u toward one and have a more responsive, faster-converging adaptive system? As we have seen, the answer is that in practical systems with noisy estimates of the gradient and/or the autocorrelation matrix, increasing u means increasing the *excess* MSE after convergence. The excess MSE is the increase in the MSE above MSE_{min} after convergence due to these noisy estimates. From (9.19), the excess MSE may be written as

$$\begin{aligned} \text{Excess MSE} &= \text{MSE} - \text{MSE}_{min} \text{ after convergence} \\ &= E\left[c'\Phi_{ff}c\right] \end{aligned} \tag{9.73}$$

For example, the noisy LMS gradient estimate in (9.53) causes the weight vector to be adjusted in the vicinity of MSE_{min} at each iteration after convergence and thus contributes to the excess MSE through variations in the translated weight vector, c, in accordance with (9.73).

To derive a simple and practical estimation of the excess MSE for the LMS algorithm, it is necessary to assume that these variations in c consist of small, independent jumps away from the optimal point at $c = 0$. With this approximation, under certain conditions, the excess MSE is

$$\text{Excess MSE} \approx u\,\text{MSE}_{min}; \quad 0 < u << 1 \tag{9.74}$$

The conditions are that u must be small, the eigenvalues of Φ_{ff} must be approximately equal, and the excess MSE is small (on the order of 0.1 or less). The derivation of this useful approximation, which is lengthy and involves a number of assumptions that must be supported, is not included here. Derivations are provided in Widrow et al.[10] and in Chapters 5 and 6 of Widrow and Stearns.[5]

Instead of the excess MSE, we use a normalized measure of the excess MSE called *misadjustment* (M), defined as follows:

$$\boxed{\text{Misadjustment:} \quad M = \frac{\text{Excess MSE}}{\text{MSE}_{min}} = u} \tag{9.75}$$

The misadjustment and the convergence time constants (9.72) constitute the principal performance measures for adaptive system performance.

The approximation in (9.75) is said to be good for values of M up to 0.1.[26] This assumption is investigated in Exercises 9.15 and 9.16. Both exercises investigate (9.75) using the example we have been using throughout this chapter, which is the one-step predictor in Chapter 8, Figure 8.7, with two weights and a sinusoidal input. The example is modified as shown in Figure 9.9 by adding white noise to d_k in order to produce a nonzero minimum MSE. Notice that the adaptive filter can cancel the sine wave, but it is not able to cancel the noise because the noise does not appear in f_k.

Exercise 9.15 consists of starting the weight vector at b_{opt} and running 2000 iterations of the LMS algorithm with adaptive gain $u = 0.05$. The result is illustrated in Figure 9.10, where 2000 iterations of the weight vector are plotted as points on the weight plane. The plot illustrates the typical random excursions of the converged weight vector in the vicinity of the minimum MSE, which are the cause of misadjustment.

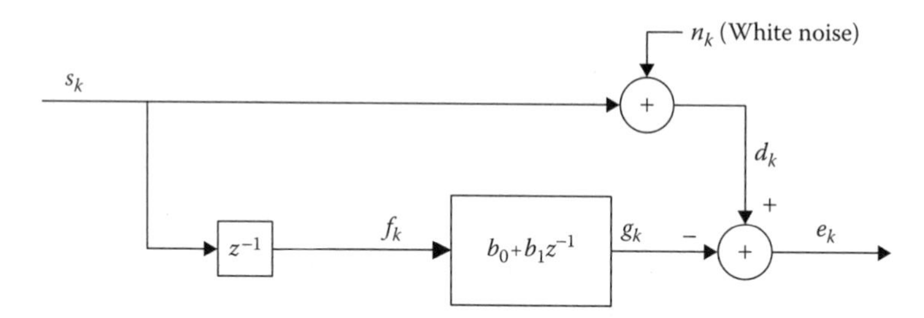

FIGURE 9.9
White noise (n_k) is added to d_k in the adaptive predictor used in Section 9.5 in order to produce a nonzero minimum mean-squared error.

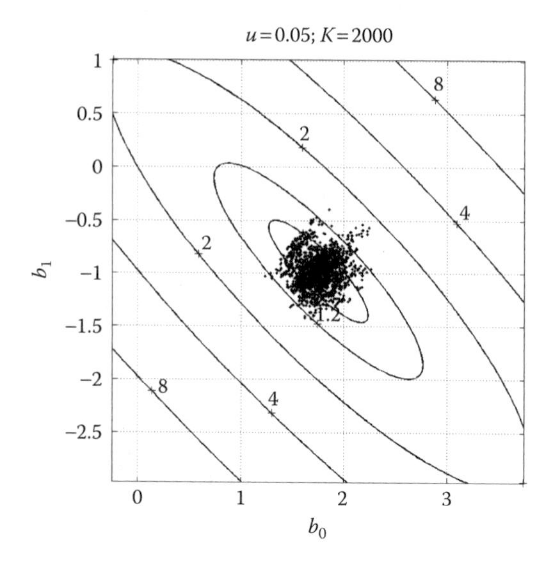

FIGURE 9.10
Result of Exercise 9.15, illustrating how the least-mean-square (LMS) algorithm drives the converged weight vector in the vicinity of minimum mean-squared error and causes misadjustment.

Exercise 9.16 consists of starting the weight vector at b_{opt} and making several independent runs of the LMS algorithm for each of several values of u, measuring the misadjustment in each case by averaging the squared prediction error, e_k^2. The result is plotted in Figure 9.11, which shows the experimental points with the approximation in (9.75). Even though the eigenvalues of Φ_{ff}, which are [0.1340, 1.8660] in this example, are quite different and do not fulfill the second condition for the validity of (9.75), the result verifies this equation as a useful approximation. However, as the misadjustment increases above 0.1, M becomes increasingly larger than the approximation in (9.75).

If we assume the validity of (9.75) as well as the time constants of (9.72) with $\sigma^2 = \lambda_{\text{min}}$, which are only approximations for LMS and RLS convergence as we have seen, we may combine the formulas for M and τ_{MSE} to obtain

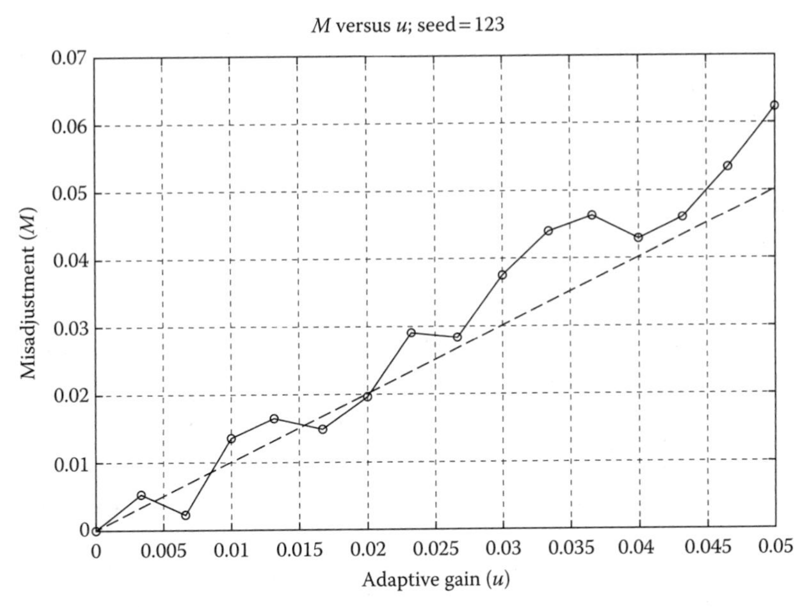

FIGURE 9.11
Result of Exercise 9.16, in which misadjustment is measured experimentally under stationary conditions with different adaptive gain values: The dashed line represents the theoretical approximation in (9.74).

$$\boxed{M \approx \frac{N}{4\tau_{\mathrm{MSE}}}; \quad 0 < u << 1,\ 1 < M < 0.1} \tag{9.76}$$

Although this formula does not apply in all situations, it nevertheless serves to summarize adaptive system performance. It demonstrates that in adaptive signal processing, low misadjustment and fast convergence are desirable but conflicting requirements. Faster convergence means noisier performance with a greater excess MSE. In fact, noting that approximately four time constants ($4\tau_{\mathrm{MSE}}$) are required for convergence from an arbitrary MSE to $\mathrm{MSE}_{\min}$, we can restate (9.76) as follows:[10]

$$\boxed{\text{Misadjustment } (M) \approx \frac{\text{number of weights } (N)}{\text{convergence time } (4\tau_{\mathrm{MSE}})}; \quad \begin{cases} 0 < u << 1 \\ 1 < M < 0.1 \end{cases}} \tag{9.77}$$

This is perhaps the simplest and most general statement one can make about the performance of linear adaptive systems that use algorithms like those we have been discussing in this chapter.

Beyond these considerations, at least at the time of this writing, the design of adaptive systems is less than an exact science. The number of weights (N), the adaptive gain (u), and the forgetting factor (α) must be chosen with the performance requirements in mind; but once they are chosen, performance in nonstationary conditions also is affected by unpredictable changes in signal statistics. The fact that adaptive signal processing systems work and survive in so many different applications, like the fact that biological systems do the same, is quite remarkable.

9.8 Other Adaptive Structures and Algorithms

In this chapter, we introduced adaptive signal processing using an FIR filter as the adaptive filter and discussed the LMS and RLS adaptive algorithms. These are the simplest and most widely applicable components of adaptive systems today, but there are other kinds of components and applications discussed in the literature.

One type of application using the FIR filter, which we have not mentioned in the chapter, is the *adaptive array*,[3,5,9] a form of which is illustrated in Figure 9.12. (A simple form of this filter was used in Chapter 8, Exercise 8.24.) The array consists of M sensors, s_1 through s_M, each s_m feeding its adaptive filter, $B_m(z)$. The desired signal, d_k, is derived in a manner that points the array at a signal source, or away from an interfering source, or in some other manner in a process known as *adaptive beamforming*.[3,5] Systems like this, with more than one adaptive filter driven by a single error, are known as *multiple-input* adaptive systems.

When each of the adaptive filters in Figure 9.12 consists of just a single weight, the system is called an *adaptive linear combiner*.[5,6] The adaptive linear combiner adapts in the same manner as the FIR filter, except that the elements of the input vector f are available simultaneously, that is, at the same

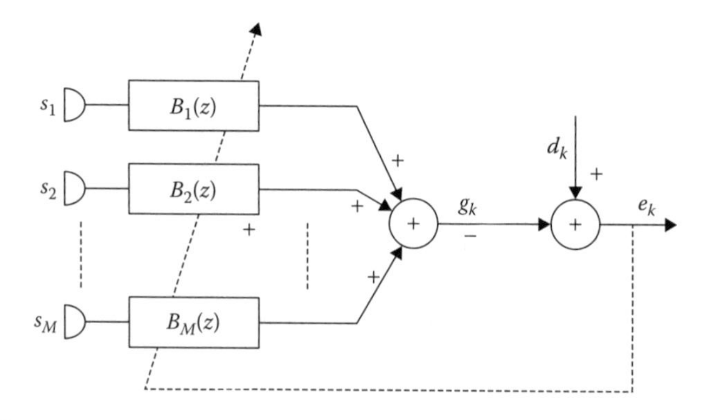

FIGURE 9.12
An adaptive array, consisting of a multiple-input adaptive system: To achieve a common goal, the adaptive filters are adjusted using a single error, e_k.

time step. For example, the adaptive linear combiner could take the form of a linear array of photosensors that moves relative to (scans) a visual image, or it could take the form of a planar array of sensors viewing a changing scene.

Other types of adaptive filter structures have been proposed. One is the *adaptive lattice*,[17–19,21,22,26] which is developed by allowing the weights, κ and λ, of one of the lattice structures described in Chapter 4, Section 4.11, to be adapted with a version of the LMS algorithm.

Recursive filters as well as nonlinear filter structures have been applied in adaptive signal processing.[5,6,11–13] Some of the nonlinear structures, for example, use *neural networks* consisting of networks of threshold elements that act like neurons. Recursive adaptive filters have not found a large number of applications, because the associated MSE functions are generally not quadratic and may contain local minima. But, recursive filters have poles, and their operations are, in general, reversible, so they remain candidates for adaptive systems in compression applications and system identification.

Other adaptive algorithms include special algorithms for adapting recursive and nonlinear systems,[6] including *genetic* algorithms,[5,15] which use the evolutionary or "survival of the fittest" approach to minimize the MSE.

There is also a class of algorithms known as *blind* algorithms used in communications systems that adapt to changing signal or channel conditions. In *blind equalization*,[6] the desired signal, d, which is known at the source but not at the receiver, is approximated (guessed) at the receiver using portions or properties of the received signal and is then used to modify the received signal. Blind equalization is used in adaptive communications receivers to reduce channel distortion as well as interfering noise and other signals, including echoes of the desired signal.

Exercises

9.1 In Section 9.3, the weight convergence formula (9.12) was proven by taking the z-transform of the recursion formula (9.10). Instead of using the z-transform, prove (9.12) by induction.

9.2 Examine the validity of the weight convergence time constant (9.16) by making 4 subplots in the same figure with $u = 0.05$ in the first plot and then 0.1, 0.2, and 0.4 in the rest of the subplots. In each subplot, complete the following:

a. Plot $1 - e^{-k/\tau_w}$ versus k for $k = [0, 5\tau_w]$ using a dashed line.

b. Plot symbols at $k = \tau_w, 2\tau_w, \ldots, 5\tau_w$.

c. Plot the exact convergence $1 - (1 - 2u)^k$ as a connected discrete plot. Comment on the validity of (9.16) with respect to each plot.

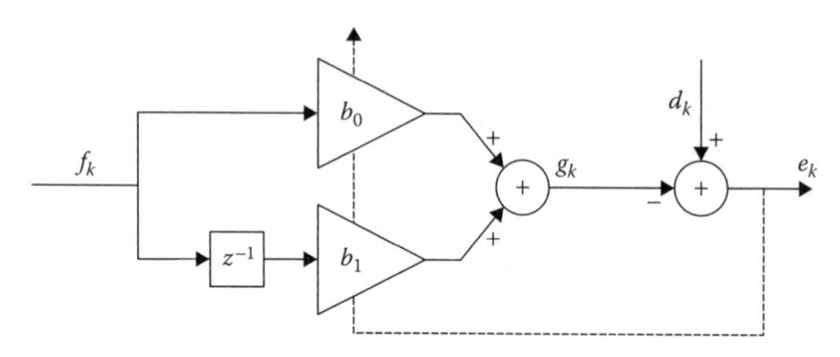

FIGURE 9.13
Adaptive filter with two weights.

9.3 A simple adaptive filter with two weights is illustrated in Figure 9.13. The signals f and d are as follows:

$$f_k = \sin(2\pi k/N); \quad d_k = 2\cos(2\pi k/N)$$

a. Express the autocorrelation matrix Φ_{ff} as a function of N, the number of samples per cycle.

b. Express the cross-correlation vector, φ_{fd}, as a function of N.

c. Express the MSE as a quadratic function of b_0 and b_1.

9.4 For the adaptive filter described in Exercise 9.3, complete the following:

a. Write an expression for the gradient of the error surface.

b. Write an expression for the optimal weight vector.

9.5 For the adaptive filter described in Exercise 9.3, assume $N = 10$ and complete the following:

a. Calculate the optimal weight vector, b_{opt}.

b. On the $b_0 - b_1$ plane, plot contours (curves of constant MSE) for $N = 10$ and the following MSE values: 0.0001, 1.0, 5.0, and 10.0. Use $b \pm 6$ to define the limits of the b-plane. (Note: This plot is most easily done using the MATLAB *contour* function.)

9.6 Suppose we modify Figure 9.13 to produce the adaptive linear combiner in Figure 9.14. Now there are two input signals instead of one, and no delays are involved. We assume stationary signals with the following statistics:

$$E\left|d_k^2\right| = 36, \; E\left|f_{1k}^2\right| = 2, \; E\left|f_{2k}^2\right| = 3, \; E\left|f_{1k}f_{2k}\right| = 1,$$

$$E[f_{1k}d_k] = 6, \; E[f_{2k}d_k] = 4$$

a. Express Φ_{ff} and φ_{fd} for this case.

b. Write an expression for MSE as a function of b_0 and b_1.

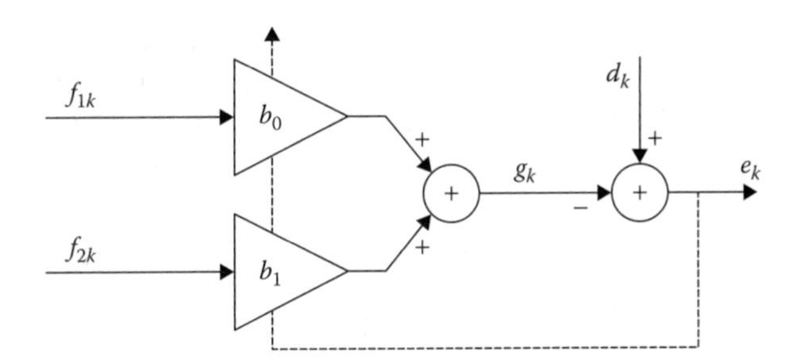

FIGURE 9.14
Adaptive linear combiner.

c. Find the optimal weight vector, b_{opt}.

d. Determine the minimum MSE.

9.7 For the adaptive linear combiner in Figure 9.14,

a. Find the eigenvalues, λ_0 and λ_1.

b. Find the eigenvector matrix, Q.

c. Plot MSE contours on the $b_0 - b_1$ plane using MSE values [10, 50, 100, 200].

d. In the same figure, plot translated (c) axes and rotated (v) axes. Note that the v axes are the principal axes of the elliptical MSE contours and thus indicate the steepest and the least-steep descent paths on the MSE surface.

9.8 For the adaptive linear combiner in Figure 9.14,

a. Write an expression for the gradient vector as a function of the weight vector.

b. Starting with initial weight vector $b(0)$ = [10; 10] and using parameters $u = 0.1$ and σ_f^2 equal to the average input signal power in Exercise 9.6, compute the first 5 weight values on a path that converges directly on b_{opt}.

c. Repeat Exercise 9.8b using the path of steepest descent instead of the direct path.

9.9 For the adaptive linear combiner in Exercise 9.3, complete the following:

a. Write an expression for the gradient vector as a function of the weight vector.

b. Starting with initial weight vector $b(0) = [2; -8]$ and using parameters $u = 0.1$ and σ_f^2 equal to $\varphi_{ff}(0, 0)$, compute the first 5 weight values on a path that converges directly to b_{opt}.

c. Repeat Exercise 9.9b using the path of steepest descent instead of the direct path.

9.10 In the prediction example in Section 9.5 and Figure 9.6, the nonstationary input is described as a unit sine wave, that is, stationary at 0.05 Hz-s for a long time and then doubles in frequency.

a. What is the optimal weight vector, b_{opt}, for predicting the initial part of the signal?

b. What is the minimum MSE associated with Exercise 9.10a?

c. In the upper of three subplots, plot the complete data frame as it appears in the upper plot in Figure 9.6.

d. Operate the LMS algorithm on the data frame with 1000 samples with b initialized to b_{opt} and $u = 0.2$. Plot the residue after $k = 0$ in the center subplot and observe that your plot looks the same as the lower plot in Figure 9.6.

e. Operate the RLS algorithm on the data frame with 1000 samples with b initialized to b_{opt}, $u = 0.2$, and $\alpha = 0.001$. Plot the residue after $k = 0$ in the lower subplot, and compare the result with the center plot.

9.11 An adaptive filter with 2 weights is used as in Figure 9.15 to minimize the squared error, e_k^2, with the following inputs at frequency $\nu = 0.1$ Hz-s:

$$f_k = \cos(2\pi\nu k); \quad d_k = 3\cos(2\pi\nu k + \pi/5); \quad 0 \le k < \infty$$

a. Find the autocorrelation matrix Φ_{ff}.

b. Find the eigenvalues of Φ_{ff} by hand using (9.24).

c. Find the time constant for convergence of the MSE using the LMS algorithm with $u = 0.1$.

d. Plot a learning curve (MSE versus k) for the LMS algorithm that begins at the mean-squared value of d_k and converges to zero over $k = 0{:}99$.

e. On the same graph, plot a learning curve using the average eigenvalue in place of λ_{min}.

f. Run the LMS algorithm with signals d and f, beginning with weight vector $b(0) = [0, 0]'$, and plot e_k^2 versus k, again on the

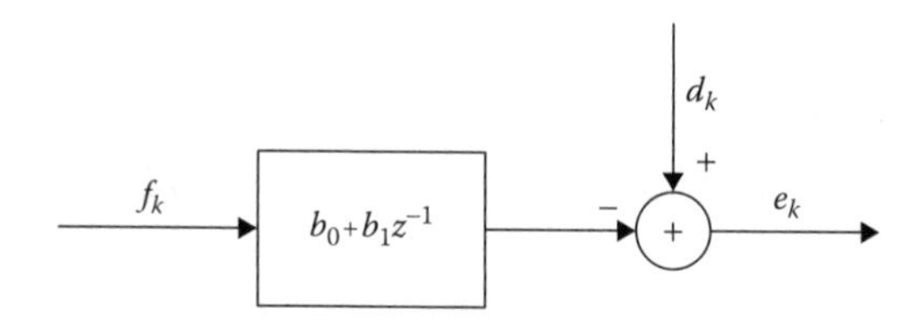

FIGURE 9.15
Adaptive filter with two weights.

same graph. Note how a running average of this plot would run between the two theoretical curves.

9.12 Repeat Exercise 9.11, but this time use the RLS algorithm with $u = \alpha = 0.1$ and $b(0) = [0, 0]'$. Plot theoretical learning curves for ideal convergence and convergence using the LMS algorithm. This time, use the range $k = 0{:}50$. Note the faster convergence of the RLS algorithm, and comment on the comparison of the squared error plot with the learning curves.

9.13 Suppose, as in Exercises 9.11 and 9.12, that the input to an adaptive filter with two weights is an arbitrary sinusoid at frequency ν Hz-s, say

$$f_k = A\cos(2\pi\nu k + \gamma); \quad 0 \le k < \infty$$

a. Write an expression for the autocorrelation matrix Φ_{ff}.

b. Write a quadratic expression and solve for the eigenvalues of Φ_{ff}.

c. Prove that the ratio of the minimum to the average eigenvalue is

$$\lambda_{min}/\lambda_{av} = 1 - \cos(2\pi\nu)$$

d. Show (in terms of time constants) why the aforementioned expression implies increasingly nonideal convergence of the LMS algorithm as ν decreases.

9.14 Verify the result of Exercise 9.13 by plotting LMS learning curves for frequencies $\nu = 0.1$ and $\nu = 0.2$ Hz-s and $k = 0{:}99$, using the signals in Exercise 9.11 in the configuration of Figure 9.15.

9.15 This exercise illustrates misadjustment using the LMS algorithm with $u = 0.05$ and the configuration in Figure 9.9 with the following signals:

$$s_k = \sqrt{2}\sin(2\pi[0:2000]/12)'; \quad n_k = \text{white Gaussian noise}; \quad \sigma_n^2 = 1$$

a. Set the random number seed to 123, and find the optimum weight vector using actual signals. Compare with the theoretical optimum in Figure 9.2, which was $b_{opt} = [\sqrt{3}, -1]$.

b. Plot the error contours on the weight plane as shown in Figure 9.10.

c. Run 2000 iterations of the LMS algorithm with $u = 0.05$, using b_{opt} for the initial weight vector. Plot a point for the weight vector at each iteration, and complete the plot in Figure 9.10.

d. Compute the theoretical misadjustment for this case.

e. Compute the misadjustment using the error signal, e_k, in this example, and compare the result with the answer to part (d).

9.16 This exercise tests the theoretical misadjustment formula (9.75) using the LMS algorithm in a manner similar to Exercise 9.15, that is, with the configuration in Figure 9.9 and with the following signals:

$$s_k = \sqrt{2}\,\sin(2\pi[0:2000]/12)';$$

$$n_k = \text{white uniform noise}; \quad \sigma_n^2 = 0.05, \ \mu_n = 0$$

(Note: This exercise involves a total of 1.5 million LMS iterations. If your computer is too slow, use a compiled program to obtain the 16 data points, and then plot these using MATLAB.)

The objective is to start the LMS algorithm at b_{opt} and obtain misadjustments experimentally at each of 16 values of u:

$$u = \frac{0.05n}{15}; \quad n = 0, 1, \ldots, 15$$

The experimental misadjustment at $u = 0$ is zero, because no adjustment is allowed in this case. At each value of u greater than zero, make 10 runs, each with 10^4 LMS iterations. At the beginning of the mth run, set the weights to $b_{opt} = [\sqrt{3}, -1]$ and add a phase shift of $2m\pi/10$ to the argument of s_k so the signal begins at a different point in its cycle. After all 10 runs, compute the MSE and the misadjustment (M) based on the theoretical minimum MSE. Finally, plot M versus u, and obtain a result similar to Figure 9.11.

References

1. Widrow, B., and M. E. Hoff Jr. 1960. Adaptive switching circuits 1960. *IRE WESCON Conv Rec* 4:96–104.
2. Widrow, B., et al. 1975. Adaptive noise canceling: Principles and applications. *Proc IEEE* 63(12):1692–1716.
3. Widrow, B., et al. 1967. Adaptive antenna systems. *Proc IEEE* 55(12):2143–59.
4. Haykin, S. S. 1991. *Adaptive Filter Theory.* 2nd ed. Englewood Cliffs, NJ: Prentice Hall.
5. Widrow, B., and S. D. Stearns. 1985. *Adaptive Signal Processing*. Englewood Cliffs, NJ: Prentice Hall.
6. Treichler, J. R., C. R. Johnson Jr., and M. G. Larimore. 1987. *Theory and Design of Adaptive Filters.* New York: Texas Instruments, Inc. and John Wiley & Sons.
7. Widrow, B., J. M. McCool, and B. Medoff. 1978. Adaptive control by inverse modeling. *12th Asilomar Conference on Circuits, Systems, and Computers,* p.90. IEEE Press.
8. Widrow, B., D. Shur, and S. Shaffer. 1981. On adaptive inverse control. *15th Asilomar Conference on Circuits, Systems, and Computers,* p.185. IEEE Press.
9. Monzingo, R. A., and T. W. Miller. 1980. *Introduction to Adaptive Arrays.* New York: John Wiley & Sons.

10. Widrow, B., et al. 1976. Stationary and nonstationary learning characteristics of the LMS adaptive filter. *Proc IEEE* 64(8):1151–62.
11. Feintuch, P. L. 1976. An adaptive recursive LMS filter. *Proc IEEE* 64(11):1622–4.
12. Larimore, M. G., J. R. Treichler, and C. R. Johnson Jr. 1980. SHARF: An algorithm for adapting IIR digital filters. *IEEE Trans Acoust Speech Signal Process* 28(4):428–40.
13. White, S. A. 1975. An adaptive recursive digital filter. *9th Asilomar Conference on Circuits, Systems, and Computers*, p.21. IEEE Press.
14. Widrow, B., and J. M. McCool. 1976. A comparison of adaptive algorithms based on the methods of steepest descent and random search. *IEEE Trans Antennas Propag* 24(5):615–37.
15. Etter, D. M., and M. M. Masakawa. 1981. A comparison of algorithms for adaptive estimation of the time delay between sampled signals. *Proc ICASSP-81* 1253–6.
16. Dentino, M. J., J. McCool, and B. Widrow. 1978. Adaptive filtering in the frequency domain. *Proc IEEE* 66(12):1658–9.
17. Makhoul, J. 1978. A class of all-zero lattice digital filters: Properties and applications. *IEEE Trans Acoust Speech Signal Process* 26(4):304–14.
18. Makhoul, J. 1977. Stable and efficient lattice methods for linear prediction. *IEEE Trans Acoust Speech Signal Process* 25(5):423–8.
19. Griffiths, L. J. 1978. An adaptive lattice structure for noise-canceling applications. *Proc ICASSP-78* 87–90.
20. Griffiths, L. J., F. R. Smolka, and L. D. Trembly. 1977. Adaptive deconvolution: A new technique for processing time-varying seismic data. *Geophysics* 42(4):742.
21. Makhoul, J., and R. Viswanathan. 1978. Adaptive lattice methods for linear prediction. *Proc ICASSP-78* 83–6.
22. Makhoul, J. L., and L. K. Cosell. 1981. Adaptive lattice analysis of speech. *IEEE Trans Acoust Speech Signal Process* 29(3):654–9.
23. Widrow, B., and E. Walach. 1984. On the statistical efficiency of the LMS algorithm with nonstationary inputs. *IEEE Trans Inf Theory* 30(2):211–21.
24. Thomas Jr., G. B. 1968. *Calculus and Analytic Geometry*. 4th ed. Chap. 10. Reading, MA: Addison-Wesley.
25. Luenberger, D. G. 1973. *Introduction to Linear and Nonlinear Programming*. Chap. 7. Reading, MA: Addison-Wesley.
26. Lim, J. S., and A. V. Oppenheim, eds. 1988. *Advanced Topics in Signal Processing*. Chap. 5. Englewood Cliffs, NJ: Prentice Hall.
27. Johnson Jr., C. R. 1988. *Lectures on Adaptive Parameter Estimation*. Englewood Cliffs, NJ: Prentice Hall.

Further Reading

Ahmed, N., et al. 1979. A short-term sequential regression algorithm. *IEEE Trans Acoust Speech Signal Process* 27(5):453–7.

Atal, B. S., and S. L. Hanauer. 1971. Speech analysis and synthesis by linear prediction of the speech wave. *J Acoust Soc Am* 50(2):637–55.

Bershad, N. J., and P. L. Feintuch. 1979. Analysis of the frequency domain adaptive filter. *Proc IEEE* 67(12):1658–9.

Clark, G. A., and P. W. Rodgers. 1981. Adaptive prediction applied to seismic event detection. *Proc IEEE* 69(9):1166–8.

Clark, G. A., S. K. Mitra, and S. R. Parker. 1981. Block implementation of adaptive digital filters. *IEEE Trans Circuits Syst* 28:584–92.

Clark, G. A., S. R. Parker, and S. K. Mitra. 1983. A unified approach to time- and frequency-domain realization of FIR adaptive digital filters. *IEEE Trans Acoust Speech Signal Process* 31(5):1073–83.

Cowan, C. F. N., and P. M. Grant, eds. 1985. *Adaptive Filters*. Englewood Cliffs, NJ: Prentice Hall.

David, R. A. 1981. *IIR Adaptive Algorithms Based on Gradient Search Techniques*, Ph.D. thesis, Stanford University, Stanford, CA.

Ferrara, E. R. 1980. Fast implementation of LMS adaptive filters. *IEEE Trans Acoust Speech Signal Process* 28:474–5.

Gersho, A. 1969. Adaptive equalization of highly dispersive channels for data transmission. *Bell Syst Tech J* 48:55.

Honig, M. L., and D. G. Messerschmitt. 1984. *Adaptive Filters*. Norwell, MA: Kluwer Academic Publishers.

Itakura, F., and S. Saito. 1971. Digital filtering techniques for speech analysis and synthesis. *Proc 7th International Congress on Acoustics*, p.261. Budapest.

Landau, A. I. 1979. *Adaptive Control: The Model Reference Approach*. New York: Marcel Dekker, Inc.

Lucky, R. W. 1966. Techniques for adaptive equalization of digital communication systems. *Bell Syst Tech J* 45:255–6.

Makhoul, J. 1973. Spectral analysis of speech by linear prediction. *IEEE Trans Audio Electroacoust* 21(3):140–8.

Poularikas, A. D., and Z. M. Ramadan. 2006. *Adaptive Filtering Primer with Matlab*. Boca Raton, FL: CRC Press.

White, S. A. 1980. A nonlinear digital adaptive filter. *14th Asilomar Conference on Circuits, Systems and Computers*, p.350. IEEE Press.

Youn, D. H., N. Ahmed, and G. C. Carter. 1983. Magnitude-squared coherence function estimation: An adaptive approach. *IEEE Trans Acoust Speech Signal Process* 31(1):137–42.

Zeidler, J. R., et al. 1978. Adaptive enhancement of multiple sinusoids in uncorrelated noise. *IEEE Trans Acoust Speech Signal Process* 26(3):240–54.

10

Signal Information, Coding, and Compression

10.1 Introduction

In this chapter, we first discuss the measurement of signal *information content* as a basis for the rest of the subjects in the chapter. Then, we will examine topics applicable to the encoding and compression of signals, that is, waveforms and images. Encoding and compression[1–7] are broad subjects, and our goal here is to introduce some of the basic techniques and show how they apply.

Signal coding and compression seem destined always to be necessary parts of the world of digital signal processing (DSP). Every time the technology offers twice the amount of storage capacity, users discover they have more than twice the amount of information to store. When the capacity of a communication link is increased, the demand for channel capacity seems to increase in proportion. Thus, there is always the requirement to render signals in ways that reduce the related storage capacity or communication bandwidth.

The objective here is to *reduce* or *compress* the number of *bits* (binary digits) in the representation of a signal or image or, in other words, to remove unnecessary or redundant information from a signal in ways that reduce its size and yet allow its restoration.

Note that just the opposite is done when signals are encoded for *error detection* and *error correction*. In these cases, redundant information is deliberately added to the signal in order to allow its verification. For example, you could append the sum of all the samples in a frame of waveform data before sending the frame, thus allowing the receiver to detect an error in one of the samples. Simple and practical coding methods for error detection and correction such as Hamming and cyclic codes[3,5] are well known, but these are subjects usually not covered in a course on DSP, and again, our interest here is in *removing* redundant information from a signal rather than adding it.

Another reason for our interest in coding and compression lies in the close relationship between these subjects and the *encryption* of signals and other data. Encryption is a process where data is transformed in ways such that recovery is (ideally) not possible except for those who know the transformation. Generally, a transformation is easier to discover when the data contains redundant information, such as a repeated sequence of samples. Thus, although not always done, it is generally best to remove redundant information from data before encrypting it, and then restore that same information after decryption.

When we discuss coding and compression, special terms are commonly used. In coding theory, a sample of a waveform or a pixel in an image becomes a *symbol*, and in some kinds of coding, there is a *symbol code* associated with each symbol. Furthermore, the waveform or image must be *discrete*, with a finite range of values. The symbol set for a *continuous* waveform, however small its range of values, would be infinite. Symbol codes are introduced in Section 10.4.

10.2 Measuring Information

The bit (binary digit) is the basic unit we use to measure information. The bit has two possible *states* (zero and one). Other familiar measures are the *byte* (8 bits, $2^8 = 256$ states), which is the basic element in many storage and computing systems, and the computer *word*, which may consist of 16, 32, or 64 bits.

The common measure of the amount of information in a waveform or image (or in any similar set of discrete data) is essentially the log base 2 ($\log_2$) of its a priori probability of occurrence. Consider the example in Figure 10.1,

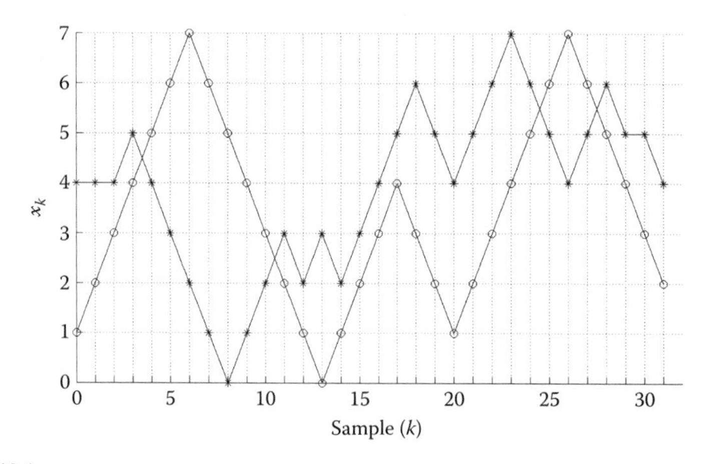

FIGURE 10.1
Two possible waveform vectors out of a large ensemble of different vectors: Each sample in a vector has 8 possible values, and there are 32 samples in a vector; so the ensemble consists of 8^{32} vectors.

which illustrates two possible discrete waveforms in a particular time frame. At each time step (k) in the frame, a given waveform may have any of 8 different amplitudes (states), that is, any sample value from 0 through 7. It follows that any pair of samples may exist in $8^2 = 64$ different states, any vector of three samples in 8^3 states, and so on. Therefore, the number of different waveform vectors that could occupy the frame in Figure 10.1 with 32 states is

$$\text{Number of possible vectors} = 8^{32} \tag{10.1}$$

Suppose all 8^{32} waveforms are equally likely. Then, the probability of each possible waveform is 8^{-32}, and as soon as we are given one waveform vector, such as one of the two shown in Figure 10.1, we have gained the amount of information in the vector, which is

$$\text{Information in a vector} = -\log_2(8^{-32}) = 96 \text{ bits} \tag{10.2}$$

The term used for this information measure is *entropy* (H). It is used in the sense that, in this example, the identification of a particular waveform results in the entropy (uncertainty) being reduced by 96 bits. Thus, our basic measure of information, or entropy, which is due to C.E. Shannon,[1] is

$$\boxed{\text{Entropy of event with probability } P\text{:} \quad H = -\log_2 P} \tag{10.3}$$

Now suppose, instead of defining an "event" to be an occurrence of a complete vector, we define an event to be the selection of one of the $K = 8$ possible symbols (values of x_k) in Figure 10.1 and let P_n represent the probability of the n-th event, as discussed in Chapter 7, Section 7.2. Then it is reasonable to extend (10.3) and define the average or expected symbol entropy:

$$\text{Average entropy (bits/symbol):} \quad H = -\sum_{n=0}^{N-1} P_n \log_2 P_n \tag{10.4}$$

If all waveform vectors are equally likely, as we have assumed so far, then all 8 symbols in Figure 10.1 must also be equally likely with $P_n = 1/8$ and, hence, $H = 3$ bits/symbol. Thus, the entropy of a vector with 32 equally likely symbols is 3(32) = 96 bits, and we see that the entropy measure (10.4) agrees with our idea of accumulating "information." As each successive symbol becomes known, our uncertainty about the waveform decreases by 3 bits until, after 32 symbols are known, we have a total of 96 bits of information.

There are different ways to apply the entropy measure (10.4). First, the measure is applicable to all kinds of symbol sets (letters, numbers, etc.), and all kinds of information besides just waveforms and images, which are our primary interests here.

Second, when the entropy measure is applied generally to the measurement of information and specifically to the design of codes for signal compression, the determination of the symbol probabilities (P_n) depends on the application. For example, suppose we are measuring the information in the single sentence you are now reading, and P_{101} stands for the probability of occurrence of the 101st ASCII symbol *e*. We could estimate P_{101} in two ways: (1) as the relative occurrence of *e* in this book or in all literature and (2) as the relative occurrence of *e* in the sentence. If we wanted a valid measure of information, we would choose the first way; but if we were interested in compressing just the sentence as an isolated message, we would choose the second, as explained further in Sections 10.3 and 10.4.

10.3 Two Ways to Compress Signals

In this section, we discuss briefly two different and essentially independent ways to compress a signal or image. For this discussion, we take one of the waveforms in Figure 10.1, say the waveform beginning with $x_0 = 1$, and assume this waveform is to be encoded with fewer than 96 bits and stored in a database. Exact retrieval must be possible knowing only the general rules used for encoding. The waveform is plotted in Figure 10.2.

We first notice that the symbols, that is, waveform values in Figure 10.2, are not equally likely. Using the *freq* function defined in Chapter 7 (Section 7.9), we have the following:

Symbol (n) in x:	0	1	2	3	4	5	6	7
Frequency $[f_x(n)]$:	$\frac{1}{32}$	$\frac{4}{32}$	$\frac{6}{32}$	$\frac{6}{32}$	$\frac{5}{32}$	$\frac{4}{32}$	$\frac{4}{32}$	$\frac{2}{32}$

(10.5)

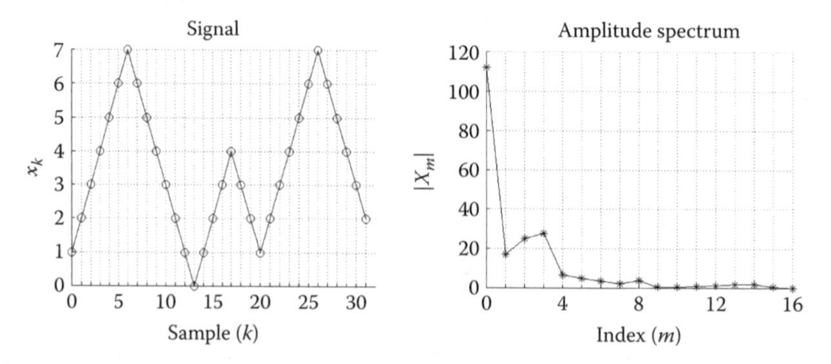

FIGURE 10.2
One of the signal vectors in Figure 10.1 and its amplitude spectrum. Note that the amplitude *distribution* is given in (10.5).

Using $f_x(n)$ in place of P_n in (10.4), the average symbol entropy in this case is

$$H_x = -\sum_{n=0}^{7} f_x(n)\log_2 f_x(n) = 2.8553 \text{ bits/symbol} \tag{10.6}$$

Because the entropy in bits per symbol is less than 3, the number of bits per symbol required for the ordinary binary codes (000 through 111), we know that *entropy coding*, which is discussed in Section 10.4, may provide a way to store the signal with fewer than 32(3) = 96 bits.

Second, we notice in the amplitude spectrum of x, which is also plotted in Figure 10.2, that not all *spectral* elements are equal. As we noted in Chapter 7, Section 7.5, this implies that the signal is *correlated*, that is, the autocorrelation function, $\varphi_{xx}(m)$, is not just an impulse at $m = 0$. In other words, the samples of x are not independent and therefore, *linear predictive coding* as it is described in Chapter 8, Section 8.10, could also be applied to compress the waveform. In Figure 10.3, we have a plot of the original vector, x, with the discrete prediction error vector, e. In this case, the symbol frequencies, similar to (10.5), are as follows:

Symbol (n) in e:	-1	0	1
Frequency $[f_e(n)]$:	$\frac{11}{32}$	$\frac{6}{32}$	$\frac{15}{32}$

(10.7)

The corresponding entropy, H_e, is less than H_x in (10.6):

$$H_e = -\sum_{n=0}^{2} f_e(n)\log_2 f_e(n) = 1.4948 \text{ bits/symbol} \tag{10.8}$$

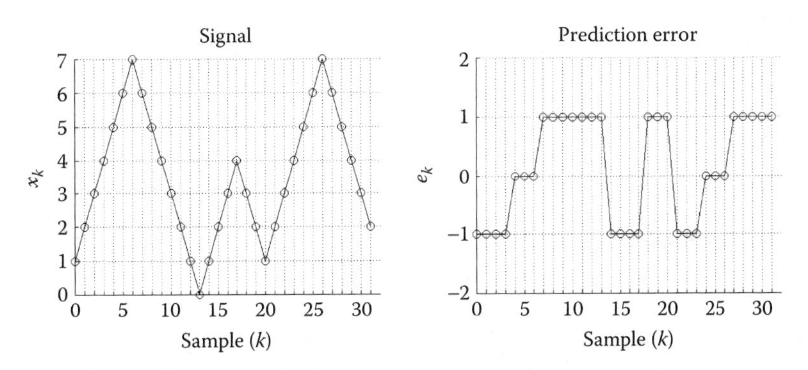

FIGURE 10.3
The signal vector in Figure 10.2 and its prediction error rounded to the nearest integer: The one-step predictor has two weights derived using covariance functions.

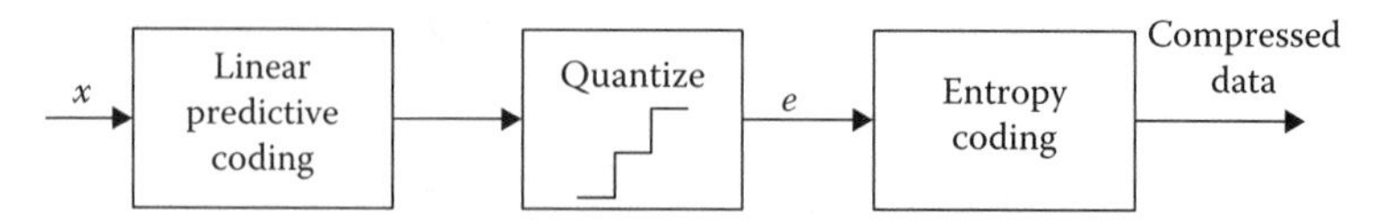

FIGURE 10.4
Compression of a signal (x) in two stages: The first stage reduces intersymbol dependence, and the second stage applies coding to reduce the number of bits per symbol in e, the quantized prediction error.

Thus, in comparison with (10.6), further compression is possible due to the lowered signal entropy. Note that the predictor weights (two weights in this case) must be stored with the error vector as described in Chapter 8, but this would not be significant with a long signal vector.

Thus, we see that there are two ways to compress a signal. One aims at encoding the signal in a way such that the number of bits is close to the signal entropy times the signal length. The second aims at reducing the range of the signal amplitude, and thus the number of bits per symbol, by removing intersymbol dependence and whitening the spectrum of the signal.

Regarding the second method, predictive coding is not the only way to reduce intersymbol dependence. Other methods include *string coding*,[5,8–10] which is particularly applicable to text and certain kinds of images; *vector quantization*;[11] and some types of *transform coding*.[7] Transform coding methods are described in Sections 10.5 through 10.7.

A final important point is that the two methods may be used *together* in a single compression operation. In the example we have been considering, a symbol code could be applied to vector e to replace the binary code of two bits per symbol and approach the entropy limit in (10.8), and we would have the scheme shown in Figure 10.4, that is, decorrelation followed by entropy coding of the quantized prediction error. In the illustration we have given here, the prediction error is simply rounded to the nearest integer, which allows exact recovery of the original vector as demonstrated in Chapter 8. One could produce further compression by increasing or varying the size of the quantizer steps, but then exact recovery would no longer be possible. This is known as *lossy* compression, as opposed to *lossless* compression, and is applicable in situations such as narrow-band telephony, or image and video compression, where exact recovery is not a requirement.

10.4 Adaptive Predictive Coding

Before we discuss entropy coding in more detail, *adaptive predictive coding* provides a useful example of the two-stage compression process just described in Section 10.3. The general idea of linear prediction and recovery was discussed in Chapter 8, Section 8.10, and illustrated in Figure 8.22. In order to

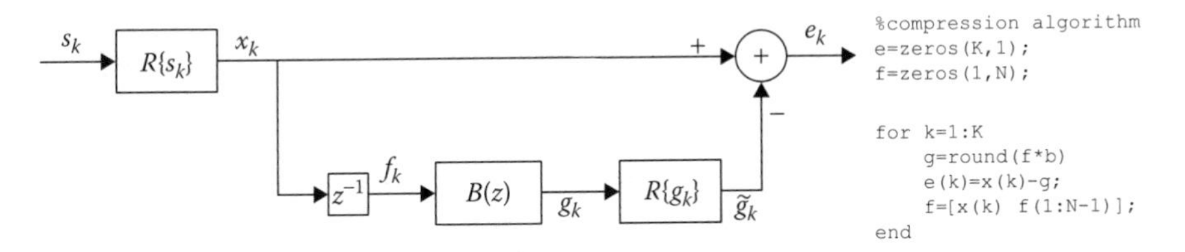

FIGURE 10.5
Nonadaptive prediction structure with rounding to produce a discrete prediction error.

make the process lossless, that is, have exact recovery of the original signal, we must use a configuration such as the one shown in Figure 10.5, which shows the process beginning by *rounding* each input sample, x_k. If the samples are from an analog-to-digital converter, then of course they are already rounded and have the overall finite range of the converter. But if x_k is coming from a floating-point processor, then it is first rounded to produce samples with a finite set of possible values, like those in Figures 10.2 and 10.3.

The prediction process in Figure 10.5 is like that in Chapter 8, Figure 8.22, but with two rounding operations, each indicated by $R\{\}$. First, each signal sample, s_k, is rounded (if necessary) to produce a discrete version, x_k. Second, the filter output is also rounded (if necessary) to produce $\tilde{g}_k$. Then the prediction error, e_k, is therefore an integer available for entropy coding. (If e_k were a continuous variable with an infinite number of possible values, entropy coding would obviously not be possible.) The algorithm in the figure assumes a fixed, nonadaptive filter, *B(z)*, implemented in a floating-point processor. The rounded input signal, *x*, is a column vector of length *K*, and *b* is the filter weight vector of length *N*. Note that *f* is a row vector and *b* is a column, so the product $g = \text{round}(f * b)$ is a scalar.

With the design in Figure 10.5, lossless recovery of the discrete signal, x_k, is possible. Assume first that a *fixed* filter, $B(z)$, is used to compress a given signal vector, *x*, of finite length *K*. Then, as discussed briefly in Chapter 4, Section 4.8, lossless recovery is possible as shown in Figure 10.6, which is similar to the recovery operation of Chapter 8, Figure 8.22. Note that only the weight vector, *b*, and the error signal, *e*, are used in the recovery process. As a practical consideration, the first *N* samples of *e* may be stored separately because in general they tend to be large compared with the remaining samples of *e*.

The main purpose of this section is to show that the prediction and recovery processes in Figures 10.5 and 10.6 can be made *adaptive* and remain lossless. We replace the fixed finite impulse response (FIR) filter, $B(z)$, with an adaptive filter running the least-mean-square (LMS) algorithm in Chapter 9, (9.54). In the latter, let

$$\mu_k = \frac{2u}{N\sigma_f^2(k)} \tag{10.9}$$

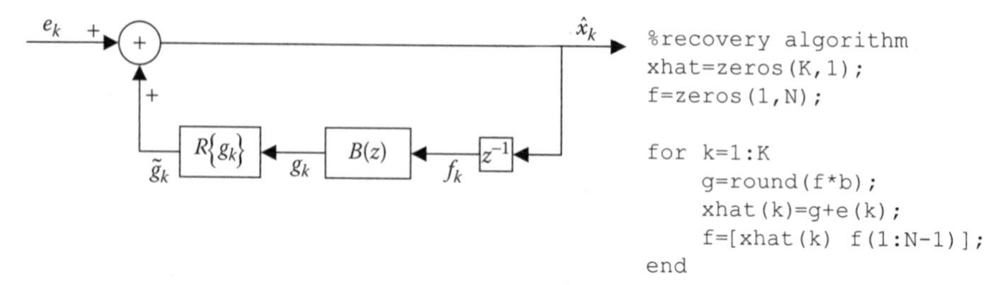

FIGURE 10.6
Lossless recovery of x_k in Figure 10.5.

and assume $\sigma_f^2(k)$ is initially nonzero and updated in terms of present and past signal values, x_k, x_{k-1}, and so on, in the manner of Chapter 9, (9.69), for example. Then the LMS algorithm (9.54) in Chapter 9 is

$$b_{k+1} = b_k + \mu_k e_k f(k) \tag{10.10}$$

where $f(k)$ is the f vector at the k-th time step. Now we can make the predictor adaptive by modifying the algorithms in Figures 10.5 and 10.6 as follows:

```
%adaptive compression            %adaptive recovery
b = zeros (N, 1);                xhat = zeros (K, 1);
e = zeros (K, 1);                f = zeros (1, N);
f = zeros (1, N);                b = zeros (N, 1);

for  k = 1:K                     for  k = 1:K
     g = round (f*b);                 g = round (f*b);
     e (k) = x (k) - g;               xhat (k) = g + e (k);
     f = [x (k), f (1:N-1)];          f = [xhat (k), f (1:N-1)];
     b = b + mu*e (k)*f';             b = b + mu*e (k)*f';
end                              end
```
(10.11)

Here, μ ("mu") is held constant for simplicity. Note however that b is updated after $\hat{x}_k$ is computed in the recovery process; so μ_k, which is used in the update of b, could be computed in terms of present as well as past signal values using (10.9). And, since $\hat{x}_k = x_k$ in the lossless case, b is updated during recovery exactly as it is updated during compression. Note also that b is updated in both algorithms using f' instead of f, because b is a column vector.

Many applications, especially in speech, audio, and video, as well as with some types of instrumentation, allow *lossy* compression with inexact recovery of the original signal, that is, where the mean-squared error (MSE) between the original and recovered signals is controlled to be a small but finite number. One way to achieve this is to reduce the range of the prediction error via *quantization*. Suppose, for example, that the discrete prediction error, e_k in Figure 10.5, is in the range

$$-e_{max} \geq e_k \geq e_{max}, \quad \text{where} \quad e_{max} = \max(|e_k|) \tag{10.12}$$

[If the predictor is to be designed before $\max(|e_k|)$ is known, e_{max} may be set to some overall maximum such as the maximum absolute analog-to-digital converter output.] Suppose we have a long vector, *e*, of prediction error samples. Each sample in the vector then has a value in the aforementioned range, that is, one of $N = 2e_{max} + 1$ possible integer values. Let f_n now represent the frequency of occurrence of the *n*-th of these *N* possible values in the vector, as described in Section 10.3. Then the average entropy in the prediction error is

$$H_e = -\sum_{n=1}^{N} f_n \log_2 f_n \text{ bits/sample} \tag{10.13}$$

If the length of *e* is, say, *K* samples, then, with entropy coding described in Section 10.5, a minimum of KH_e bits will be required to encode the prediction error.

The vector *e*, with each discrete sample having only *N* possible values, is said to be *quantized* in its original form. Suppose we now define a "quantizing" operation that multiplies *e* by a number less than one and rounds the result as follows, in which $R\{\}$ stands for "round":

$$\tilde{e}_k = R\{qe_k\}; \quad q < 1 \quad \text{and} \quad k = 0, 1, \ldots, K-1 \tag{10.14}$$

Then the new vector, $\tilde{e}$, is quantized with $\tilde{N} = 2\tilde{e}_{max} + 1$ possible sample values, smaller than the original number, and so the sum in (10.13) now has fewer terms. Furthermore, the new sample values are generally close to the original values, so the frequencies in (10.13) are generally distributed as they were originally. Therefore the new entropy, $H_{\tilde{e}}$, is generally less than the original entropy, H_e.

An illustration of this case is given in Figure 10.7. The original vector is in the upper plot, in which e_{max} in (10.12) is indicated. Then *q* in (10.14) is chosen to reduce the prediction error to a 4-bit binary number and the result, $\tilde{e}_k$ in (10.14), is shown in the center plot. Finally, a recovered version of *e* is shown in the lower plot together with the original vector. The recovered version is obtained, as one might expect, as follows:

$$\hat{e}_k = R\left\{\frac{\tilde{e}_k}{q}\right\} \tag{10.15}$$

As indicated in Figure 10.7, the entropy in this illustration was reduced to about ¾ of the original entropy, and quantizing in this case also allows each quantized sample, $\tilde{e}_k$, to be represented with a 4-bit binary number. The samples of the recovered vector, $\hat{e}$, are close enough to those of *e* to make this approach viable in many DSP applications.

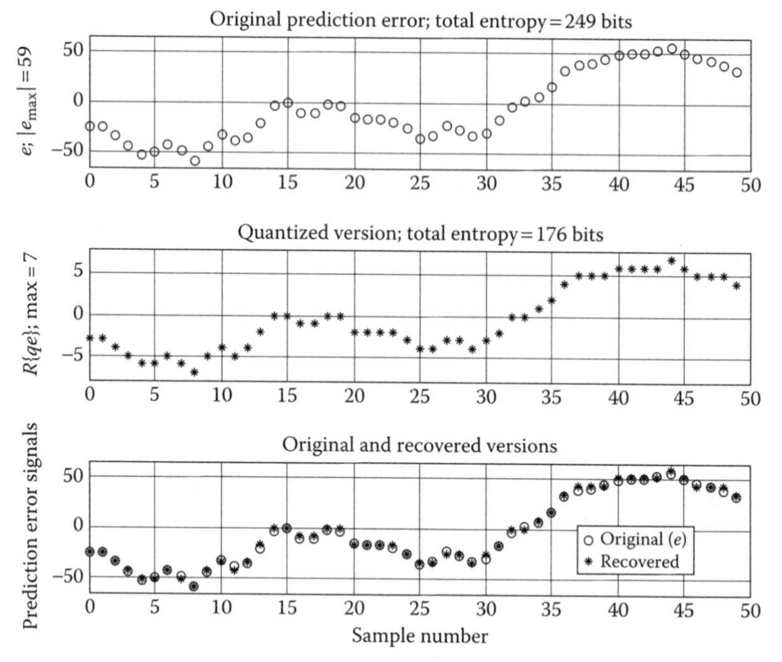

FIGURE 10.7
Quantizing the prediction error in Figure 10.5 for lossy compression: The value of q in this case is 0.1186.

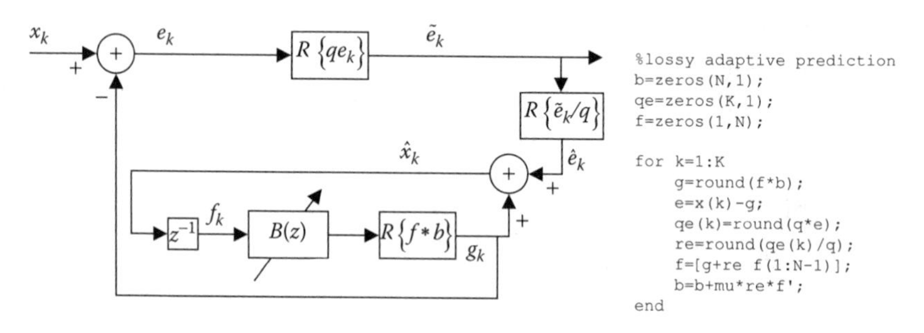

FIGURE 10.8
Lossy adaptive prediction: Similar to Figure 10.1 except the input to the adaptive filter is the (delayed) recovered version of the original signal.

If the prediction error is quantized to reduce its entropy, the overall design of the adaptive predictor is affected. Unless the *same* prediction error is used to adjust the filter weights in both prediction and recovery, the recovery process will usually become unstable. This, in turn, means *reconstructed* signal and error values, $\hat{x}_k$ and $\hat{e}_k$, must be used to adjust the adaptive filter at both ends of the process. The encoder must therefore be configured as in Figure 10.8,

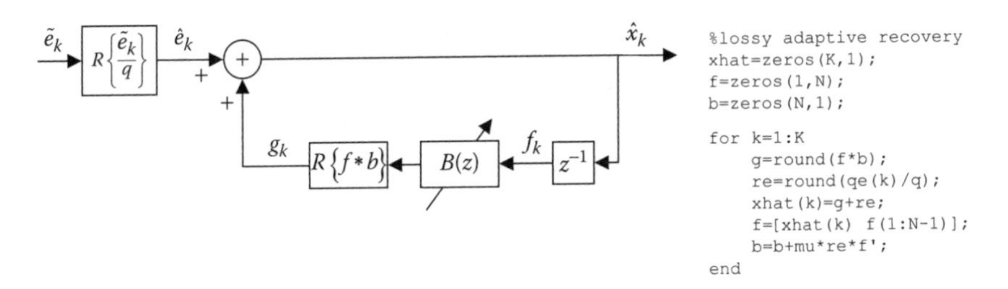

FIGURE 10.9
Lossy recovery of $\hat{x}_k$ in Figure 10.8.

which also contains the encoding algorithm. In the algorithm, "qe" is the quantized prediction error, which is $\tilde{e}_k$ in (10.14), and "re" is the recovered error, which is $\hat{e}_k$ in (10.15). The recovery process, similar to Figure 10.6, is shown in Figure 10.9. The process just described has the desired effect of having the weight vector values change during recovery exactly as they change during prediction, assuming the same floating-point processor is used in both situations.

10.5 Entropy Coding

The term *entropy coding* refers to the first kind of compression we discussed in Section 10.3. In this compression method, we encode a vector or image in a way such that the number of bits in the encoded version approaches the average entropy in bits/symbol times the number of symbols. In this section, we look at two coding schemes that are practical and also come close to being optimal in the sense just described.

Entropy coding begins with a set of symbols such as the eight possible waveform values in Figures 10.1 through 10.3. The coding schemes work well when the number of symbols is small, but are not directly applicable when the symbol set is large. Consider the example in Figure 10.10. The upper plot is a segment consisting of 1,000 samples of the seismic signal in Chapter 8, Figure 8.23, and we note that the range of symbol values in this segment is over 20,000. The distribution of signal amplitudes in the segment is shown in the center plot of Figure 10.10. Entropy coding affords essentially no improvement in this case, because even the most frequent samples appear only twice (with frequency 0.002) in the entire segment. This leads us to the lower plot, in which $f_c(n)$ is the frequency of the n-th category of symbols, each category having a range of 2048. Entropy coding then consists of encoding the category sequence rather than the sample sequence and, for each

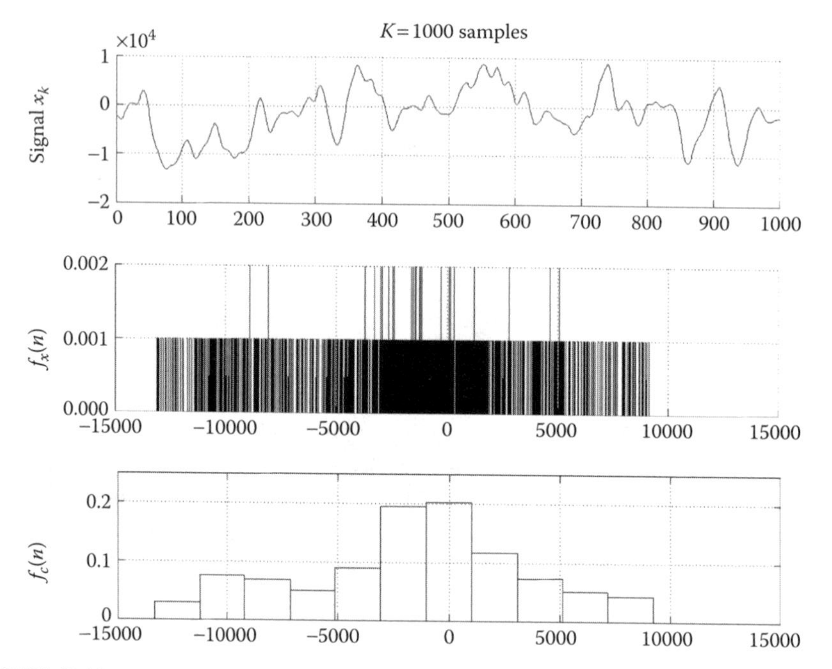

FIGURE 10.10
The seismic waveform of Chapter 8, Figure 8.23: (Upper) A segment, x, of the seismic waveform in Figure 8.23; (center) amplitude distribution of x, showing that entropy coding will not work; (lower) amplitude distribution of categories, which may be encoded as described in this section.

symbol, appending an 11-bit offset to locate the symbol within the category. Using this approach, entropy coding is applicable to nearly all kinds of waveforms and images.[12]

The first coding scheme is called *Huffman coding* after D.A. Huffman.[2,5,13] Huffman coding is a type of *fixed-symbol coding*, in which a specific code is assigned to each possible symbol. Our intent here is only to give a simple example of Huffman coding that shows how to construct Huffman codes. The formal theoretical background is left to texts on coding and information theory,[2,5] which are recommended for further study.

For our illustration, we use the second waveform in Figure 10.1, that is, the waveform shown in Figure 10.11. Next to the waveform plot in Figure 10.11 are the waveform vector (x), the frequency vector multiplied by $K = 32$ (which shows the number of times each symbol appears in x), and the average entropy in bits/symbol computed using (10.4) with $P_k = f_x(k)$.

The rules for Huffman code production are simple. They involve the production of a *binary code tree*, the structure of which contains the codes for all the possible symbols. An algorithm that produces a Huffman code tree is shown next, along with an illustration of its use with the data in Figure 10.11:

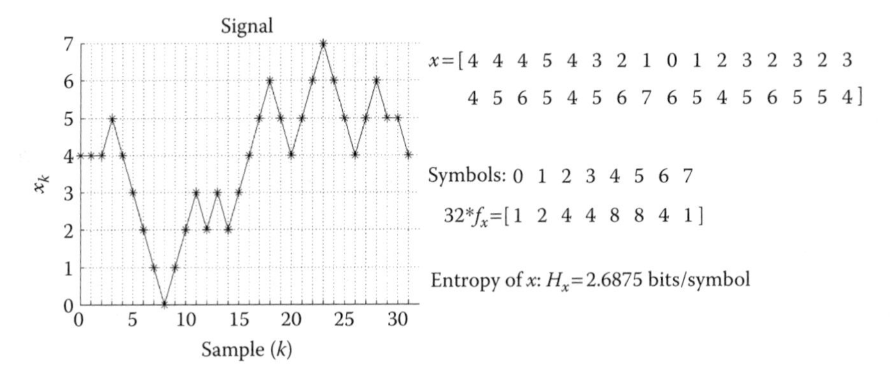

FIGURE 10.11
One of the signal vectors in Figure 10.1.

Huffman code production

Assign a symbol and its frequency in terms of (symbol, frequency) to each of N nodes.

Then perform the following:

1. List the N nodes in the order of decreasing frequency.
2. Combine nodes N and $N-1$ into a new node.
3. Record the link from these nodes to the new node.
4. There are now $N-1$ nodes; so decrease N by 1.
5. Repeat steps 1–4 until $N=1$.

Tree information is now complete. (10.16)

Iteration								
Initial	1	2	3	4	5	6	7	8
(0,1)	(4,8)	(4,8)	(4,8)	(10,8)	(11,8)	(12,16)	$(13,16)_{14}$	(14,32)
(1,2)	(5,8)	(5,8)	(5,8)	(4,8)	(10,8)	$(11,8)_{13}$	$(12,16)_{14}$	
(2,4)	(2,4)	(2,4)	(9,4)	(5,8)	$(4,8)_{12}$	$(10,8)_{13}$		
(3,4)	(3,4)	(3,4)	(2,4)	$(9,4)_{11}$	$(5,8)_{12}$			
(4,8)	(6,4)	(6,4)	$(3,4)_{10}$	$(2,4)_{11}$				
(5,8)	(1,2)	$(8,2)_9$	$(6,4)_{10}$					
(6,4)	$(0,1)_8$	$(1,2)_9$						
(7,1)	$(7,1)_8$							

(10.17)

Initially, in the first column in (10.17) each symbol is assigned to a *node*, and the nodes are listed with corresponding symbol frequencies as shown. For convenience, we use $32f_x$ instead of f_x, so the frequencies in (10.17) sum to 32 instead of 1.

Tree production begins in the second column (column 1), where the two least-frequent nodes, 0 and 7, are attached to node 8. The attachment is denoted by the subscript (8) on these two nodes in column 1, thus satisfying step 3 in the algorithm (10.16). In this manner, we move from column 1 to the second iteration in column 2, which has $N - 1 = 7$ nodes, and so on, until after 8 iterations there is a single node, which represents the top of the code tree. In the fourth iteration, notice that step 1 is necessary, that is, node 10 is not at the bottom of column 4, because there are less-frequent nodes.

All the Huffman code information for the signal in Figure 10.11 is contained in (10.17); but it is best illustrated in the drawing of the tree in Figure 10.12. The tree is formed by connecting the nodes as indicated in (10.17), and then the codes are formed by assigning binary labels (arbitrarily) to the branches emerging from each node. The resulting Huffman codes are listed in the figure.

The first thing to note about codes generated in this manner is the assignment of *shorter codes to more-frequent symbols*, which of course will lead to a shorter representation of the waveform in Figure 10.11. Second, the tree

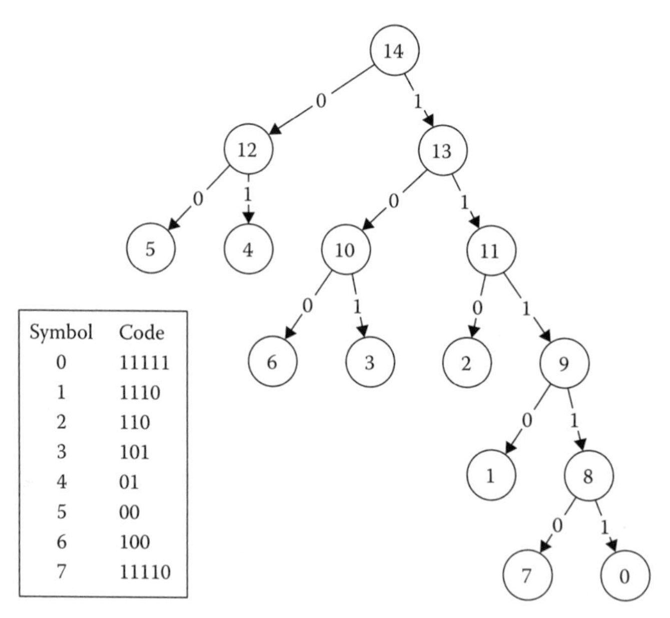

Symbol	Code
0	11111
1	1110
2	110
3	101
4	01
5	00
6	100
7	11110

FIGURE 10.12
Huffman codes and tree for the vector in Figure 10.11.

structure implies an important property called the *prefix property*, without which the codes would be useless for signal compression:

$$\boxed{\text{Prefix property: No code begins with another code.}} \tag{10.18}$$

We can easily see why Huffman codes have this property. If, say, 100 were the prefix of the code for any symbol besides 6, then node 6 would have emerging branches, which is impossible because node 6 represents a symbol and not an internal node in the tree.

The prefix property implies that Huffman codes can be concatenated (strung together) to represent the complete signal vector in Figure 10.11, thereby producing a representation shorter than the original 96-bit binary representation and yet allowing recovery of the original data. The recovery process consists (in principal) of beginning with the first bit of the binary string and proceeding down from the top of the tree (node 14 in Figure 10.12) until a symbol node is reached, recording the symbol, and repeating the process until all symbols are decoded.

The corresponding binary and Huffman encoded bit streams for the vector in Figure 10.11 are as follows, the Huffman string being encoded with the codes in Figure 10.12:

```
symbol: 4  4  4  5  4  3  2  1  0  1  2  3  2  3  2  3
        4  5  6  5  4  5  6  7  6  5  4  5  6  5  5  4

binary: 100100100101100011010001000001010011010011010011
        100101110101100101110111110101100101110101101100

Huffman:010101000110111011101111111101101011101011101010
        10010000010010011110100000100100000001
```
(10.19)

Note how the continuous Huffman string would be decoded without ambiguity using the tree or the table in Figure 10.12.

Due to the small number (32) of samples, which is not realistic, the Huffman vector with 86 bits is only 10 bits shorter than the binary vector with 96 bits, but it illustrates the principal. Furthermore, the Huffman vector in this case is optimal. It matches the entropy of the waveform vector in Figure 10.11, that is,

$$\text{Total entropy} = 32H_x = 32(2.6875) = 86 \text{ bits} \tag{10.20}$$

This fortuitous result is not true in general. It is true, as one may observe from the nature of the binary tree, only when the symbol frequencies are all negative powers of two. Otherwise, some of the possible codes in the tree will not be used. See Exercise 10.7, for example, which provides such a case. The type of entropy coding discussed next gets around this constraint on the symbol probabilities.

Two final points on Huffman coding: first, there is not just one unique set of Huffman codes for a given signal, that is, for a given set of symbols and symbol frequencies. This is easily seen in Figure 10.12 where we could exchange the "0" and "1" emerging from any node and change the code without changing any of the code lengths, thus producing a comparable but different Huffman code. In fact, given an ordered set of symbols and a generally understood rule for assigning zeros and ones to branches, we could construct the code tree knowing only the lengths of the symbol codes. Thus, in Huffman coding applications, *the only data we must store or communicate with the encoded signal vector consists of the ordered set of code lengths.* Because of this, Huffman coding is used widely in signal coding and compression.

Second, MATLAB® functions for Huffman coding are available from many sources, including the Internet, at the time of this writing. A simple function included on the publisher's website for this text that generates Huffman codes is

```
[H, L]=h_codes(x)
```
(10.21)

Given a signal x, H is a vector of Huffman codes for the symbols in x and L is a vector giving the length of each code. (Note that L is necessary, because some codes begin with zero.) The signal vector is assumed to consist of integers in the range $[0, x_{max}]$. The function calls two other functions: (1) `code_length(x)`, which does as its name implies; and (2) `freq(x)`, which computes a symbol frequency table. The latter function was introduced in Chapter 7, Section 7.9, and is useful for computing amplitude distributions in general. Its description is

```
[f,xmin,xmax]=freq(x)
```
(10.22)

When `freq` is executed, f becomes a vector of symbol frequencies, that is, the amplitude distribution of x, in the range $[x_{min}, x_{max}]$.

The next type of entropy coding we discuss is called *arithmetic coding*.[5,12,14,15] Arithmetic coding differs from Huffman coding in that the encoded version of the signal vector or array does not consist of a sequence of codes for individual signal elements (symbols). In arithmetic coding, the signal is encoded by processing the symbols one at a time in order, and the end result is a single long binary fraction rather than a sequence of symbol codes.

To illustrate the production of an arithmetic code, we again use the waveform vector in Figure 10.11, which is shown again in Figure 10.13 with its table of symbol frequencies. The frequencies sum to one and represent the symbol probabilities for this particular vector.

The concept (but not exactly the method, as we shall see) of arithmetic coding is illustrated in Figure 10.14. The encoding process begins by establishing a line from zero to (in this case) $2^{15} = 32{,}768$, with 8 partitions with lengths that are proportional to the 8 symbol frequencies. This is the vertical line at the left of the diagram. Notice that any one of the eight possible

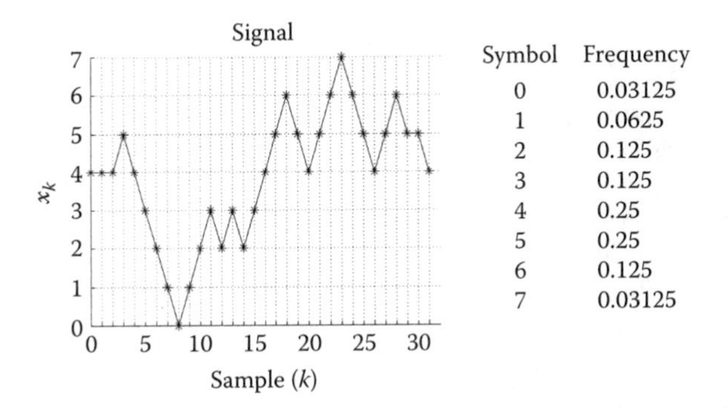

Symbol	Frequency
0	0.03125
1	0.0625
2	0.125
3	0.125
4	0.25
5	0.25
6	0.125
7	0.03125

FIGURE 10.13
Signal in Figure 10.11 with frequency table.

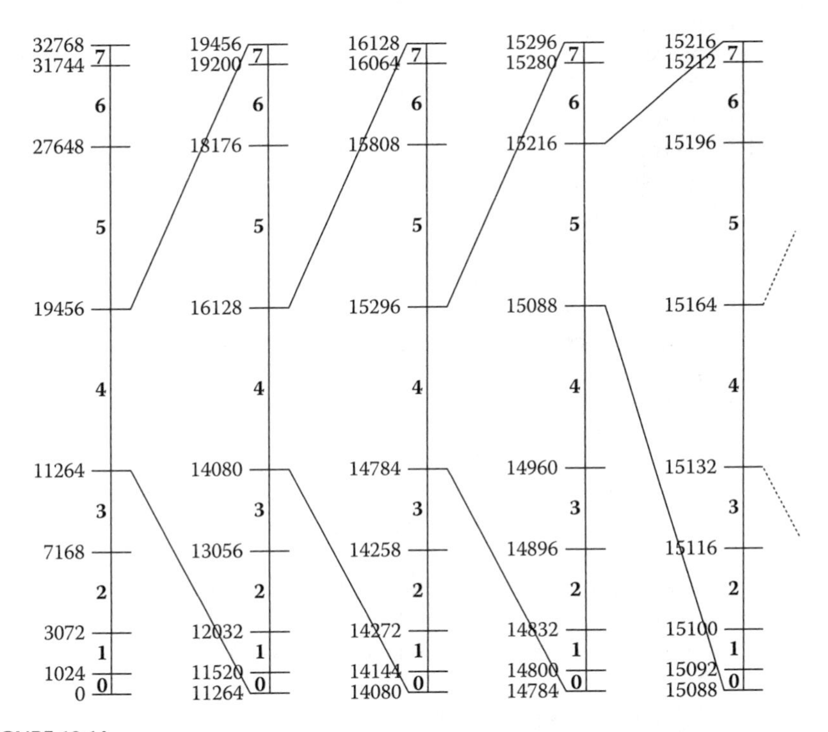

FIGURE 10.14
Arithmetic encoding of the first five symbols of the signal in Figure 10.13: Each symbol is used in turn to select its interval in an overall range of integers, and thus produce a smaller interval to be used by the following symbol.

symbols specifies a unique piece of the line, with the length of the piece being proportional to the symbol frequency. Thus, for example, the interval from 27,648 to 31,744 stands for the symbol 6, and so on. Now instead of the entire interval, we agree (arbitrarily) to represent the interval by its *lower bound*. Thus, the symbol 6 would be represented by 27,648, and so on.

The idea in arithmetic coding is to represent the entire message as a small interval on this line. The encoding process begins in this illustration with the interval [0, 32768) and proceeds with the definition of an interval that grows smaller with each symbol. As the sequence of symbols increases in length, it requires a diminishing fraction of the line to represent it, that is, to distinguish it from other sequences of the same length.

The process is illustrated in Figure 10.14, which shows the encoding of just the first 5 symbols in the vector in Figure 10.13, that is, [4, 4, 4, 5, 4]. With the first symbol (4), the range is reduced from [0, 32768) to the interval [11264, 19456), which is represented by the second line from the left. With the second symbol (4 again), the interval is reduced to [14080, 16128). With the third (4), the interval is reduced again to [14784, 15296). The next symbol (5) selects a different part of the line and reduces the interval to [15088, 15216). Notice that the number of significant bits required to distinguish the upper and lower limits of the interval (and thus the final symbol from other symbols) increases at each stage. The final symbol in Figure 10.14 (4) causes the interval to shrink again to [15132, 15164) and, thus, the number of bits needed to distinguish the interval grows with each symbol.

In fact, we can see in the example of Figure 10.14 that to process all 32 symbols in the vector and still represent the intervals with integers, we would have to begin with a range much larger than 2^{15}. Because the total signal entropy is 86 bits, we would require an initial integer range of $2^{86} \approx 10^{26}$ for the entire signal. However, the range is not necessary, because the most significant bits of the interval limits can be stripped (and saved) as encoding proceeds. For example, in Figure 10.14, the binary representations of the range limits on the third line are

$$(14080)_{10} = (011011100000000)_2$$

$$(16128)_{10} = (011111100000000)_2$$

We could represent the interval just as well without the two leading bits in each binary number. Practical arithmetic coding algorithms work by stripping and saving these unnecessary leading bits as symbols are processed, thus maintaining a practical integer word size. The MATLAB arithmetic encoding and decoding functions that use this technique are included in the *functions* folder on the publisher's website:

```
[y,Nbits]=a_encode(x)
          x=a_decode(y,f,K)          (10.23)
```

The functions are useful with the exercises, but probably will need modification for other applications. The encoding function, *a_encode*, encodes an integer vector or array, x, into another vector, y, with 8 bits per element of y, and also supplies the total number of output bits. If *a_encode* is used with the signal in Figure 10.13, the output bit stream is

```
Arithemtic coding of signal in Figure 10.13:
01110110  01000110  11000001  01011010  10110011  11100111
01010100  11011000  01001111  11011110  011101
```
(10.24)

Comparing this with (10.19), we see that the bit stream length (86 bits) is the same as the Huffman length in this example but that there are no individual symbol codes in this case. Instead, there is a single code representing the entire signal.

The `a_encode` function makes use internally of the symbol frequencies computed with the `freq` function in (10.22). The frequency data must be included with the code in (10.24) or in some way made available to the decoder. The decoding function, `a_decode`, uses the encoding algorithm, that is, enough bits are read to discern the first symbol, which in turn is used to compute the interval for the second symbol just as in encoding, and so on. The arguments f and K of `a_decode` in (10.23) are the frequency table and the length of x, respectively.

As we have seen, Huffman and arithmetic encoding both result in an encoded vector length that approaches or matches the average symbol entropy times the number of symbols. Both encoding methods also have the overhead associated with including the symbol lengths (Huffman) or symbol frequencies (arithmetic).

Our final point regarding these entropy coding methods is that either method can be made adaptive. *Adaptive arithmetic coding* is especially easy, assuming that the range of symbols is known globally. We begin with the assumption that all symbol probabilities are equal or with any globally understood assumption about the symbol frequencies. After the first symbol, say x_1, is encoded with this assumption, the frequency table is modified by incrementing the frequency of x_1, and so on. As encoding proceeds, in general, frequencies become more accurate and, thus, the entropy limit is approached.

Adaptive coding has two advantages: (1) It eliminates the overhead. There is no need to include code lengths or frequencies with the encoded signal. (2) In cases where the symbol frequencies change in a long signal, the encoder can adapt to these changes by continuously modifying the frequency table.

As we leave this section, we note that the entropy limit, that is, bits per symbol times number of symbols, is not usually the limit of compressibility. We saw in Section 10.4 that there are two kinds of redundancy in signals usually encountered in engineering, and entropy coding addresses only one of these. In Exercise 10.24, we apply the process in Figure 10.4 to the signal in Figure 10.8 and find that the encoded message is thereby reduced from the previous entropy limit (86 bits) to a lower entropy limit (24 bits).

10.6 Transform Coding and the Discrete Cosine Transform

In Section 10.3, we discussed the distinction between lossless and lossy signal compression. Lossy compression is appropriate in applications where some kinds of errors in the reconstructed signal do not matter to the end user. In speech communications, for example, audio signals are routinely encoded in a lossy manner in order to conserve bandwidth.[16] In image and video processing, the JPEG and MPEG standards[7] involve lossy compression techniques. These also involve a form of encoding known as *transform coding*.[7,17]

The objective in transform coding is to transform the signal vector or array into another domain, usually the frequency domain, in which most of the essential information in the signal is contained in a relatively small part of the transform domain, that is, in a relatively few elements of the transform vector or array. Almost any kind of transform that provides mapping of vectors or arrays to another domain and allows recovery via an inverse transform is a possible candidate for accomplishing this objective, depending on the data being transformed. In this section, we discuss the *Discrete Cosine Transform* (DCT), which has turned out to be especially useful in waveform and image compression. In Section 10.8, we discuss *multirate processing*, which produces transforms used for similar purposes.

These transforms are applied to signal compression essentially in the manner shown in Figure 10.4; that is, transform coding followed by entropy coding. The concept is shown with more detail in Figure 10.15. The signal is processed in segments, with each segment being transformed independently. In current image-processing practice, for example, 8×8 pixel segments are processed one at a time, until the entire image is processed. Then the transform components are (or may be) quantized selectively, so that information in the more important components is kept in more detail. The definition of "more important" is both signal dependent and subjective and, therefore, the selective quantizing scheme that works best for all waveforms or all images does not exist. Different kinds of data, and different uses of data, require different designs of this stage of the operation.

Next, in Figure 10.15, the quantized transform data is encoded and other essential data, such as the Huffman code lengths or the arithmetic code

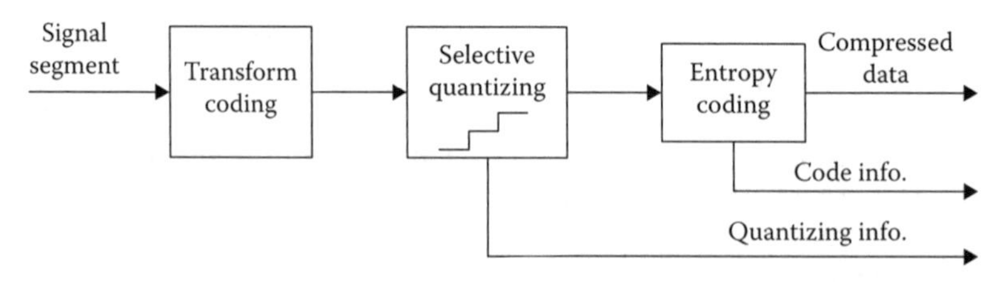

FIGURE 10.15
Signal compression using transform coding: Depending on relative importance, transform components may be quantized with different quantizing step sizes. The step sizes, along with the information needed for entropy decoding, must be included with the compressed signal data.

frequencies, are attached to the compressed data. Unless it is known globally, the quantizing information, that is, the number of steps used to quantize each transform value, must also be contained in the compressed data.

Transform coding and predictive coding are competing candidates for waveform and image compression in the sense that both remove the same kinds of redundancy in a signal. For example, a predictor reduces the information in narrowband waveform components by subtraction, and a transform coder reduces the information in these same components by concentrating them into fewer transform components.

With this idea in mind, it is natural to think first of the discrete Fourier transform (DFT) as a candidate for transform coding. The DCT[7,17,18] is related to the DFT and, in fact, may be written as a function of the DFT, as we shall see. The DCT and the DFT are similar, because they transform a signal component at a given frequency into essentially the same transform component and also because both transforms have unique, nearly symmetric inverses. However, unlike the DFT, the DCT has real components, which are easier to use and quantize.

The one-dimensional DCT of a vector x with N elements is defined as follows:

Discrete Cosine Transform (DCT):

$$X_m^{DCT} = c_m \sum_{n=0}^{N-1} x_n \cos\left(\frac{(2n+1)m\pi}{2N}\right); \quad m = 0, 1, \ldots, N-1$$

$$c_m = \begin{cases} \sqrt{1/N}; & m = 0 \\ \sqrt{2/N}; & m > 0 \end{cases} \qquad (10.25)$$

We use X_m^{DCT} to represent the DCT here and to distinguish the DCT from the DFT. There are variations of the form in (10.25),[7,17,18] but this form is used especially in compression, because, as we shall see, the scaling factors (c_m) produce transform components generally in a range of values similar to that of the signal components. Note that unlike the DFT, the DCT produces real components, which may be positive or negative, but not complex.

As with the DFT, there is a unique inverse DCT that recovers the original vector or array. When the DCT is expressed as in (10.25), the inverse has the following symmetric form:

Inverse DCT:

$$x_n = \sum_{m=0}^{N-1} c_m X_m^{DCT} \cos\left(\frac{(2n+1)m\pi}{2N}\right); \quad n = 0, 1, \ldots, N-1$$

$$c_m = \begin{cases} \sqrt{1/N}; & m = 0 \\ \sqrt{2/N}; & m > 0 \end{cases} \qquad (10.26)$$

Because the scaling factors are the same in (10.25) and (10.26), we conclude that a signal and transform elements will have comparable ranges of values. The demonstration that (10.26) is the inverse of (10.25) is similar to the demonstration in Chapter 3, Equation 3.22, for the DFT inverse and is left for Exercise 10.25 at the end of this chapter.

In fact, the DCT may be expressed as a function of the DFT. We do this by constructing a vector consisting of any signal vector, x, appended to itself. To begin, we use X_m^{DFT} to represent the DFT of x as defined in Chapter 3, (3.7), that is,

$$X_m^{DFT} = \sum_{n=0}^{N-1} x_n e^{-j\frac{2\pi nm}{N}}; \quad m = 0, 1, \ldots, N-1 \tag{10.27}$$

Now X_m^{DFT} is, in general, complex. But we have seen (e.g., in Chapter 8, Equation 8.22) that if x is reversed, the DFT of x^r is the conjugate of X_m^{DFT}. Therefore, we should be able to combine x and x^r and produce a vector with imaginary DFT components that cancel, that is, a vector with a real DFT. One such vector is the following vector with length $2N$:

$$u = [u_0 \ldots u_{2N-1}] = [x_{N-1} \ldots x_1 \; x_0 \; x_0 \; x_1 \ldots x_{N-1}] \tag{10.28}$$

The DFT of u is then

$$\begin{aligned} U_m^{DFT} &= \sum_{n=0}^{N-1} u_n e^{-j\frac{2\pi mn}{2N}} \\ &= \sum_{k=N-1}^{0} x_k e^{-j\frac{2\pi m(k-N+1)}{2N}} + \sum_{i=0}^{N-1} x_i e^{-j\frac{2\pi m(i+N)}{2N}}; \quad m = 0, 1, \ldots, N-1 \end{aligned}$$

(In the final sums, the indices are $k = N - n - 1$ and $i = n - N$.)

To satisfy the conjugate relationship between vector x and its reversal, u must now be centered at the origin, that is, halfway between the x_0 samples in (10.28). Thus, we imagine a discrete domain in which the indices are aligned as follows:

$$\begin{array}{rccccccc} \text{Element of } u\text{:} & x_{N-1} & \cdots & x_0 & x_0 & \cdots & x_{N-1} \\ \text{Index:} & -N+\frac{1}{2} & \cdots & -\frac{1}{2} & \frac{1}{2} & \cdots & N-\frac{1}{2} \end{array} \tag{10.29}$$

We recall from Chapter 3, Table 3.3, property 8, that the time shift described in (10.29), that is, the left shift of $N - 1/2$ steps, causes the DFT, U_m^{DFT}, to be multiplied by the factor $e^{j[2\pi m(N-1/2)/2N]}$. The transform, V_m^{DFT}, of the shifted time series is given by

$$
\begin{aligned}
V_m^{DFT} &= U_m^{\text{DFT}} e^{j\frac{2\pi m(N-1/2)}{2N}} \\
&= \sum_{k=0}^{N-1} x_k e^{j\frac{2\pi m(k+1/2)}{2N}} + \sum_{i=0}^{N-1} x_i e^{-j\frac{2\pi m(i+1/2)}{2N}}; \quad m = 0,1,\ldots,2N-1
\end{aligned} \tag{10.30}
$$

The two sums in the last line are conjugates of each other, so their imaginary parts cancel. Therefore, by limiting m in (10.30) to its first N values, we can write an expression that contains all the DCT components in (10.25):

$$
\begin{aligned}
V_m^{DFT} &= 2\,\text{Re}\left\{\sum_{i=0}^{N-1} x_i e^{-j\frac{2\pi m(i+1/2)}{2N}}\right\} \\
&= 2\sum_{n=0}^{N-1} x_n \cos\left(\frac{(2n+1)m\pi}{2N}\right) = \frac{2X_m^{DCT}}{c_m}; \quad m = 0,1,\ldots,N-1
\end{aligned} \tag{10.31}
$$

Thus, we have a derivation of the DCT in terms of a related DFT. We note that the upper sum in (10.31) is the DFT of x appended with N zeros and delayed (shifted right) by 1/2 sample. There are two consequences of this. First, we see that the components of the DCT of a vector, x, are nearly the same as the components of the DFT of the vector $[x^r\ x]$ and, thus, we would expect the DCT and the DFT to produce similar spectra. The similarity is illustrated in Figure 10.16, where DFT and DCT amplitudes are plotted for comparison. The signal in this

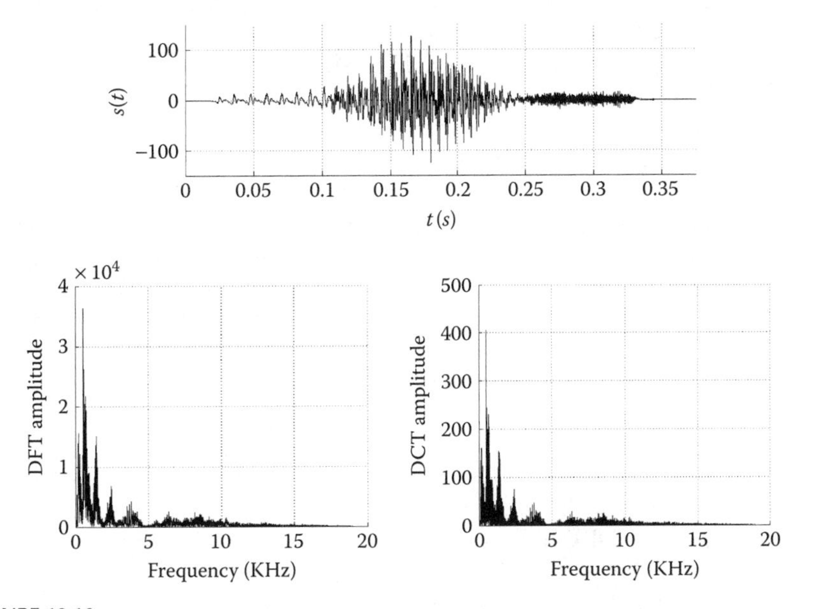

FIGURE 10.16
A segment of the speech signal, *s(t)*, in Chapter 8, Figure 8.17, with discrete Fourier transform and discrete cosine transform amplitudes, showing the similarity between the two transforms.

case is a portion of the speech waveform in Chapter 8, Figure 8.17. Note that although the DCT has twice as many components, the two amplitude spectra are very similar in form.

Second, we have in the first line of (10.31) a way to compute the DCT using the fast Fourier transform (FFT) algorithm, that is, an equivalent expression for X, the DCT of x, is

$$
\begin{aligned}
& y = [x_0 \;\; x_1 \;\; \ldots \;\; x_{N-1} \;\; 0 \;\; 0 \;\; \ldots \;\; 0] \\
& X_m^{DCT} = c_m \operatorname{Re}\left\{ FFT\{y\} e^{-j\frac{m\pi}{2N}} \right\}; \quad m = 0, 1, \ldots, N-1
\end{aligned} \tag{10.32}
$$

The zeros in (10.32) increase the FFT size from N to $2N$, and there are more efficient algorithms for computing the DCT[7]; but the form in (10.32) is useful if only the FFT is available.

By the same line of reasoning, a formula equivalent to (10.32) for the inverse DCT in (10.26) is

$$
\begin{aligned}
& Y_m = c_m e^{j\frac{m\pi}{2N}} X_m^{DCT}; \quad m = 0, 1, \ldots, N-1; \\
& x_n = 2N \operatorname{Re}\left\{ IFFT\left\{ [Y_0 \;\; Y_1 \;\; \ldots \;\; Y_{N-1} \;\; 0 \;\; 0 \;\; \ldots \;\; 0] \right\} \right\}; \quad n = 0, 1, \ldots, N-1
\end{aligned} \tag{10.33}
$$

where *IFFT* stands for the inverse DFT. Two MATLAB functions, `y=sp_dct(x)` and `x=sp_idct(y)`, that compute the DCT and its inverse using (10.32) and (10.33) are included on the publisher's website for this text to demonstrate these methods. The derivation of (10.33) is similar to the derivation of (10.32) and is left for Exercise 10.26.

In image coding and spectral analysis, we use a *two-dimensional* transform. In general, the two-dimensional transform of an array consists of transforming the *columns* of the array one at a time and then transforming the *rows* of the result. Thus, the two-dimensional transform is quite different from the one-dimensional transform. Recall that the DFT of an array, for example, is given as the DFT of each column. Thus, the two-dimensional DFT, which we call "DFT2," is produced as follows, using ".′" to indicate the transpose:

$$
\text{(Two-dimensional) DFT2}(x) = \text{DFT}\{\text{DFT}\{x\}.'\}.' \tag{10.34}
$$

The MATLAB `fft2` function accomplishes this computation.

An illustration of the two-dimensional DCT in image coding is given in Figure 10.17. The two-dimensional DCT is computed as in (10.34) with the DCT in place of the DFT. In the illustration, we use the image known as Boats, which appears often in the DSP literature. The complete image, a 256 × 256 array of 8-bit grayscale pixels, is shown on the upper left. On the lower left, we extracted a 32 × 32–pixel segment consisting of the top of the tower in the upper left quadrant of the complete image. The DCT of this segment is shown

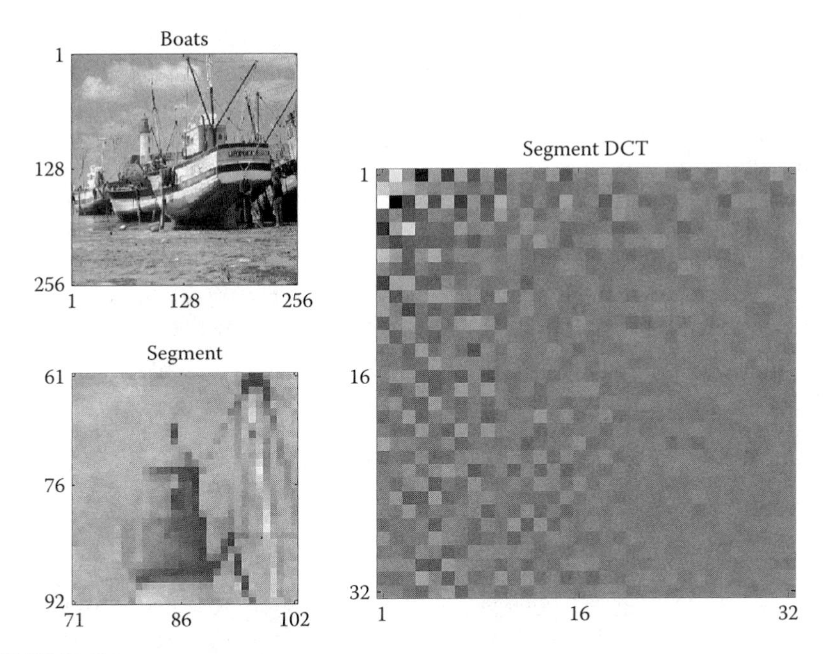

FIGURE 10.17
Two-dimensional discrete cosine transform of a segment of the Boats image: The segment includes the top of the tower in the upper left quadrant of the complete image.

on the right in Figure 10.17. To plot the DCT segment, the DCT components were scaled to the image pixel range, that is, [0, 255]. In this representation of the DCT, *low-frequency* components appear at the upper left, *high-frequency* components are at the lower right, and component *values* increase from black to white.

Compared with the original image, the rounded DFT and the rounded DCT in this case have lower entropies per element, as seen in the following data (10.35). (For DFT entropy, we used a comparable array consisting of separate real and imaginary parts with 8 bits/element.) These entropies are for the complete image, not the segment, in Figure 10.17:

$$\begin{array}{ll} \text{Entropy of Boats image:} & 7.19 \text{ bits/pixel} \\ \text{Entropy of DFT components:} & 1.50 \text{ bits/component} \\ \text{Entropy of DCT components:} & 1.76 \text{ bits/component} \end{array} \tag{10.35}$$

Thus, transform coding reduces the image entropy. Furthermore, as seen here, the choice of the DCT over the DFT is due not to entropy but to the real component values in the DCT, which are more amenable to quantization as well as to other procedures that lead to lossy image compression.

To produce the aforementioned entropies, the mean value of the image was subtracted before transforming, the transforms were scaled to the range [0, 255], and the transform components were rounded to the nearest integer. In image compression, the information needed to undo these steps must be included with the data.

The rounding operation is a simple example of the quantizing step in Figure 10.15, which, in this case, results in a compression in terms of the entropy ratio of 7.19/1.76 or about 4:1. With this degree of quantization, the quality of the reconstruction via the inverse DCT, which is shown in Figure 10.18, is good.

When the image transform is *scanned* to convert the array to a vector, as it is in some compression schemes, the *zigzag scan* (Figure 10.19) is preferable, because it places components at similar frequencies near each other in the

FIGURE 10.18
Original image and reconstruction after 4:1 compression using two-dimensional discrete cosine transform quantization.

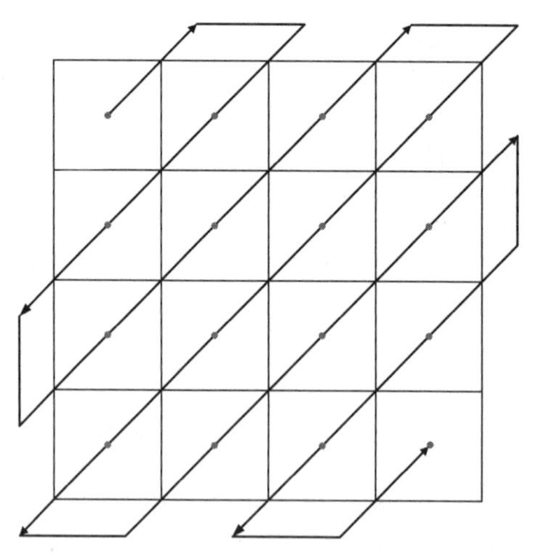

FIGURE 10.19
Zigzag scanning of two-dimensional pixels or transform components.

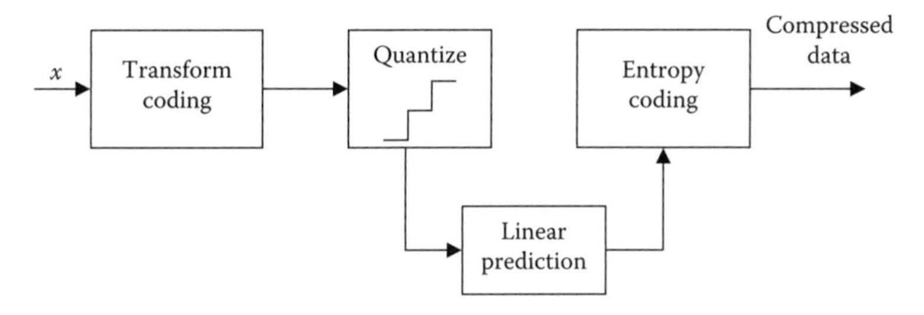

FIGURE 10.20
A linear prediction stage added to the compression scheme in Figure 10.15.

scanning sequence. A MATLAB function called $y = image\_scan(x)$, for which x is any rectangular array and y is the zigzag scan vector, is included to accomplish this type of scan.

Because the DCT components are likely to be correlated along the scan line, linear predictive coding of these components prior to entropy coding, illustrated in Figure 10.20, may be advantageous. The result of adding this step, that is, zigzag scanning followed by linear predictive coding, to the compression process for the Boats image reduces the DCT entropy slightly from the result in (10.35).

10.7 The Discrete Sine Transform

A discrete sine transform (DST) exists* and is derived in much the same way as the DCT in Section 10.7. Due mainly to the JPEG and MPEG standards, both of which use the DCT, the DST is applied less often in image compression. However, the DCT and the DST together give us a better understanding of the nature of these transforms and, besides, the DST is preferable in certain kinds of waveform compression.

In (10.29), we constructed the vector u in a way such that the shifted version of u was an *even* function of its index, and we derived a formula for the DCT using U^{DFT}, the DFT of u. For the DST, we construct a similar vector, s, such that the shifted version of s is an *odd* function of its index:

$$\begin{aligned} s &= \begin{bmatrix} s_0 & s_1 & \cdots & s_N & s_{N+1} & s_{N+2} & \cdots & s_{2N+1} \end{bmatrix} \\ &= \begin{bmatrix} 0 & x_{N-1} & \cdots & x_0 & 0 & -x_0 & \cdots & -x_{N-1} \end{bmatrix} \end{aligned} \tag{10.36}$$

In this case, unlike u in (10.28), zeros are inserted into the sequence. Then the elements of s are shifted left in order to align the element s_{N+1} with index

* The Discrete Sine Transform, *Wikipedia, the free encyclopedia*, en.wikipedia.org.

zero, to provide odd symmetry around the origin. Compared with (10.29), the index alignment now looks like this:

$$\begin{aligned} &v_k = s_{k-(N+1)} \\ &v = \begin{bmatrix} 0 & x_{N-1} & \cdots & x_0 & 0 & -x_0 & \cdots & -x_{N-1} \end{bmatrix} \\ &k = \begin{bmatrix} (-N-1) & -N & \cdots & -1 & 0 & 1 & \cdots & N \end{bmatrix} \end{aligned} \quad (10.37)$$

An example comparing the DCT signal in (10.29) with the DST signal in (10.37) is given in Figure 10.21.

At this point, notice the similarity between s in (10.36) and v in (3.60) in Chapter 3, Section 3.12. Both vectors if extended periodically are odd functions, continuous with continuous first derivatives at the end points, with Fourier series containing only sine terms. Thus, the DST we are about to describe is directly applicable in the resampling processes described in Chapter 3.

The shift from (10.36) to (10.37) is $N+1$ time steps. According to property 8 in Chapter 3, Table 3.3, the shift of $N+1$ time steps corresponds with a phase shift in the m-th DFT component of $2\pi(N+1)m/(2N+2) = m\pi$ radians. [Note that the length of s here is $2N+2$ instead of $2N$ in (10.29)]. Thus, in terms of the DFT of s, the DFT of the left-shifted version of s is

$$V_m^{DFT} = S_m^{DFT} e^{jm\pi} = (-1)^m S_m^{DFT}; \quad m = 0, 1, \ldots, 2N+1 \quad (10.38)$$

This version of V^{DFT} differs from (10.30); since our objective here is to derive the DST itself in addition to relating it to the DFT, we now formulate the DFT of v, the shifted version of s, by summing over the range of k in (10.37) noting that $v_{-N-1} = v_0 = 0$:

$$V_m^{DFT} = \sum_{k=-N-1}^{N} v_k e^{-j\frac{2\pi mk}{2N+2}} = \sum_{k=-N}^{-1} v_k e^{-j\frac{2\pi mk}{2N+2}} + \sum_{n=1}^{N} v_n e^{-j\frac{2\pi mn}{2N+2}}; \quad m = 0, 1, \ldots, 2N+1$$

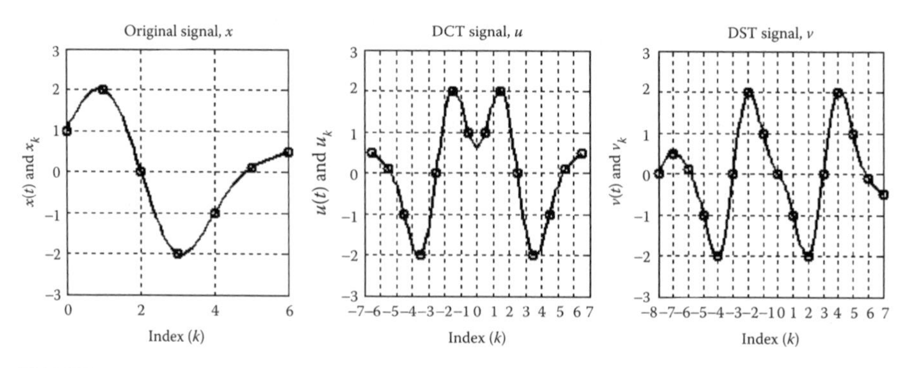

FIGURE 10.21
Left: original signal, x; center: even signal, u, used in the discrete cosine transform; and right: odd signal, v, used in the discrete sine transform.

In the first of the two sums on the right, the index (i) of x corresponding with v_k in (10.37) is $i = -k - 1$, and in the second sum the index of x is $i = n - 1$. Thus we have

$$\begin{aligned} V_m^{DFT} &= \sum_{i=N-1}^{0} x_i e^{j\frac{\pi(i+1)m}{N+1}} + \sum_{i=0}^{N-1} -x_i e^{-j\frac{\pi(i+1)m}{N+1}} \\ &= 2j\sum_{i=0}^{N-1} x_i \sin\frac{\pi(i+1)m}{N+1}; \quad m = 0, 1, \ldots, 2N+1 \end{aligned} \tag{10.39}$$

Noting that $V_0^{\mathrm{DFT}} = 0$ and also, as in any DFT of length $2N + 2$ where x is a real signal vector, $V_{2N+2-m}^{\mathrm{DFT}}$ and V_m^{DFT} are complex conjugates, we now express the DST of x in terms of the independent, nonzero components of V_m^{DFT} as follows:

Discrete Sine Transform (DST):

$$\begin{aligned} X_m^{DST} &= c\sum_{k=0}^{N-1} x_k \sin\frac{(k+1)(m+1)\pi}{N+1}; \quad m = 0, 1, \ldots, N-1 \\ c &= \sqrt{\frac{2}{N+1}} \end{aligned} \tag{10.40}$$

[Note that the range of m in (10.40)], that is, from 0 to $N - 1$, corresponds with the range 1 to N in (10.39). The coefficient c has the same effect as with the DCT. It generally scales the DST components into the range of the signal.

From the definitions of the DST here and V^{DFT} in (10.38) and (10.39), we now have a formula for the computation of the DST in terms of an FFT, similar to (10.32) that computes the DCT in terms of an FFT:

$$\begin{aligned} S_m^{DFT} &= \mathrm{FFT}\{s\}; \quad m = 0, 1, \ldots, 2N+1 \\ X_m^{DST} &= c\frac{(-1)^{m+1}}{2j} S_{m+1}; \quad m = 0, 1, \ldots, N-1 \end{aligned} \tag{10.41}$$

In this result, note that the periodic extension of s is an odd function; therefore, only the imaginary parts of the DFT components (i.e., the sine terms in the FFT) can be nonzero. The reason for using s instead of v for the signal vector here is that the MATLAB function *fft.m* is written for the indices of s in (10.36) rather than those of v.

As with the DCT, the inverse DST is the same as the DST itself:

Inverse DST:

$$\begin{aligned} x_k &= c\sum_{m=0}^{N-1} X_m^{DST} \sin\frac{(m+1)(k+1)\pi}{N+1}; \quad k = 0, 1, \ldots, N-1 \\ c &= \sqrt{\frac{2}{N+1}} \end{aligned} \tag{10.42}$$

The proof of this is not difficult, and we will give it here since the DST is less well-known than the DCT. When we substitute (10.40) in (10.42), the result is

$$x_k = c^2 \sum_{m=0}^{N-1} \sum_{i=0}^{N-1} x_i \sin\frac{(i+1)(m+1)\pi}{N+1} \sin\frac{(m+1)(k+1)\pi}{N+1} \tag{10.43}$$

Let $n = m+1$. Then, since the term for $n = 0$ is zero, we can include the latter, change the order of summation, and write

$$x_k = c^2 \sum_{i=0}^{N-1} x_i \sum_{n=0}^{N} \sin\frac{(i+1)n\pi}{N+1} \sin\frac{(k+1)n\pi}{N+1} \tag{10.44}$$

In Chapter 2, (2.27), we saw that the product of two sine functions summed over an integral number of cycles is zero, and in the inner sum in (10.44), unless $i = k$, at least one of the sine functions is summed over one or more cycles, that is, either $i > 0$ or $k > 0$. Thus, the inner sum is nonzero only when $i = k$ and, in this case, using the relationship for $\sin^2\theta$ from Chapter 1, Table 1.2, the inner sum is

$$\sum_{n=0}^{N} \sin^2\frac{(k+1)n\pi}{N+1} = \frac{1}{2}\sum_{n=0}^{N}\left(1+\cos\frac{2(k+1)n\pi}{N+1}\right) = \frac{N+1}{2} \tag{10.45}$$

(Again, the cosine sum over an integral number of cycles is zero.) Using these results in (10.44), we have

$$x_k = c^2\left(\frac{N+1}{2}\right)x_k = x_k \tag{10.46}$$

Thus the substitution results in an identity, which proves the inverse DST formula in (10.42).

Since the DST in (10.40) and its inverse in (10.42) are identical functions, the inverse can also be computed in terms of the FFT as in (10.41). However, the DST components must first be expanded into a series in the same way components of the signal x are expanded to form the vector s in (10.36). That is,

$$\begin{aligned}
S &= \begin{bmatrix} S_0 & S_1 & \cdots & S_N & S_{N+1} & S_{N+2} & \cdots & S_{2N+1} \end{bmatrix} \\
&= \begin{bmatrix} 0 & X_{N-1}^{DST} & \cdots & X_0^{DST} & 0 & -X_0^{DST} & \cdots & -X_{N-1}^{DST} \end{bmatrix}; \\
s_k &= \mathrm{FFT}\{S\}; \quad m = 0,1,\ldots,2N+1 \\
x_k &= c\frac{(-1)^{k+1}}{2j} s_{k+1}; \quad k = 0,1,\ldots,N-1
\end{aligned} \tag{10.47}$$

Functions for computing the DST and its inverse are included in the *functions* folder available on the publisher's website. The functions are

$$X = dst(x);\\ x = idst(X) \tag{10.48}$$

These implement, respectively, the DST formula (10.41) and the inverse DST formula in (10.47). Both functions use the MATLAB *fft* function in the aforementioned manner, that is, by first constructing the periodic sequences s and S. When x is an array of vectors in columns instead of a single vector, $X = dst(x)$ is the array of DSTs of the columns of x, and $idst(X)$ computes the inverse DST of the columns of X.

The DST as well as the DCT has been applied to the lossy compression of successive short segments of a continuous time series. An example is discussed here in which a long waveform is broken into segments. Our objective is to compress the segments separately using transform coding and be able to reconnect the segments accurately when the segments are decompressed and attached to reconstruct the long waveform.

Example 10.1 Waveform Compression Using the DST

Figure 10.22 is a plot of a portion of a long waveform partitioned into segments. The idea is to compress and encode each segment separately for storage or transmission with negligible loss of accuracy, and be able to reattach the segments and recover a version of the original waveform without the kind of noise caused by regularly spaced steps in the recovered waveform where the segments are reconnected. In other words, the original segments are continuous with continuous derivatives from one segment to the next, and the recovered segments should be likewise to a given degree of accuracy.

Given one of the segment vectors x, the first step in applying the DST or DCT is to construct a ramp function from x_0 to x_{K-1}, where K is the segment length.

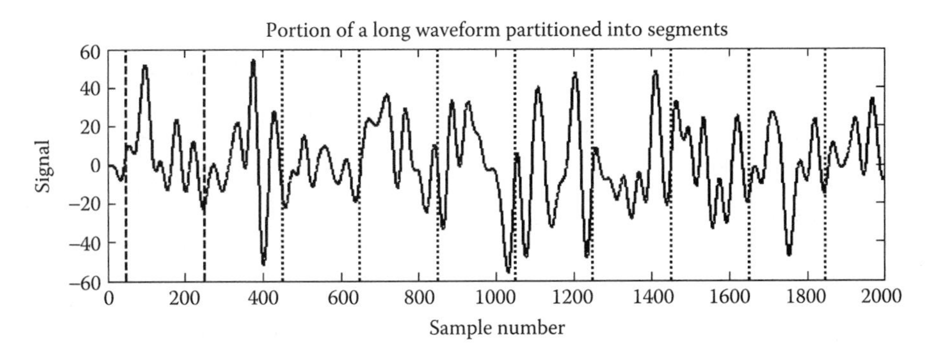

FIGURE 10.22
Portion of a long waveform partitioned into segments: The segment used in Example 10.1, within the dashed lines on the left, consists of sample numbers 50 through 250.

An example with $K = 200$ is shown in Figure 10.23, in which the segment (x) is the first segment, between the dashed lines, in Figure 10.22. The ramp function is the line of K samples from x_0 to x_{K-1}. Next, the ramp is subtracted from x to form the vector v shown in Figure 10.24, with $v_0 = v_{K-1} = 0$. The idea in this procedure is to encode v_1 through v_{K-2} using DST and encode the end points of x, that is, x_0 and x_{K-1}, separately for the following two reasons: (1) to keep the total number of degrees of freedom at K and (2) to allow separate encoding of the segment end points so the segments can be reconnected accurately.

When the transform components are to be encoded in a lossy manner, typically by ignoring or inaccurately representing high-frequency components, the DST offers a possible advantage. Recall how the vector s was formed from the signal vector in (10.36). In the present case, for the DST of $[v_1 \cdots v_{K-2}]$, s would be constructed as follows:

$$\begin{aligned} s &= \begin{bmatrix} s_0 & s_1 & \cdots & s_{K-2} & s_{K-1} & s_K & \cdots & s_{2K-3} \end{bmatrix} \\ &= \begin{bmatrix} 0 & v_{K-2} & \cdots & v_1 & 0 & -v_1 & \cdots & -v_{K-2} \end{bmatrix} \end{aligned} \tag{10.49}$$

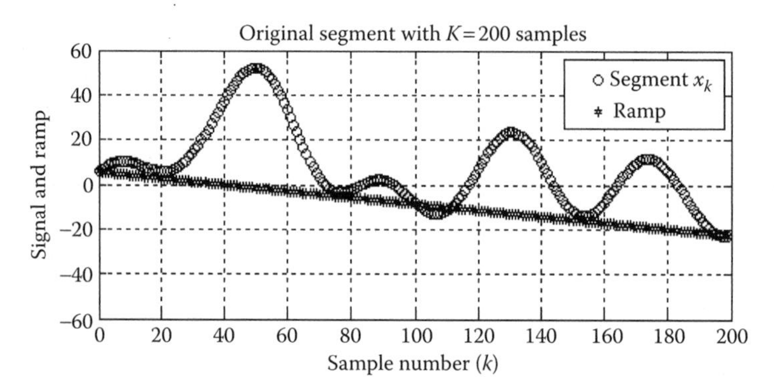

FIGURE 10.23
Original waveform segment, x, and ramp function from x_0 to x_{K-1}.

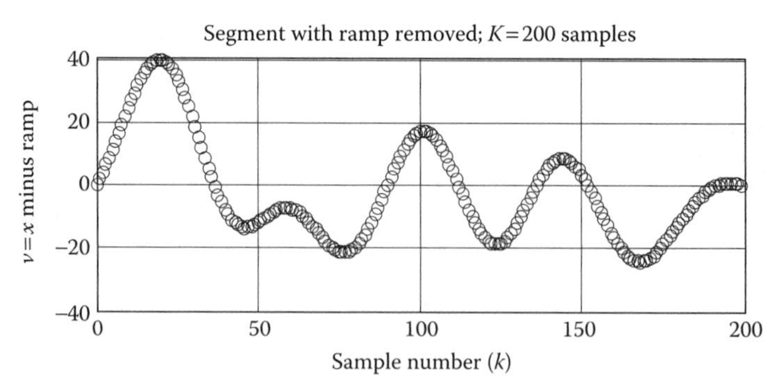

FIGURE 10.24
Original waveform segment, x, with ramp function removed and $v_0 = v_{K-1} = 0$.

A plot showing one period of s in (10.49) is shown in Figure 10.25. Note that the periodic extension of s is continuous and has a continuous first derivative (first difference) both at the end points and at the halfway points, that is, at $k = 0$, $k = K - 1$, $k = 2K - 2$, and so on. Since, according to (10.41), the DST is proportional to the DFT of s, this means the DST will not have artificial high-frequency components produced by discontinuities in the periodic extension of s and its first derivative. Comparing s in (10.36) with u in (10.29), we can see that this is not generally true of the DCT. The result is a slight but possibly significant advantage in using the DST in this type of application. In this example, both transforms map the signal power into a small part of the spectrum, as shown in Figure 10.26. When x is reconstructed from just 16 of the 198 transform components along with accurately encoded end points, the maximum DST reconstruction error is just over 1% of the maximum signal magnitude, $|x_{max}|$, and the maximum DCT reconstruction error is just over 8% of $|x_{max}|$.

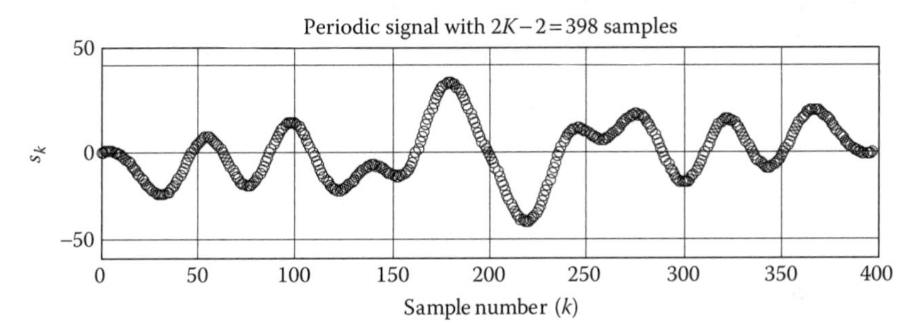

FIGURE 10.25
Periodic signal, s. The discrete Fourier transform of s is used to compute the discrete sine transform as in (10.41).

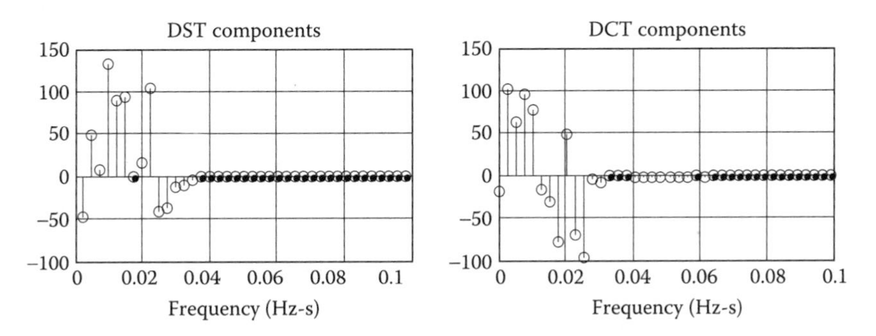

FIGURE 10.26
Discrete sine transform and discrete cosine transform components in Example 10.1: Only the first 20% of the overall spectrum is shown here.

10.8 Multirate Signal Decomposition and Subband Coding

Another method used to analyze and compress waveforms and images involves a process known as multirate signal processing, or *multirate signal decomposition,* combined with *subband coding.*[19,20] These terms are more or less self-descriptive, as we shall see in this section. Multirate processing is also related to signal transformation using *wavelets.* These are all broad subjects, and there are several good reference texts for the reader wishing to pursue them.[19–25] Our goal is to introduce the concepts and show how they apply to signal and image analysis and compression.

Because filtering is a familiar subject in DSP, it is easiest to describe these concepts in terms of filtering operations. The key process in this approach is known as *decimation,* or *downsampling.* A simple filtering and decimation process is illustrated in Figure 10.27. The principle is the same, as we shall see, for both waveforms and images, and although only two filters and downsampling by a factor of 2 are shown in Figure 10.27, the same analysis holds for any number (M) of filters and downsampling by the factor M.

The downsampling operations shown in Figure 10.27 are similar to the downsampling discussed in Chapter 5, Section 5.3. For our discussion, the signals in the lowpass channel are labeled as follows:

$$\begin{aligned} x &= [x_0\ x_1\ x_2\ \ldots] \\ u^1 &= \left[u_0^1\ u_1^1\ u_2^1\ \ldots\right];\quad u^{1D} = \left[u_0^1\ u_2^1\ u_4^1\ \ldots\right] \end{aligned} \tag{10.50}$$

Thus u^{1D}, the downsampled or decimated version of u^1, is a sequence consisting of every other sample of u^1.

If the decimated sequences u^{1D} and u^{2D} are defined in this manner, then clearly these sequences together have the same number of samples as their

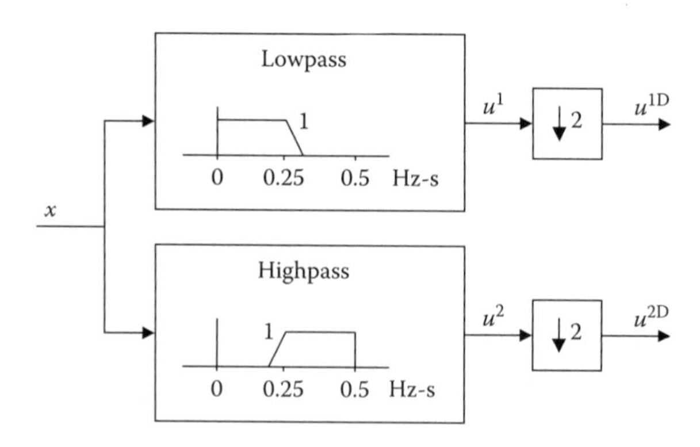

FIGURE 10.27
Single stage of a multirate signal-processing system.

parent vector x. (We assume for this discussion that the length of x is even.) Furthermore, if, say, u^{2D} contains only a small amount of useful information compared with u^{1D}, then it is useful to be able to subject the two signals separately to quantization and entropy coding, that is, *subband* coding. An encoding and recovery concept is illustrated in Figure 10.28. L and H stand for the lowpass and highpass filters, respectively, in Figure 10.28, Q stands for quantizing, EC stands for entropy coding, and ED stands for entropy decoding. We assume EC–ED is a lossless process. We view this concept as encoding and recovery in a *multirate processor with two components* produced by the filters L and H, each component the result of convolving x with the impulse response of the respective filter, as described in Chapter 4.

Notice that multirate processing differs from transforms like the DCT in that it involves digital filters that may be used to process a signal or image continuously rather than one vector at a time. This continuous type of processing in lossy compression eliminates the *blocking effects* of the DCT in the recovered signal, which may be objectionable in waveform and image compression.

Multirate processing with more components involves the application of the same type of filtering and downsampling in cascade arrangements, as we will discuss shortly. The main subject in multirate theory, however, lies in the design of lowpass and highpass filters, which produce useful transformations and allow exact or at least acceptable recovery of the original signal.

The first question we might ask is whether the decimated sequences, u^{1D} and u^{2D} in Figure 10.27, which contain the same number of samples as x, contain enough information to reconstruct x, even with ideal filters. Let us assume small (lossless) quantizing steps and lossless entropy coding, resulting in the version shown in Figure 10.29. Then the answer to our question lies in the relationship between the spectrum of a continuous signal and the spectrum

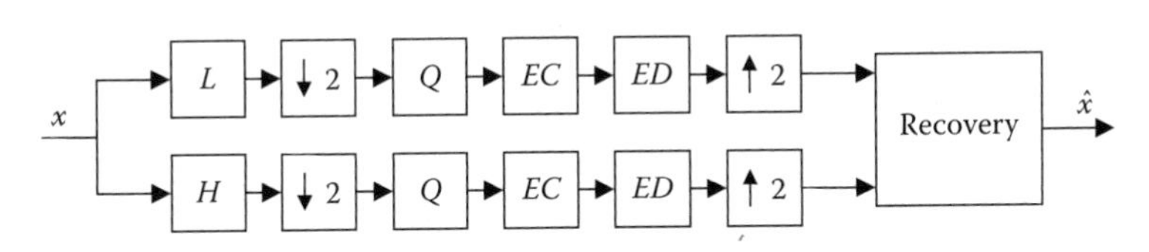

FIGURE 10.28
Encoding and recovery involving a wavelet transform with two components: Symbols Q, EC, and ED stand for quantization, entropy coding, and entropy decoding, respectively.

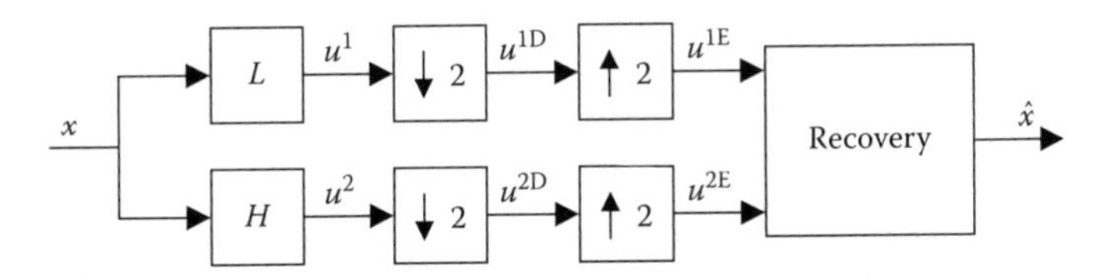

FIGURE 10.29
Equivalent of Figure 10.28 when the quantizing and entropy coding operations are lossless.

of its sample vector. This relationship is given in Chapter 3, (3.47), which gives the spectrum of the samples as a superposition of shifted spectra of the continuous signal.

Suppose now that we have a waveform, $x(t)$, sampled with time step T that is small enough to prevent aliasing. Let $X(\omega)$ represent the continuous DFT of $x(t)$ (e.g., as in (3.25)). We use this representation here instead of the more correct notation, $X(e^{j\omega T})$, for the sake of brevity. Because the sampling theorem is satisfied, we note from (3.47) that $TX(\omega)$ in the range $|\omega| \le \pi/T$ also represents the Fourier transform of $x(t)$. Suppose the Fourier transform of $x(t)$ has the form shown at the upper left in Figure 10.30. Note that the spectrum of $x(t)$ occupies the entire range from zero to half the sampling rate; so any attempt to downsample x would result in an aliased reconstruction, as described in Chapter 3, Section 3.10, and particularly in (3.47).

The DFT of a long vector x, consisting of samples of $x(t)$ with time step T, is shown at the upper right in Figure 10.30. This is one cycle of the DFT, which we designate $X(\omega)$. The DFT consists of repetitions of $X(\omega)$ over all frequencies.

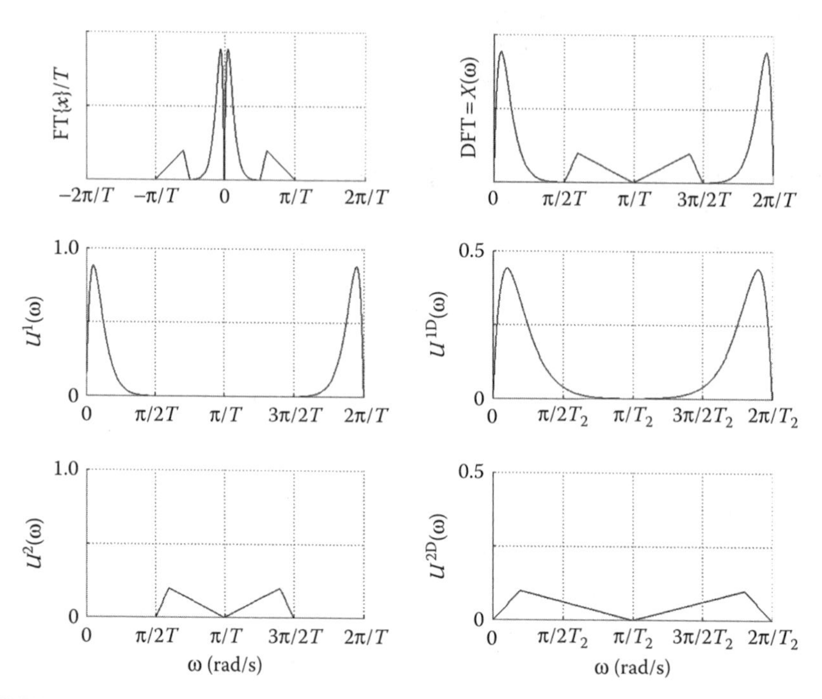

FIGURE 10.30
Upper left: continuous transform of $x(t)$ scaled by $1/T$; upper right: discrete Fourier transform (DFT) of sample vector x with time step T; middle left: DFT of u^1, which is x after lowpass filtering; middle right: DFT of u^1 after downsampling with $T_2 = 2T$; and lower plots: same as middle plots for u^2, which is x after highpass filtering.

We also note that these transforms are usually complex and that we are using *real* transforms here for the sake of the illustration.

For this discussion, we assume the filters in Figure 10.29 are ideal and produce u^1 and u^2 with DFTs $U^1(\omega)$ and $U^2(\omega)$ shown at the middle and lower left in Figure 10.30. That is, the filters extract the low- and high-frequency portions of $X(\omega)$ without overlap. Because the spectra of u^1 and u^2 are (ideally) band limited, the DFTs of these downsampled signals, $U^{1D}(\omega)$ and $U^{2D}(\omega)$, which are shown at the middle and lower right, are not aliased. Note that the time step after downsampling is

$$T_2 = 2T \tag{10.51}$$

Therefore, the frequency range in the plots of $U^{1D}(\omega)$ and $U^{2D}(\omega)$ is half the range in the plots of $U^1(\omega)$ and $U^2(\omega)$. Also, the spectral values are only half as large for the decimated sequences, because these have only half as many samples. Thus, all the information in $X(\omega)$ (and thus in x) is contained unambiguously in the downsampled signals u^{1D} and u^{2D} and recovery of x is at least theoretically possible.

We now consider the question of recovery. As shown in Figure 10.29, recovery is preceded by *upsampling*, or *expansion*. In general, expansion by a factor M consists of inserting $M-1$ zeros between each pair of samples and is thus an application of property 6 in Chapter 3, Table 3.3, which states that the DFT of the expanded sequence consists of M repetitions of the DFT of the sequence before expansion.

In the case of Figure 10.29 where the factor is $M=2$, we have the following sequences:

$$\text{Expansion: } M=2: \begin{cases} u^{1E} = [u_0^{1D} \quad 0 \quad u_1^{1D} \quad 0 \quad u_2^{1D} \ \ldots \\ u^{2E} = [u_0^{2D} \quad 0 \quad u_1^{2D} \quad 0 \quad u_2^{2D} \ \ldots \end{cases} \tag{10.52}$$

In accordance with property 6, u^{1E} has a DFT, U^{1E}, consisting of two repetitions of U^{1D}, and likewise for u^{2E}. This result is illustrated in Figure 10.31. Again, because $T_2 = 2T$, the frequency range on the right in Figure 10.31 is twice that on the left. We can also see that an ideal lowpass filter would succeed in extracting the portion of U^{1E} from 0 to $\pi/2T$ rad/s, and an ideal highpass filter would succeed in extracting the portion of U^{2E} from $\pi/2T$ to π/T rad/s. Thus, the ideal recovery of x from its decimated components would be accomplished as in Figure 10.32, and we may view Figure 10.32 as a simple example of a wavelet transform (in the left half of the figure) and its inverse (in the right half).

Practical multirate processors, however, use real digital filters (both FIR and infinite impulse response [IIR]) in place of the ideal filters (L and H) and actually *allow aliasing* to occur during the transformation and recovery. Aliasing causes signal spectra to be modified as described in Chapter 3,

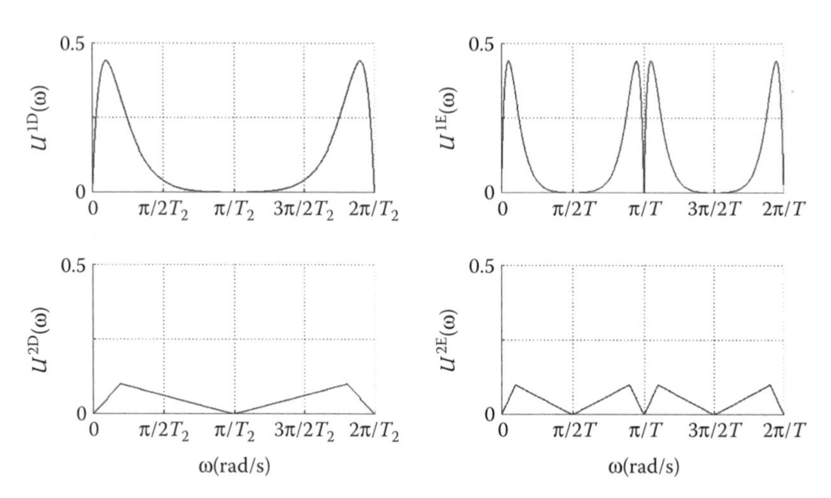

FIGURE 10.31
Discrete Fourier transforms (DFTs) of the downsampled sequences on the left and DFTs of the expanded sequences on the right: With $T_2 = 2T$, expansion consists of inserting a 0 between each pair of samples in the downsampled sequences. Frequency ranges on the right are twice those on the left.

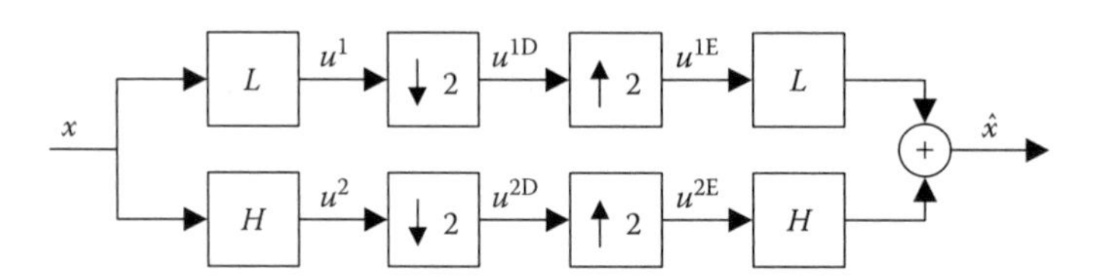

FIGURE 10.32
Ideal decomposition and recovery of a signal using decimation with $M = 2$.

Sections 3.10 and 3.11, and as illustrated, for example, in Figure 3.16. In the present case, aliasing would cause the spectral components in Figure 10.31 to overlap and add together in the vicinity of $\omega = \pi/T_2$ in the left-hand plots and in the vicinity of $\omega = \pi/2T$ and $\omega = 3\pi/2T$ in the right-hand plots, as described by (3.47) in Chapter 3.

With aliasing present, we now consider the configuration in Figure 10.33, in which the input and recovery filters (L^1 and L^2, or H^1 and H^2) are not necessarily the same. In this configuration, we have the following DFT relationships for the *forward transform*:

$$\begin{aligned}
U^1(\omega) &= X(\omega)L^1(\omega); \quad U^{1D}(\omega) = \frac{1}{2}\left(U^1(\omega) + U^1\left(\omega - \frac{\pi}{T}\right)\right) \\
U^2(\omega) &= X(\omega)H^1(\omega); \quad U^{2D}(\omega) = \frac{1}{2}\left(U^2(\omega) + U^2\left(\omega - \frac{\pi}{T}\right)\right)
\end{aligned} \tag{10.53}$$

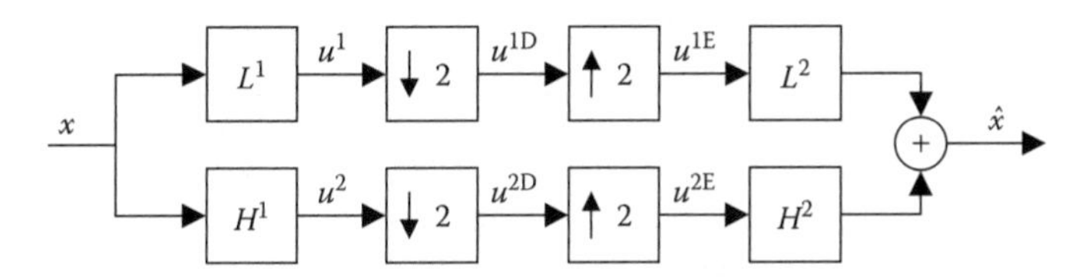

FIGURE 10.33
Decomposition and recovery of a signal using real filters that allow aliasing.

In each line, the first expression describes the input filtering operation. The second expression is due to aliasing in the vicinity of $\omega = \pi/T_2$ and amounts to an application of (3.47) with two terms (converted to DFTs), one for $m = 0$ and the second for $m = 1$.

For the inverse transform, that is, the second half of Figure 10.33, the following relationships are applicable in the range $|\omega| \le \pi/T$:

$$\left.\begin{array}{l} U^{1E}(\omega) = U^{1D}(\omega); \\ U^{2E}(\omega) = U^{2D}(\omega); \end{array}\right\} \hat{X}(\omega) = L^2(\omega)U^{1E}(\omega) + H^2(\omega)U^{2E}(\omega) \qquad (10.54)$$

The first two equations here are illustrated in Figure 10.31, and the third is evident from Figure 10.33.

Before proceeding to see whether $\hat{X}(\omega)$ can be made equal to $X(\omega)$ in the presence of aliasing, we express the DFTs in (10.53) and (10.54) using z-transform notation, that is, with

$$z = e^{j\omega T} \qquad (10.55)$$

as in Chapter 4, (4.5). Then each DFT of the form $X(\omega)$, or more precisely, $X(e^{j\omega T})$, becomes $X(z)$, and each DFT of the form $U(\omega - \pi/T)$ becomes $U(-z)$ via (10.55). If we make these substitutions in (10.54) and (10.55) and combine terms, we obtain the following result:

$$\begin{aligned} 2\hat{X}(z) &= L^2(z)[X(z)L^1(z) + X(-z)L^1(-z)] \\ &\quad + H^2(z)[X(z)H^1(z) + X(-z)H^1(-z)] \\ &= X(z)[L^1(z)L^2(z) + H^1(z)H^2(z)] \\ &\quad + X(-z)[L^1(-z)L^2(z) + H^1(-z)H^2(z)] \end{aligned} \qquad (10.56)$$

With $\hat{X}(z)$ expressed in this form, it is now easy to proceed.

To prevent aliasing in the output, we will not allow $X(-z)$ in (10.56) to be part of the reconstruction. Thus, in the last line of (10.56), we set

$$L^1(-z)L^2(z) + H^1(-z)H^2(z) = 0 \qquad (10.57)$$

With this constraint on the filters, the second term on the second line in (10.56) vanishes, and we can see the *additional* requirement for exact recovery of the input signal:

$$\text{For exact recovery:}\ \ L^1(z)L^2(z) + H^1(z)H^2(z) = 2 \tag{10.58}$$

We consider now the choice of the input filters L^1 and H^1. These are often chosen by first designing $L^1(z)$ as a linear-phase FIR filter and then specifying

$$H^1(z) = L^1(-z),\ \text{that is,}\ H^1(\omega) = L^1(\omega - \pi/T) \tag{10.59}$$

With this specification, we create (in terms of the DFT) $H^1(\omega)$ as the *mirror image* of $L^1(\omega)$, as illustrated in Figure 10.34. Again, note that the plots show DFT amplitudes over the range $0 \le \omega \le 2\pi/T$, rather than the usual range $0 \le \omega \le \pi/T$, in order to emphasize the frequency shift in (10.58). Thus, with (10.58), we assure that if $L^1(z)$ is a good lowpass filter, then $H^1(z)$ is a good highpass filter.

If we now impose the constraint in (10.58), we can eliminate $H^1(z)$ in (10.56) and (10.57), with the following result:

$$\begin{aligned}\text{Prevent aliasing:}\ \ & L^2(z)L^1(-z) + H^2(z)L^1(z) = 0\\ \text{Exact recovery:}\ \ & L^2(z)L^1(z) + H^2(z)L^1(-z) = 0\end{aligned} \tag{10.60}$$

With these two constraints, we have now specified the recovery filters in terms of the single filter, $L^1(z)$, that is,

$$L^2(z) = L^1(z);\quad H^2(z) = -L^1(-z) = -H^1(z) \tag{10.61}$$

The result is shown in Figure 10.35, where there is now only a single choice of a lowpass filter, $L(z)$. Because the input filter gains are mirror images with respect to the quadrature frequency $\pi/4T$, the filters in this system are known as *quadrature mirror filters* (qmfs).

Figure 10.35 contains the basic elements of this introductory discussion of multirate processing. The first half of the figure represents a *signal decomposition*,

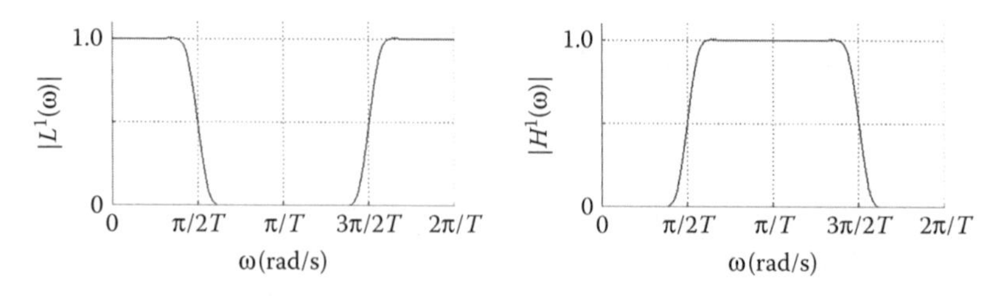

FIGURE 10.34
Amplitude gains of lowpass and highpass finite impulse response mirror-image filters, with $H^1(\omega)$ equal to $L^1(\omega - \pi/T)$.

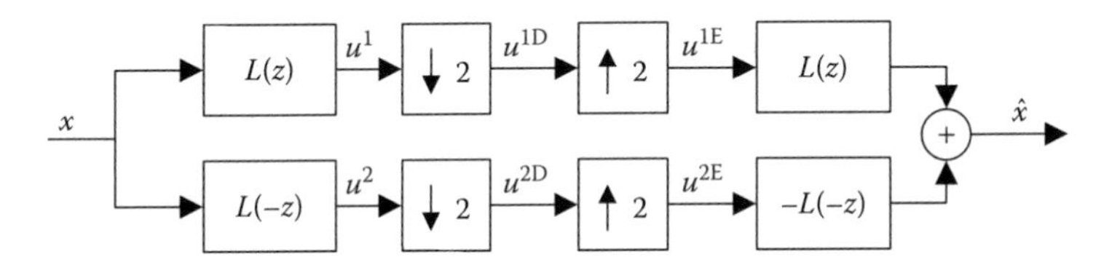

FIGURE 10.35
Signal decomposition and exact recovery with quadrature mirror filters: $L(z)$ is any lowpass filter with cutoff at $\omega_c = \pi/4T$, and the other filters are determined from $L(z)$.

and the second half represents an expansion, or recovery, of the signal. There are other considerations in the design and use of filters for multirate processing, the selection of the factor M, and so on, including the use of adaptive filters in multirate systems, which are beyond our scope here. The reference by P.P. Vaidyanathan[21] is particularly recommended for further study.

We can, however, make some observations about Figure 10.35 and its use in signal analysis and compression, as discussed, for example, with Figure 10.28. First, each filter in Figure 10.35 introduces a phase shift that we have ignored by plotting real spectra in Figures 10.30 and 10.31. If the filters are linear-phase FIR filters such as those described in Chapter 5, the phase shift from x to $\hat{x}$ in Figure 10.35 amounts to a signal delay of $N - 1$ samples, where N is the number of weights in each filter. Thus, the entire process can be made causal without causing distortion. Exercises 10.22 and 10.23 illustrate this point.

Second, as we noted in Section 10.6, any transform that processes a waveform vector can be made two dimensional, as discussed previously in connection with the FFT and the DCT. Let TR(x) represent any transformation of the signal vector x. When x is an array, TR(x) is defined to operate on the *columns* of x. Then the two-dimensional transform, TR2, is defined always as follows:

$$\mathrm{TR2}\{x\} = \mathrm{TR}\{\mathrm{TR}\{x\}.'\}.' \tag{10.62}$$

In the case of the multirate processing, the "transform" may be formed by ordering the output sequences from low to high frequency to form the output vector in each column of the transformed array.

Third, we also observe that the configuration in Figure 10.35 may be expanded in different ways to form a complete transform. Two of the most common ways are illustrated in Figures 10.36 and 10.37. In Figure 10.36, we see the first two stages of a decomposition with *equal* bands. Each successive stage decomposes its input vectors into twice as many output vectors, each output vector having half the number of samples in one of the input vectors. Thus, the frequency domain of x is partitioned into twice as many equal parts with each stage, and a transform with N stages has 2^N *components*, that is, decimated waveform vectors. Note the *superscript* labeling of

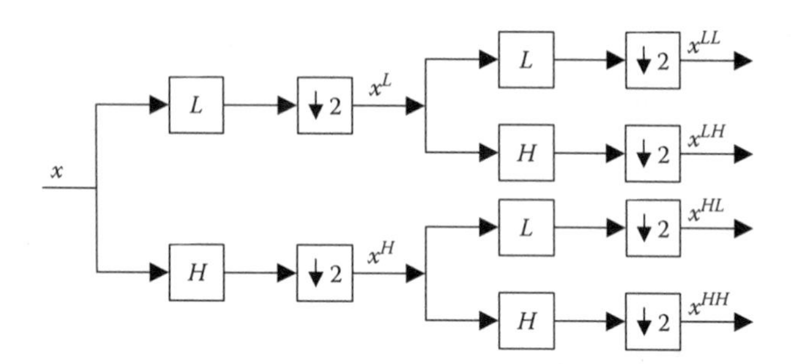

FIGURE 10.36
Decomposition with equal bands.

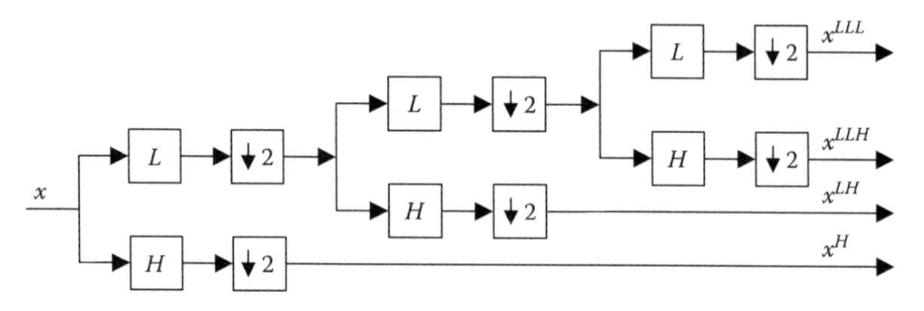

FIGURE 10.37
Decomposition with octave bands.

the downsampled vectors in Figure 10.36 (and Figure 10.37 as well). These indicate the relative frequency bands, and the *number* of superscripts tells us how many times the length of the input has been reduced by a factor of 2 due to downsampling.

Decomposition with *octave* bands is illustrated in Figure 10.37. In this case, only the low-frequency component is decomposed at each stage. The applicability of this approach may be seen by examining the speech spectra in Figure 10.16 and the image spectrum in Figure 10.17. In general, the spectra of naturally occurring waveforms and images tend to be concentrated at low frequencies, and the amplitude spectrum tends to change at a faster rate at lower frequencies, thus justifying the uneven bandwidths in the transform. There is an advantage with the octave bands, because decomposition with N stages produces $N+1$ components instead of the 2^N components produced in the equal-band transform.

Using the speech signal in Figure 10.16, the equal- and octave-band decompositions are illustrated in Figure 10.38, which shows all the signals in the system, and Figure 10.39, which shows just the input and output signals. In the octave-band decomposition, we can observe that the signal energy is distributed more uniformly in the three high-frequency bands, and we note that more stages in the system would decompose x^{LLL} without changing the other three signals.

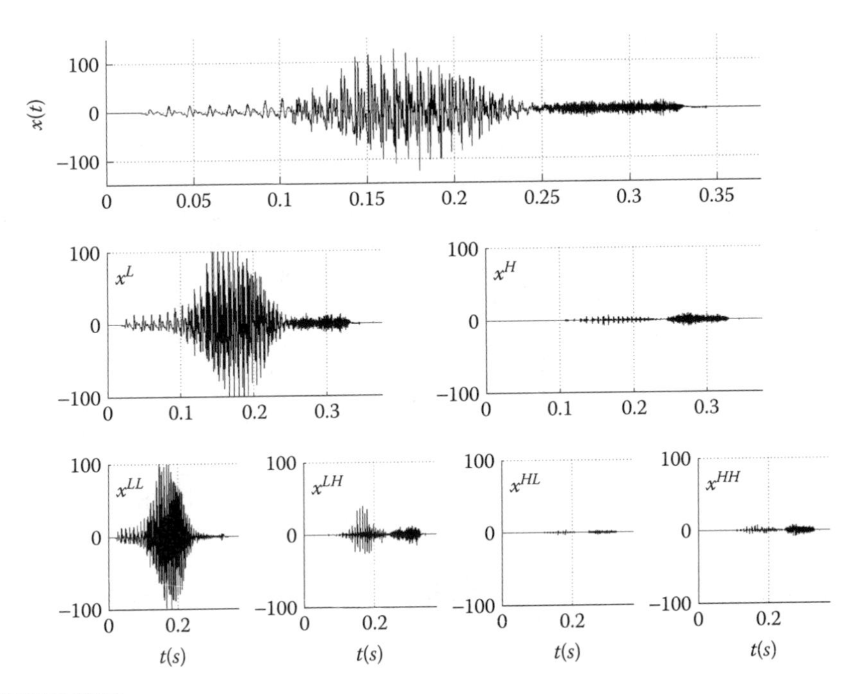

FIGURE 10.38
Decomposition of a short speech segment into equal bands in accordance with Figure 10.36.

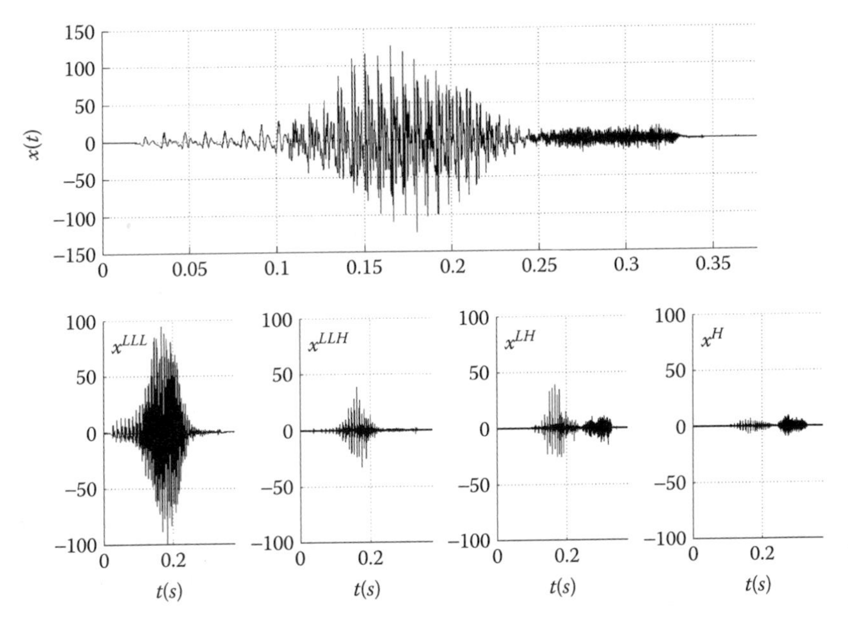

FIGURE 10.39
Decomposition of a short speech segment into octave bands in accordance with Figure 10.37.

10.9 Time–Frequency Analysis and Wavelet Transforms

In this final section, we introduce the general concept known as *time–frequency analysis* as a method for analyzing and transforming signals. Time–frequency analysis is closely related to all the transforms we discussed, that is, the DFT and the DCT, and also to multirate processing and filtering. To begin, let us reconsider the *spectrogram*, which was introduced in Chapter 6. In the illustration in Chapter 6, Figure 6.20, we had an example of partitioning the time–frequency plane as might be represented by Figure 10.40. The plane is divided into segments such that the time increment is the time step, T, and the frequency increment is π/NT, where N is the number of bandpass filters, that is, teeth in the comb filter. The height of each rectangle in Figure 10.40, which is π/NT rad/s, then represents the *frequency resolution*, that is, the passband of the corresponding tooth in the comb filter. The width represents the *time resolution*, that is, the single step from kT to $(k + 1)T$ in the output of the filter.

Now instead of the comb filter in Chapter 6, suppose we define a *short-time discrete transform* using a scheme similar to that used for spectral analysis in Chapter 7. That is, we use a sliding window, ω_{k-nN}, of length N and take (say) the DFT or the DCT at intervals of NT seconds (s) in time, resulting in a different set of N transform components every N time steps and partitioning the time–frequency plane as illustrated in Figure 10.41. (In this case, the transform components are saved rather than averaged as they were in

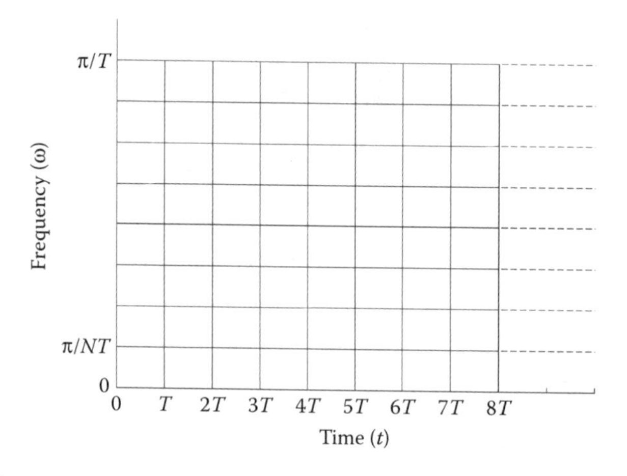

FIGURE 10.40
Representation of how the spectrogram resolves time and frequency: Each waveform in the spectrogram is within a bandwidth of π/NT rad/s, and the points on each waveform are spaced T seconds apart.

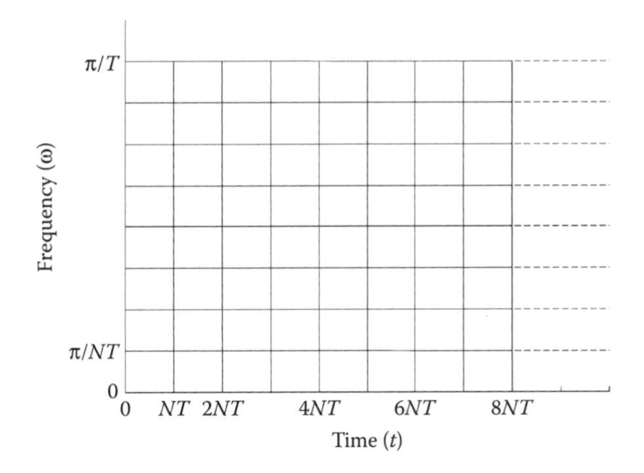

FIGURE 10.41
A short-time transform, such as the short-time discrete cosine transform, partitions the time–frequency plane into equal parts. The frequency resolution is π/NT radians per second, and the time resolution is NT seconds, where N is the length of the time-domain window.

Chapter 7.) The short-time DCT may be described by modifying (10.25) as follows:

$$X_m^{DCT} = c_m \sum_{k=0}^{\infty} x_k \omega_{k-nN} \cos\left(\frac{(2k+1)m\pi}{2N}\right); \quad m = 0,1,\ldots,N-1; \quad n = 0,1,\ldots,\infty \quad (10.63)$$

In this description, ω_{k-nN} is the window that covers the time domain from nNT to $(N + 1)NT$ s and selects the sequence in the range $[x_{nNT} \cdots x_{(n+1)NT})$. We now have a frequency resolution equal to π/NT rad/s and a time resolution equal to NT s, as shown in Figure 10.41. Note that the implied constant area (π) of each rectangle illustrates the *uncertainty principle.* If we increase N in order to increase frequency resolution, we must accept a proportionate decrease in time resolution.

In the case of the short-time discrete transform, we have something quite similar to the equal-band multirate filter described previously in this section. Downsampling by the factor N, as illustrated, for example, for $N = 4$ in Figure 10.36, increases the time step from T to NT, as we have seen, and produces signals with frequencies at multiples of π/NT rad/s. Thus, the time–frequency resolution for equal-band multirate processing is also depicted in Figure 10.41 and is the same as that for the short-time DCT. To emphasize the time resolution produced by downsampling, we use Figure 10.42, which illustrates the same resolution in frequency and time with 3 stages, that is, $N = 8$.

Short-time transforms and equal-band coding do not, however, have the advantage of *octave-band* coding with its nonuniform frequency resolution.

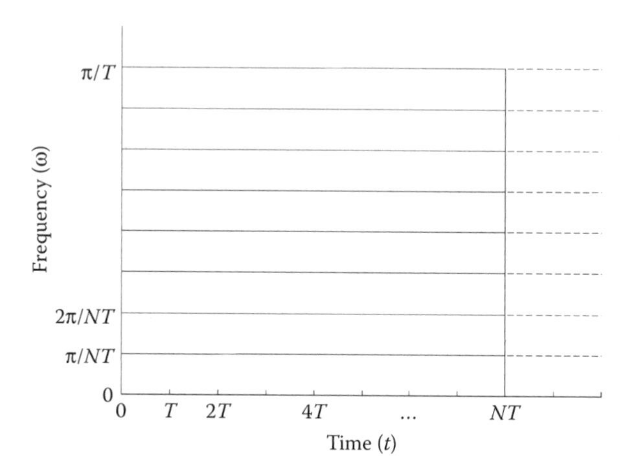

FIGURE 10.42
Time–frequency resolution for the equal-band multirate filter with 3 stages and $N = 8$ output signals: Frequency resolution is π/NT, and with downsampling the time resolution becomes NT seconds.

This is an important advantage to have with the encoding of data from natural sources, because, as we mentioned previously, the spectra of these data generally tend to decrease in proportion to frequency, that is, as $1/\omega$. In the example used in Figure 10.37, octave-band coding partitions the time–frequency plane as shown in Figure 10.43. Note that the area of each rectangle (time–frequency resolution) is constant (π).

Wavelet transforms[20,21,23–26] may be viewed as a generalization of the octave-band coding concept. Suppose we view the three-stage system in Figure 10.37 as a single filter bank with varying bandwidths, as in Figure 10.44. Each transfer function, $H^m(z)$, is the overall transfer function of the corresponding channel in Figure 10.37, with downsampling occurring all at once instead of one stage at a time. For convenience, we assume the signal is unbounded and (as before) the filters are noncausal, with impulse responses centered around zero and also unbounded, so in the time domain we may express each filter output as follows:

$$U^m(z) = X(z)H^m(z); \quad u_k^m = \sum_{n=-\infty}^{\infty} x_n h_{k-n}^m; \quad m = 0,1,2,3; \quad -\infty < k < \infty \tag{10.64}$$

Each output, y^m, is downsampled according to its position in Figure 10.44, and so we have

$$Y_k^m = \sum_{n=-\infty}^{\infty} x_n h_{ik-n}^m; \quad \begin{cases} m = 0,1,2,3 \\ i = 8,8,4,2 \\ -\infty < k < \infty \end{cases} \tag{10.65}$$

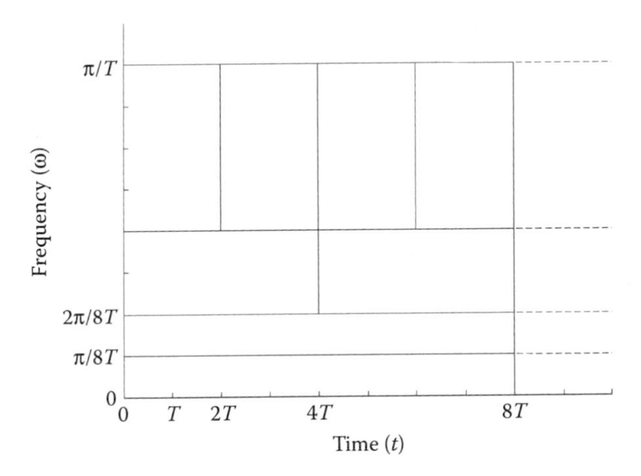

FIGURE 10.43
Time–frequency resolution for the octave-band multirate filter with three stages and four output signals: Frequency resolution decreases with frequency, time resolution increases with frequency, and the product of the two remains constant and equal to π.

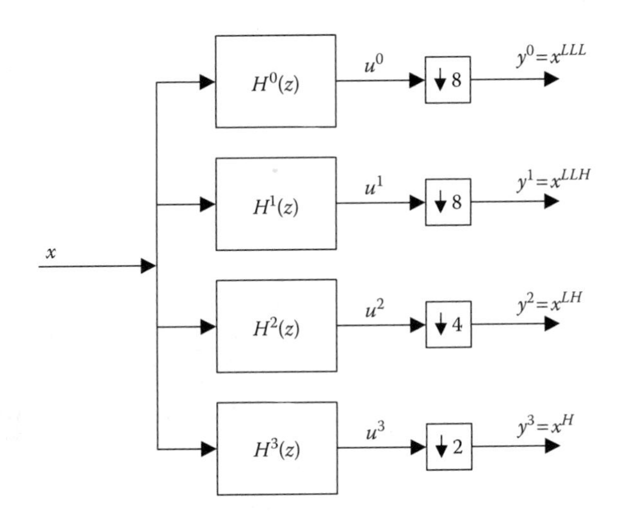

FIGURE 10.44
Filter structure for the wavelet transform, which is the same as the octave-band structure in Figure 10.37 with filtering and downsampling combined into single operations: The wavelet transform components are y^0 through y^3.

It is easy to see how to extend this and add stages to Figure 10.37, for example, m = [0 1 2 3 4] and i = [16 16 8 4 2], and so on. The decimated signals, y_k^m, are called *wavelet coefficients*, and together they comprise the *wavelet transform* of x. The *wavelets*, h_n^m, may be derived in different ways, and their derivation is the subject of texts on wavelet transforms, several of which are listed in the References section. At the time of this writing, MATLAB also offers a special toolbox of wavelet transforms.

Exercises

General instructions: A few of the exercises in this section involve processing a signal from the *dsp data* folder, which may be downloaded from the publisher's website for this text. When this is the case, you may use the MATLAB file *display_data* to display all the signals in a single figure, and then use the expressions in *display_data* to read a specific file. In doing so, you may need to modify the path to the folder. Executing *display_data* should result in a color version of the two signals and the image in Figure 10.45.

10.1 The grayscale image in Figure 10.18 has 256 pixels in each of 256 rows. Each pixel has a range of 256 levels of gray, ranging from black (0) to white (255).

a. How many different images are possible?

b. If all possible images are equally likely, how much information is contained in a single image?

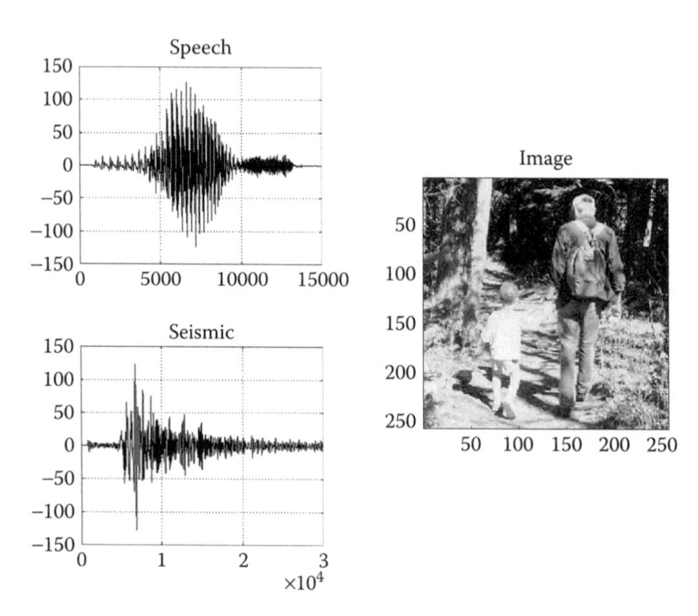

FIGURE 10.45
Speech waveform, seismic waveform, and image used in several of the exercises.

10.2 You receive the following binary message from a source that sends one or zero with equal probability:

$$x = [01000011011010000110100101110000]$$

a. What is the entropy of the message, based on the source probabilities?

b. What is the entropy measure that would be used for encoding x, based on its content?

c. The message actually consists of consecutive 8-bit ASCII symbols for the word "chip." Compute the information received in these four symbols, assuming all ASCII symbols are equally likely.

10.3 Suppose we wish to encode the following message using a Huffman or arithmetic code:

```
Hatred stirs up strife, but love covers all offenses.
```

a. How many symbols are in the message, including spaces and punctuation?

b. How many *different* symbols are in the message, including spaces and punctuation?

c. Write all the different symbols in a row vector, s. Underneath s, write a vector f containing the symbol frequencies expressed as fractions. Assume that s and f are to be included with the encoded information.

d. Show how to compute the average entropy of the symbols in the message, and compute it.

10.4 Suppose the sample vector, x, plotted with the circular (∘) symbol in Figure 10.1, is to be entropy-encoded using its own statistics. What is the smallest number of bits you could have in the encoded version, not counting the frequency table and other similar extra information?

10.5 What is the average entropy in bits per symbol of the two waveforms in Figure 10.1 taken together?

10.6 Given the sample vector x, plotted with the circular (∘) symbol in Figure 10.1, complete the following:

a. Describe the symbol set.

b. Develop a frequency table in terms of $[Nf_x(n)]$.

c. Develop a Huffman code tree following the procedure in (10.16), and draw the tree.

d. Using the code tree, tabulate the set of symbol codes and code lengths.

e. Encode the entire message, and express the result as a binary string.

f. Compute the total entropy of x, and compare it with the length of the result of Exercise 10.6e.

10.7 (See "General instructions" above Exercise 10.1.) Quantize the speech signal in Figure 10.45 to integers in the range [0, 7], plot the result, and do Exercise 10.6a through 6f.

10.8 (See "General instructions" above Exercise 10.1.) For the speech signal in Figure 10.45, complete the following:

a. Compute the minimum number of bits required for entropy coding.

b. Use the *h_codes* function to produce the complete set of binary symbol codes. Print the codes in binary. Find the shortest code, and verify its prefix property.

c. Compute the length of the encoded signal and compare with the result of Exercise 10.8a. (You do not need to construct the encoded signal.)

10.9 Given the sample vector x, plotted with the circular (∘) symbol in Figure 10.1, complete the following:

a. Describe the symbol set.

b. Develop and print a frequency table.

c. Use function *a_encode* to encode the entire message. Express the result as a binary string.

d. Compute the total entropy of x and compare it with the length of the result of Exercise 10.9c.

10.10 Given the message vector in Exercise 10.3, assume the symbol set consists only of the ASCII symbols actually used in the message, and do Exercise 10.9a through 9.d.

10.11 (See "General instructions" above Exercise 10.1.) Let $x(1{:}30000)$ represent the seismic signal in Figure 10.45.

a. Compute the minimum number of bits required for entropy coding of $x(9000{:}9999)$.

b. Use the *a_encode* function to encode $x(9000{:}9999)$. Print the first 32 bits of the output.

c. Compute the length of the encoded signal and compare with the result of Exercise 10.11a.

10.12 (See "General instructions" above Exercise 10.1.) Let x represent the seismic signal in Figure 10.45.

a. Filter x using a least-squares one-step predictor with six weights. Plot x and prediction error, e, in separate subplots. Do not include start-up values e_0–e_5.

b. Encode e with e_0–e_5 excluded. Compare the length of the Huffman-coded version with the entropy of x and the entropy of e.

10.13 (See "General instructions" above Exercise 10.1.) Let x represent the seismic signal in Figure 10.45.

a. Compute the DFT vector, X. Append the imaginary part of X to the real part. Scale the resulting real vector (Y) to the range [0, 255], and plot the result.

b. Using Huffman coding, compare the length of the encoded version of Y with the entropy of Y and the entropy of x.

c. Perform Exercise 10.13b using arithmetic coding. Comment on the results of Exercises 10.13b and 10.13c.

10.14–10.19 **NOTE:** Exercises 10.14 through 10.19 are specified in the table below. In each case, a waveform is processed in the manner described. The note above Exercise 10.1 explains how to obtain and plot the signals. In each exercise, do the following:

a. Process the signal before coding as described, except in Exercises 10.14 and 10.15.

b. Plot the signals (and processed signals, except in Exercises 10.14 and 10.15).

c. Shift the values so the minimum is 0 and the range is [0, 255]. Construct a Huffman code list consisting of [symbol, code length] for each symbol, and print the list.

d. Without actually creating the encoded message, compute the message length and the resulting compression ratio assuming 8 bits/symbol in the original signal, and compare it with the results of the other exercises you have worked in the same column of the table.

	Speech Signal	**Seismic Signal**
Huffman coding only.	Exercise 10.14	Exercise 10.15
Predictive coding quantized to integers; six weights Huffman coding.	Exercise 10.16	Exercise 10.17
DCT of signal–mean value, quantized to 8-bits Huffman coding.	Exercise 10.18	Exercise 10.19

10.20 Process the speech signal as in Exercise 10.18, but in place of the DCT use a four-stage octave-band encoder with FIR filters having seven weights each and using the Hamming window.

a. Plot the concatenated encoder output, from low to high frequency.

b. Translate the output to integers in the range [0, 255]. Find and print a set of Huffman codes.

c. Compute the length of the encoded message and the compression ratio.

10.21 (See "General instructions" above Exercise 10.1.) For the color version of the image in Figure 10.45, note that the array dimensions are [256, 256, 3], that is, there are 3 arrays containing the red, green, and blue pixels of the image. Compute the two-dimensional DCT of each of the 3 arrays and combine these into the array X with dimensions [256, 256, 3]. Delete the direct current (d.c.) component, that is, X(1, 1, :). Create a color figure with two images, each about 3.5" on a side. Put the original color image on the left, and on the right, display the segment X(1:10,1:10,:) as a color image. Comment on how the two-dimensional DCT components change with the spatial frequency.

10.22 Refer to the diagram in Figure 10.35, which illustrates a qmf stage with recovery. Make a figure with the following 8 subplots (4×2), in order:

1. Chirping sinusoidal signal, x_n versus n, with 128 samples given by x = sin(2*π*[0:127].*linspace(.1,.4,128)/4).
2. Amplitude gain of $L(z)$, a lowpass, linear-phase FIR filter designed with N = 13 weights and the Hamming window, and also the amplitude gain of the mirror filter, $L(-z)$.
3. Signal u_n^{1D} versus n.
4. Signal u_n^{2D} versus n.
5. Signal u_n^{1E} versus n.
6. Signal u_n^{2E} versus n.
7. Reconstructed signal $\hat{x}_n$ versus n.
8. Signals x_n and $\hat{x}_{n-N}$ together versus n. These two signals should be nearly the same. Explain why the shift of N samples is necessary to align the two signals.

10.23 Refer to the three-stage, octave-band filter in Figure 10.37. In this exercise, we will use a similar filter with $M = 4$ stages to partition the speech signal, x, described in "General instructions."

a. First, truncate x so its length, K, is the largest possible multiple of 2^M. What is the revised value of K?

b. Filter x through 4 stages with each stage using N = 21 weights and the Hamming window, producing 8 signals similar to the

4 signals in Figure 10.38. Concatenate the 8 signals into a single vector, e, with K samples.

c. Process e through the mirror image of the octave-band filter, and sum the result to produce the reconstructed signal, $\hat{x}$.

10.24 In Section 10.3 we saw how predictive coding could be applied to reduce the entropy of a signal prior to entropy coding, following the scheme in Figure 10.4. Apply this scheme to the waveform in Figure 10.13.

a. Apply the one-step predictor used in Figure 10.3, and make a plot similar to Figure 10.3 of the signal and the prediction error.

b. Compute the entropy of the original signal.

c. Compute and compare the entropy of the prediction error, e, excluding e_0.

10.25 Prove by substitution in (10.25) that the inverse DCT formula in (10.26) is correct.

10.26 By using the inverse DFT formula (Chapter 3, 3.23), prove that the fast version of the inverse DCT in (10.33) is identical to the original formula in (10.26).

References

1. Shannon, C. E. 1948. A mathematical theory of communication. *Bell Syst Tech J* 27:379–423 and 623–56.
2. Ingels, P. M. 1971. *Information and Coding Theory*. Scranton, PA: Intext.
3. Hamming, R. W. 1980. *Coding and Information Theory.* Englewood Cliffs, NJ: Prentice Hall.
4. Held, G., and T. Marshall. 1991. *Data Compression: Techniques and Applications, Hardware and Software Considerations.* New York: John Wiley & Sons.
5. Golomb, S. W., R. E. Peile, and R. A. Schotz. 1994. *Basic Concepts in Information Theory and Coding.* New York: Plenum Press.
6. Sayood, K. 1996. *Introduction to Data Compression.* San Francisco, CA: Morgan Kaufman Publishers, Inc.
7. Rao, K. R., and J. J. Hwang. 1996. *Techniques and Standards for Image, Video, and Audio Coding.* Chap. 5. Upper Saddle River, NJ: Prentice Hall.
8. Ziv, J., and A. Lempel. 1977. A universal algorithm for sequential data compression. *IEEE Trans Inf Theory* 23(3):337–43.
9. Nelson, M. R. 1989. LZW data compression. *Dr Dobb's J* 29:86.
10. Regan, S. M. 1990. LZW revisited. *Dr Dobb's J* 15(6):126–7.
11. Gersho, A., and R. M. Gray. 1992. *Vector Quantization and Signal Compression.* Norwell, MA: Kluwer Academic Publishers.
12. Stearns, S. D. 1995. Arithmetic coding in lossless waveform compression. *IEEE Trans Signal Process* 43(8):1874–9.

13. Huffman, D. A. 1952. A method for the construction of minimum-redundancy codes. *Proc IRE* 40(9):1098–1101.
14. Rissanen, J., and G. G. Langdon Jr. 1979. Arithmetic coding. *IBM J Res Dev* 23(2):149.
15. Witten, I. H., R. M. Neal, and J. G. Cleary. 1987. Arithmetic coding for data compression. *Commun ACM* 30(6):520–40.
16. Jayant, N. S., and P. Noll. 1984. *Digital Coding of Waveforms—Principles and Applications to Speech and Video*. Englewood Cliffs, NJ: Prentice Hall.
17. Rao, K. R., and P. Yip. 1990. *Discrete Cosine Transform: Algorithms, Advantages, Applications*. Boston, MA: Academic Press.
18. Ahmed, N., and K. R. Rao. 1975. *Orthogonal Transforms for Digital Signal Processing*. New York: Springer-Verlag.
19. Crochiere, R. E., and L. R. Rabiner. 1983. *Multirate Digital Signal Processing*. Englewood Cliffs, NJ: Prentice Hall.
20. Vetterli, M., and J. Kovacevic. 1995. *Wavelets and Subband Coding*. Englewood Cliffs, NJ: Prentice Hall.
21. Vaidyanathan, P. P. 1993. *Multirate Systems and Filter Banks*. Englewood Cliffs, NJ: Prentice Hall.
22. Shynk, J. 1992. Frequency-domain and multirate adaptive filtering. *IEEE SP Mag* 9(1):14–37.
23. Burrus, C. S., R. A. Gopinath, and H. Guo. 1998. *Introduction to Wavelets and Wavelet Transforms*. Upper Saddle River, NJ: Prentice Hall.
24. Boggess, A., and F. J. Narcowich. 2001. *A First Course in Wavelets with Fourier Analysis*. Upper Saddle River, NJ: Prentice Hall.
25. Strang, G., and T. Nguyen. 1996. *Wavelets and Filter Banks*. Wellesley, MA: Wellesley-Cambridge Press.
26. Saha, S. 2002. Image compression—from DCT to wavelets: A review. *ACM Crossroads* (ACM electronic publication) www.acm.org/crossroads.

11

Models of Analog Systems

11.1 Introduction

Digital models of analog filters and other continuous systems are useful in situations where we wish to duplicate or replace the continuous system with an "equivalent" discrete system. As suggested in Figure 11.1, the accuracy of the model is measured in terms of a discrete error, e_k, at regularly spaced sample points. The analog-to-digital converters (ADCs) in Figure 11.1 are assumed to produce impulse samples at synchronized points in time. In Chapter 9, we discussed the design of digital filters that minimize the mean-squared value of e_k. In this chapter, we wish to arrive at models that produce a small or vanishing value of e_k under specified input conditions.

A simple linear approximation scheme, called the *impulse-invariant approximation*, is introduced first. It is a generalization of the procedure given in Chapter 5, (5.17).

To begin, suppose the digital filter is to be designed to approximate the linear system in Figure 11.1 for an arbitrary input, $x(t)$, that is, for any finite-valued input function whose sample set $[x_k]$ is provided in sequence. In a typical digital model of, say, a production process, a guidance system, or a vehicle, there are many linear subsystems as well as nonlinear subsystems, and many different input and output variables that cannot be predetermined. The digital filtering concepts here generally are applicable when the linear subsystems can be described in terms of transfer functions or when one is interested in the spectral characteristics of the output variables.

The digital approximation or *model* of the system in Figure 11.1 will be discussed for the case in which $H_a(s)$, expressed as a ratio of polynomials in s, is a *proper fraction*, that is, the degree of the numerator is less than that of the denominator. In this case the inverse transform $h_a(t)$ is more easily defined, and the approximation scheme can include the use of $h_a(t)$. Section 11.2 describes first the impulse-invariant approximation to $H_a(s)$ and the related approximation to the convolution integral.

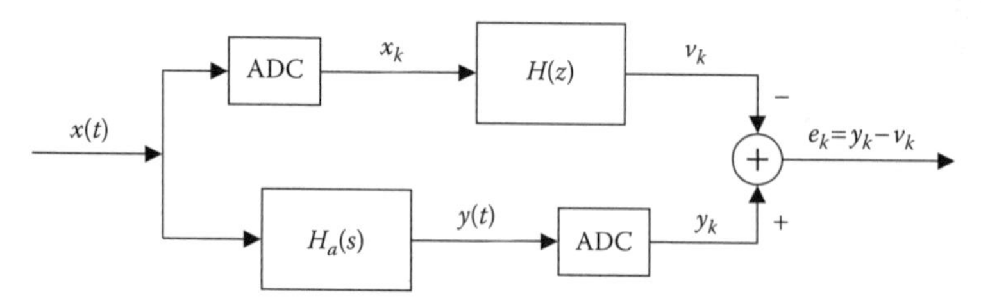

FIGURE 11.1
Discrete model, $H(z)$, of a continuous linear system, $H_a(s)$: $H(z)$ and $H_a(s)$ are different functions.

11.2 Impulse-Invariant Approximation

This type of approximation has been described by Kaiser[8] and by Gold and Rader.[3] Let $H_a(j\omega)$ be the desired transfer function expressed as a Fourier transform. Since the transfer function of a digital filter is a discrete Fourier transform (DFT), in the absence of aliasing, a model of $H_a(j\omega)$, $H_0(e^{j\omega T})$, is clearly possible in accordance with (3.47) and (3.48) in Chapter 3 simply by using the digital filter having the transfer function $TH_a(j\omega)$. Thus, with no aliasing,

$$H_a(j\omega) = TH_0(e^{j\omega T}); \quad |\omega T| < \pi \tag{11.1}$$

In other words, assuming a finite impulse response vector, h_a, T times the DFT of h_a is a discrete approximation to the Fourier integral of h_a in the sense of (3.47):

$$\sum_{k=0}^{K-1} h_{ak} e^{-j\omega kT} T \approx \int_{t=0}^{KT} h_a(t) e^{-j\omega t} dt \tag{11.2}$$

Thus, if $H(z)$ in Figure 11.1 is made to be the z-transform of Th_a, the two filters would have the same response to the unit impulse function, provided we define the analog "impulse" as a rectangular pulse with amplitude T and duration $1/T$.

Since the world of continuous signals does not consist of impulse functions or combinations thereof, the result in (11.1) is not usually of practical value. But it does at least show that linear continuous systems have digital counterparts, that is, digital filters that behave the same in this restricted sense.

On account of (11.1), the discrete and analog responses must be equal at the sample points. Therefore, we must assume the sampling theorem is satisfied, and we must also assume the analog transfer function, $H_a(s)$, is expressed as a proper rational function of s, that is, a rational function of s in which the degree of the denominator is less than the degree of the numerator. The reason for the latter is the *initial value theorem* (Truxal[17]), which states that

$$\lim_{t \to 0} h_a(t) = \lim_{s \to \infty} sH_a(s) \tag{11.3}$$

Therefore, if $H_a(s)$ is not proper, the discrete sample h_0 is not finite and the z-transform in Chapter 4, (4.6), is not defined. If $H_a(s)$ is not proper, a partial-fraction expansion can be used.

Summarizing and somewhat restating the above, the following steps would be taken to obtain an impulse-invariant approximation to a proper, rational, continuous transfer function:

1. Begin with the desired transfer function, $H_a(s)$.
2. Determine $h_a(t)$ from a Laplace transform table such as the Appendix.
3. In accordance with Chapter 3, (3.47), find $H_0(z)$ as the z-transform of the sampled version of $Th_a(t)$. The Appendix eliminates step 2 by giving $H_0(z)$ in terms of $H_a(s)$.
4. The corresponding difference equation is then given as in Chapter 4, (4.10).

When the input signal, $x(t)$ in Figure 11.1, is the unit impulse function, the aforementioned procedure produces transforms of the same signal, and therefore, v and y are equal at the sample points. Examples 11.1 and 11.2 illustrate the simple nature of the impulse-invariant model as well as its practical limitations.

EXAMPLE 11.1

Derive the impulse-invariant model of $H_a(s) = 1/(s+1)$. In steps 2 and 3 above we obtain the following:

$$h_a(t) = e^{-t}$$

$$H_0(z) = \frac{T}{1 - e^{-T} z^{-1}} \tag{11.4}$$

The z-transform of e^{-t} is on line 151 in the Appendix. As with any rational function of $j\omega$, $H_a(j\omega)$ does not vanish beyond any finite frequency. Therefore, as suggested by Figure 11.2, aliasing occurs in the DFT of $h_a(t) = e^{-t}$ at any finite sampling rate. Figure 11.2 illustrates the case when $\pi/T = 10$ rad/s, with the solid line being a plot of $|H_a(j\omega)|$ and the dashed line a plot of $|H_0(e^{j\omega T})|$ found by taking the magnitude of $H_0(z)$ in (11.4) with $z = e^{j\omega T}$.

Once $H_0(z)$ has been derived as in this example, the digital response to an impulse or step function can be examined in closed form.

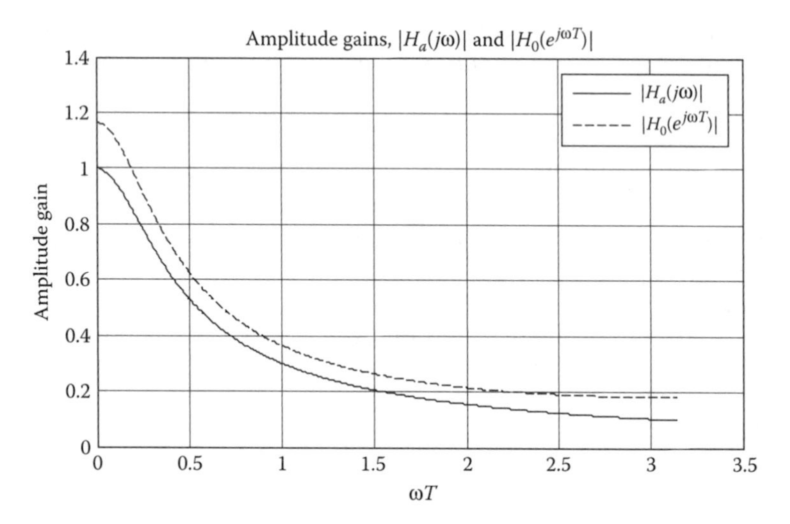

FIGURE 11.2
Continuous (solid) and impulse-invariant digital (dashed) amplitude gains for $H_a(s) = 1/(s + 1)$ and $T = \pi/10$.

EXAMPLE 11.2

Compare the step-function responses of the analog and digital filters in Example 11.1. On line 101 in the Appendix, the Laplace transform of the unit step function is $1/s$. Thus, the analog response is

$$Y(s) = X(s)H_a(s) = \frac{1}{s(s+1)}; \quad y(t) = 1 - e^{-t} \tag{11.5}$$

Similarly, the digital response from line 101 in the Appendix and (11.4) is

$$V(z) = \left(\frac{z}{z-1}\right)\frac{T}{1-e^{-T}z^{-1}} = \frac{Tz^2}{(z-1)(z-e^{-T})} \tag{11.6}$$

The inverse transform of $V(z)$ here can be found by putting $V(z)$ in the form of line 155 in the Appendix:

$$V(z) = \frac{Tz}{1-e^{-T}}\left(\frac{(1-e^{-T})z}{(z-1)(z-e^{-T})}\right) \tag{11.7}$$

The inverse transform of (11.7) is then

$$v_k = \frac{T(1-e^{-(k+1)T})}{1-e^{-T}} \tag{11.8}$$

The continuous and digital responses for two values of T are compared in Figure 11.3. The impulse-invariant approximation, v_k, approaches samples of $y(t)$ as T diminishes.

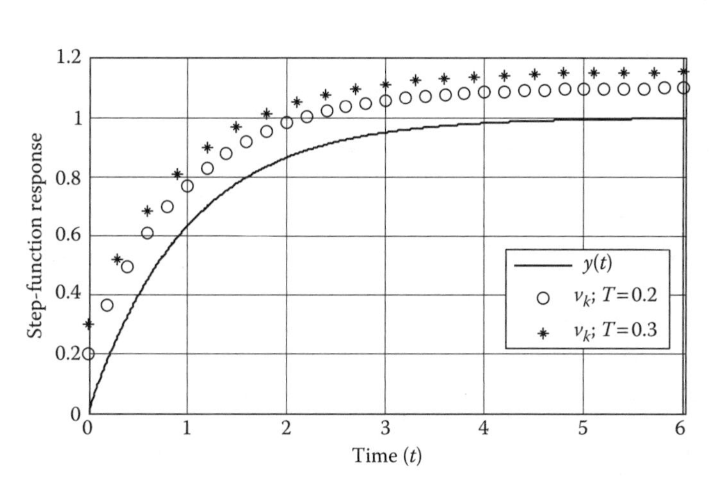

FIGURE 11.3
Step-function response, v_k, using the impulse-invariant model of $H_a(s) = 1/(s + 1)$.

Also, it appears that most of the error could be eliminated simply by reducing the d.c. gain of $H(z)$. The reason for this observation is discussed in Section 11.7.

When $x(t)$ is an arbitrary finite input signal, $y(t)$ is the convolution of $x(t)$ with the impulse response $h_a(t)$. In this case, the impulse-invariant approximation applies as in (11.2), that is,

$$v_k = \sum_{n=0}^{\infty} h_{an} x_{k-n} T \approx \int_{t=0}^{\infty} h_a(t) x(kT - t) dt = y_k \tag{11.9}$$

Thus, when we use the digital model with impulse response equal to samples of $Th_a(t)$, we have a discrete approximation to the continuous convolution integral.

For example, suppose $x(t)$ is the unit step function in Example 11.2, that is, x is zero for $t < 0$ and one for $t \geq 0$. Then the discrete convolution in (11.9) is

$$v_k = T \sum_{n=0}^{\infty} h_{an} x_{k-n} = T \sum_{n=0}^{k} e^{-nT}(1) = \frac{T\left(1 - e^{-(k+1)T}\right)}{1 - e^{-T}} \tag{11.10}$$

The upper limit changes to k in the second sum because the step function begins at $k - n = 0$. The final expression is found from the geometric series formula (1.19) in Chapter 1. Thus we arrive at the same expression as in (11.8), which is plotted in Figure 11.3. The form in (11.10) illustrates more clearly the nature of the impulse-invariant approximation. Consider a specific case, say, with $k = 6$ and $t = 6T$. The effect in Figure 11.3 is illustrated in Figure 11.4, which shows the components of the continuous and discrete convolution integrals at $t = 6T$.

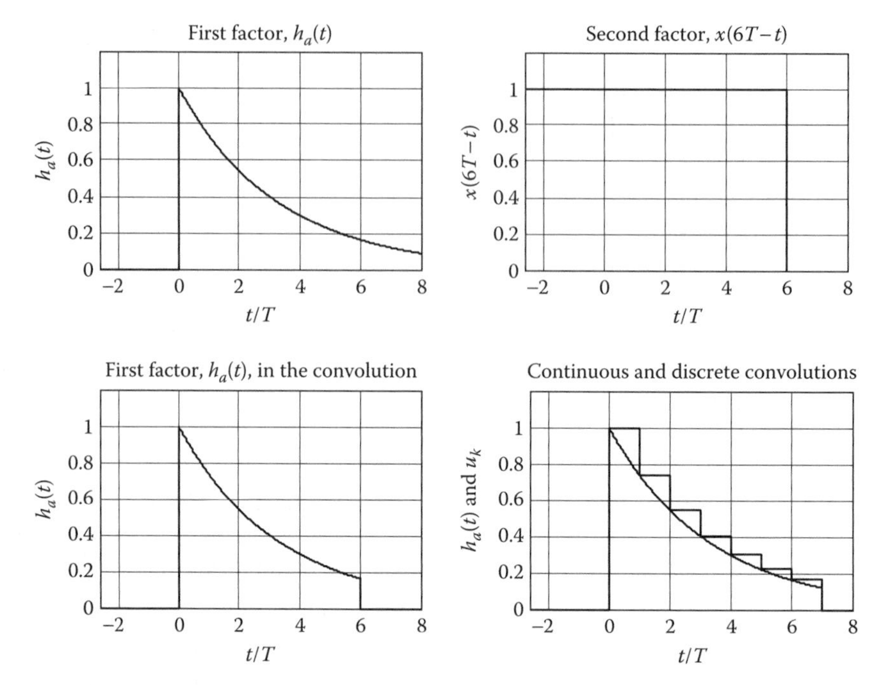

FIGURE 11.4
Illustration of the convolution in (11.9) at $t = 6T$.

The product of the two upper plots produces the lower left plot. The convolution is the area under the plot. The discrete convolution is the area under the steps in the lower right plot, and we can observe the effect in Figure 11.3, namely, that the discrete convolution exceeds the continuous convolution and approaches the latter as $T \to 0$.

The convolution analysis in Figure 11.4 shows that the impulse-invariant approximation leaves something to be desired in the way of accuracy. The question of how to achieve greater accuracy is addressed in Section 11.6. For now, the impulse-invariant model offers at least the argument that there is a direct comparison between digital and continuous systems, and that digital systems can be designed with frequency response functions and convolutions similar to those of continuous systems.

11.3 Final Value Theorem

In the analysis of continuous linear systems, if the final value of a system response to a given input exists, it can be found by letting t approach infinity in the expression for the output, $y(t)$. It can often be found more conveniently from the transform, $Y(s)$, via the following argument from Truxal,[17] Chapter 1. If $y(t)$ reaches a constant final value, say L, then $y(t)$ can be expressed as the sum of a step function of amplitude L plus some other decaying function, that is,

$$y(t) = Lu(t) + y_1(t)e^{-at} \tag{11.11}$$

where $u(t)$ is the unit step function, $y_1(t)$ is finite-valued, and a is greater than zero. Using this representation of $y(t)$ and applying lines G and 101 in the Appendix, the product $sY(s)$ becomes

$$sY(s) = L + sY_1(s+a) \tag{11.12}$$

Now we can see that letting $s \to 0$ in (11.12) and $t \to \infty$ in (11.11) both produce the same result, namely $y(\infty) = L$. Therefore,

$$\boxed{\lim_{t\to\infty} y(t) = \lim_{s\to 0} sY(s)} \tag{11.13}$$

is the final value theorem for continuous systems. Note the duality between (11.13) and the initial value theorem in (11.3).

There is a similar theorem for discrete systems. Again, suppose y_k is always equal to L for k greater than some integer. Then, as in (11.11), one could write

$$y_k = L + y_{1k}e^{-kaT} \tag{11.14}$$

with the same constraints on y_1 and a. In this case, taking the product $(z-1)Y(z)$, the result from the Appendix, lines G and 101, is

$$(z-1)Y(z) = zL + (z-1)Y_1(ze^{aT}) \tag{11.15}$$

And again, the final value L is obtained either by letting $k \to \infty$ in (11.14) or $z \to 1$ in (11.15). Therefore,

$$\boxed{\lim_{k\to\infty} y_k = \lim_{z\to 1}(z-1)Y(z)} \tag{11.16}$$

is the final value theorem for discrete systems.

EXAMPLE 11.3

Determine the final value of the step-function response of $H_a(s) = 1/(s+1)$. Equation 11.13 gives the result immediately:

$$\lim_{t\to\infty} y(t) = \lim_{s\to 0} sX(s)H(s) = \lim_{s\to 0} s\left(\frac{1}{s}\right)\left(\frac{1}{s+1}\right) = 1 \tag{11.17}$$

Similarly, (11.16) gives the final value of u in Figure 11.1 when the impulse-invariant model of $H_a(s)$ is used for $H(z)$:

$$\lim_{k\to\infty} u_k = \lim_{z\to 1}(z-1)U(z)$$
$$= \lim_{z\to 1} X(z)H_0(z) = \lim_{z\to 1}(z-1)\left(\frac{z}{z-1}\right)\left(\frac{Tz}{z-e^{-T}}\right) = \frac{T}{1-e^{-T}} \qquad (11.18)$$

11.4 Pole–Zero Comparisons

In the general case of the discrete model in Figure 11.1, it is instructive to compare the digital poles and zeros of $H(z)$ with the analog poles and zeros of $H_a(s)$. As discussed in Chapter 4, Sections 4.5 and 4.6, the digital zeros are the zeros of the numerator of $H(z)$ and the digital poles are zeros of the denominator. We saw that $H(z)$ was *stable* (in the sense of having a decaying impulse response) when all its poles were inside the unit circle, that is, $|z|<1$.

Analog poles and zeros were also discussed briefly in Chapter 6, where we mentioned that $H_a(s)$ is stable when its poles are in the left half of the s-plane, that is, $\text{Re}(s)<0$, where $\text{Re}(s)$ is the real part of s.

One way to compare the model, $H(z)$, with the original, $H_a(s)$, suggested by Jury (1964), is to compare the poles and zeros of both on the s-plane. Doing so illustrates the differences in frequency response as well as the effect of the sampling theorem. We already know that the frequency response of $H(z)$ is found by substituting $z=\exp(j\omega T)$ in $H(z)$. Suppose, therefore, that we substitute $z=\exp(sT)$ in $H(z)$. Then the frequency response can be viewed on the s-plane in the same way we view the response of $H_a(s)$, using vectors drawn between a point on the positive $j\omega$ axis and each of the poles and zeros in the system.

When we use $z=\exp(sT)$, the mapping of poles or zeros from the z-plane to the s-plane is not singular. First, we express z in polar form:

$$z = re^{j(\theta\pm 2n\pi)}; \quad 0\le n<\infty \qquad (11.19)$$

That is, z is at radius r and angle θ on the z-plane for all values of n. If we now let $z=\exp(sT)$ and express s as a function of z, the result is

$$s = \frac{1}{T}\log z = \frac{1}{T}[\log r + j(\theta \pm 2\pi n)]; \quad 0\le n<\infty \qquad (11.20)$$

Thus, a single point on the z-plane maps into an infinite number of points on the s-plane. For example, letting $n=0$ and $s=\alpha+j\omega$, the infinite strip on the s-plane bounded by $\omega=-j\pi/T$ and $\omega=+j\pi/T$ maps in accordance with (11.19) onto the entire z-plane, with the left half s-plane mapping to the interior of the unit circle on the z-plane. The general idea is illustrated in Figure 11.5. The semi-infinite strip on the left half of the s-plane represents the case with α, the real part of s, less than zero, and $n = 0$ in (11.20).

In (11.19), we also see that, depending on the value of n, any point on the left half of the s-plane maps to a single point inside the unit circle on the z-plane.

Conversely, a single pole or zero inside $|z| = 1$ maps in accordance with (11.20) to an infinite number of poles or zeros evenly spaced at $2\pi/T$ intervals along a vertical line on the left half of the s-plane, as illustrated in Figure 11.6. The figure shows a single zero on the z-plane at $z = 1$, which maps to $s = 0$ and to multiples of $s = j\pi/T$ along the $j\omega$ axis of the s-plane, as well as a

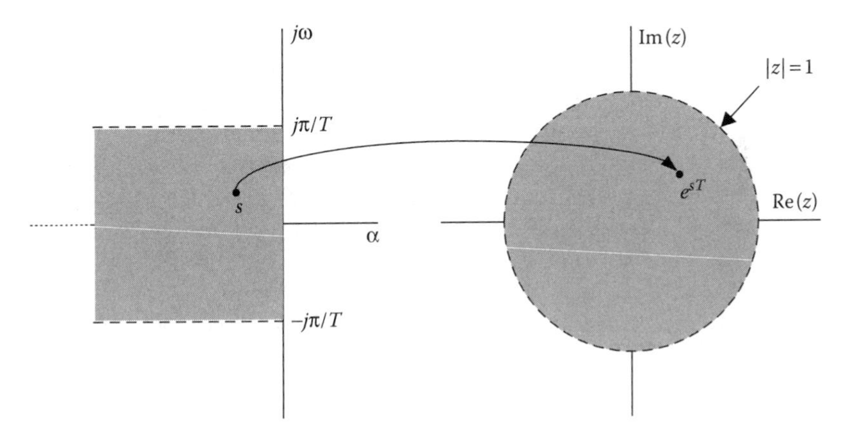

FIGURE 11.5
Mapping from the s-plane (left) to the z-plane (right): The semi-infinite strip shown on the left half of the s-plane maps to the interior of the unit circle on the z-plane.

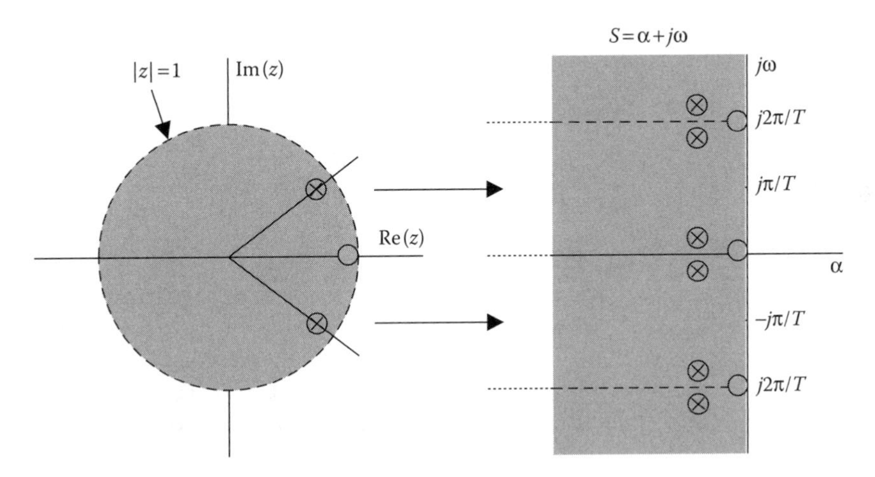

FIGURE 11.6
Mapping from the z-plane (left) to the s-plane (right), in accordance with (11.20).

similar mapping of the conjugate pair of poles on the z-plane. The mappings in Figures 11.5 and 11.6 are helpful (to some) in understanding the relationship of the Fourier transform to the DFT described in Chapter 4, and also the sampling theorem. When spectral content of the signal is entirely within the frequency range $|\omega| < \pi/T$, the mapping is one-to-one in both directions. Therefore, as we discussed in Chapter 3, the continuous signal can be recovered from the sample vector.

Suppose now that the signals are all within $|\omega| < \pi/T$ and we are modeling an analog system, $H_a(s)$, using the impulse-invariant approximation. Then, in accordance with the procedure in Section 11.2, the impulse-invariant model of $H_a(s)$ is

$$H_0(z) = TZ\{L^{-1}[H_a(s)]\} \tag{11.21}$$

where $Z\{\}$ means "z-transform of" and $L^{-1}[\,]$ means "inverse Laplace transform of." In other words, the mappings of the poles and zeros of $TH_a(s)$ and $H_0(z)$, which are transforms of the same signal, are as illustrated in Figures 11.5 and 11.6, and the model is therefore "impulse invariant."

The emphasis so far has been on comparing digital and continuous systems with the assumption that a digital model of a continuous system is the objective. We have noticed that the impulse-invariant models do not produce accurate approximations. In Figure 11.3, for example, the model was accurate only as $T \to 0$. Thus, the impulse-invariant approximation, although useful for conveying the fundamental relationships between analog and digital systems, is not accurate enough in most cases where digital simulation is the primary objective.

The remaining sections emphasize digital simulation as the main objective. Different schemes for deriving the model transfer function, $H(z)$, are introduced. For simulation purposes, the new schemes are considered as improvements over the impulse-invariant models, with greater accuracy being achieved for given values of T.

11.5 Approaches to Modeling

Digital models of continuous systems can be classified in various ways: One way is to classify models in terms of the signals being processed by the model. The discussion in this chapter is based on a deterministic approach to modeling in which the signals are essentially predetermined, that is, known in advance. A statistical approach, in which only signal statistics are used, was discussed in Chapter 8.

When the desired linear transfer function $H_a(s)$ is itself a rational function of s, as it must be for any realizable linear transfer function, the methods described here will provide simple and accurate digital models. In general, by decreasing the sampling interval T, the error in the model can be made as small as desired in the interval $\omega < \pi/T$ without making the model unduly complex.

Digital modeling plays an important role in modern systems analysis. It is used to test complicated and expensive system designs on the computer, so that construction costs are limited to software until the design is completed. There are plenty of applications of this in areas like navigation, guidance, and control systems for aerospace vehicles, nuclear reactors, radar tracking, and communication systems. There are other applications in areas like biology and medicine, economics, geology, and demography, and, in short, in any situation where a continuous system can be described appropriately. The References section contains several texts on the general subject of digital simulation.[1,2,4,5,9–14,16]

When the model involves inputs, outputs, and transfer functions, the frequency-domain approach is generally applicable, and the designer of the digital model is concerned with designing an accurate approximation to an idealized "real world." The methods in this chapter apply particularly to this concern, and are also useful in establishing a general connection between linear continuous and discrete systems.

A digital model of a continuous system, as defined here, involves a transfer function, $H(z)$, representing each linear portion of the system given by an $H_a(s)$ and also accounting for any limits or other nonlinearities that may exist in the continuous system. Nonlinearities are discussed later in Section 11.9.

The principal transfer functions are those illustrated in Figure 11.7, which is the same as Figure 11.1 and emphasizes that the ADCs are synchronized and produce impulse samples. The system to be modeled is described by $H_a(s)$, with input $x(t)$ and output $y(t)$. The modeling problem consists of obtaining the digital transfer function, $H(z)$, so that the digital output, v_k, is a good representation of y_k, that is, $y(kT)$ at the sample points. The resulting frequency response function, $H(j\omega)$, is obtained by substituting $e^{j\omega T}$ for z in $H(z)$ and represents the response of the digital model. Note that $H(j\omega)$ here is not necessarily the DFT of $1/T$ times the sampled inverse transform of

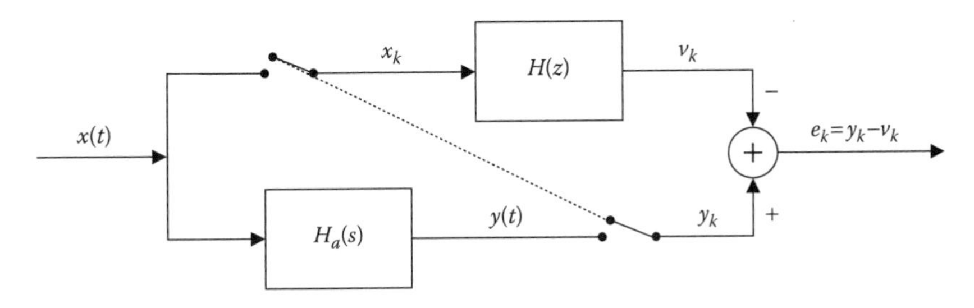

FIGURE 11.7
Discrete model, $H(z)$, of a continuous linear system, $H_a(s)$, emphasizing that the samples of x and y are impulse samples synchronized in time.

$H_a(s)$ as it was in the impulse-invariant approximation. The exact relationship between $H(j\omega)$ and $H_a(s)$ depends on the simulation method, that is, on how $H(z)$ is derived from $H_a(s)$.

The error model in Figure 11.7 is a *discrete* error model because the error, e_k, is measured only at the sample points. The model assumes one is interested in accuracy only at these points rather than in a continuous approximation to $y(t)$. The discrete error model has been used by several authors.[1,4,12,13] The discrete error and its DFT in Figure 11.7 are

$$\begin{aligned} e_k &= y_k - v_k \\ E(j\omega) &= Y(j\omega) - V(j\omega) = X(j\omega)H_a(j\omega) - V(j\omega) \end{aligned} \tag{11.22}$$

The discrete spectrum of e is seen to depend on the DFTs of both x and y, that is, on how well these functions represent their continuous counterparts. Since the modeling error depends on the input signal x and yet the model $H(z)$ is generally produced with only $H_a(j\omega)$ known, it is reasonable to classify simulation methods according to properties of $x(t)$. In particular, simulation methods are likely to differ according to whether the spectrum of $x(t)$ falls into one or the other of two broad classes:

1. $|X(j\omega)|$ is significantly greater than zero for $\omega \geq \pi/T$.
2. $|X(j\omega)|$ is negligible for $\omega \geq \pi/T$.

In class 1 models, one cannot hope to obtain a completely satisfactory general-purpose model because aliasing is present. In effect, the sampling process does not convey enough information about $x(t)$ to allow the simulator to distinguish it from other functions, no matter how good the simulator. The approach in Section 11.6, therefore, is to first make the model accurate for a particular known input and to then examine the accuracy of this model for other inputs.

For class 2 models, in which the spectrum of $x(t)$ is limited to frequencies below π/T rad/s, one can attain models of arbitrarily good accuracy. The problem in this case is to make $H(j\omega)$ approach $H_a(j\omega)$ in the interval $|\omega| < \pi/T$ so the error, e, will approach zero regardless of the exact form of $x(t)$. Methods for accomplishing this are discussed in Section 11.7.

11.6 Input-Invariant Models

Input-invariant models apply generally to either of the two classes described in Section 11.5. They are error-free for specific input functions and more or less accurate for others, depending on how much the other functions differ from the specified function. They are discussed first because one of them, the impulse-invariant method, has been discussed previously. The procedure for generalizing the impulse-invariant method described in Section 11.2 and producing models that are invariant for other input signals can be summarized in the following six steps, in which, as mentioned previously, $L\{\}$ signifies the Laplace transform of the continuous signal and $Z\{\}$ signifies the z-transform of the sampled signal:

1. Determine the desired transfer function, $H_a(s)$.
2. Specify an input signal, $i(t)$, other than the impulse function used in Section 11.2.
3. Find the Laplace and z transforms of $i(t)$, $I(s) = L\{i(t)\}$ and $I(z) = Z\{i(t)\}$.
4. The continuous output signal is the inverse transform, $y(t) = L^{-1}\{I(s)H_a(s)\}$.
5. The z-transform of $y(t)$ in step 4 is then $Y(z) = Z\{L^{-1}\{I(s)H_a(s)\}\}$.
6. The invariant model is

$$H(z) = Y(z)/I(z) = Z\{L^{-1}\{I(s)H_a(s)\}\}/I(z) \tag{11.23}$$

Since $Y(z)$ is by definition $Z\{y(t)\}$, y_k and u_k in Figure 11.1 must be the same, $e_k = 0$, and the model is exact when the input is $x(t) = i(t)$. These steps are illustrated in Example 11.4, which develops the step-invariant model (rather than the impulse-invariant model in Example 11.1) for $H_a(s) = 1/(s+1)$.

EXAMPLE 11.4

Develop the step-invariant model of $H_a(s) = 1/(s+1)$. The six steps given in Section 11.6 yield the following result in this example:

1. $H_a(s) = 1/(s+1)$
2. $i(t) = u(t)$, the unit step at $t = 0$
3. From the Appendix, line 101, $I(s) = 1/s$ and $I(z) = z/(z-1)$
4. $y(t) = L^{-1}\{I(s)H_a(s)\} = L^{-1}\left\{\dfrac{1}{s(s+1)}\right\} = 1 - e^{-t}$
5. $Y(z) = Z\{1 - e^{-t}\} = \dfrac{z}{z-1} - \dfrac{z}{z-e^{-t}} = \dfrac{z(1-e^{-T})}{(z-1)(z-e^{-T})}$
6. $H_s(z) = \dfrac{Y(z)}{I(z)} = \dfrac{1-e^{-T}}{z-e^{-T}}$

$H_s(z)$ is the step-invariant model of $H_a(s)$ in the sense that in Figure 11.7, if $x(t)$ is the unit step function $u(t)$, then $v_k = y_k$ and the model is exact.

We have seen that the impulse-invariant model is perfectly accurate when the input $x(t)$ is the unit impulse with amplitude $1/T$ and the step-invariant model is perfect when $x(t)$ is the unit step function with samples $x_k = 1;\ k \geq 0$. The same is obviously true if we design the model for any specified input $x(t)$.

If these models were accurate only for their specified inputs, they would be of rather limited practical interest. However, with linear superposition, any input-invariant model can be shown to give zero error in response to *any linear combination* of the specified input function.

For example, consider the stepwise or "zero-order hold" version of $x(t)$ in Figure 11.8, consisting of a superposition of step functions. If the stepwise version is used for the continuous signal in place of $x(t)$, then, as suggested in Figure 11.9, the model is exact provided $H_s(z)$ is the step-invariant model of $H_a(s)$. The error, e_k, is always zero for the stepwise version of any input signal.

This example illustrates the following theorem:

Zero-order hold theorem

Let $H_a(s)$ be any linear transfer function of s, and let

$$H_s(z) = \frac{z-1}{z} Z\left\{ L^{-1}\left\{ \frac{H_a(s)}{s} \right\} \right\} \tag{11.24}$$

Then $H_s(z)$ is an exact model of $H_a(s)$ for any input signal consisting of steps occurring at the sample points.

The zero-order hold theorem is just a restatement of step 6 of Section 11.6 (11.23) with $i(t)$ specified as the unit step function $u(t)$.

A similar result can be obtained for the first-order hold situation by replacing the step invariant $H_s(z)$ with a ramp invariant $H_r(z)$. The discrete error

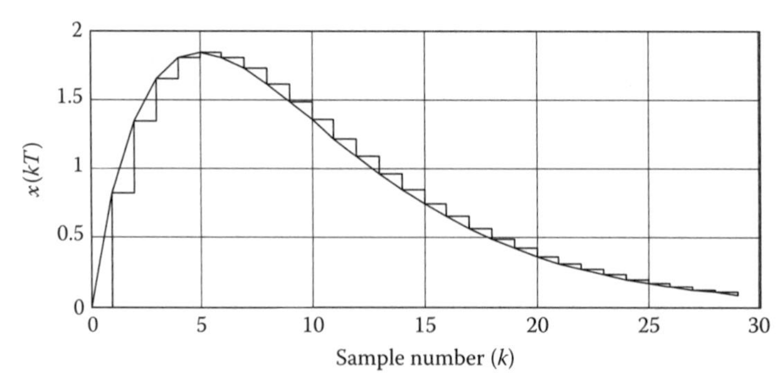

FIGURE 11.8
Continuous signal $x(t)$, with zero-order hold version.

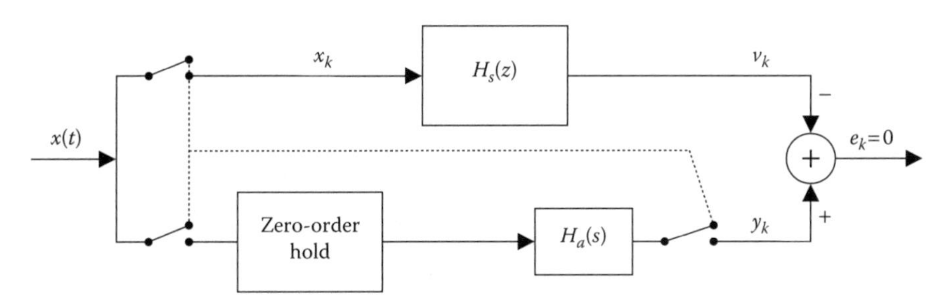

FIGURE 11.9
Step-invariant model of $H_a(s)$: The discrete error, e_k, is 0 for any finite input signal, $x(t)$.

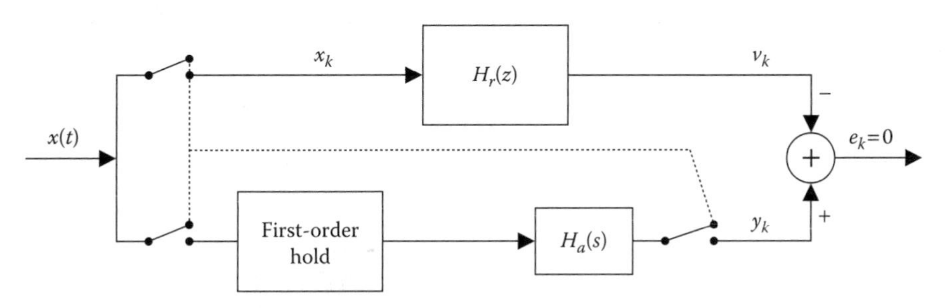

FIGURE 11.10
Ramp-invariant model of $H_a(s)$: The discrete error, e_k, is 0 for any finite input signal, $x(t)$.

model is shown in Figure 11.10. Again the error is zero, this time provided that the input to $H_a(s)$ is the first-order hold version of $x(t)$ (i.e., a sequence of straight lines connecting the sample points).

The first-order hold theorem corresponding with Figure 11.5 is as follows:

First-order hold theorem

Let $H_a(s)$ be any linear transfer function of s, and let

$$H_r(z) = \frac{(z-1)^2}{Tz} Z\left\{L^{-1}\left\{\frac{H_a(s)}{s^2}\right\}\right\} \tag{11.25}$$

Then $H_r(z)$ is an exact model of $H_a(s)$ for any input signal consisting of straight lines between the sample points.

Again, the first-order hold theorem is a restatement of step 6 of Section 11.6 (11.23) with $i(t)$ specified as the unit ramp function $i(t) = t$ for $t \geq 0$.

It is also instructive to prove (11.25) by expressing explicitly the input signal described in the theorem, an example of which is shown in Figure 11.11.

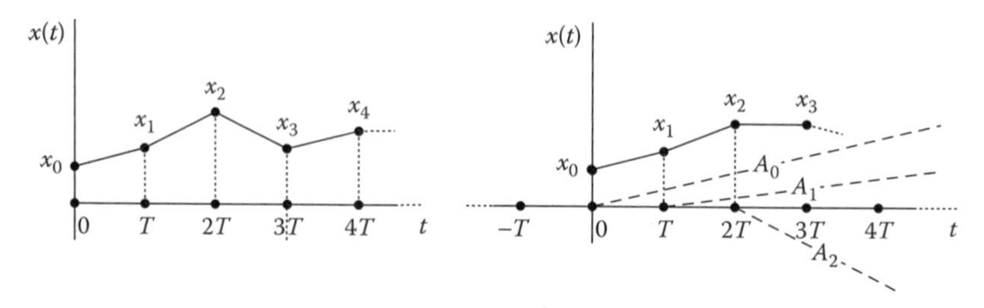

FIGURE 11.11
Signal, $x(t)$, consisting of straight lines between samples, formed by adding successive ramp functions.

The signal, $x(t)$, is expressed as a sum of ramp functions beginning at successive time steps,

$$x(t) = x_0 + \sum_{k=0}^{\infty} A_k(t - kT)u(t - kT) \tag{11.26}$$

where $u(t - kT)$ is a unit step at time kT and A_k is the slope of the ramp beginning at time $t = kT$, which, when added to the sum, brings $x(t)$ on a line from x_k to x_{k+1}. The successive ramp functions are illustrated on the right in Figure 11.11. The sum of ramp functions at any sample point $t = kT$ causes $x(t)$ to equal the sample value x_k. Between sample points, the sum of ramps produces a line from one point to the next.

Now, suppose a single term in the sum is the input to the analog transfer function, $H_a(s)$. Then, according to lines H and 102 in the Appendix, we have a term of the form

$$i(t) = A(t - kT)u(t - kT); \quad I(s) = \frac{Ae^{-kTs}}{s^2}; \quad \text{and} \quad I(z) = \frac{ATz}{z^k(z-1)^2} \tag{11.27}$$

The invariant model transform in step 6 of Section 11.6 (11.23) is then

$$H_r(z) = \frac{Z\{L^{-1}\{I(s)H_a(s)\}\}}{I(z)} = \frac{z^k(z-1)^2}{ATz} Z\left\{L^{-1}\left\{\frac{Ae^{-kTs}H_a(s)}{s^2}\right\}\right\} \tag{11.28}$$

In this result, A is a constant, and e^{-kTs}, which translates to z^{-k} in the z-transform, produces a delay in the impulse response of $H_a(s)/s^2$. Thus, (11.28) becomes

$$H_r(z) = \frac{z^k(z-1)^2 Az^{-k}}{ATz} Z\left\{L^{-1}\left\{\frac{H_a(s)}{s^2}\right\}\right\} = \frac{(z-1)^2}{Tz} Z\left\{L^{-1}\left\{\frac{H_a(s)}{s^2}\right\}\right\} \quad (11.29)$$

which is the same as $H_r(z)$ in (11.25). This again proves the theorem, because if $x(t)$ in (11.26) is a sum of ramp terms and if the model is exact for each term, the model must also be exact for the sum, that is, for any input of the form of $x(t)$ with lines between the sample points.

EXAMPLE 11.5

Develop the ramp-invariant model of $H_a(s) = 1/(s+1)$ and illustrate its response to $x(t) = 2te^{-t}$. According to the first-order hold theorem (11.25), using the transforms of line 156 in the Appendix, $H_r(z)$ is calculated as follows:

$$\begin{aligned} H_r(z) &= \frac{(z-1)^2}{Tz} Z\left\{L^{-1}\left\{\frac{H_a(s)}{s^2}\right\}\right\} = \frac{(z-1)^2}{Tz} Z\left\{L^{-1}\left\{\frac{1}{s^2(s+1)}\right\}\right\} \\ &= \frac{(z-1)^2}{Tz}\left(\frac{Tz}{(z-1)^2} - \frac{(1-e^{-T})z}{(z-1)(z-e^{-T})}\right) = 1 - \frac{(z-1)^2(1-e^{-T})z}{Tz(z-1)(z-e^{-T})} \quad (11.30) \\ &= \frac{T(z-e^{-T})-(z-1)(1-e^{-T})}{T(z-e^{-T})} = \frac{(T-1+e^{-T})z - Te^{-T}+1-e^{-T}}{T(z-e^{-T})} \end{aligned}$$

The difference equation corresponding with this result is

$$y_k = \frac{1}{T}\left((T-1+e^{-T})x_k - (Te^{-T}-1+e^{-T})x_{k-1}\right) + e^{-T}y_{k-1} \quad (11.31)$$

The effect of this formula is illustrated in Figure 11.12 using $T = 0.5$. On the left is $x(t)$ and $s(t)$, the latter being composed of straight lines between sample points. On the right are $y(t)$ and the sample set $[v_k]$ in Figure 11.10, which are obtained using the model in (11.31). In this case, T is large enough to cause a visible error; nevertheless, $s(t)$ is a fairly accurate approximation to $x(t)$, and the model gives a useful result.

It is instructive to compare the model in this example with the model obtained via the equation error method described in Chapter 8. To obtain a comparable model, we restrict the a and b vectors to length 2 as in (11.31), that is, $a = [1\ a_1]$ and $b = [b_0\ b_1]$. Then the vectors are adjusted iteratively to minimize the mean-squared error (MSE) in the manner illustrated in Figure 11.13. That is, y is the vector of output samples of $H_a(s)$ in Figure 11.1, and x is the input sample vector. In the equation-error method as described in Chapter 8, weight vector b is first optimized using autocovariance matrix R_{xx} and cross-covariance vector r_{xd}, that is, $[r_{xd}(0)\ r_{xd}(1)]$. Then, using the labels in Figure 11.13, the model error vector is

$$e = d - g = y - f - g \quad (11.32)$$

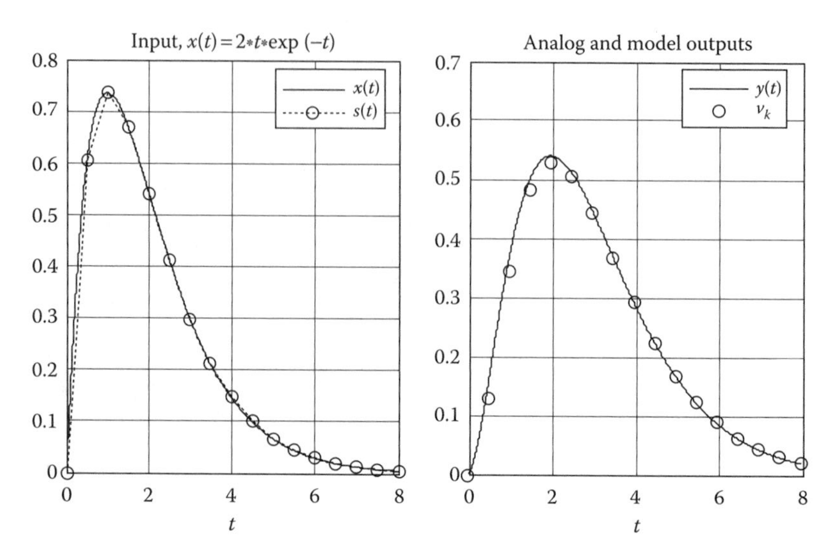

FIGURE 11.12
Signals in Example 11.5, illustrating the ramp-invariant model.

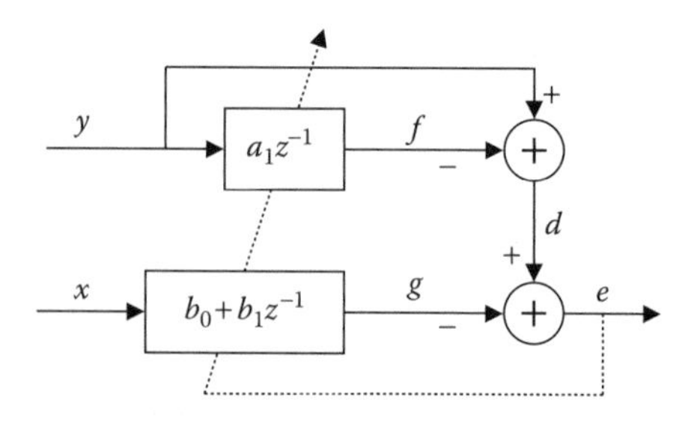

FIGURE 11.13
Equation error model for Example 11.5.

To minimize e, we must therefore adjust a_1 to make f as near as possible to

$$h = y - g \tag{11.33}$$

Thus, after optimizing b and computing g, we then optimize a (which is just a_1 in this case) using R_{xx} and r_{yh}, that is, $r_{yh}(1)$ to account for the delay associated with a_1. Then we repeat these two steps, alternately adjusting b and a_1 until MSE ceases to decrease. The process is illustrated in Figure 11.14. Twenty iterations are enough to make the MSE negligible in this case. The weight adjustments are plotted on the left in Figure 11.14, the resulting MSE improvement is shown in the center, and the analog output signal is plotted with the model outputs on the right. Compared with the ramp-invariant result

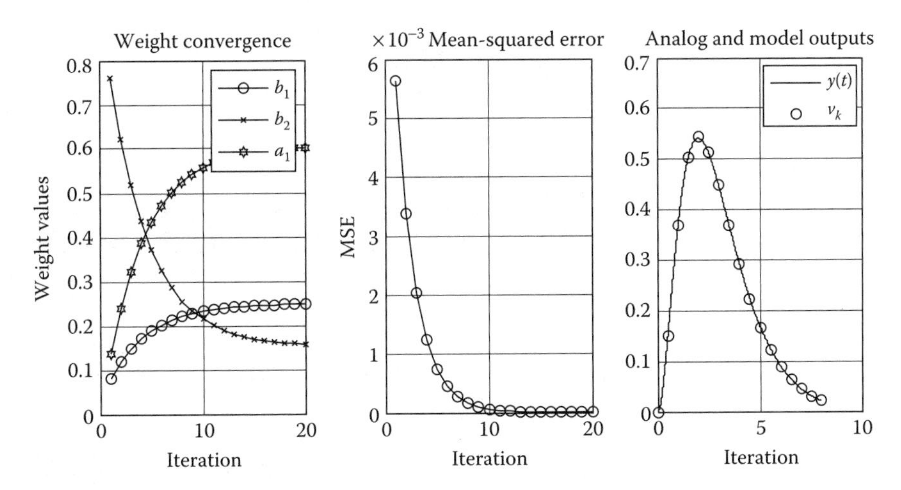

FIGURE 11.14
Equation error results for Example 11.5, illustrating the accuracy of the equation error model for a specified signal.

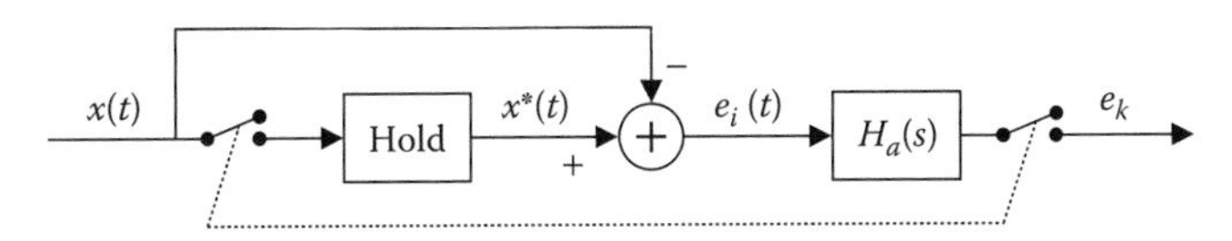

FIGURE 11.15
The model error, e, as the result of passing the input error, $e_i = x^* - x$, through $H_a(s)$.

in Figure 11.12, the equation error model produces a slightly more accurate result, but of course the equation error model is specific for the input signal, x, specified for this example, whereas the ramp-invariant model is specified for *any* input signal consisting of straight lines between sample points.

Invariant models have been described by Jury[7] and also by Rosko,[12] who provide examples of their application. They are useful when input signals are not frequency limited because they provide an indication of the error in terms of how well the signal, $x(t)$, is represented by a similar signal, $x^*(t)$, with steps or straight lines between the sample points. If the representation is good, as it is in Figure 11.12 for example, then one can usually expect satisfactory results from the model.

In fact, the exactness of the model for $x^*(t)$ can be used to establish a bound on the output error, which is due to the difference between $x(t)$ and $x^*(t)$. The discrete error model for this is shown in Figure 11.15. If we view the error as noise added to $x(t)$, as suggested in the figure, then the output error is simply the result of passing the noise, $e_i(t) = x^*(t) - x(t)$, through $H_a(s)$ to produce the model error, e.

The Laplace transform of the continuous error, $e(t)$, from which the sample vector e is taken, is seen in Figure 11.15 to be

$$E(s) = H_a(s)E_i(s) \tag{11.34}$$

The inverse transform of the product (11.34), similar to the case for z-transforms in Chapter 4, is the convolution of the two inverse transforms, $h_a(t)$ and $e_i(t)$, that is,

$$e(t) = L^{-1}\{H_a(s)E_i(s)\} = \int_0^t h_a(\tau)\, e_i(t-\tau)d\tau \tag{11.35}$$

Now, an upper bound on the magnitude of the modeling error, $e_k = e(kT)$, can be established by letting $e_{i\max}$ represent the maximum magnitude of $e_i(t)$, that is, the maximum discrepancy (magnitude) between $x(t)$ and $x^*(t)$, in some interval, say $0 \le t < mT$. From (11.35), this upper bound is

$$|e_k| \le e_{i\max} \int_0^{mT} |h_a(\tau)d\tau|; \quad k \le m \tag{11.36}$$

In most physical modeling situations, $h_a(t)$ will have finite energy, and the bound in (11.36) will be good for all samples, that is, for $m \le \infty$. Note that the bound is directly proportional to the maximum input error, $e_{i\max}$, and can, therefore, usually be made as small as desired by reducing the time step T and thus the discrepancy between $x(t)$ and $x^*(t)$. In Figure 11.12 for example, when T is reduced from 0.5 to 0.25, the MSE is reduced by more than an order of magnitude. Error bounds for other first- and second-order systems are found in the Exercises section at the end of this chapter (see Exercises 11.23 and 11.24).

In conclusion, the invariant models described in this section are easy to obtain and useful in situations where the signals can be approximated with steps or lines between sample points. They do not require $x(t)$ to be band limited, and the simulation error can be assessed in terms of the discrepancy between $x(t)$ and its approximation, $x*(t)$.

11.7 Other Linear Models

Two other simple models are described in this section: (1) an *adjusted version of the impulse-invariant approximation* and (2) a substitutional method called *bilinear approximation*, in which $H(z)$ is obtained by substituting a function of

z for s in $H_a(s)$. Recall that the latter was used in Chapter 6 to convert analog Butterworth and Chebyshev filters to corresponding digital filters.

Both methods are easiest to discuss and analyze if applied in situations where there is no aliasing. Then, matching the continuous transfer function, $H_a(j\omega)$, with the model transfer function, $H(e^{j\omega T})$, in the interval $|\omega| < \pi/T$ is a reasonable objective. If it can be achieved, then the model will be accurate in the sense that y_k and v_k will be equal in Figure 11.1.

The simplest approximation to $H_a(s)$ is the impulse-invariant approximation, $H_0(z)$, described in Section 11.2. In Example 11.1 and Figure 11.2, we observed that in the case of $H_a(s) = 1/(s+1)$, the difference between $H_0(e^{j\omega T})$ and $H_a(j\omega)$ could be substantially reduced if H_0 is simply multiplied by a constant. This approach, in which the adjusted impulse-invariant approximation, $H_A(z)$, is obtained by scaling $H_0(z)$, was suggested by Fowler.[2] The constant scaling factor is obtained by adjusting $H_0(z)$ to make its final value match that of $H_a(s)$ for a step-function input. In this case, the final value theorems in (11.13) and (11.16) result in the following equation for the unit step function, in which $y(t)$ is the output of $H_a(s)$ and v_k is the output of $H_0(z)$:

$$\begin{aligned} \lim_{t\to\infty} y(t) &= \lim_{s\to 0} sY(s) = \lim_{s\to 0} s\left(\frac{1}{s}\right)H_a(s) = \lim_{s\to 0} H_a(s) \\ \lim_{k\to\infty} v_k &= \lim_{z\to 1}(z-1)V(z) = \lim_{z\to 1}(z-1)\left(\frac{1}{z-1}\right)H_0(z) = \lim_{z\to 1} H_0(z) \end{aligned} \tag{11.37}$$

We therefore scale $H_0(z)$ to obtain the adjusted version, $H_A(z)$, as follows:

$$H_A(z) = \frac{H_a(0)}{H_0(1)} H_0(z) \tag{11.38}$$

With this scaled version of $H_0(z)$, the model will have the correct final value for a step or any input that settles to a constant final value. In other words, the d.c. response of the model is exact and $H_A(e^{j0}) = H_a(j0)$. For this reason, $H_A(z)$ is also called the "d.c.-adjusted" impulse-invariant model of $H_a(s)$.

If either limit in (11.37) is zero or infinite, $H_0(z)$ may be adjusted to have the correct final value for some other input, such as an impulse or ramp. However, in these cases, it is often best to use one of the other models.

To illustrate the adjusted impulse-invariant model, suppose we redo the examples in Figures 11.2 and 11.3 using the d.c.-adjusted version of $H_0(z)$. In Example 11.1, we had

$$H_a(s) = \frac{1}{s+1} \quad \text{and} \quad H_0(z) = \frac{T}{1-e^{-T}z^{-1}} \tag{11.39}$$

Therefore,

$$H_A(z) = \frac{H_a(0)}{H_0(1)} H_0(z) = \frac{1}{T/(1-e^{-T})} \frac{T}{1-e^{-T}z^{-1}} = \frac{(1-e^{-T})}{1-e^{-T}z^{-1}} \tag{11.40}$$

The results are compared with Figures 11.2 and 11.3 in Figure 11.16, which illustrates the advantage of adjusting the impulse-invariant model in accordance with (11.38).

The second simple approximation method we will discuss here is a substitutional method called the bilinear approximation. The use of this method in the design of recursive filters is described in Chapter 6. In this method, a simple ratio of linear functions of z is substituted for s in $H_a(s)$ to produce the digital model, $H_b(z)$. The substitution we will use is

$$s \leftarrow A\frac{z-1}{z+1} \tag{11.41}$$

in which A is a constant. This bilinear form has several properties of interest in digital simulation. It is rational and therefore $H_b(z)$ must be rational if $H_a(s)$ is a rational function. The bilinear substitution also has the effect of mapping the interior of the unit circle in the z-plane onto the primary strip of the left half of the s-plane, a property common to $s = (1/T)\log z$ discussed in Section 11.3. In fact, with $A = 2T$, (11.41) is a crude approximation to $s = (1/T)\log z$, which accounts in a rough way for the validity of the bilinear model: The poles and zeros of $H_a(s)$ in the primary strip are roughly those of $H_b(e^{st})$ in the primary strip. The bilinear substitution with $A = 2T$ is called "Tustin's approximation," after A. Tustin, who showed that with this value of A, sT is a simple approximation to $\log z$ when the bilinear substitution is made.

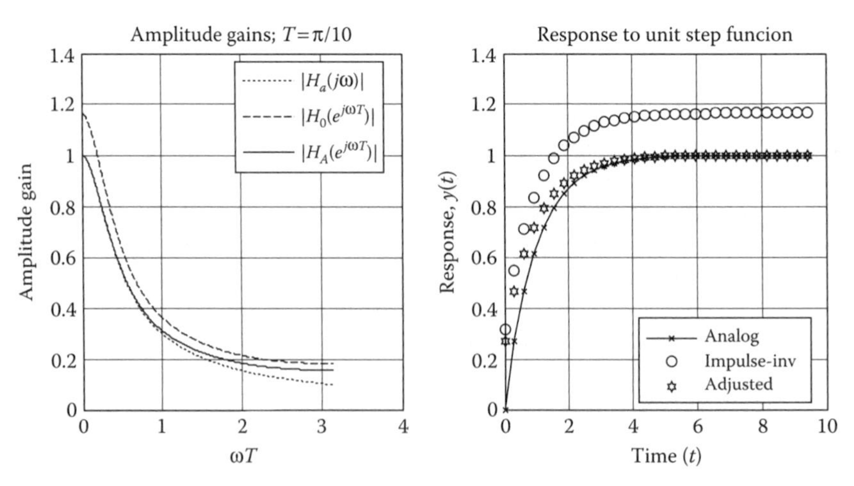

FIGURE 11.16
Comparing the impulse-invariant model of $1/(s + 1)$ with its adjusted version.

Of greater interest in modeling, however, is the warping of the frequency axis produced by the bilinear substitution. In Chapter 6, (6.41) and (6.42), the warping with $A = 1$ was given by

$$H_b(e^{j2\pi\nu}) = H_a[j\tan(\pi\nu)]; \quad \text{that is,} \quad H_b(e^{j\omega T}) = H_a[j\tan(\omega T/2)] \tag{11.42}$$

With $A \neq 1$, it is easy to modify (11.42) and see that the warping produced by the substitution in (11.41) produces the model with

$$H_b(e^{j\omega T}) = H_a[jA\tan(\omega T/2)] \tag{11.43}$$

With this result one can see, first, that the functions H_a and H_b are equal at $\omega = 0$, which means no d.c. adjustment is needed here. Second, $H_b(e^{j\omega T})$ at $\omega = \pi/T$ is equal to $H_a(\omega)$ at $\omega = \infty$. In other words, H_a in the range $0 \le \omega < \infty$ is warped into the interval $0 \le \omega < \pi/T$ to produce the model H_b. And finally, the constant A can be adjusted to make H_b equal to H_a at any single frequency less than π/T rad/s, say ω_0. That is, if we make H_b in (11.43) equal to $H_a(j\omega_0)$ in (11.43), then $H_a(j\omega_0) = H_a[A\tan(\omega_0/2)]$ and, therefore,

$$A = \omega_0 \cot(\omega_0 T/2) \tag{11.44}$$

These properties of bilinear models are illustrated in Example 11.6.

EXAMPLE 11.6

We again use $H_a(s) = 1/(s+1)$ to develop a model using the bilinear substitution, and make the model exact at $\omega_0 = \pi/2T$ rad/s. With this value, A in (11.44) is

$$A = \frac{\pi}{2T}\cot\left(\frac{\pi}{2T}\left(\frac{T}{2}\right)\right) = \frac{\pi}{2T} \tag{11.45}$$

Using (11.41) in $H_a(s)$, the bilinear model is then

$$H_b(z) = \frac{1}{\left(\frac{\pi}{2T}\right)\left(\frac{z-1}{z+1}\right)+1} = \frac{2T(z+1)}{\pi(z-1)+2T(z+1)} = \frac{2T(z+1)}{(2T+\pi)z+(2T-\pi)} \tag{11.46}$$

The amplitude spectrum and unit step response of $H_b(z)$ are shown in Figure 11.17 for comparison with Figure 11.16. The comparison illustrates how the warping of the frequency axis caused by the bilinear substitution, which was beneficial for the design of IIR digital filters in Chapter 6, results generally in models that are less accurate with wideband input signals, because the bilinear model generally can

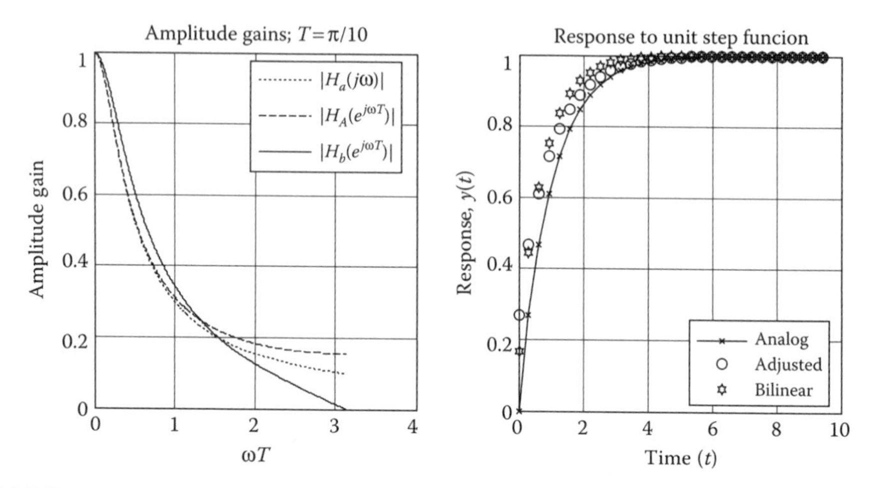

FIGURE 11.17
Comparing the adjusted impulse-invariant model of $1/(s+1)$ with the bilinear model.

be made exact only at zero and one other predetermined frequency. On the other hand, the bilinear transformation involves only the simple substitution for s to $H_a(s)$ in (11.41), rather than the z-transform of an inverse Laplace transform.

11.8 Comparison of Linear Models

It is difficult to compare these models in a general way because accuracy depends on the ensemble of input signals and on $H_a(s)$, as well as on the sampling interval, T.

To give the reader an idea of the quality of the modeling methods we have discussed, we compare the following four models of first- and second-degree versions of $H_a(s)$ in the frequency domain:

1. Step-invariant model
2. Ramp-invariant model
3. Adjusted zero-order model
4. Bilinear model with $A=2/T$ (Tustin's approximation described in Section 11.7)

Our basis for comparison will be the absolute gain error, that is, the absolute difference between $H_a(j\omega)$ and the model gain, $H(e^{j\omega T})$, at ω rad/s:

$$E(j\omega)=\left|H_a(j\omega)-H(e^{j\omega T})\right|;\quad 0\le\omega<\pi/T \tag{11.47}$$

For our first-degree version of $H_a(s)$, we use the system in Examples 11.1 through 11.6:

$$H_a(s) = \frac{1}{s+1} \tag{11.48}$$

The four digital models, each referred to its derivation, are:

1. $H_s(z)$ (Example 11.4) $= \dfrac{1-e^{-T}}{z-e^{-T}}$ (11.49)

2. $H_r(z)$ (11.30) $= \dfrac{(T-1+e^{-T})z - Te^{-T} + 1 - e^{-T}}{T(z-e^{-T})}$ (11.50)

3. $H_A(z)$ (11.40) $= \dfrac{1-e^{-T}}{1-e^{-T}z^{-1}} = \dfrac{(1-e^{-T})z}{z-e^{-T}}$ (11.51)

4. $H_b(z)$ (11.41 with $A = 2/T$) $= \dfrac{T(z+1)}{(T+2)z+(T-2)}$ (11.52)

The absolute gain error in (11.47) is plotted in Figure 11.18 in terms of dB versus ωT. Note that this is the absolute gain error, not the error in absolute gain. In other words, the measure in (11.47) is affected by both amplitude gain and phase differences. In Figure 11.16, the two models with the lower absolute gain errors are the ramp-invariant and bilinear models.

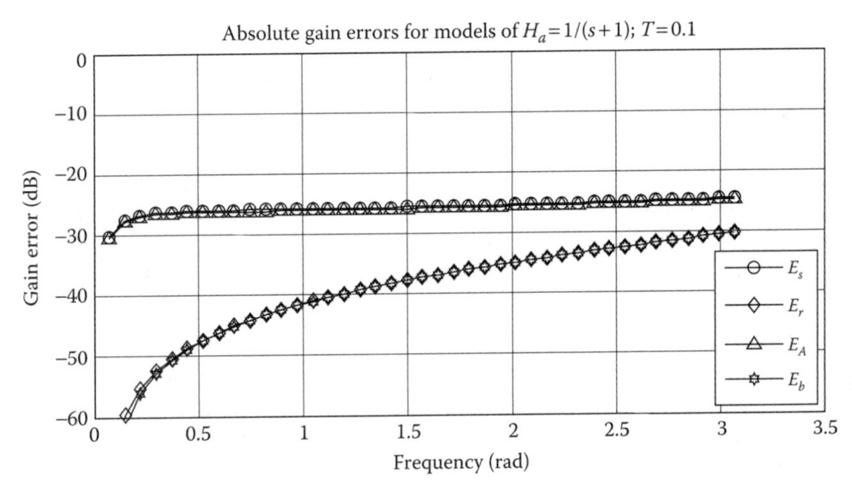

FIGURE 11.18
Absolute gain errors for four models of $1/(s + 1)$ with $T = 0.1$.

For our second-degree version of $H_a(s)$, we use a transfer function with complex conjugate poles at $-1 \pm j2$ on the s-plane:

$$H_a(s) = \frac{1}{(s+1)^2 + 2^2}; \quad h_a(t) = \frac{1}{2} e^{-t} \sin 2t \tag{11.53}$$

with the impulse response, $h_a(t)$, given in the Appendix on line 206. The step-invariant and ramp-invariant models, $H_s(z)$ and $H_r(z)$, respectively, are derived from lines 311 and 407 in the Appendix, and $H_b(z)$ is determined as before in accordance with (11.41) with the coefficient $A = 2/T$. To describe the four digital models in this case, we define the following coefficients:

$$\begin{aligned} &c_1 = 5T + 2; \quad c_2 = e^{-T} \sec\theta \cos(2T - \theta); \quad c_3 = 2e^{-T}\cos(2T); \\ &c_4 = e^{-2T}; \quad c_5 = 5T^2 \end{aligned} \tag{11.54}$$

In the definition of c_2, $\theta = \tan^{-1}(-1/2)$ in $H_s(z)$ and $\tan^{-1}(3/4)$ in $H_r(z)$. The models may then be expressed as follows:

1. $$H_s(z) = \frac{1}{5}\left(1 - \frac{(z-1)(z-c_2)}{z^2 - c_3 z + c_4}\right) \tag{11.55}$$

2. $$H_r(z) = \frac{(c_1 - 2c_2 + 2c_3 - 4)z^2 + (2 + 4c_2 - c_3 c_1 - 2c_4)z + (c_1 c_4 - 2c_2)}{25T(z^2 - c_3 z + c_4)} \tag{11.56}$$

3. $$H_A(z) = \frac{(1 - c_3 + c_4)z}{5(z^2 - c_3 z + c_4)} \tag{11.57}$$

4. $$H_b(z) = \frac{T^2(z+1)^2}{(c_5 + 4T + 4)z^2 + 2(c_5 - 4)z + (c_5 - 4T + 4)} \tag{11.58}$$

Again the absolute gain error in (11.47) is plotted in Figure 11.19 in terms of dB versus ωT. In Figure 11.19 in this case, the step-invariant model has a somewhat greater absolute gain error when compared with the other three models. The ramp-invariant error, E_r, is the smallest over most of the range of ω. Note, however, that in the bilinear simulation, the constant A could have been chosen to force the model in (11.58) to produce a gain error of zero at any single frequency. These examples are of course neither conclusive nor inclusive and only illustrate the use of the absolute gain error to compare digital models.

Modeling errors using measures similar to $E(j\omega)$ in (11.47) or a specified time-domain error, e_k, have been documented by several authors. Jury[7] and

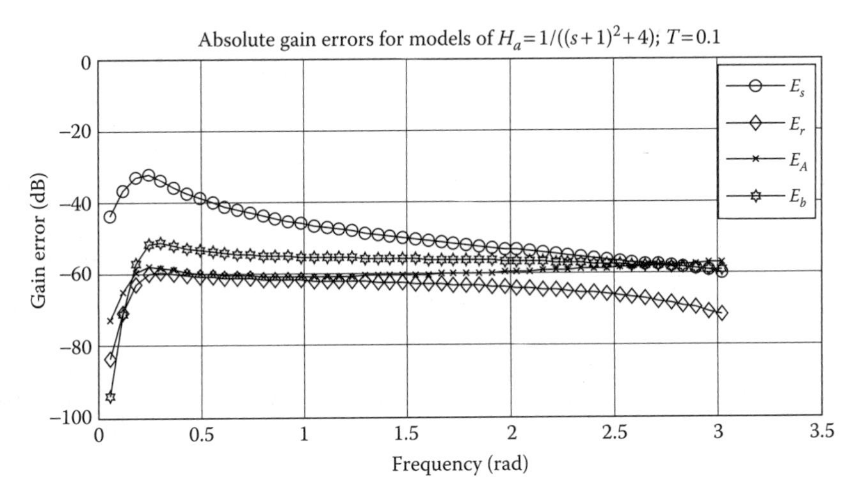

FIGURE 11.19
Absolute gain errors for four models of $1/((s+1)^2+2^2)$ with $T=0.1$.

Rosko[12] present plots of the frequency-domain errors for the integration operators $1/s^n$. Wait[15] gives examples of frequency-domain errors in various models of first- and second-order systems, and Rosko[12] provides a number of both time- and frequency-domain errors in models of various systems.

11.9 Models of Multiple and Nonlinear Systems

Continuous electrical and mechanical systems often involve more than one input signal as well as a number of different transfer functions that are sometimes connected in feedback loops. Furthermore, nonlinearities may be present in the form of limits, thresholds, and so on. The control system in Figure 11.20 is an example. A single model of the entire system would not usually be attempted because: (1) the limiter is a nonlinear component of the system for large signals; (2) the feedback signal, $f(t)$, may be required explicitly; and (3) there are two overall transfer functions, one from $x_1(t)$ to $y(t)$ and the other from $x_2(t)$ to $y(t)$. There is no single approach to the simulation of such a complex structure that is best for all systems. The most straightforward method is to derive models for $H_1(s)$, $H_2(s)$, and $H_3(s)$, as well as the limiter, that is, to model each block in the diagram individually. This could be less accurate than an overall model. In general, the model of a product, say $H(z)$ derived from $H(s)=H_1(s)H_2(s)$, is different from the product of models, that is, $H_1(z)H_2(z)$, depending on how the z-transfer functions are derived. An important exception to this

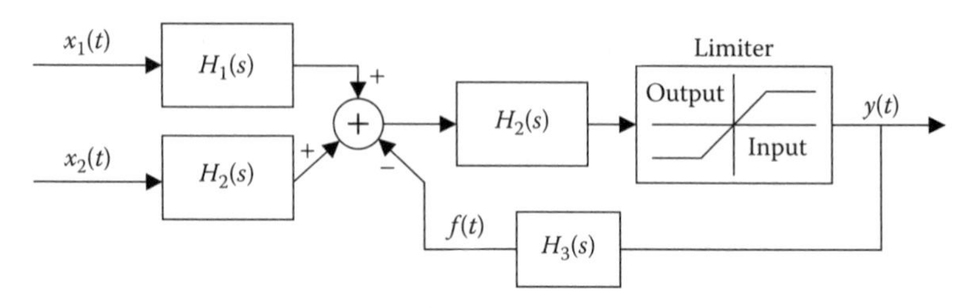

FIGURE 11.20
System with feedback as well as a nonlinear element.

general rule occurs with the bilinear approximation. Here, since $H(z)$ is found by replacing s with $A(z-1)/(z+1)$ in $H(s)$, $H(z)=H_1(z)H_2(z)$ follows from $H(s)=H_1(s)H_2(s)$. Example 11.7 illustrates some of the features of this discussion.

EXAMPLE 11.7

For the system in Figure 11.21, derive a model that gives an exact step-function response. Show how the model differs from the product of individual step-invariant transfer functions. The individual step-invariant models are essentially derived in Example 11.4, in which we can easily see that the step-invariant model of $H(s)=c/(s+c)$ is

$$H_c(z)=\frac{z-1}{z}Z\left\{L^{-1}\left[\frac{c}{s(s+c)}\right]\right\}=\frac{1-e^{-cT}}{z-e^{-cT}} \tag{11.59}$$

Thus, the product of the models of the individual stages in Figure 11.20 is

$$H(z)=H_1(z)H_2(z)=\frac{(1-e^{-T})}{(z-e^{-T})}\frac{(1-e^{-2T})}{(z-e^{-2T})}=\frac{1-e^{-T}-e^{-2T}+e^{-3T}}{z^2-(e^{-T}+e^{-2T})z+e^{-3T}} \tag{11.60}$$

The step-invariant model of the cascade combination in Figure 11.21 differs from this result. The model is found in the same manner, that is,

$$H(z)=\frac{z-1}{z}Z\left\{L^{-1}\left[\frac{2}{s(s+1)(s+2)}\right]\right\} \tag{11.61}$$

This time, the transforms are on line 300 in the Appendix. With $a=1$ and $b=2$ on line 300, the cascade model becomes

$$\begin{aligned} H(z)&=\frac{z-1}{z}\left(\frac{z}{z-1}+\frac{2z}{(-1)(z-e^{-T})}-\frac{z}{(-1)(z-e^{-2T})}\right) \\ &=\frac{(1-e^{-T})^2(z+e^{-T})}{(z-e^{-T})(z-e^{-2T})} \end{aligned} \tag{11.62}$$

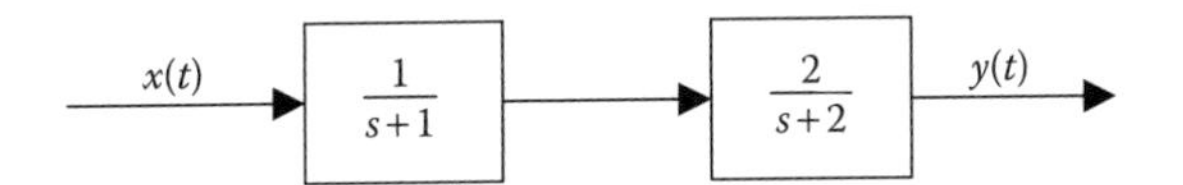

FIGURE 11.21
System used in Example 11.7.

Thus, the overall model in (11.62) is not the same as the product of individual models in (11.60), and (11.60) is, therefore, not a step-invariant model of the overall system in Figure 11.21. This fact is easily demonstrated by comparing the responses given by (11.60) and (11.62) with the analog response, $h(t)$, on line 300 in the Appendix.

Besides the limiter in Figure 11.20, other nonlinearities are common in continuous systems. Hard limits, saturation effects, threshold devices, logarithmic responses, and so on are found or may be approximated in various circumstances. All of these in general are easy to simulate with one or a few MATLAB® expressions; but they generally need to be implemented as individual components that operate explicitly at each time step. For example, a device that limits a signal, x, to a maximum value of L would be modeled at each time step (k) by an expression such as

```
y(k)=min(x(k),L)
```
(11.63)

Methods for simulating closed-loop systems with nonlinearities have been described by Hurt,[6] Sage and Burt,[13] Fowler,[2] and Rosko.[12] Generally, these methods involve simulating the individual blocks in the block diagram, adding a unit delay in the feedback loop if necessary for realizability, and then adding blocks either inside or outside the loop to make the overall transfer function more accurate. In this manner, the difficulties mentioned in this section can be alleviated, at least to some extent.

To follow this general plan, some or all of the following six steps can be taken to simulate a closed-loop system with nonlinearities:

1. Replace the nonlinear elements temporarily with linear components.
2. After considering the known properties of the input signals, construct a digital model of each block in the block diagram.
3. Add a unit delay inside the closed loop if (and only if) it is necessary to make the closed-loop model realizable.
4. If possible, add a constant gain inside the closed loop to make the (overall) closed-loop poles agree with those of a model of the closed-loop $H(s)$.
5. Add a block at the input to make the overall $H(z)$ a desired (i.e., impulse-invariant, step-invariant, ramp-invariant, etc.) model of $H(s)$.
6. Replace the nonlinear elements with suitable simulations or expressions, and test the model.

We conclude this chapter with Example 11.8, which amounts to a case study of a system with feedback and a nonlinear element. The example is meant to show how, with some patience, the engineer can design a useful and reasonably accurate model.

EXAMPLE 11.8

The block diagram in Figure 11.22 includes a device, $H_1(s)$, whose output, $y(t)$, is limited such that its magnitude cannot exceed one. An integrator, $H_2(s)$, is in the feedback leg.

The feedback signal, $f(t)$, causes $g(t)$ to oscillate in response to a transient input and to reach a final value of zero for a constant input, as seen in the form of the overall (linear) $H_a(s)$ for small signals (Chapter 4, Section 4.8):

$$H_a(s) = \frac{H_1(s)}{1 + H_1(s)H_2(s)} = \frac{s}{(s+1)^2 + 9} \tag{11.64}$$

Step 1 of the procedure detailed following Example 11.7 is accomplished by modeling the system for small signals and letting the limiter have unit gain. For step 2, we select the step-invariant form and construct the individual block models, again using the appropriate lines in the Appendix:

$$H_1(z) = \frac{z-1}{z} Z\left\{L^{-1}\left[\frac{1}{s(s+2)}\right]\right\} = \frac{1-e^{-2T}}{2(z-e^{-2T})} \tag{11.65}$$

This equation is similar to (11.59), and

$$H_2(z) = \frac{z-1}{z} Z\left\{L^{-1}\left[\frac{10}{s^2}\right]\right\} = \frac{10_T}{z-1} \tag{11.66}$$

The step-invariant model of the integrator in (11.66) has a useful property in the closed-loop model: There are more poles than zeros. Thus, in the digital version of Figure 11.22,

$$f_k = 10Ty_{k-1} + f_{k-1} \tag{11.67}$$

and the simulation is realizable without the need of step 3 (of the procedure described following Example 11.7), because f_k and e_k can be calculated in terms of past values of loop variables. (A counterexample is discussed later in this section.)

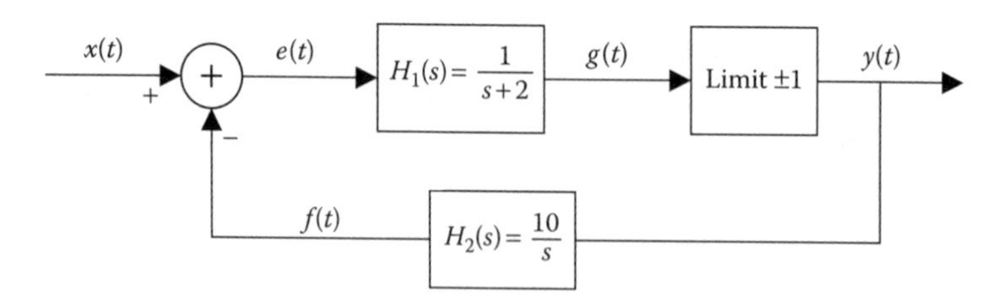

FIGURE 11.22
System used in Example 11.8.

Next, for step 4, the step-invariant model of the overall $H_a(s)$, which is obtained using (11.64) and line 206 in the Appendix, is

$$H_{ov}(z) = \frac{z-1}{z} Z\left\{L^{-1}\left[\frac{s}{s[(s+1)^2+9]}\right]\right\} = \frac{z-1}{z} Z\left\{L^{-1}\left[\frac{1}{(s+1)^2+9}\right]\right\}$$

$$= \frac{z-1}{z}\left(\frac{\left(\frac{1}{3}\right) z e^{-T}\sin 3T}{z^2 - 2ze^{-T}\cos(3T) + e^{-2T}}\right) = \frac{\left(\frac{1}{3}\right)(z-1)e^{-T}\sin 3T}{z^2 - 2ze^{-T}\cos(3T) + e^{-2T}} \tag{11.68}$$

On the other hand, the *denominator* of the transfer function of the feedback model, $H_f(z)$, with constant gain K inside the closed loop is

$$D = 1 + KH_1(z)H_2(z) = 1 + K\left(\frac{1-e^{-2T}}{2(z-e^{-2T})}\right)\left(\frac{10T}{z-1}\right) \tag{11.69}$$

which, when $H_f(z)$ is rationalized, becomes

$$D_r = z^2 - (1+e^{-2T})z + e^{-2T} + 5KT(1-e^{-2T}) \tag{11.70}$$

In this case, there is no value of K that will make the roots of (11.70) equal to the poles of (11.68); so step 4 is not possible in this example. (Again, a counterexample is discussed later in this section.)

To accomplish step 5, an input block, $H_3(z)$, is appended to the front end to make the overall feedback model transfer function step invariant, that is, equal to $H_{ov}(z)$ in (11.68). Thus, using (11.65) and (11.66) with (11.68), we have

$$H_{ov}(z) = \frac{\left(\frac{1}{3}\right)e^{-T}\sin 3T(z-1)}{z^2 - 2ze^{-T}\cos(3T) + e^{-2T}} = \frac{H_3(z)H_1(z)}{1 + H_1(z)H_2(z)};$$

$$H_3(z) = H_{ov}(z)\left(\frac{1}{H_1(z)} + H_2(z)\right) \tag{11.71}$$

$$= \left(\frac{2e^{-T}\sin 3T}{3(1-e^{-2T})}\right)\left(\frac{z^2 - (1+e^{-2T})z + e^{-2T} + 5T(1-e^{-2T})}{z^2 - 2ze^{-T}\cos(3T) + e^{-2T}}\right)$$

Finally, step 6 (of the procedure described in Example 11.7) is accomplished by placing the limiter in the digital model. The completed step-invariant model is shown in Figure 11.23. The model obviously can be realized if the computations

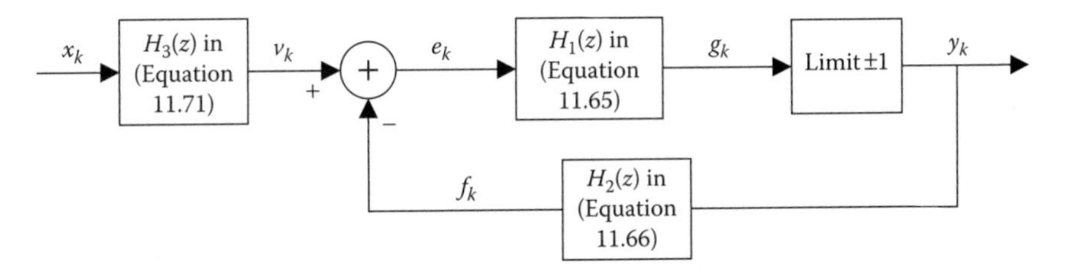

FIGURE 11.23
Step-invariant model of the system in Figure 11.22.

are made in order, that is, first v and f, the latter in accordance with (11.67); then e; then g; and finally y. Initial values x_{-2}, x_{-1}, e_{-2}, etc., may be set to zero in most situations.

Before giving an illustration using the model in Example 11.8, it is instructive to reconstruct the model with another choice for the blocks in step 2. Using the adjusted impulse-invariant form described in Section 11.2 and (11.38) for each block in Figure 11.22, the individual block model of $H_1(s)$ is

$$H_{01}(z) = Z\left\{L^{-1}\left[\frac{1}{s+2}\right]\right\} = \frac{z}{z-e^{-2T}};$$
$$H_{A1}(z) = \frac{H_{a1}(0)}{H_{01}(1)} H_{01}(z) = \frac{(1-e^{-2T})z}{2(z-e^{-2T})} \tag{11.72}$$

The adjusted impulse-invariant model of $H_2(s)$ is found by noting that the final value of $h_2(t)$, given by (11.13), is the same as the final value of h_{2k} given by (11.16). That is, no d.c. adjustment is needed in this case and

$$H_{A2}(z) = Z\left\{L^{-1}\left[\frac{10}{s}\right]\right\} = \frac{10z}{z-1} \tag{11.73}$$

The important difference in this case is that in the digital model of Figure 11.22, $H_{A2}(z)$ requires the current value of y, that is,

$$f_k = 10y_k + f_{k-1} \tag{11.74}$$

Therefore, step 3 (of the procedure described in Example 11.7) requires that we place a delay somewhere in the loop to make the model realizable. Choosing (arbitrarily) to place the delay in the feedback portion has the effect of multiplying $H_{A2}(z)$ in (11.73) by z^{-1}. Now the feedback signal, f_k, is computed in terms of y_{k-1} instead of y_k, and the closed-loop simulation is realizable.

For step 4, we need the overall transfer function of the model with the feedback delay, comparable to but different from (11.68) for the step-invariant model. Using either the steps in Section 11.2 or the general formula, (11.23), as well as line 208 in the Appendix, the overall impulse-invariant model is

$$H_{ov}(z) = TZ\{L^{-1}[H_a(s)]\} = TZ\left\{L^{-1}\left[\frac{s}{(s+1)^2+9}\right]\right\}$$
$$= \frac{\left(\frac{T}{3}\right)[3z^2 - ze^{-T}(3\cos 3T + \sin 3T)]}{z^2 - 2ze^{-T}\cos 3T + e^{-2T}} \tag{11.75}$$

In this case, since $H_a(0) = 0$, $H_{ov}(z)$ cannot be adjusted in accordance with (11.30). With a constant gain K in the feedback loop, the transfer function of the feedback model becomes

$$H_f(z) = \frac{H_{A1}(z)}{1 + Kz^{-1}H_{A1}(z)H_{A2}(z)} \tag{11.76}$$

and we can see that after using (11.72) and (11.73) in this expression and rationalizing the result, the denominator of $H_f(z)$ becomes

$$D_r = z^2 - [1 + e^{-2T} - 5K(1 - e^{-2T})]z + e^{-2T} \tag{11.77}$$

Equating this result to the denominator in (11.75) and solving for K, we have

$$K = \frac{1 + e^{-2T} - 2e^{-T}\cos 3T}{5(1 - e^{-2T})} \tag{11.78}$$

Finally, to accomplish step 5 of the procedure described in Example 11.7, we derive an input transfer function, $H_3(z)$, that will cause the overall model to be step invariant. The derivation of $H_3(z)$ is similar to (11.71):

$$H_3(z) = H_{ov}(z)\left(\frac{1}{H_{A1}(z)} + Kz^{-1}H_{A2}(z)\right) \tag{11.79}$$

If we substitute $H_{ov}(z)$ in (11.68) for the step-invariant model, $H_{A1}(z)$ in (11.72), and $H_{A2}(z)$ in (11.73) into (11.79), the expression reduces to

$$H_3(z) = \frac{2e^{-T}\sin(3T)}{3(1 - e^{-2T})z} \tag{11.80}$$

With $H_3(z)$, the model is now step invariant even though its components are not. After performing step 6 of the procedure described in Example 11.7 and inserting the limiter, the completed step-invariant simulation using the impulse-invariant blocks is as shown in Figure 11.24.

Having the two step-invariant models in Figures 11.23 and 11.24, it is instructive to compare them. Although they are different, the two models are guaranteed via the construction of $H_3(z)$ to give identical, exact results when the continuous input $x(t)$ is composed of steps and is small so the operation is linear. This behavior is illustrated in Figure 11.25, where $x(t)$ is a unit

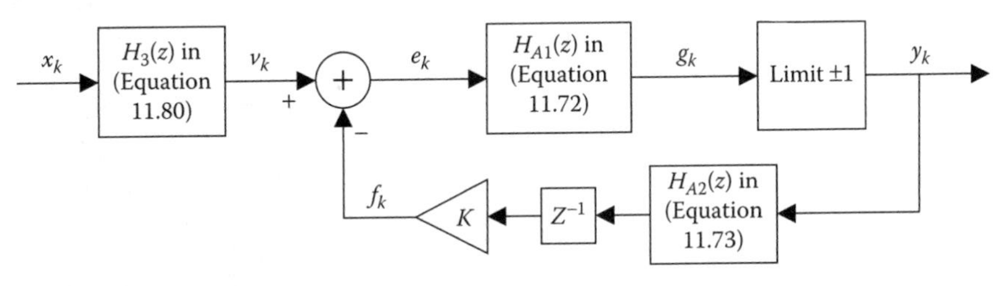

FIGURE 11.24
Step-invariant model of the system in Figure 11.22 with impulse-invariant components.

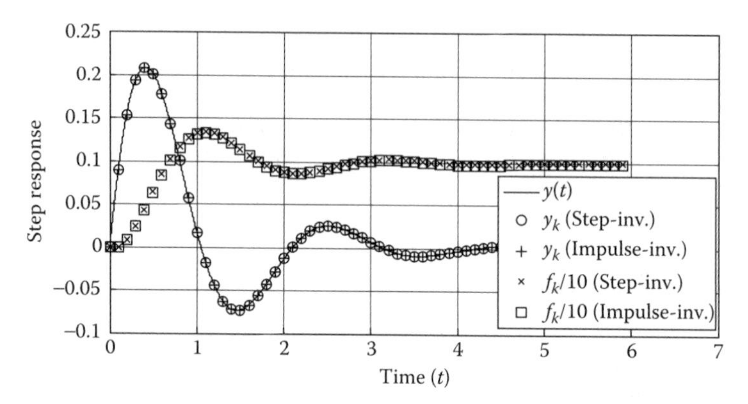

FIGURE 11.25
Linear operation of models for Example 11.8 (Figures 11.22, 11.23, and 11.24) with time step $T = 0.1$: The input signal is a unit step function at $t = 0$. The output signals (v_k and f_k) are exact in this case.

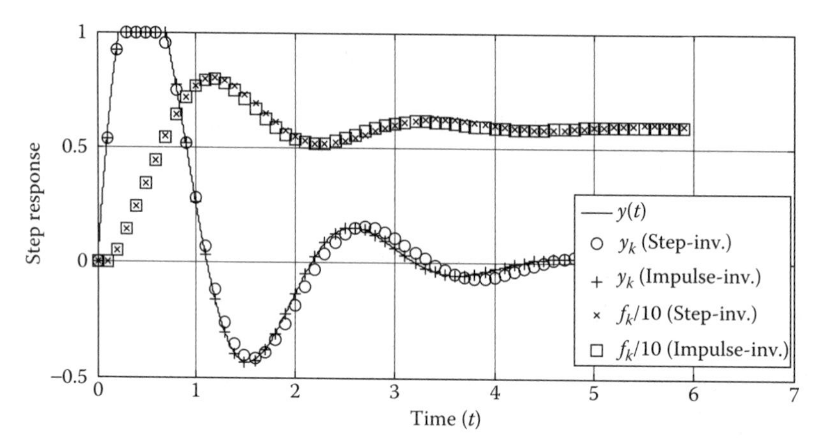

FIGURE 11.26
Nonlinear operation of models for Example 11.8 (Figures 11.22, 11.23, and 11.24) with time step $T = 0.1$: The input signal is a step function with amplitude 6 at $t = 0$. The output signals (v_k and f_k) are nearly exact in this case.

step function and, consequently, the limit is not reached. In this case, the outputs (y_k) of both models are exact at the sample points. A scaled version of the feedback signal, $f_k/10$, is also shown in Figure 11.25.

Figure 11.26 illustrates nonlinear operation in response to a step input with amplitude 6 at $t = 0$. Both models are exact until g_k reaches 1.0 and operation becomes nonlinear. After the limit is reached, there are slight inaccuracies, which should be considered when the designer selects the time step to be used in operating the models.

Regarding the latter, and assuming the sampling theorem is satisfied when analog inputs are digitized, it is always possible to resample the digitized input using a shorter time step before applying one of these models. Then, for example, the accuracy of the step-invariant model is improved because the stepwise version of the input is generally closer to the continuous version.

11.10 Concluding Remarks

In this chapter, some of the simplest methods for simulating continuous transfer functions have been introduced. Simulation *per se* has been only part of our objective here; the remainder has been to develop some of the interesting relationships between digital and continuous systems. The usefulness of these relationships is not limited to modeling analog systems. For example, the bilinear substitution was used in Chapter 6 to design digital filters.

The question of the overall accuracy of the different simulation methods is difficult to address in any general way. As shown in this chapter, errors can be measured in the time domain, in the frequency domain, under specified input conditions with nonlinear operation, and so on.

In the final analysis, simulation always involves a mixture of art, science, luck, and varying degrees of honesty. It is generally possible to model a complex process so that the outcome is as the engineer wishes to see it rather than a faithful reproduction of reality, and one must always be careful not to let misleading results go unnoticed on account of the complexity of the model. "Reality checks" in any form are always desirable.

Exercises

11.1 Derive a model of $H_a(s) = 1/(s+a)$ that is exact when the input is the sampled version of $f(t) = Ae^{-at}$. Find the appropriate difference equation for the model.

11.2 Repeat Exercise 11.1 with $H(s)$ changed to $H(s) = 1/(s+b)$.

11.3 Using the answer to Exercise 11.2, compute the output for $f_k = Ae^{-akT}$ and confirm analytically that it is correct.

11.4 Determine the transfer function, $H_r(z)$, of the ramp-invariant model of $H_a(s) = 1/(s+a)$.

11.5 Describe the step-invariant model of $H_a(s) = 1/(s^2+2s+5)$ in terms of a difference equation that computes output samples (y_k) in terms of input samples (x_k).

11.6 In Exercise 11.5, show that the continuous system, $H_a(s)$, and its step-invariant model are both stable by plotting poles on the s- and z-planes, respectively.

11.7 Produce a figure like Figure 11.12 using a step-invariant model of $H_a(s) = 1/(s+1)$ instead of the ramp-invariant model.

11.8 Prove that the final value of the d.c.-adjusted impulse-invariant model of $H_a(s) = 1/(s+a)$ is correct when the input to the model is a step function with amplitude A.

11.9 Derive the adjusted impulse-invariant model, $H_0(z)$, for the analog integrator, $H_a(s) = 1/s$.

11.10 Derive $H_0(z)$ for the adjusted impulse-invariant model of $H_a(s) = 1/(s+1)^2$ and describe the model in terms of a difference equation.

11.11 Using the bilinear approximation, derive a model of $H_a(s) = 1/(s+1)^2$, which is accurate at frequencies $f = 0$ and 2 KHz, assuming a sampling frequency of 16 KHz.

11.12 Plot the power gain of $H_a(s) = 1/[s(s+1)]$ together with the power gain of a bilinear model accurate at 1 Hz, with time step $T = 0.25$.

11.13 Using the bilinear approximation, derive a model of the system in Figure 11.21. Make the model accurate at 4 Hz, assuming a sampling frequency of 20 Hz. Demonstrate the model by plotting analog and digital responses to a unit rectangular pulse lasting from $t = 0$ to $t = 0.5$ s.

11.14 Derive a step-invariant model of the system in Figure 11.27. Then with time step $T = 0.1$ s, demonstrate the model by plotting analog and digital responses to the signal in Exercise 11.13, that is, to a unit rectangular pulse lasting from $t = 0$ to $t = 0.5$ s.

11.15 Develop an impulse-invariant model of the system in Figure 11.28. Adjust the d.c. gain appropriately and demonstrate the model by comparing analog and digital responses to the signal in Exercise 11.13,

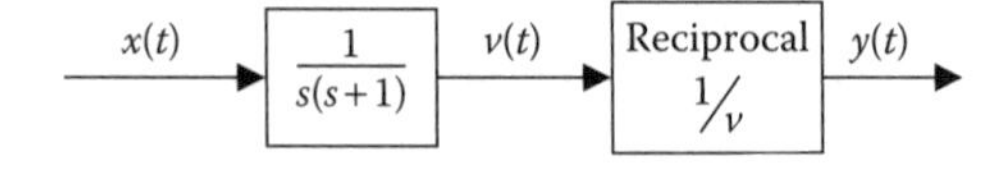

FIGURE 11.27
System for Exercise 11.14.

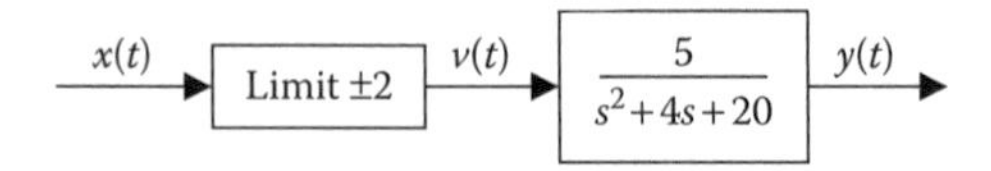

FIGURE 11.28
System for Exercise 11.15.

that is, to a unit rectangular pulse lasting from $t = 0$ to $t = 0.5$ s. Assume the sampling frequency is $f_s = 20$ Hz.

11.16 Express the response of $H(s)$ in (11.64) to a unit step function at $t = 0$.

11.17 For the step-invariant model in Figure 11.23, write a difference equation that expresses v in terms of x. Then write a second expression for y in terms of v, assuming the signals are small so that the limiting component has no effect.

11.18 In Figure 11.24, if the unit delay is moved up to delay the input to $H_{A1}(z)$, show (1) that the model is realizable, and (2) how the gain constant (K) is affected by doing so.

11.19 Modify Figure 11.24 by placing both K and z^{-1} in the forward part of the loop ahead of $H_{A1}(z)$, and determine $H_3(z)$ in the modified model.

11.20 Determine the bilinear blocks needed to simulate Figure 11.22. Use Tustin's approximation, that is, the bilinear substitution with $A = 2T$.

11.21 Using the bilinear blocks from Exercise 11.20 and a feedback delay, construct a complete bilinear model of Figure 11.22. Demonstrate the model using input signal $x(t) = 2 * t * \exp(-t)$ and time step $T = 0.1$.

11.22 Develop ramp-invariant blocks for the nonlinear system in Figure 11.29. Express each transfer function, $H_n(z)$, as a rational function of z and the time step T.

11.23 The two most basic configurations of two or more linear systems, which we discussed in Chapter 4, are shown in Figure 11.30 in terms

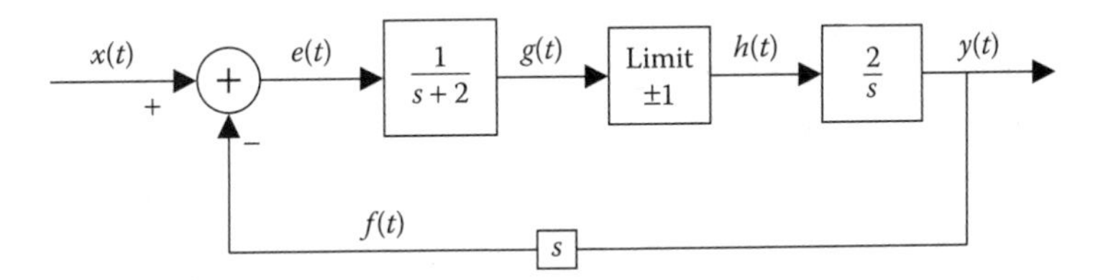

FIGURE 11.29
System for Exercise 11.22.

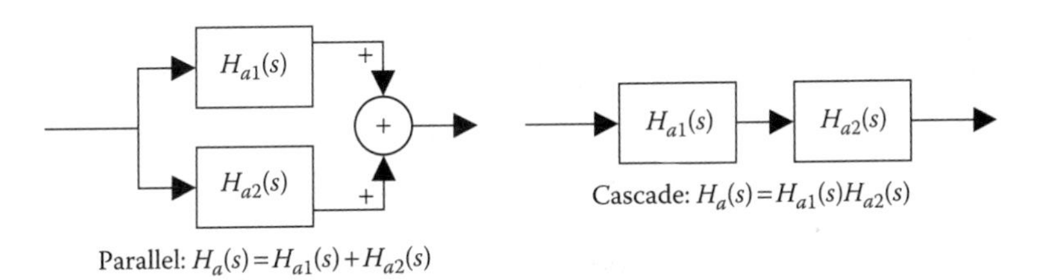

FIGURE 11.30
Parallel and cascade configurations for Exercises 11.23 and 11.24.

of continuous analog systems, $H_{a1}(s)$ and $H_{a2}(s)$. Only two analog systems are illustrated in each configuration, but of course any number of analog systems could be appended in parallel or in cascade. The question we wish to answer is as follows: Is an invariant model [in the form of (11.23)] of either of the *analog* configurations the same as the corresponding *digital* configuration of individual invariant models?

In this exercise, consider just the *parallel* configuration on the left in Figure 11.30. Prove that the answer must be "yes" for any invariant model by letting

$$H_a = H_{a1} + H_{a2}$$

in (11.23) and applying line C in the Appendix to obtain the following:

$$H(z) = \frac{1}{I(z)} Z\{L^{-1}\{I(s)H_a(s)\}\} = \frac{1}{I(z)} Z\{L^{-1}\{I(s)H_{a1}(s)\}\} + \frac{1}{I(z)} Z\{L^{-1}\{I(s)H_{a2}(s)\}\}$$

and explain the meaning of all the terms in this result.

Demonstrate the aforementioned conclusion by constructing a step-invariant model, $H(z)$, of the following analog system:

$$H_a(s) = \frac{2(s+1.5)}{(s+1)(s+2)}$$

Use line 301 in the Appendix to do this. Next, express $H_a(s)$ as the sum of two analog functions such that $H_a(s) = H_{a1}(s) + H_{a2}(s)$. Construct the two step-invariant models $H_1(z)$ and $H_2(z)$, and show that $H(z) = H_1(z) + H_2(z)$.

11.24 Consider now the *cascade* system in Figure 11.30. We learned in Chapter 4 that although the inverse transform of a sum is the sum of individual inverse transforms, the same is not true for the inverse transform of a product. Use this fact to argue that, in general, the invariant model of a cascade system is not the same as the product of the two individual invariant models.

Demonstrate this by constructing a step-invariant model, $H(z)$, of the following analog system:

$$H_a(s) = \frac{2}{(s+1)(s+2)}$$

Use line 300 in the Appendix to do this. Next, express $H_a(s)$ as the product of two analog functions such that $H_a(s) = H_{a1}(s)H_{a2}(s)$. Then construct the individual step-invariant models $H_1(z)$ and $H_2(z)$, and show that $H(z)$ and $H_1(z)H_2(z)$ have the same denominators, but not the same numerator.

11.25 Suppose $e_{i\max}$ is the maximum difference between $x(t)$ and its held version, $x^*(t)$, in Figure 11.15. What is the upper bound on the modeling error, e_k, when $H_a(s) = A/(s+a)$ with $a > 0$?

11.26 Repeat Exercise 11.25, this time with the expression $H_a(s) = A/[(s+a)^2 + b^2]$.

11.27 Solve Exercise 11.25 using the expression $H_a(s) = As/[(s+a)^2 + b^2]$.

References

1. Defigueiredo, R. J. P., and A. N. Netravali. 1971. Optimal spline digital simulators of analog filters. *IEEE Trans* CT-18(6):711–17.
2. Fowler, M. E. 1965. A new numerical method for simulation. *Simulation* 6:324–30.
3. Gold, B., and C. M. Rader. 1969. *Digital Processing of Signals*. New York: McGraw-Hill.
4. Greaves, C. J., and J. A. Cadzow. 1967. The optimal discrete filter corresponding to a given analog filter. *IEEE Trans* AC-12(3):304–7.
5. Hildebrand, F. B. 1968. *Finite-Difference Equations and Simulations*. Englewood Cliffs, NJ: Prentice Hall.
6. Hurt, J. M. 1964. New difference-equation technique for solving nonlinear differential equations. *AFIPS Conf Proc* 25:169.
7. Jury, E. I. 1964. *Theory and Application of the z-Transform Method*. New York: John Wiley & Sons.
8. Kaiser, J. F. 1966. Digital filters. In *System Analysis by Digital Computer*, ed. J. F. Kaiser and F. F. Kuo. Chap. 7. New York: John Wiley & Sons.
9. Lapidus, L., and J. H. Seinfeld. 1971. *Numerical Solution of Ordinary Differential Equations*. New York: Academic Press.
10. Rea, J. L. 1972. z-Transformation techniques in digital realization of coaxial equalizers. *Sandia Lab* SC-RR-72:0524.
11. Reitman, J. 1971. *Computer Simulation Applications*. New York: Wiley.
12. Rosko, J. S. 1972. *Digital Simulation of Physical Systems*. Reading, MA: Addison-Wesley.
13. Sage, A. P., and R. W. Burt. 1965. Optimum design and error analysis of digital integrators for discrete system simulation. *AFIPS Conf Proc* 27(1):903.
14. Schroeder, D. H. 1972. *A New Optimization Procedure for Digital Simulation*. Ph.D. dissertation, University of New Mexico, Albuquerque, New Mexico.
15. Wait, J. V. 1970. Digital filters. In *Active Filters: Lumped, Distributed, Integrated, Digital, and Parametric*, ed. L. P. Huelsman. Chap. 5. New York: McGraw-Hill.
16. Chaturvedi, D. K. 2010. *Modeling and Simulation of Systems Using Matlab and Simulink*. Boca Raton, FL: CRC Press.
17. Truxal, John, G. 1955. *Automatic Feedback Control System Synthesis*. New York: McGraw-Hill.

Further Reading

Ceschino, F., and J. Kuntzmann. 1966. *Numerical Solution of Initial Value Problems.* Englewood Cliffs, NJ: Prentice-Hall.

Gibson, J. E. 1963. *Nonlinear Automatic Control.* Chap. 4. New York: McGraw-Hill.

Hamming, R. W. 1973. *Numerical Methods for Scientists and Engineers.* 2nd ed. New York: McGraw Hill.

Kelly, L. G. 1967. *Handbook of Numerical Methods and Applications.* Chap. 19. Reading, MA: Addison Wesley.

Ragazzini, J. R., and G. F. Franklin. 1958. *Sampled-Data Control Systems.* Chap. 4. New York: McGraw-Hill.

12

Pattern Recognition with Support Vector Machines

12.1 Introduction

Humans possess a natural ability to recognize spatial and temporal patterns in the world around them by processing signals gathered through their senses. A long-standing goal of artificial intelligence is to duplicate this ability with man-made sensors and computing devices. In practice, however, the goal of duplicating human ability often gives way to the goal of optimizing performance for the available sensing and computing resources. Furthermore, humans are notoriously bad at certain types of pattern recognition problems where man-made sensors and computers excel (e.g., detecting subtle changes). Practical pattern recognition problems, where the goal is to optimize performance, are ubiquitous in the modern world as witnessed by this highly abbreviated list of diverse examples: speech recognition, fingerprint recognition, face recognition, iris recognition, optical character recognition, image target recognition, bar code recognition, computer network intrusion detection, nuclear event detection, medical disease detection, financial fraud detection, structural damage detection (e.g., for bridges, buildings, airplanes, and ships), and anomaly detection (e.g., for screening at airports, shipping ports, and manufacturing production lines). Although pattern recognition systems can be very complex, their basic operation is conceptually quite simple: First, convert the input data into *patterns* and then assign a *class label* to each pattern. This chapter focuses on the second step, but we start with a brief discussion of the first.

The process of converting input data into patterns relies heavily on the unique characteristics of the application and therefore varies widely from one application to the next. Nevertheless, some common themes have emerged. One of these is *locality*, namely that individual patterns are often formed from local data components. For example, with time series data it is common to form patterns from local time windows; with image data it is common to form patterns from local spatial windows; and with text data it is common to form patterns from local syntactic units such as paragraphs, sentences, words, or characters. These windows may be overlapping or not, and may have fixed or variable size (and shape in the case of images). A second theme that has emerged is the notion of *feature design*; original data patterns are converted to *feature vectors*, that is, vectors whose individual components correspond to features of the original pattern. The purpose of feature design is to extract, isolate, or amplify the information considered most relevant to the recognition task (i.e., to isolate the key features of the data). This may involve processing at a global scale before computing the local pattern features, and typically relies on signal processing tools like those developed in the previous chapters. Examples of global processing include removing the d.c. component of a time-series, registering an image or time series (i.e., aligning with a reference), or converting all characters of a text document to lowercase. Another example common to shape-based image object recognition is to apply a spatial highpass filter followed by a thresholding operation to create a so-called *edge image*. Once the global processing is complete, the local data units (like the aforementioned windows) are converted into feature vectors. For example, if we are looking for signals with a particular frequency signature (e.g., a fundamental frequency plus harmonics), we might first convert the individual time windows into spectral density estimates. On the other hand, if we are looking for a particular type of transient behavior then it may be more appropriate to convert to wavelet coefficients. Common features for image windows include fast Fourier transform (FFT) coefficients, discrete cosine transform (DCT) coefficients, wavelet coefficients, or various statistical quantities such as minimum, maximum, mean, standard deviation, skewness, and kurtosis. In the shape-based image object recognition problem, the features might include a collection of *shape descriptors* such as size, length, width, orientation, and medial axis points, which are derived from the object outline extracted from the edge image. A common feature representation for local text units is a histogram whose individual bins correspond to the frequency of occurrence of particular character sequences. For example, when the particular character sequences correspond to words from a dictionary this histogram is often called a *bag-of-words* representation. More information on the feature design process can be found in the literature, several of which are listed in the References section.[1–4] In summary, whether the pattern formation process is based on locality, feature design, or some other approach, the output patterns often take the form of finite dimensional vectors with real-valued components.

This chapter is concerned with the task of designing a pattern classifier that assigns *class labels* to patterns. For example, in the fingerprint recognition problem where patterns correspond to fingerprint images, the labels represent the identity of an individual, and in the disease detection problem where patterns correspond to a patient's test results, the labels represent *disease* or *not disease*. Pattern recognition problems with only two labels are called *two-class* problems, and they are quite common. Indeed, most so-called *detection* problems, in which we want to determine the presence or absence of a particular type of event, are of this form. Pattern recognition problems with more than two labels are called *multiclass* problems. Two-class and multiclass problems have much in common; but since the multiclass problem can be decomposed into a collection of two-class problems, the two-class problem is often considered more fundamental. Thus, we will develop the two-class problem in some detail and then in Section 12.5 briefly describe how two-class methods can be extended to the multiclass problem. A more thorough treatment can be found in one of the many excellent texts on pattern recognition.[2,5–11]

Patterns from individual classes can be highly variable. Indeed, there is often a natural variability within the pattern classes. For example, in the handwritten digit recognition problem, there are a wide variety of two-dimensional spatial patterns (i.e., scribbles on a piece of paper) that correspond to each of the 10 pattern classes "0," "1," …, "9." But patterns are also subject to extraneous variabilities that are irrelevant to the recognition task such as measurement noise, environmental noise, background events in the measurement environment, and imperfections in the pattern formation step. To simultaneously account for all these variabilities, it is common to cast pattern recognition problems into a probabilistic setting where patterns are modeled as samples of a random variable, and the probability distribution of the random variable characterizes the relative frequency of occurrence of all possible variations of the pattern due to both natural and extraneous causes. In this setting, the probability distributions for the individual classes are combined to form a *mixture* distribution that describes the overall data synthesis process.

Complete knowledge of this mixture distribution is sufficient to design the optimal pattern classifier and compute precisely how well it works. In practice, however, we (almost) never have complete distribution knowledge. Compensation for this lack of knowledge can be achieved by using empirical observations, that is, data samples drawn from the distribution, to help design the classifier and estimate its performance. This process of using existing data samples to design a classifier that will be used to make predictions about future data samples is called *learning*, and it is performed by a computer *algorithm* that accepts *training* data as its input and produces a pattern classifier as its output. This is conceptually the same as the approaches described in Chapters 8 and 9, except here we are designing pattern classifiers instead of filters.

In this chapter, we make a distinction between learning problems where the distribution belongs to a *known model class*, for example, Gaussian, and those where it does not. In problems in which the distribution model class is unknown, the learning problem can be quite complicated. This is often the case in practice. Ideally we would like a learning method that works well for arbitrary distributions, but no such method exists.[12] Consequently, we might settle for a method that simply outperforms other methods across all distributions, but again no such method exists.[13] This situation has led to a proliferation of learning methods, including neural networks, decision trees, nearest neighbors, kernel methods, and many others,[2,5–12] whose strengths and weaknesses are quite varied. The *support vector machine* (SVM) described in this chapter is a learning method that incorporates elements from many of these other methods, and is roughly able to combine their strengths and mitigate their weaknesses. More specifically, the SVM learning method is able to incorporate a wide range in the type and amount of prior distribution knowledge, is practical and easy to use, is computationally efficient, and is robust to the unknown aspects of the distribution and to the uncertainties associated with how well the finite training data represent the underlying distribution.

Before proceeding, we briefly mention some notational conventions unique to this chapter. We use the *big-oh* notation $O(\cdot)$ to express upper bounds. Formally, for problems of size n, the notation $O(g(n))$ defines a set of functions $\{f(n)\}$ that satisfy $0 \le f(n) \le cg(n)$ for all $n \ge n_0$, for some positive constants c and n_0. Informally we say that an algorithm that performs a total of $f(n) = c_2 n \log n + c_1 n + c_0$ computations for some finite constants (c_0, c_1, c_2) has a run time $O(n \log n)$, since we can always bound $f(n)$ by $cn \log n$ for a suitable choice of c. Also, as a general rule we will use bold variables such as $\mathbf{x}$ to represent vectors; subscripted regular variables x_i to represent the corresponding vector components, for example, $\mathbf{x} = (x_1, x_2, \ldots, x_d)^T$; and subscripted bold variables such as $\mathbf{x}_i$ to represent the ith vector from a collection of vectors. Furthermore, we use $\mathbb{R}$ to denote the space of real-valued vectors in one dimension (i.e., real-valued scalars), and $\mathbb{R}^d$ to denote the space of real-valued vectors in d dimensions. This chapter makes heavy use of probability density functions which were first described in Chapter 7. The main difference here is that these density functions are defined over vectors $\mathbf{x}$ instead of scalars x. We also make heavy use of the expected value operator E, which is defined as an integral with respect to a density function as described in Chapter 7. This chapter uses several different density functions/expected value operators and, therefore, uses subscripts to distinguish between the different cases. Finally, we use $\min_{x\in\chi} r(x)$ to denote the minimum value of a function r over all x in a set χ, and $\arg\min_{x\in\chi} r(x)$ to denote the subset of χ where the minimum is achieved.

12.2 Pattern Recognition Principles

We now describe the standard two-class pattern recognition problem. First we assume that the pattern formation process produces real-valued pattern vectors of the form $\mathbf{x} = (x_1, x_2, \ldots, x_d)^T \in \mathbb{R}^d$. The pattern class labels are denoted by the variable y, which takes the values +1 or –1. Our probabilistic model assumes that labeled data points ($\mathbf{x}$, y) are generated as samples of a random variable (X, Y) with a probability density $p_{X,Y}$. This density function is the same kind of probability function that is introduced in Chapter 7, but this time with two variables—the random vector X and the random scalar Y—instead of just one. This density can be decomposed into a mixture of densities as follows:

$$\begin{aligned} p_{X,Y}(\mathbf{x},y) &= \delta(y-1)p_{X,Y}(\mathbf{x},1)+\delta(y+1)p_{X,Y}(\mathbf{x},-1) \\ &= \delta(y-1)P_1 p_{X|1}(\mathbf{x})+\delta(y+1)P_{-1}p_{X|-1}(\mathbf{x}) \end{aligned}$$

where δ is the unit impulse function,

$$P_1 = \int p_{X,Y}(\mathbf{x},1)\,d\mathbf{x} \quad \text{and} \quad P_{-1} = \int p_{X,Y}(\mathbf{x},-1)\,d\mathbf{x} \tag{12.1}$$

are the class probabilities, and the conditional densities $p_{X|1}$ and $p_{X|-1}$ are the densities for class 1 and –1, respectively. The class probabilities (P_1, P_{-1}) satisfy $P_1 = 1 - P_{-1}$ and represent the probability that a sample is generated from class (1, –1). Equivalently, they represent the fraction of samples generated from class (1, –1) over the long run. The density $p_{X,Y}$ characterizes labeled data points ($\mathbf{x}$, y), but in the real-world environment where we plan to deploy our classifier, we do not get to see the labels. Indeed, this is what we want to predict. The density that characterizes the unlabeled data $\mathbf{x}$ that we see in the deployed environment is the X-marginal probability density

$$p_X(\mathbf{x}) = \int p_{X,Y}(\mathbf{x},y)\,dy = P_1 p_{X|1}(\mathbf{x}) + P_{-1}p_{X|-1}(\mathbf{x})$$

which is obtained by integrating out the y variable. An example of this density and its two components is shown in Figure 12.1a for a hypothetical one-dimensional problem.

A *pattern classifier* is a real-valued function f that assigns a label

$$\texttt{sign}\, f(\mathbf{x}) = \begin{cases} 1, & f(\mathbf{x}) > 0 \\ -1, & f(\mathbf{x}) \le 0 \end{cases}$$

to every point $\mathbf{x} \in \mathbb{R}^d$. A classifier f commits an error for sample ($\mathbf{x}$, y) when $\texttt{sign}\, f(\mathbf{x}) \neq y$. The goal is to construct a classifier whose expected error is as small as possible. If we define the *indicator* function

$$I(a) = \begin{cases} 1, & a \text{ is true} \\ 0, & a \text{ is false} \end{cases}$$

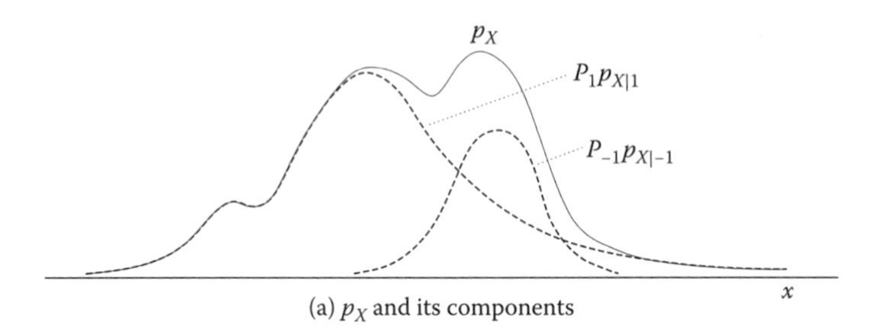

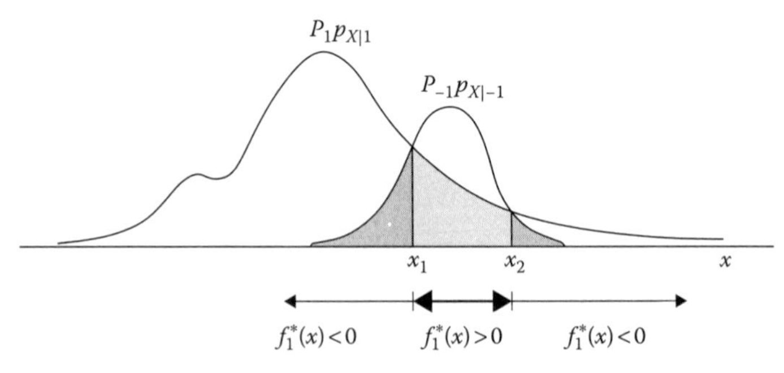

FIGURE 12.1
Class errors and Bayes error. (a) p_X and its components $P_1 p_{X|1}$, $P_{-1} p_{X|-1}$; (b) an optimal classifier f_1^* partitions the values of **x**. The class errors ($P_1 e_1$, $P_{-1} e_{-1}$) correspond to the (light and dark) shaded regions and the Bayes error e^* is the total area of the shaded regions.

then the expected error e can be written as

$$\begin{aligned} e(f) &= E_{X,Y}[I(\mathtt{sign}\, f(\mathbf{x}) \neq y)] = \int I(\mathtt{sign}\, f(\mathbf{x}) \neq y) p_{X,Y}(\mathbf{x}, y)\, d\mathbf{x}\, dy \\ &= P_1 \int I(f(\mathbf{x}) \leq 0) p_{X|1}(\mathbf{x})\, d\mathbf{x} + P_{-1} \int I(f(\mathbf{x}) > 0) p_{X|-1}(\mathbf{x})\, d\mathbf{x} \\ &= P_1 e_1(f) + P_{-1} e_{-1}(f) \end{aligned} \tag{12.2}$$

where

$$\begin{aligned} e_1(f) &= \int I(f(\mathbf{x}) \leq 0) p_{X|1}(\mathbf{x})\, d\mathbf{x} \\ e_{-1}(f) &= \int I(f(\mathbf{x}) > 0) p_{X|-1}(\mathbf{x})\, d\mathbf{x} \end{aligned} \tag{12.3}$$

are the expected errors for class 1 and –1. These expected errors are also called *error rates* because they represent the rate at which the classifier f will make errors when presented with a continuous stream of samples from $p_{X,Y}$. The class error rates e_1 and e_{-1} go by many different names in the literature.

In statistics, they are often referred to as *type I* and *type II* errors, and in detection problems where class labels (1, –1) are used to represent (*target, background*) the errors (e_1, e_{-1}) are often referred to as the *missed detection* and *false alarm* rates, respectively.

The minimum error rate over all possible functions is defined as

$$e^* = \min_{\text{all } f} e(f)$$

This is the so-called *Bayes error*, and it represents a lower bound on the error that can be achieved for data generated by $p_{X,Y}$. Classifiers f^* that achieve the Bayes error can be expressed in many different, but equivalent, forms. Indeed, if f_0^* is optimal, then any function f_i^* that has the same sign at all values of $\mathbf{x}$ is also optimal. In particular, any scaled function $c(\mathbf{x})f_0^*(\mathbf{x})$ for $c(\mathbf{x}) > 0, \forall \mathbf{x}$ is also optimal.

An obvious way to minimize the average error over all $\mathbf{x}$ is to minimize the average error at each value of $\mathbf{x}$. This can be accomplished by assigning the most likely label at each $\mathbf{x}$. This rule can be realized by the optimal classifier

$$f_0^*(\mathbf{x}) = P_{1|X}(\mathbf{x}) - P_{-1|X}(\mathbf{x}) \tag{12.4}$$

where the function $P_{y|X}(\mathbf{x})$ computes the probability that the true label is y for $X = \mathbf{x}$. This classifier simply assigns the label of the class with the largest probability at $\mathbf{x}$. Using Bayes rule, we can write $P_{y|X}(\mathbf{x}) = P_y p_{X|y}(\mathbf{x})/p_X(\mathbf{x})$. Substituting this expression for $P_{y|X}(\mathbf{x})$ and then applying the aforementioned scaling property leads to an equivalent optimal form:

$$f_1^*(\mathbf{x}) = P_1 p_{X|1}(\mathbf{x}) - P_{-1} p_{X|-1}(\mathbf{x}) \tag{12.5}$$

This form is illustrated in Figure 12.1b; the figure shows how $f_1^*(\mathbf{x})$ partitions the $\mathbf{x}$ space, which in this case is the real line, into $y = 1$ and $y = -1$ regions according to the largest mixture component. Multiplying $f_1^*(\mathbf{x})$ by $1/P_1 p_{X|-1}(\mathbf{x})$ and applying the scaling property gives the optimal form

$$f_2^*(\mathbf{x}) = \underbrace{\frac{p_{X|1}(\mathbf{x})}{p_{X|-1}(\mathbf{x})}}_{\text{Likelihood ratio}} - \frac{P_{-1}}{P_1} \tag{12.6}$$

which is expressed in terms of the so-called likelihood ratio. Finally, since the log is a monotonic function and comparing two ratios is equivalent to comparing their log ratios, we obtain the optimal form

$$f_3^*(\mathbf{x}) = \log[p_{X|1}(\mathbf{x})] - \log[p_{X|-1}(\mathbf{x})] - \log(P_{-1}/P_1) \tag{12.7}$$

which often leads to considerable simplifications when the conditional densities $p_{X|Y}$ are members of the exponential family.

EXAMPLE 12.1

Consider a two-class pattern recognition problem in which the class densities are multivariate Gaussians of the form [i.e., multivariate versions of the density function in Chapter 7, (7.11)]

$$p_{x|y}(\mathbf{x}) = G_y(\mathbf{x}) = (2\pi)^{-d/2}\left|\Sigma_y\right|^{-1/2}\exp\left\{-\frac{1}{2}(\mathbf{x}-\boldsymbol{\mu}_y)^T\Sigma_y^{-1}(\mathbf{x}-\boldsymbol{\mu}_y)\right\}, \quad y \in \{-1, 1\}$$

with class means and covariances given by

$$\begin{aligned}\boldsymbol{\mu}_y &= E_{X|y}[\mathbf{x}]\\ \Sigma_y &= E_{X|y}[(\mathbf{x}-\boldsymbol{\mu}_y)(\mathbf{x}-\boldsymbol{\mu}_y)^T]\end{aligned}$$

This is the so-called two-class Gaussian problem. The mean vectors $\boldsymbol{\mu}_y$ represent the average values for vectors from each class, and the covariance matrices Σ_y represent deviance from the average. The astute reader will recognize a similarity between the covariance matrices Σ_y here and the autocorrelation matrix Φ_{ff} in Chapter 8, (8.6), and the autocovariance matrix R_{ff} in Section 8.5. All of these matrices capture the second-order statistics for a particular random vector. In fact, the covariance definition here can be viewed as a generalization of the definitions in Chapter 8 in that the random vectors here are not necessarily derived from a time series signal. Substituting into the log-likelihood optimal form in (12.7) and simplifying gives

$$f_q^*(\mathbf{x}) = \mathbf{x}^T Q\mathbf{x} + \mathbf{w}^T\mathbf{x} + b \tag{12.8}$$

where

$$\begin{aligned}Q &= \frac{1}{2}(\Sigma_{-1}^{-1} - \Sigma_1^{-1})\\ \mathbf{w} &= \Sigma_1^{-1}\boldsymbol{\mu}_1 - \Sigma_{-1}^{-1}\boldsymbol{\mu}_{-1}\\ b &= \frac{1}{2}\boldsymbol{\mu}_{-1}^T\Sigma_{-1}^{-1}\boldsymbol{\mu}_{-1} - \frac{1}{2}\boldsymbol{\mu}_1^T\Sigma_1^{-1}\boldsymbol{\mu}_1 + \frac{1}{2}\log(|\Sigma_{-1}|) - \frac{1}{2}\log(|\Sigma_1|) - \log\left(\frac{P_{-1}}{P_1}\right)\end{aligned} \tag{12.9}$$

Since f_q^* is a quadratic polynomial in $\mathbf{x}$, it is called a "quadratic" classifier, hence the subscript q. The parameter settings in (12.9), which are optimal for the two-class Gaussian problem, are only one way to realize a quadratic classifier and others will be discussed in Section 12.3.2. When $\Sigma_{-1} = \Sigma_1 = \Sigma$, the quadratic term cancels and the optimal form simplifies to a linear classifier

$$f_l^*(\mathbf{x}) = \mathbf{w}^T\mathbf{x} + b \tag{12.10}$$

where

$$\begin{aligned}\mathbf{w} &= \Sigma^{-1}(\boldsymbol{\mu}_1 - \boldsymbol{\mu}_{-1})\\ b &= \frac{1}{2}\boldsymbol{\mu}_{-1}^T\Sigma^{-1}\boldsymbol{\mu}_{-1} - \tfrac{1}{2}\boldsymbol{\mu}_1^T\Sigma^{-1}\boldsymbol{\mu}_1 - \log\left(\frac{P_{-1}}{P_1}\right)\end{aligned} \tag{12.11}$$

Again, the parameter settings in (12.11) are only one way to realize a linear classifier, and others will be discussed in Section 12.3.2.

12.3 Learning

In practice, the optimal classifiers in (12.4) through (12.7) cannot be implemented directly because the probability functions are not known. Even in the two-class Gaussian problem where the distribution *model class* is known, we almost never have complete knowledge of the *model parameters*, that is, the means ($\boldsymbol{\mu}_1$, $\boldsymbol{\mu}_{-1}$), covariances (Σ_1, Σ_{-1}), and class probabilities (P_1, P_{-1}). To compensate, we use *empirical data* to help "fill in" the missing information. In particular, we consider the case in which the empirical data consists of a collection of *labeled* samples ($\mathbf{x}_1$, y_1), …, ($\mathbf{x}_n$, y_n) called the *training data*. For example, in the target recognition problem we assume that we can gather a collection of both targets and nontargets to form a training set that can be used to help design a classifier. The corresponding learning problem can be stated as follows: *Given a collection* $D_n = ((\mathbf{x}_1, y_1), \ldots, (\mathbf{x}_n, y_n))$ *of samples from an unknown density* $p_{X,Y}$, *determine a classifier* $\hat{f}$ *whose error* $e(\hat{f})$ *is as close to* e^* *as possible*. In learning theory, this is called the *supervised classification problem*, where the word *supervised* refers to the fact that the training samples contain ground truth labels. An example is shown in Figure 12.2 for some two-dimensional data. Here, we display a scatterplot of samples $\mathbf{x}_i$ whose labels are indicated by different plotting symbols. A classifier $\hat{f}$ will partition the data space into two sets, $\{\mathbf{x}: f(\mathbf{x}) \leq 0\}$ and $\{\mathbf{x}: f(\mathbf{x}) > 0\}$, and assign the labels –1 and +1, respectively. This is often visualized by overlaying the *decision boundary* $\{\mathbf{x}: \hat{f}(\mathbf{x}) = 0\}$ on the scatterplot. The decision boundaries for two hypothetical classifiers are shown in Figure 12.2. In the supervised classification problem, we must use the labeled training data to design a classifier $\hat{f}$ that partitions the data space so that the labels assigned to *future data* are as accurate as possible. Can you determine which of the two classifiers illustrated in Figure 12.2 will most accurately classify future data samples? What strategy would you use to design such a classifier?

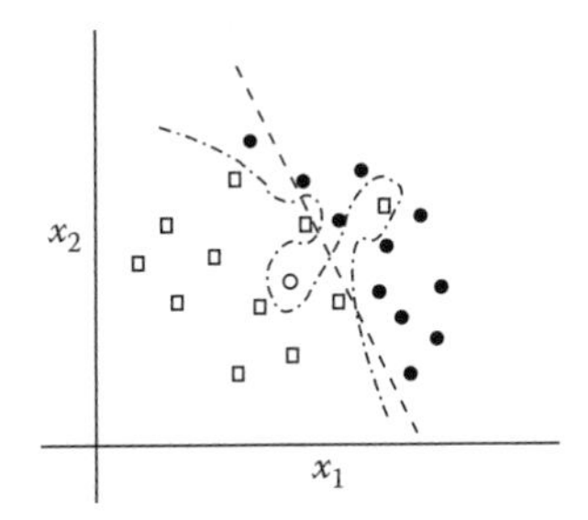

FIGURE 12.2
Scatterplot of labeled training data for a supervised classification problem, and two hypothetical classifier decision boundaries.

Incomplete distribution knowledge implies that, probabilistically, we can never design an optimal classifier regardless of the learning method, and we can never know precisely how well our designed classifier works regardless of the tests performed. However, with an informative *sample plan* and a well-designed *learning method*, we can design classifiers $\hat{f}$ that are *near optimal* in the sense that their *excess error* $e(\hat{f}) - e^*$ is *controlled*, that is, bounded in size and guaranteed to decrease and approach zero with larger values of n. Furthermore, we can produce reliable performance estimates $\hat{e}(\hat{f})$ in the sense that their accuracy $|\hat{e}(\hat{f}) - e(\hat{f})|$ is also controlled. Sections 12.3.1 and 12.3.2 provide a brief introduction to sample plans and learning methods.

12.3.1 The Independent and Identically Distributed Sample Plan

The *sample plan* is a description of (or a model for) how the training data are obtained. Conceptually, the sample plan tells us what kind of information is carried by the training data about the future data that we want to classify. In this chapter, we employ the most commonly used and heavily studied *independent and identically distributed* (iid) sample plan. This sample plan assumes that the training data $D_n = ((\mathbf{x}_1, y_1), \ldots, (\mathbf{x}_n, y_n))$ are iid samples from $p_{X,Y}$, where $p_{X,Y}$ is the probability density of the future data that we want to classify. There are actually two assumptions here: independence and identical distributions. The independence assumption states that the sample values are all mutually independent, that is, no sample values depend on any other sample values. This assumption can be violated, for example, if the *order* of the samples contains information about the individual sample values. In particular when samples are gathered from *overlapping* image or time series windows, then successive samples contain some of the exact same data values (although in different positions) and this is an obvious violation of the independence assumption. This violation can sometimes be mitigated by collecting training data from nonoverlapping windows, but this is no guarantee of independence. It turns out that independence is often not only a strong assumption to satisfy, but also a stronger assumption than is actually needed. That is, the development and analysis carried out under the independence assumption is often extremely useful in cases where the assumption is violated. With this in mind, independence is often assumed because it greatly simplifies development and analysis. Finally, we note that the sample independence assumption does *not* imply that the *individual components* of $\mathbf{x}$ are independent (and typically they are not). The *identically distributed* assumption simply states that all the samples are drawn from the same probability distribution. This assumption is violated if the data statistics are changing over time/space. Although in many pattern recognition problems this assumption is not strictly true, it is often considered a reasonable assumption over an appropriate time/space window.

The iid sample plan allows us to establish a rigorous link between the *generalization errors* in (12.2) and (12.3), which express expectations as integrals with respect to probability densities, and their *empirical* (finite sample) approximations, which express expectations as summations with respect to data samples. Specifically, the empirical versions of (12.1) through (12.3) take the form

$$\hat{P}_1 = \frac{1}{n}\sum_{i=1}^{n} I(y_i = 1) = \frac{n_1}{n}, \quad \hat{P}_{-1} = \frac{1}{n}\sum_{i=1}^{n} I(y_i = -1) = \frac{n_{-1}}{n}$$
$$\hat{e}(f) = \frac{1}{n}\sum_{i=1}^{n} I(\mathrm{sign}\, f(\mathbf{x}_i) \neq y_i)$$
$$\hat{e}_1(f) = \frac{1}{n_1}\sum_{i:y_i=1} I(\mathrm{sign}\, f(\mathbf{x}_i) \neq 1), \quad \hat{e}_{-1}(f) = \frac{1}{n_{-1}}\sum_{i:y_i=-1} I(\mathrm{sign}\, f(\mathbf{x}_i) \neq -1) \tag{12.12}$$

Under the iid assumption, these are Monte Carlo approximations of the integrals in (12.1) through (12.3) and are, therefore, known to have accuracy $O(1/\sqrt{n})$[12,14] (see Chapter 8 of the book given by reference 12). More specifically, since they are all sums of indicator functions (i.e., *binary* functions) over iid samples, each sum is a binomial random variable. This means that $\hat{e}$ is a random variable with mean e and variance $e(1-e)$, where e is the generalization error value (similarly for $\hat{P}_1, \hat{P}_{-1}, \hat{e}_1$, and $\hat{e}_{-1}$). Thus, it is straightforward to show that $\hat{e}$ is within approximately $1.64\sqrt{\frac{\hat{e}(1-\hat{e})}{n}}$ of the generalization error value e with probability 0.95. However, a word of caution is in order when these estimates are used to evaluate functions obtained through a learning process. In particular, if D_n is used to design a function $\hat{f}$ and the same data are then used to compute the empirical error $\hat{e}(\hat{f})$, then $\hat{e}(\hat{f})$ will typically be a *biased* estimate of the generalization error $e(\hat{f})$ and will generally not satisfy the aforementioned accuracies. For example, if $\hat{f}$ is chosen to minimize $\hat{e}$ then it is almost obvious that $\hat{e}(\hat{f})$ will be an overly optimistic estimate of $e(\hat{f})$. A common procedure for obtaining unbiased estimates is to partition D_n randomly into two sets, a new (smaller) training set D_{n_a} and a *test* set T_{n_b}, and then use D_{n_a} to design the function $\hat{f}$ and T_{n_b} to estimate the error of $\hat{f}$. The accuracy of the unbiased holdout estimate is then $O(1/\sqrt{n_b})$.

12.3.2 Learning Methods

Many learning methods use the training data to approximate one of the optimal forms $f_0^* - f_3^*$ in (12.4) through (12.7). We briefly review two classes of methods: (1) one based on approximating $f_1^* - f_3^*$ and (2) the other based on approximating f_0^*.

The first class of methods is called *plug-in* methods because it uses the training data to estimate the class probabilities and densities and then substitutes these estimates into one of $f_1^* - f_3^*$. The details depend on the distribution knowledge. When the distribution model class is known and the class densities belong to a parametric family (e.g., Gaussians) then the parameters θ of the density $p_{X,Y}$ can be chosen to maximize the so-called *likelihood function L*, which takes the form $L(\theta) = \prod_{i=1}^{n} p_{X,Y}[(\mathbf{x}_i, y_i)|\theta]$ under the iid sample plan. Equivalently, θ can be chosen to maximize the log-likelihood function $\log(L)$. This is called the *maximum likelihood* (ML) method. In particular when the class densities are Gaussian, the maximum likelihood estimates take the form of sample averages (see Exercise 12.1)

$$\hat{\boldsymbol{\mu}}_1 = \frac{1}{n_1}\sum_{i:y_i=1} \mathbf{x}_i, \quad \hat{\boldsymbol{\mu}}_{-1} = \frac{1}{n_{-1}}\sum_{i:y_i=-1} \mathbf{x}_i$$
$$\hat{\Sigma}_1 = \frac{1}{n_1}\sum_{i:y_i=1} (\mathbf{x}_i - \hat{\boldsymbol{\mu}}_1)(\mathbf{x}_i - \hat{\boldsymbol{\mu}}_1)^T, \quad \hat{\Sigma}_{-1} = \frac{1}{n_{-1}}\sum_{i:y_i=-1} (\mathbf{x}_i - \hat{\boldsymbol{\mu}}_{-1})(\mathbf{x}_i - \hat{\boldsymbol{\mu}}_{-1})^T$$
$$\hat{P}_1 = \frac{n_1}{n}, \quad \hat{P}_{-1} = \frac{n_{-1}}{n} \tag{12.13}$$

The classifier is realized by substituting these estimates into the simplified form in (12.8) and (12.9) [or (12.10) and (12.11)]. This is called the *Gaussian Maximum Likelihood* (GML) method.

When the distribution model class is unknown, it is common to use the sample count estimates $(\hat{P}_1, \hat{P}_{-1}) = (\frac{n_1}{n}, \frac{n_{-1}}{n})$ for the class probabilities and adopt a so-called *nonparametric* form for the class densities. For example, the *Parzen-kernel density* estimate takes the form

$$\hat{p}_{X|y}(\mathbf{x}) = \frac{1}{n_y}\sum_{i:y_i=y} k(\mathbf{x}_i, \mathbf{x})$$

where k is a so-called the *kernel function* that is positive and typically chosen to give larger values when $\mathbf{x}$ is closer to $\mathbf{x}_i$. A popular choice for k is the so-called *Gaussian kernel*

$$k(\mathbf{x}_i, \mathbf{x}) = \frac{1}{\sqrt{2\pi\sigma^d}} e^{-\frac{\|\mathbf{x}-\mathbf{x}_i\|^2}{2\sigma^2}}$$

With this choice, the kernel density estimate is simply a sum of Gaussians with equal covariance $\sigma^2 I$ centered at the training samples $\mathbf{x}_i$. The classifier is realized by substituting these estimates into (12.5), (12.6), or (12.7).

The second class of methods uses the training data to approximate $f_0^* = P_{1|X} - P_{-1|X}$. One common path is to exploit the observation that the mean-squared error $E_X[[f_0^*(\mathbf{x}) - f(\mathbf{x})]^2]$ can be minimized by minimizing $E_{X,Y}[(y - f(\mathbf{x}))^2]$, that is, the binary labels y are an acceptable surrogate for the true (unknown) target values f_0^* (see Exercise 12.2). This motivates the *least-squares* (LS) method, which chooses f to minimize

$$\frac{1}{n}\sum_{i=1}^{n}[y_i - f(\mathbf{x}_i)]^2.$$

When f takes the form of a linear classifier, that is, $f(\mathbf{x}) = \mathbf{w}^T\mathbf{x} + b$, then the LS solution takes the familiar form seen earlier in Chapter 8, (8.6) and Table 8.1. In particular, if we let $\acute{\mathbf{x}}^T = (\mathbf{x}^T, 1)$ and $\acute{\mathbf{w}}^T = (\mathbf{w}^T, b)$, then the LS solution takes the following form:

$$\acute{\mathbf{w}}^* = \begin{bmatrix} \mathbf{w}^* \\ b^* \end{bmatrix} = \left[\frac{1}{n}\sum_{i=1}^{n}\acute{\mathbf{x}}_i\acute{\mathbf{x}}_i^T\right]^{-1}\left[\frac{1}{n}\sum_{i=1}^{n}y_i\acute{\mathbf{x}}_i\right]. \tag{12.14}$$

This is called the LS-LINEAR learning method. If the data statistics satisfy $\Sigma_1 = \Sigma_{-1} = \Sigma$ and $P_1 = P_{-1} = .5$, then it is easy to show that the LS-LINEAR solution is identical to the GML solution (see Exercise 12.3). Furthermore, $\mathbf{w}^*$ remains the same (up to a positive scalar) when $P_1 \neq P_{-1}$. In addition, this $\mathbf{w}^*$ is equivalent to the well-known *Fisher's linear discriminant*.[6] The fact that the same $\mathbf{w}^*$ is obtained from multiple perspectives suggests that it may be valid for a more general class of distributions. Indeed, in practice $\mathbf{w}^*$ often performs well in cases where the data are not Gaussian.

The LS method is also used to design the multilayer perceptron (MLP) neural network whose functions take the form $f(\mathbf{x}) = \psi_0 + \sum_{m=1}^{M}\psi_m \tanh(\mathbf{w}_m^T\mathbf{x} + b_m)$ where $\psi_m \in \mathbb{R}$ and $\mathbf{w}_m \in \mathbb{R}^d$. In this case, the most common algorithm for determining the values of $(\psi, \mathbf{w}_1, \ldots, \mathbf{w}_M)$ is the celebrated *backpropagation* algorithm, which is an iterative algorithm based on the same type of gradient search that is employed by the least-mean-square (LMS) algorithm in Chapter 9.

Although we have described only a few specific methods so far, they are representative of the larger collection of methods found in the literature. In general, if the distribution model class is known, then it is common to pursue a maximum likelihood method. The GML method is one example. In many cases, however, this approach leads to a computationally intractable learning problem whose approximate realization introduces an uncertainty that makes it difficult to control the excess error. In cases where the distribution model class is unknown, the choice of learning method becomes more complicated. One approach is to choose a method that incorporates a rich function class whose members can provide arbitrarily close approximations to the optimal classifier regardless of its form. The Parzen-kernel density estimate with Gaussian kernel and MLP neural network are examples in this direction. However, this approach also often leads to a computationally intractable learning problem whose approximate realization introduces an uncertainty that makes it difficult to control the excess error. Consequently, an alternate approach is to choose a simple function class with a computationally efficient learning strategy like the aforementioned LS-LINEAR method. However, the potentially poor approximations provided by this approach once again make it difficult to control the excess error. In summary, when the distribution model class is complicated or unknown, it is often difficult to find a learning method whose practical realization is both computationally efficient and guaranteed to control the excess error.

In general, there are three fundamental barriers that limit our ability to control the excess error: (1) incomplete distribution knowledge, (2) limited training data, and (3) limited computational resources. The learning process contains three components that can be used to confront these barriers: (1) the choice of function class $\mathcal{F}$, (2) the *learning strategy* $\mathcal{L}$ that chooses a function f from $\mathcal{F}$ based on the training data D_n, and (3) the *algorithm* that attempts to carry out this strategy in practice. The development of a learning method that guarantees control of the excess error requires a basic understanding of how these components interact with the fundamental barriers to influence the approximation error, estimation error, and computation error, which we now describe.

Incomplete distribution knowledge may lead to a choice $\mathcal{F}$ that does not include a Bayes optimal classifier. This contributes a component to the excess error called *approximation error.* Incomplete distribution knowledge may also lead to a choice of learning strategy $\mathcal{L}$ that contributes to the approximation error by failing to choose a classifier from $\mathcal{F}$ that is best for the distribution. For example, the LS-LINEAR learning strategy fails to choose the best linear classifier for many distributions.[12] Limited training data implies that D_n contains only partial information about $p_{X,Y}$ and this contributes a component to the excess error called *estimation error.* The estimation error is heavily influenced by the choice of $(\mathcal{L}, \mathcal{F})$. The core issue here is how well the characteristics of $p_{X,Y}$ represented by D_n can be exploited by $(\mathcal{L}, \mathcal{F})$. If $(\mathcal{L}, \mathcal{F})$ is unable to exploit the finer details of the data, then a relatively small number of samples may be sufficient to determine the best solution, regardless of the distribution complexity. On the other hand, if $(\mathcal{L}, \mathcal{F})$ is able to exploit the intricate details of the data, then a relatively large number of samples may be needed to prevent the learning strategy from choosing an overly complex function that performs well on the training data but poorly on future data. In this case, we say that $(\mathcal{L}, \mathcal{F})$ has *overfit* the training data. In practice, the goal is to match the complexity of $(\mathcal{L}, \mathcal{F})$ to the complexity of $p_{X,Y}$ and the size of D_n. The best match will be one that strikes an optimal balance between estimation error and approximation error. This *approximation–estimation error trade-off* is at the heart of all learning problems, and designing learning methods that strike the proper balance is one of the most challenging (and intriguing) aspects of classifier design. Finally, limited computational resources mean that it may be impossible to implement the learning strategy exactly, especially when $(\mathcal{L}, \mathcal{F})$ is complex. This contributes a component to the excess error called *computation error.* The SVM, described in Section 12.4, is a learning method that can simultaneously guarantee control of the approximation, estimation, and classification errors in problems where the distribution model class is unknown.

12.4 Support Vector Machines

Support vector machines were pioneered by Vladimir Vapnik[15–18] and have been developed extensively in the literature.[19–27] The SVM learning strategy is based on an approach championed by Vapnik called the *direct method*.[17,18] The direct method, which is more a philosophy than a method, can be simply stated as follows: *Since our performance criterion is e, we should choose a learning strategy that most directly minimizes e*. Of course, the meaning of *most directly* is somewhat subjective, but as a general rule Vapnik suggests avoiding *indirect* methods that end up solving a harder problem than necessary. For example, it is widely accepted that density estimation is harder than pattern classification, so the direct method suggests avoiding the use of density estimation techniques to solve the the pattern classification problem. The most common example of a direct method is the *empirical error minimization* strategy that selects a classifier $\hat{f}$ from $\mathcal{F}$ that directly minimizes the empirical error $\hat{e}$. This strategy completely avoids the intermediate probability functions in (12.4) through (12.7) and has the powerful advantage that it is robust to distribution.[12,18] Its major disadvantage is that it is nearly always computationally intractable, even for the simplest function classes. The SVM learning strategy overcomes this weakness by constructing a surrogate optimization criterion that is both computationally attractive and *calibrated* with respect to e in the sense that minimizing the surrogate criterion guarantees control of the excess error. Furthermore, the SVM learning method allows the user to choose a wide variety of function classes, from simple to complex, including many of the function classes employed by traditional learning methods, and uses the same core learning algorithm, which is guaranteed to be computationally efficient and to provide a global optimum, in each case. Sections 12.4.1 through 12.4.3 develop the function class $\mathcal{F}$, the learning strategy $\mathcal{L}$, and a computationally efficient algorithm for the SVM method.

12.4.1 The Support Vector Machine Function Class

The SVM function class $\mathcal{F}$ is defined by functions of the form

$$f(\mathbf{x}) = \mathbf{w}^T \boldsymbol{\phi}(\mathbf{x}) + b$$

where $\boldsymbol{\phi}: \mathbb{R}^d \rightarrow \mathbb{R}^m$ is a map from d-dimensional vectors $\mathbf{x}$ to m-dimensional vectors

$$\boldsymbol{\phi}(\mathbf{x}) = [\phi_1(\mathbf{x}), \phi_2(\mathbf{x}), \ldots, \phi_m(\mathbf{x})]^T$$

where usually $m \gg d$. Thus, f is simply a linear classifier in a higher dimensional space. The idea is that with the appropriate choice of $\boldsymbol{\phi}$, we may be able

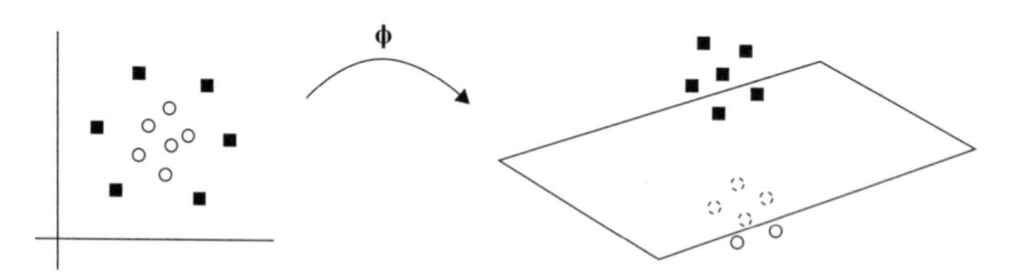

FIGURE 12.3
Illustration of mapping nonlinearly separable data in two dimensions to linearly separable data in three dimensions.

to better separate the two classes with a simple linear classifier as shown in Figure 12.3. Actually, there is no requirement that $\boldsymbol{\phi}$ map to a *higher* dimensional space: It could map to a *lower* dimensional space if one can be found where the two classes are well separated by a linear classifier. But to represent function classes with good universal approximation capabilities we must generally map to a higher dimensional space, often an *infinite*-dimensional space, in which case the inner product $\mathbf{w}^T\boldsymbol{\phi}(\mathbf{x})$ is actually an integral. Direct evaluation of $\mathbf{w}^T\boldsymbol{\phi}(\mathbf{x})$ in higher dimensional spaces, whether finite or infinite, is computationally prohibitive. However, by placing a rather mild restriction on the choice of $\boldsymbol{\phi}$ the inner product can be efficiently evaluated in the original data space without direct use of either $\boldsymbol{\phi}$ or $\mathbf{w}$. In the infinite-dimensional case, this means that the integral can be performed *without integration*. The restriction is that $\boldsymbol{\phi}$ map to a vector space called a *Reproducing Kernel Hilbert Space* (RKHS).[28] In an RKHS, the inner product $\boldsymbol{\phi}(\mathbf{x})^T\boldsymbol{\phi}(\acute{\mathbf{x}})$ can be computed using

$$\boldsymbol{\phi}(\mathbf{x})^T\boldsymbol{\phi}(\acute{\mathbf{x}}) = k(\mathbf{x},\acute{\mathbf{x}}) \tag{12.15}$$

where k is the *kernel function* for the RKHS. Consider the following simple example: Let $\mathbf{x} = (x_1, x_2)^T$ be a two-dimensional vector. Then the function $k(\mathbf{x},\acute{\mathbf{x}}) = (\mathbf{x}^T\acute{\mathbf{x}})^2$ can be represented as $k(\mathbf{x},\acute{\mathbf{x}}) = \boldsymbol{\phi}(\mathbf{x})^T\boldsymbol{\phi}(\acute{\mathbf{x}})$ where

$$\boldsymbol{\phi}(\mathbf{x}) = (x_1^2, \sqrt{2}x_1x_2, x_2^2)^T$$

since

$$k(\mathbf{x},\acute{\mathbf{x}}) = (\mathbf{x}^T\acute{\mathbf{x}})^2 = (x_1\acute{x}_1 + x_2\acute{x}_2)^2 = x_1^2\acute{x}_1^2 + 2x_1\acute{x}_1x_2\acute{x}_2 + x_2^2\acute{x}_2^2 = \boldsymbol{\phi}(\mathbf{x})^T\boldsymbol{\phi}(\acute{\mathbf{x}})$$

The relation in (12.15) allows us to implement f without direct evaluation of $\mathbf{w}^T\boldsymbol{\phi}(\mathbf{x})$ because, as we show in Section 12.4.3.1, the SVM solution for $\mathbf{w}$ takes the form $\mathbf{w} = \Sigma_i a_i\boldsymbol{\phi}(\mathbf{x}_i)$, and so the inner product can be written as

$$\mathbf{w}^T\boldsymbol{\phi}(\mathbf{x}) = \sum_i a_i\boldsymbol{\phi}(\mathbf{x}_i)^T\boldsymbol{\phi}(\mathbf{x}) = \sum_i a_i k(\mathbf{x}_i,\mathbf{x}) \tag{12.16}$$

We will also show in Section 12.4.3 that (12.15) allows us to solve the SVM design problem (i.e., to find the coefficients a_i) without direct computation in the RKHS. Thus, (12.15) allows us to completely avoid direct computation in the RKHS space and, therefore, makes the SVM approach computationally feasible. In fact, once k is known we never need to know the map $\boldsymbol{\phi}$, only that such a map exists. Furthermore, the kernel k uniquely defines the RKHS, but the map $\boldsymbol{\phi}$ may not be unique as illustrated by the fact that the maps

$$\boldsymbol{\phi}(\mathbf{x}) = (x_1^2 - x_2^2, 2x_1x_2, x_1^2 + x_2^2)^T$$

and

$$\boldsymbol{\phi}(\mathbf{x}) = (x_1^2, x_1x_2, x_1x_2, x_2^2)^T$$

are also valid for the aforementioned simple kernel $k(\mathbf{x}, \acute{\mathbf{x}}) = (\mathbf{x}^T\acute{\mathbf{x}})^2$. Thus, in practice, rather than choosing a map $\boldsymbol{\phi}$ we choose a kernel k. The only restriction is that k be a valid RKHS kernel. A sufficient condition is given by the Moore–Aronszajn theorem,[29] which loosely states that every symmetric positive semidefinite function k defines a unique RKHS, that is, every k such that $k(\mathbf{x}, \acute{\mathbf{x}}) = k(\acute{\mathbf{x}}, \mathbf{x})$, $\forall \mathbf{x}, \acute{\mathbf{x}}$ and

$$\int\int k(\mathbf{x}, \acute{\mathbf{x}})g(\mathbf{x})g(\acute{\mathbf{x}}) \geq 0$$

for all square integrable functions g. Different choices of k lead to different function classes $\mathcal{F}$, and a great deal of effort has been devoted to the design of specialized kernel functions that exploit specific domain and distribution knowledge.[5,30–34] Four common kernel functions that are motivated by the functional forms encountered earlier in Section 12.3.2 are as follows:

$$k(\mathbf{x}, \acute{\mathbf{x}}) \in \begin{cases} \mathbf{x}^T\acute{\mathbf{x}} & \text{Linear} \\ (c_1\mathbf{x}^T\acute{\mathbf{x}} + c_0)^q & \text{Polynomial} \\ e^{-\gamma\|\mathbf{x}-\acute{\mathbf{x}}\|^2} & \text{Radial basis function (RBF)} \\ \tanh(c_1\mathbf{x}^T\acute{\mathbf{x}} + c_0) & \text{Sigmoid} \end{cases} \tag{12.17}$$

The sigmoid kernel only satisfies the Moore-Aronszajn theorem for certain values of (c_1, c_0).[20,35] Using (12.16), it is easy to see that the first kernel yields a linear classifier $f(\mathbf{x}) = \mathbf{w}^T\mathbf{x} + b$ where $\mathbf{w} = \Sigma_i a_i \mathbf{x}_i$. Similarly, the second kernel yields a polynomial classifier or order q; the third kernel yields $f(\mathbf{x}) = \sum_i a_i e^{\gamma\|x-x_i\|^2} + b$, which has the same form as the Parzen-kernel classifier; and the fourth kernel yields $f(\mathbf{x}) = \sum_i a_i \tanh(c_1\mathbf{x}_i^T\mathbf{x} + c_0) + b$, which has the same form as the MLP neural network classifier. There is a slight difference in the way these functions are parameterized compared to Section 12.3.2, but the main point is that the SVM approach provides a *unified* setting for the design and analysis of classifiers using these (and many other) function classes. Furthermore, as we shall see in Section 12.4.2, the SVM design criterion is convex for all kernels and can, therefore, be optimized efficiently. In fact, the *same* core algorithm can be used to find a global optimum regardless of the choice of k. Thus, the SVM approach allows direct control over the approximation error through the choice of k without negative implications on computation.

12.4.2 The Support Vector Machine Learning Strategy

The SVM learning strategy chooses a function $\hat{f}$ from $\mathcal{F}$ by minimizing a surrogate optimization criterion. The two key components in constructing the SVM criterion are (1) *convex relaxation* and (2) *complexity regularization.* To derive the first component, we replace the classification error e with a surrogate risk function r that upper bounds e and satisfies two important properties, that is, convexity and calibration. We start by observing that the error condition $\texttt{sign}\, f(\mathbf{x}) \neq y$ is essentially equivalent to the condition $yf(\mathbf{x}) \leq 0$. The only difference is that $yf(\mathbf{x}) \leq 0$ contributes an extra error when $(f(\mathbf{x}) = 0, y = -1)$. Furthermore, the error counting function $I(yf(\mathbf{x}) \leq 0)$ is upper bounded by the *hinge function*

$$[1 - yf(\mathbf{x})]_+ = \begin{cases} 1 - yf(\mathbf{x}), & 1 - yf(\mathbf{x}) > 0 \\ 0, & 1 - yf(\mathbf{x}) \leq 0 \end{cases} \tag{12.18}$$

as shown in Figure 12.4. Thus, we can write

$$\begin{aligned} e(f) &= E_{X,Y}[I(\texttt{sign}\, f(\mathbf{x}) \neq y)] \\ &\leq E_{X,Y}[I(yf(\mathbf{x}) \leq 0)] \\ &\leq E_{X,Y}[[1 - yf(\mathbf{x})]_+] \end{aligned}$$

The function r defined by $r(f) = E_{X,Y}[[1 - yf(\mathbf{x})]_+]$ is called the *hinge risk,* and $r^* = \min_{\text{all } f} r(f)$ is the optimal hinge risk. The hinge risk upper bounds the

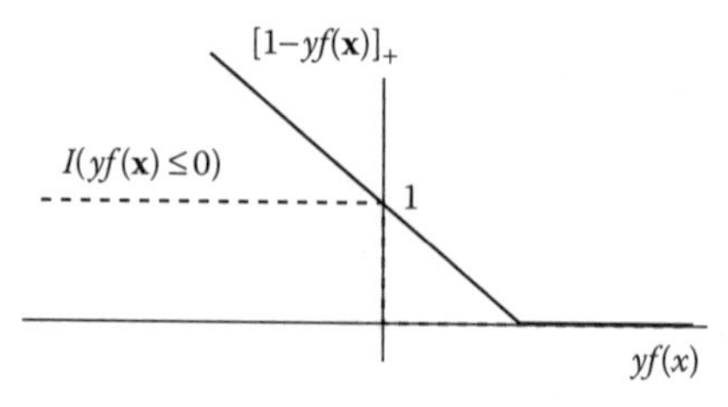

FIGURE 12.4
The hinge function $[1 - yf(\mathbf{x})]_+$ is a convex relaxation of the step function $I(yf(\mathbf{x}) \leq 0)$.

classification error for all f and all distributions. Furthermore, although it is not obvious, the hinge risk satisfies the following *calibration inequality*:

$$e(f') - e^* \leq r(f') - r^* \tag{12.19}$$

which holds for any function f'.[36–38] This inequality implies that we can control the excess classification error by controlling the excess hinge risk. This is significant because the hinge risk is a *convex* function of f, that is, $[yf(\mathbf{x})]_+ = [y(\mathbf{w}^T\boldsymbol{\phi}(\mathbf{x}) + b)]_+$ is a convex function of $(\mathbf{w}, b)$. This has the *enormous* consequence that minimizing r is computationally tractable, whereas minimizing e is not. The price we pay for this enormous benefit is not yet fully understood, but appears to be relatively mild (see the next to last paragraph of this section). The empirical hinge risk is given by

$$\hat{r}(f) = \hat{r}(\mathbf{w}, b) = \frac{1}{n}\sum_{i=1}^{n}[1 - y_i(\mathbf{w}^T\boldsymbol{\phi}(\mathbf{x}_i) + b)]_+$$

It is easy to show that minimizing $\hat{r}$ is a linear programming problem, which can be solved efficiently. However, this solution can lead to *overfitting*, that is, a lack of control on the estimation error, especially when the function class is complex.

Thus, the second component of the SVM criterion is a *complexity regularization* term that is added to the empirical hinge risk to control the estimation error. This term penalizes functions $f(\cdot) = \mathbf{w}^T\boldsymbol{\phi}(\cdot) + b$ according to the squared norm $||\mathbf{w}||^2$ of their RKHS component. Specifically, the empirical SVM criterion $\hat{s}_\lambda$ is given by

$$\hat{s}_\lambda(f) = \hat{s}_\lambda(\mathbf{w}, b) = \lambda\|\mathbf{w}\|^2 + \hat{r}(\mathbf{w}, b) = \lambda\|\mathbf{w}\|^2 + \frac{1}{n}\sum_{i=1}^{n}[1 - y_i(\mathbf{w}^T\boldsymbol{\phi}(\mathbf{x}_i) + b)]_+ \tag{12.20}$$

where the scalar $\lambda > 0$ is used to optimize the approximation–estimation error trade-off. Note that the complexity regularization term $\lambda||\mathbf{w}||^2$ maintains the convexity of the criterion. Furthermore, it plays a key role in determining the optimal solution form $\mathbf{w} = \Sigma_i a_i \boldsymbol{\phi}(\mathbf{x}_i)$ that allows the aforementioned kernel implementation (see Section 12.4.3).

The success of the SVM learning strategy depends critically on the choice of kernel k and regularization value λ. For example, if the distribution model class is unknown then a *universal* kernel should be used to control the approximation error. Roughly speaking, a universal kernel guarantees that any continuous function can be approximated arbitrarily closely by a function in $\mathcal{F}$.[26,39] Examples of universal kernels can be found in the literature.[26,39] The RBF kernel is universal, but the linear and polynomial kernels are not. Alternatively, if the distribution model class is known then it may be possible to choose a simpler (nonuniversal) kernel whose function class $\mathcal{F}$ contains functions that yield zero (or very small) excess hinge risk. Although this approach is not yet fully understood, the current rule of thumb is to choose kernels whose function class $\mathcal{F}$ contains realizations of the optimal forms obtained by substituting the model class probability functions into (12.4), (12.5), (12.6), or (12.7). For example, for the two-class Gaussian problem, we would choose a polynomial kernel with $q = 2$.

As a general rule, the choice of the regularization parameter should depend on the kernel k, the number of training samples n, and the complexity of the data distribution. For universal kernels, the conditions $\lambda_n \to 0$ and $\lambda_n \geq c\sqrt{\log n/n}$ are sufficient to guarantee asymptotic control of the approximation and estimation errors.[26,39] The lower bound may be relaxed for specific kernels. Indeed, for the RBF kernel the conditions $\lambda_n \to 0$ and $\lambda_n \geq c/n$ are sufficient.[26,39] To optimize the finite sample (i.e., nonasymptotic) performance, we must often extend our function class beyond a basic universal kernel. For example, the class $\mathcal{F}_\gamma$ that corresponds to the RBF kernel with width parameter γ must be extended to the class $\mathcal{F} = \cup_{\gamma>0}\mathcal{F}_\gamma$. This means that we must extend the learning method to include the selection of γ. Furthermore, this value must vary as a function of n (much like λ). It is intuitively obvious that γ should depend on n if we consider that larger values of γ (i.e., smaller kernel widths) are generally required to exploit the finer details of the distribution revealed by an increasing number of training samples. By studying the simultaneous effect of λ and γ on the excess error and choosing them to optimize performance, it is possible to establish specific formulas for λ and γ as a function of n.[26,40] These formulas are expressed in terms of the so-called *noise exponents*, which measure the complexity of the distribution.[26,40] In practice, the noise exponents are not known and so we must choose λ and γ by some other means. It turns out that we can achieve the same (near optimal) performance guarantees by using the simple grid search in Figure 12.5.[41] This procedure defines a finite grid of (λ, γ) values, computes a minimizer of $\hat{s}_\lambda$ for each grid point, and then computes the final solution using the (λ, γ) values from the best grid point. This procedure is based on a theorem that specifies parameter values $n_{\text{core}} = n_{\text{aux}} = n/2$ with $m_\lambda = m_\gamma = n^2$ uniformly spaced grid values over $\Lambda_{n_{\text{core}}} = (0,1], \Gamma_{n_{\text{core}}} = [1, n_{\text{core}}^{1/d}]$.[41] In practice, however, it is more common to choose $n_{\text{core}} > n_{\text{aux}}$, unless this choice of n_{core} places an excessive computational burden on the design algorithm. It is also common to choose a smaller nonuniform grid that contains roughly 10 × 10 *logarithmically* spaced values of λ and γ. Specifically, the following grids have been reported to work well: logarithmically spaced values of λ from the interval $[10^{-6} n_{\text{core}}^{-1/2}, 1]$, and logarithmically spaced values of γ from the nterval $[.1, 2n_{\text{core}}^{1/d}]$ or the interval $c\sqrt{n_{\text{core}}}/d$ where $c \in [.0001,1]$.

In summary, to control the excess error, we first make an appropriate choice of kernel, and then determine λ (and any potential kernel parameters) using a grid search. When distribution knowledge is limited, we should choose a universal kernel that can be adapted to obtain good finite sample performance. The Gaussian RBF kernel is one example. Alternatively, if the distribution model class is known, then it is sufficient to choose $\mathcal{F}$ to contain functions that yield zero (or very small) excess hinge risk. Although we gave a rule of

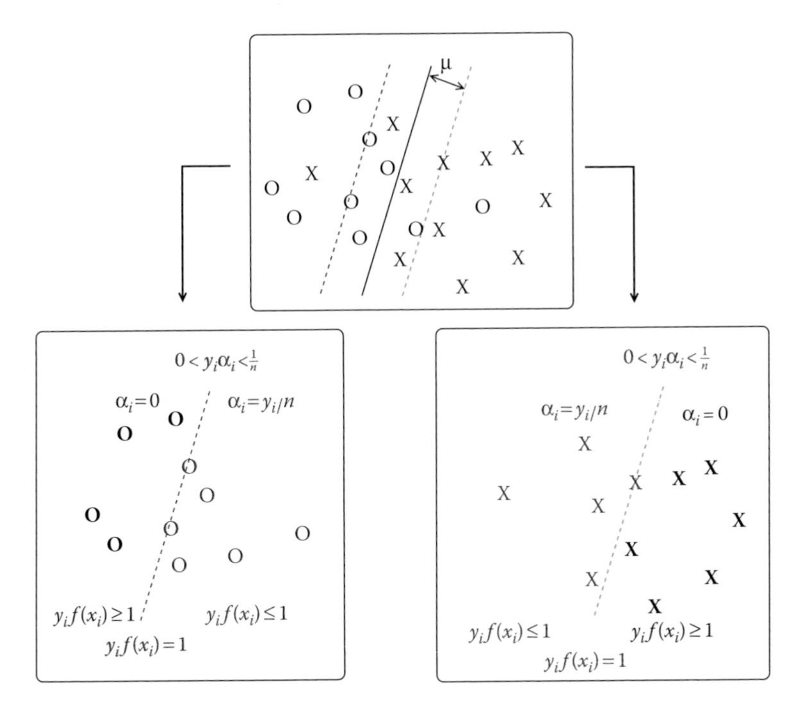

FIGURE 12.8
Optimal solution characteristics illustrated for a linear kernel.

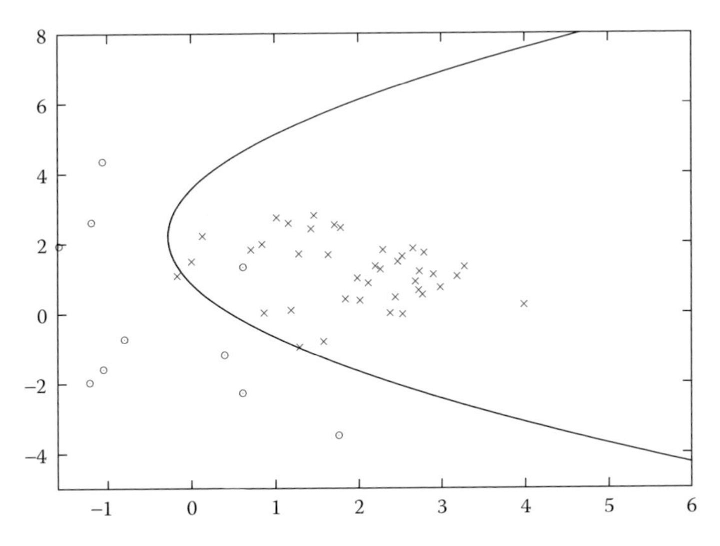

FIGURE 12.11
Results corresponding to the design of a quadratic classifier using the Gaussian maximum likelihood method: The error estimate for this result is 0.114 with an accuracy of approximately 0.005.

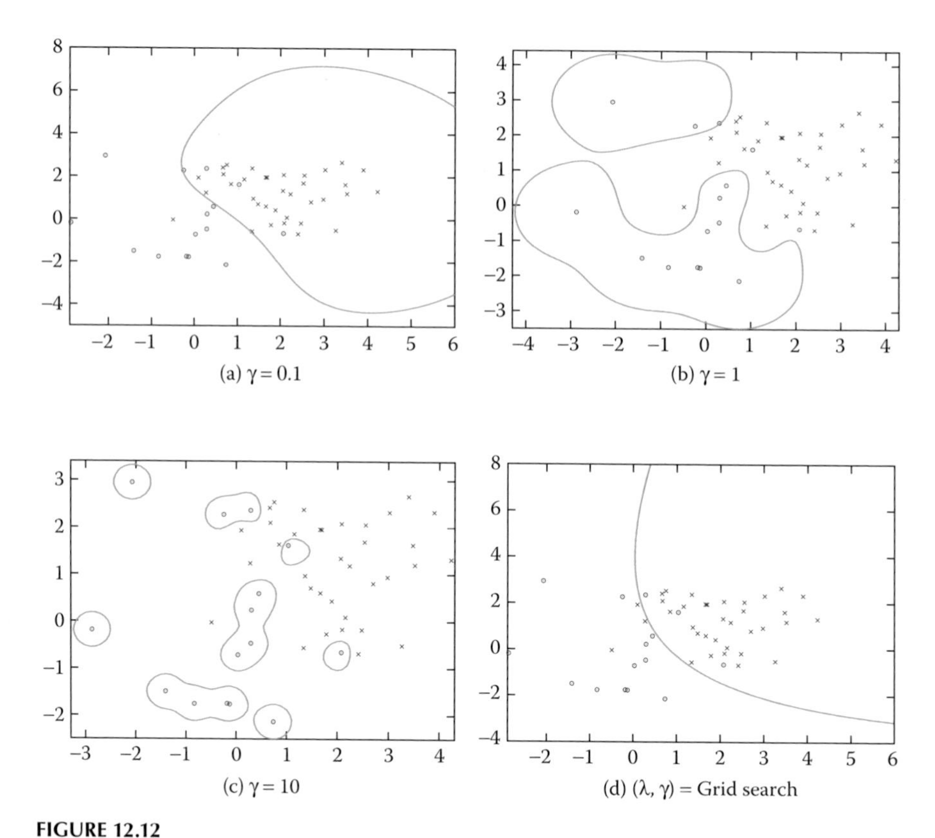

FIGURE 12.12
Decision boundaries for support vector machine–radial basis function classifiers designed with four different values of (λ, γ): (a–c): $\lambda = 0.001$ and γ is given below the figure; (d): both λ and γ are chosen using the grid search procedure.

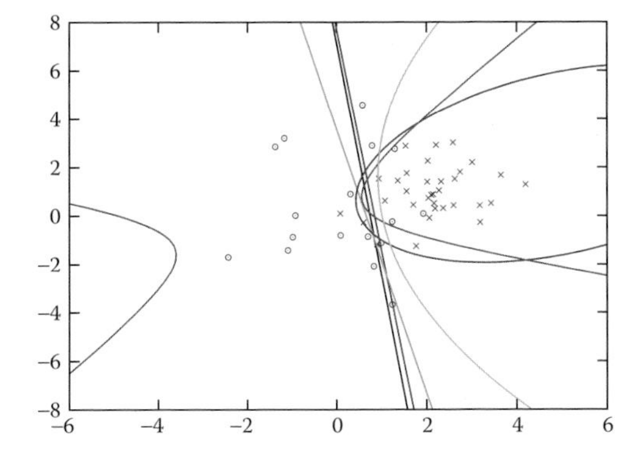

Design method	Generalization error
LS-LINEAR (black)	0.1557
GML-LINEAR (red)	0.1544
SVM-LINEAR (green)	0.1534
GML-QUAD (blue)	0.1295
SVM-QUAD (magenta)	0.1356
SVM-RBF (cyan)	0.1528
Bayes error	0.1202 (±0.0053)

FIGURE 12.13
Results of the six classifier design methods for a two-class Gaussian problem.

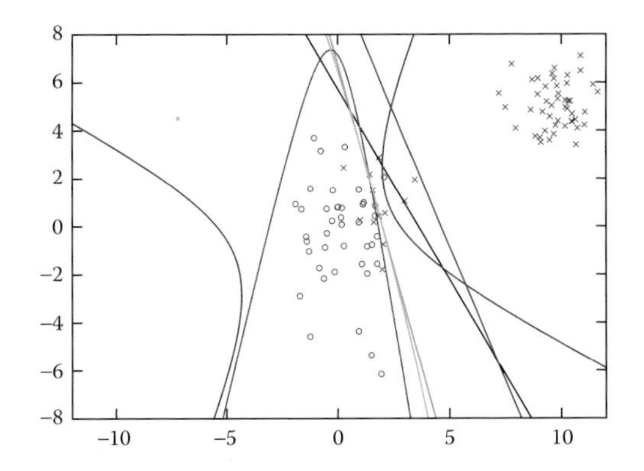

Design method	Generalization error
LS-LINEAR (black)	0.1561
GML-LINEAR (red)	0.1832
SVM-LINEAR (green)	0.0893
GML-QUAD (blue)	0.1262
SVM-QUAD (magenta)	0.0909
SVM-RBF (cyan)	0.0900

FIGURE 12.14
Comparison of classifier design methods where blue samples are drawn from a Gaussian distribution and red samples are drawn from a mixture of two Gaussians, one that overlaps with the blue and one that is far away from the overlap region.

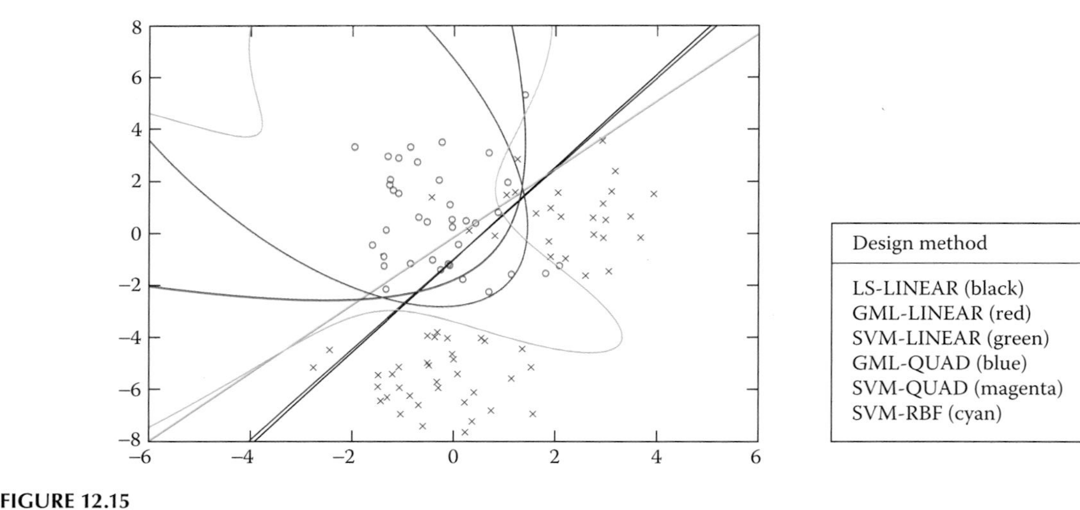

Design method	Generalization error
LS-LINEAR (black)	0.1897
GML-LINEAR (red)	0.1916
SVM-LINEAR (green)	0.2199
GML-QUAD (blue)	0.1641
SVM-QUAD (magenta)	0.1329
SVM-RBF (cyan)	0.0913

FIGURE 12.15
Comparison of classifier design methods where blue samples are drawn from a Gaussian distribution, and red samples are drawn from a mixture of two Gaussians that overlap with different regions of the blue class.

Procedure 1:

1. Partition the training data randomly into a core training set of size n_{core} and an auxiliary hold-out set of size $n_{aux} = n - n_{core}$.
2. Form a finite grid of λ values λ_i, $1 \le i \le m_\lambda$, over the interval $\Lambda_{n_{core}}$ and a finite grid of γ values γ_j, $1 \le j \le m_\gamma$ over the interval $\Gamma_{n_{core}}$.
3. Use the core training set to compute an SVM solution f_{λ_i,γ_j} for each grid point (λ_i, γ_j).
4. Choose a grid point (i^*, j^*) whose solution minimizes the empirical classification error on the auxiliary hold-out set.
5. Re-train the SVM with all the training data using and $\lambda_n = \lambda_{i*}\sqrt{n_{core}/n}$ and $\gamma_n = \gamma_{j*}(n/n_{core})^{1/d}$ to obtain the final solution.

FIGURE 12.5
Grid search algorithm for (λ, γ).

thumb earlier in this section, this approach is not yet fully understood. For example, consider the class of distributions whose Bayes optimal classifiers are linear. Choosing $\mathcal{F}$ to be the class of linear classifiers does *not* guarantee a near optimal SVM learning method, although $\mathcal{F}$ contains all possible optimal classifiers. On the other hand, simply increasing the complexity of $\mathcal{F}$ to the RBF function class *does* yield a near optimal SVM learning method. This suggests that while solutions for particular families of distributions may take a simple form, the SVM learning strategy *may* require a more complex function class to efficiently compute these solutions.

The SVM learning strategy developed in this section guarantees control of the approximation and estimation errors. In Section 12.4.3, we develop a computationally efficient algorithm that guarantees control of the computation error by producing approximate solutions $\hat{f}$ that satisfy $\hat{s}_\lambda(\hat{f}) - \min_{f\in\mathcal{F}} \hat{s}_\lambda(f) < \varepsilon$ for any user-specified value of $\epsilon > 0$.

12.4.3 The Core Support Vector Machine Algorithm

With λ and k fixed, we develop an algorithm that finds a solution that approximately minimizes $\hat{s}_\lambda$ and then use this solution to implement $\hat{f}$. In particular, we show how to do this without using $\mathbf{w}$ or $\boldsymbol{\phi}$ directly. This ability to solve and implement a possibly infinite-dimensional function using finite-dimensional computations is a trademark of the SVM approach. To develop this approach, we follow the steps in Figure 12.6. Step 1 converts the unconstrained optimization problem $[\min_{\mathbf{w},b} \hat{s}_\lambda(\mathbf{w}, b)]$ to a constrained optimization problem called a *quadratic programming* (QP) problem by adding n slack variables $\boldsymbol{\xi} = (\xi_1, \ldots, \xi_n)^T$, $\xi_i \in \mathbb{R}$. The total number of variables in this *primal* QP problem is $\dim(\mathbf{w}) + n + 1$ (where $\dim(\mathbf{w})$ is the dimension of $\mathbf{w}$, which may be infinite). Step 2 constructs a *dual* QP problem defined over $\boldsymbol{\alpha} = (\alpha_1, \alpha_2, \ldots, \alpha_n)^T$, $\alpha_i \in \mathbb{R}$. This dual QP problem has a total of n variables, regardless of the dimension of $\mathbf{w}$ and is, therefore, amenable to solution by digital computer. Thus, step 3 aims to (approximately) solve the dual QP problem. Step 4 then maps the dual solution back to the primal. This is accomplished by deriving functions

Steps to solve $\min_{\mathbf{w},b} \hat{s}_\lambda(\mathbf{w}, b)$ and implement $\hat{f}(\mathbf{x}) = \hat{\mathbf{w}}^T \boldsymbol{\phi}(\mathbf{x}) + \hat{b}$

1. Construct Primal QP

$$\begin{aligned} &\min_{\mathbf{w},b,\xi} && \lambda\|\mathbf{w}\|^2 + \frac{1}{n}\sum_{i=1}^{n}\xi_i \\ &\text{subject to} && \xi_i \geq 1 - y_i(\mathbf{w}^T\boldsymbol{\phi}(\mathbf{x}_i) + b), \quad i = 1, 2, \ldots, n \\ &&& \xi_i \geq 0, \quad i = 1, 2, \ldots, n \end{aligned}$$

2. Construct Dual QP

$$\begin{aligned} &\max_{\boldsymbol{\alpha}} && -\frac{1}{2}\boldsymbol{\alpha}^T Q\boldsymbol{\alpha} + \boldsymbol{\alpha}^T\mathbf{y} \\ &\text{subject to} && \boldsymbol{\alpha}^T\mathbf{1} = 0 \\ &&& 0 \leq y_i\alpha_i \leq \frac{1}{n}, \quad i = 1, 2, \ldots, n \end{aligned}$$

3. Solve Dual QP for $\hat{\boldsymbol{\alpha}}$

4. Map Dual Solution to Primal Solution

$$\hat{\mathbf{w}} = \mathbf{w}_{\text{map}}(\hat{\boldsymbol{\alpha}}) = \frac{1}{2\lambda}\sum_{i=1}^{n}\hat{\alpha}_i\boldsymbol{\phi}(\mathbf{x}_i)$$

$$\hat{b} = b_{\text{map}}(\hat{\boldsymbol{\alpha}}) \in \arg\min_b \frac{1}{n}\sum_{i=1}^{n}[y_i(g_i(\hat{\boldsymbol{\alpha}}) - b)]_+$$

$$\hat{\boldsymbol{\xi}} = \boldsymbol{\xi}_{\text{map}}(\hat{\boldsymbol{\alpha}}, \hat{b}) = \left(\left[y_1(g_1(\hat{\boldsymbol{\alpha}}) - \hat{b})\right]_+, \ldots, \left[y_n(g_n(\hat{\boldsymbol{\alpha}}) - \hat{b})\right]_+\right)^T$$

5. Construct $\hat{f}$ using kernel

$$\hat{f}(\mathbf{x}) = \hat{\mathbf{w}}(\mathbf{x})^T\boldsymbol{\phi}(\mathbf{x}) + \hat{b} = \frac{1}{2\lambda}\sum_{i=1}^{n}\hat{\alpha}_i k(\mathbf{x}_i, \mathbf{x}) + \hat{b}$$

FIGURE 12.6
Steps used in the derivation of the core SVM algorithm.

$\mathbf{w}_{\text{map}}$, b_{map}, $\boldsymbol{\xi}_{\text{map}}$ that map feasible values of the dual variable $\boldsymbol{\alpha}$ to feasible values of the primal variables $\mathbf{w}$, b, and $\boldsymbol{\xi}$, and then using these functions to compute $(\hat{\mathbf{w}}, \hat{b}, \hat{\boldsymbol{\xi}}) = [\mathbf{w}_{\text{map}}(\hat{\boldsymbol{\alpha}}), b_{\text{map}}(\hat{\boldsymbol{\alpha}}), \boldsymbol{\xi}_{\text{map}}(\hat{\boldsymbol{\alpha}}, \hat{b})]$ The final step is to construct $\hat{f} = \hat{\mathbf{w}}^T\boldsymbol{\phi}(\mathbf{x}) + \hat{b}$ by substituting $\mathbf{w}_{\text{map}}(\hat{\boldsymbol{\alpha}})$ for $\hat{\mathbf{w}}$ and then replacing the inner products $\boldsymbol{\phi}(\mathbf{x}_i)^T\boldsymbol{\phi}(\mathbf{x})$ with kernel evaluations $k(\mathbf{x}_i, \mathbf{x})$. Steps 1–2 and 4–5 are developed first, and then a popular algorithm for the dual QP solver in step 3 is described.

12.4.3.1 Constructing the Primal, Dual, and Dual-to-Primal Map

Consider the SVM criterion in (12.20). To derive the primal QP problem, we replace each sum term $[1 - y_i(\mathbf{w}^T\boldsymbol{\phi}(\mathbf{x}_i) + b)]_+$ with a slack variable ξ_i and then require these slack variables to satisfy $\xi_i \geq 1 - y_i(\mathbf{w}^T\boldsymbol{\phi}(\mathbf{x}_i) + b)$ and $\xi_i \geq 0$. This gives the primal QP problem as follows:

$$\begin{aligned} &\min_{\mathbf{w},b,\xi} && \lambda\|\mathbf{w}\|^2 + \frac{1}{n}\sum_{i=1}^{n}\xi_i \\ &\text{Subject to} && \xi_i \geq 1 - y_i(\mathbf{w}^T\boldsymbol{\phi}(\mathbf{x}_i) + b), \quad i = 1, 2, \ldots, n \\ &&& \xi_i \geq 0, \quad i = 1, 2, \ldots, n \end{aligned} \tag{12.21}$$

It is easy to show that for every solution $(\mathbf{w}^*, b^*, \boldsymbol{\xi}^*)$ of the primal QP the pair $(\mathbf{w}^*, b^*)$ is a minimizer of $\hat{s}_\lambda$, and that for every minimizer $(\mathbf{w}^*, b^*)$ of $\hat{s}_\lambda$ there is a feasible $\boldsymbol{\xi}^*$ such that $(\mathbf{w}^*, b^*, \boldsymbol{\xi}^*)$ is a solution of the primal QP (see Exercise 12.5). We define the primal QP criterion as follows:

$$S(\mathbf{w}, b, \xi) = \lambda \|\mathbf{w}\|^2 + \frac{1}{n}\sum_{i=1}^{n} \xi_i$$

and define $S^* = S(\mathbf{w}^*, b^*, \xi^*)$ to be its optimal value. To construct the dual QP we use the following Kuhn–Tucker theorem adapted from.[42,43]

Theorem 12.1: Consider the following convex programming problem

$$\begin{array}{ll} \min_{\psi \in \mathbb{R}^q} & r(\psi) \\ \text{Subject to} & h_i(\psi) \geq 0, \;\; i = 1, \ldots, m \end{array}$$

where r and h_i, $i = 1, \ldots, m$ are convex and there exists a ψ' such that $h_i(\psi') > 0$ for all $i = 1, 2, \ldots, m$. Define the Lagrangian

$$L(\psi, \boldsymbol{\theta}) = r(\psi) - \sum_{i=1}^{m} \theta_i h_i(\psi)$$

where $\boldsymbol{\theta} = (\theta_1, \ldots, \theta_m)^T$. If ψ^* solves the convex programming problem, then there exists Lagrange multipliers $\boldsymbol{\theta}^* = (\theta_1^*, \ldots, \theta_m^*)^T$ such that the following three conditions hold:

$$L(\psi^*, \boldsymbol{\theta}) \leq L(\psi^*, \boldsymbol{\theta}^*) \leq L(\psi, \boldsymbol{\theta}^*) \text{ for all } \psi \text{ and } \boldsymbol{\theta} \geq 0 \tag{12.22}$$

$$\theta_i^* \geq 0, i = 1, \ldots, m \tag{12.23}$$

$$\theta_i^* h_i(\psi^*) = 0, i = 1, \ldots, m \tag{12.24}$$

■

This theorem tells us that ψ^* can be found by solving the *unconstrained* optimization problem $\min_\psi L(\psi, \boldsymbol{\theta}^*)$, and that $\boldsymbol{\theta}^*$ can be found by solving the Lagrangian dual problem $\max_{\boldsymbol{\theta} \geq 0} L(\psi^*, \boldsymbol{\theta})$. Although it appears that neither problem can be solved without first knowing the solution to the other, it is often possible to solve both problems as follows: First use the optimality conditions of $[\min_\psi L(\psi, \boldsymbol{\theta})]$ to determine a map $\psi_{\text{map}}(\boldsymbol{\theta})$ that gives a value of ψ that optimizes L for any value of $\boldsymbol{\theta} \geq 0$, and then substitute this map into L to obtain a criterion $R(\boldsymbol{\theta}) = L(\psi_{\text{map}}(\boldsymbol{\theta}), \boldsymbol{\theta})$ that is expressed in terms of $\boldsymbol{\theta}$ alone. Then, after solving $\max_{\boldsymbol{\theta} \geq 0} R(\boldsymbol{\theta})$ to obtain $\boldsymbol{\theta}^*$, the solution to the primal problem is obtained using $\psi^* = \psi_{\text{map}}(\boldsymbol{\theta}^*)$. This may seem like a roundabout approach, but it is extremely useful in situations like ours where the dual problem is much easier to solve than the primal.

To apply this theorem/approach, we first form the Lagrangian of the primal QP in (12.21)

$$L(\mathbf{w},b,\boldsymbol{\xi},\boldsymbol{\alpha},\boldsymbol{\beta}) = \lambda\|\mathbf{w}\|^2 + \frac{1}{n}\sum_{i=1}^{n}\xi_i - \sum_{i=10}^{n}a_i[y_i(\boldsymbol{\phi}(\mathbf{x}_i)^T\mathbf{w}+b)-1+\xi_i] - \sum_{i=1}^{n}\beta_i\xi_i$$

where $a_i \geq 0$ and $\beta_i \geq 0$ are the Lagrange multipliers; further, recall that $\lambda > 0$. To simplify our development, we define

$$\alpha_i = y_i a_i, \quad i = 1, 2, \ldots, n$$

and use the relation $a_i = y_i\alpha_i$ (since $y_i \in \{-1, 1\}$) to make a change of variables in L. Replacing a_i with $y_i\alpha_i$ in L gives

$$L(\mathbf{w},b,\boldsymbol{\xi},\boldsymbol{\alpha},\boldsymbol{\beta}) = \lambda\|\mathbf{w}\|^2 + \frac{1}{n}\sum_{i=1}^{n}\xi_i - \sum_{i=1}^{n}\alpha_i[\boldsymbol{\phi}(\mathbf{x}_i)^T\mathbf{w}+b-y_i+y_i\xi_i] - \sum_{i=1}^{n}\beta_i\xi_i \quad (12.25)$$

and the constraints are now $y_i\alpha_i \geq 0$ and $\beta_i \geq 0$. Since L is convex and differentiable with respect to $(\mathbf{w}, b, \boldsymbol{\xi})$, the following first-order conditions are both necessary and sufficient for a minimum of L with respect to $(\mathbf{w}, b, \boldsymbol{\xi})$:

$$\left.\frac{\partial L}{\partial \mathbf{w}}\right|_{\mathbf{w}=\acute{\mathbf{w}}} = 0 \Rightarrow \acute{\mathbf{w}} = \frac{1}{2\lambda}\sum_i \alpha_i\boldsymbol{\phi}(\mathbf{x}_i) \quad (12.26)$$

$$\frac{\partial L}{\partial b} = 0 \Rightarrow \sum_i \alpha_i = 0 \quad (12.27)$$

$$\frac{\partial L}{\partial \xi_i} = 0 \Rightarrow y_i\alpha_i + \beta_i = \frac{1}{n} \quad (12.28)$$

Substituting (12.26) through (12.28) into the Lagrangian yields

$$\min_{\mathbf{w},b,\boldsymbol{\xi}} L(\mathbf{w},b,\boldsymbol{\xi},\boldsymbol{\alpha},\boldsymbol{\beta}) = \sum_{i=1}^{n} y_i\alpha_i - \frac{1}{4\lambda}\sum_{i=1}^{n}\sum_{j=1}^{n}\alpha_i\alpha_j\boldsymbol{\phi}(\mathbf{x}_i)^T\boldsymbol{\phi}(\mathbf{x}_j)$$

The Lagrangian dual problem is to maximize this criterion over $(\boldsymbol{\alpha}, \boldsymbol{\beta})$ subject to the constraints

$$y_i\alpha_i \geq 0, \beta_i \geq 0,\ y_i\alpha_i + \beta_i = \frac{1}{n}, \sum_i \alpha_i = 0$$

The first three constraints can be combined into $0 \leq y_i\alpha_i \leq 1/n$ so that $\boldsymbol{\beta}$ is eliminated from both the criterion and the constraints, leaving a dual optimization problem over $\boldsymbol{\alpha}$ alone:

$$\begin{aligned} \max_\alpha \quad & -\frac{1}{2}\boldsymbol{\alpha}^T Q \boldsymbol{\alpha} + \boldsymbol{\alpha}^T y \\ \text{Subject to} \quad & \sum_{i=1}^n \alpha_i = 0 \\ & 0 \le y_i \alpha_i \le \tfrac{1}{n}, \quad i = 1, 2, \ldots, n \end{aligned} \tag{12.29}$$

where Q is an n-by-n matrix with elements

$$Q_{ij} = [\boldsymbol{\phi}(\mathbf{x}_i)^T \boldsymbol{\phi}(\mathbf{x}_j)]/2\lambda = k(\mathbf{x}_i, \mathbf{x}_j)/2\lambda \tag{12.30}$$

Note that this dual QP problem has size n, independent of the dimension of $\mathbf{w}$, and is defined by computations that can be performed entirely in the original data space through the use of the kernel to evaluate the components of Q. Note also that the set of feasible values

$$\mathcal{A} = \left\{ \boldsymbol{\alpha} : \; (0 \le y_i \alpha_i \le 1/n) \quad \text{and} \quad \left(\sum_{i=1}^n \alpha_i = 0 \right) \right\} \tag{12.31}$$

is both convex and compact. Define the set of optimal solutions

$$\mathcal{A}^* = \arg\max_{\boldsymbol{\alpha} \in \mathcal{A}} R(\boldsymbol{\alpha}) \tag{12.32}$$

and define the dual QP criterion

$$R(\boldsymbol{\alpha}) = -\frac{1}{2}\boldsymbol{\alpha}^T Q \boldsymbol{\alpha} + \boldsymbol{\alpha}^T y$$

With $Z = [\boldsymbol{\phi}(\mathbf{x}_1), \boldsymbol{\phi}(\mathbf{x}_2), \ldots, \boldsymbol{\phi}(\mathbf{x}_n)]$, we can write $Q = Z^T Z/2\lambda$, verifying that Q is symmetric ($Q^T = (Z^T Z)^T = Q$) and positive semidefinite ($\mathbf{u}^T Q \mathbf{u} = |Z\mathbf{u}|^2 \ge 0$). Thus, $R(\boldsymbol{\alpha})$ is a concave function over $\mathcal{A}$ ensuring that $\mathcal{A}^*$ is convex and compact and that $R^* = R(\mathcal{A}^*)$ is unique.

In Section 12.4.3, we develop an algorithm that computes an approximate solution $\hat{\boldsymbol{\alpha}}$ to the dual QP. Once we have an approximate solution to the dual, we must map back to the primal. The expression in (12.26) gives a value $\acute{\mathbf{w}}$ that minimizes L for any feasible $\boldsymbol{\alpha}$ and is therefore a natural choice for the function $\mathbf{w}_{\text{map}}$. Thus, we define

$$\mathbf{w}_{\text{map}}(\boldsymbol{\alpha}) = \frac{1}{2\lambda} \sum_{i=1}^n \boldsymbol{\alpha}_i \boldsymbol{\phi}(\mathbf{x}_i) \tag{12.33}$$

which is the expression for $\mathbf{w}$ that was promised earlier in Sections 12.4.1 and 12.4.2. Next, we seek maps b_{map} and $\boldsymbol{\xi}_{\text{map}}$ that minimize the primal QP problem once $\mathbf{w}$ has been replaced by $\mathbf{w}_{map}(\boldsymbol{\alpha})$. Substituting $\mathbf{w}_{map}(\boldsymbol{\alpha})$ into the components of the primal QP in (12.21) gives

$$\lambda \left\| \mathbf{w}_{\text{map}}(\boldsymbol{\alpha}) \right\|^2 = \lambda \left\| \frac{1}{2\lambda} \sum_{i=1}^{n} \alpha_i \boldsymbol{\phi}(\mathbf{x}_i) \right\|^2 = \frac{1}{2} \boldsymbol{\alpha}^T Q \boldsymbol{\alpha}$$

and

$$\begin{aligned} 1 - y_j \mathbf{w}_{\text{map}}(\boldsymbol{\alpha})^T \boldsymbol{\phi}(\mathbf{x}_j) &= 1 - \frac{y_j}{2\lambda} \sum\nolimits_{i=1}^{n} \alpha_i \boldsymbol{\phi}(\mathbf{x}_i)^T \boldsymbol{\phi}(\mathbf{x}_j) \\ &= 1 - \frac{y_j}{2\lambda} \sum\nolimits_{i=1}^{n} \alpha_i k(\mathbf{x}_i, \mathbf{x}_j) \\ &= 1 - y_j (Q\boldsymbol{\alpha})_j \end{aligned}$$

If we define $g(\acute{\boldsymbol{\alpha}})$ to be the gradient of R evaluated at $\acute{\boldsymbol{\alpha}}$, that is,

$$g(\acute{\boldsymbol{\alpha}}) = \nabla_{\boldsymbol{\alpha}} R \big|_{\boldsymbol{\alpha}=\acute{\boldsymbol{\alpha}}} = -Q\acute{\boldsymbol{\alpha}} + y \tag{12.34}$$

then we can write

$$1 - y_j \mathbf{w}_{\text{map}}(\boldsymbol{\alpha})^T \boldsymbol{\phi}(\mathbf{x}_j) = y_j g_j(\boldsymbol{\alpha})$$

where $g_j(\boldsymbol{\alpha})$ is the jth component of the gradient vector. With these substitutions, the primal QP problem takes the form

$$\begin{aligned} &\min_{b,\xi} && \frac{1}{2}\boldsymbol{\alpha}^T Q \boldsymbol{\alpha} + \frac{1}{n}\sum\nolimits_{i=1}^{n} \xi_i \\ &\text{Subject to} && \xi_i \ge y_i[g_i(\boldsymbol{\alpha}) - b] \\ &&& \xi_i \ge 0, \quad i = 1, 2, \ldots, n \end{aligned}$$

and since $\boldsymbol{\alpha}$ is fixed, this is equivalent to the optimization problem

$$\begin{aligned} &\min_{b,\xi} && \frac{1}{n}\sum\nolimits_{i=1}^{n} \xi_i \\ &\text{Subject to} && \xi_i \ge y_i[g_i(\boldsymbol{\alpha}) - b] \\ &&& \xi_i \ge 0, \quad i = 1, 2, \ldots, n \end{aligned}$$

The obvious choice of $\acute{\xi}_i$ that makes the criterion as small as possible and still satisfies the constraints is

$$\acute{\xi}_i = \xi_{\text{map}_i}(\boldsymbol{\alpha}, b) = [y_i(g_i(\boldsymbol{\alpha}) - b)]_+ \tag{12.35}$$

Substituting ξ_{map} into the aforementioned optimization problem leaves a scalar optimization problem for b':

$$b' = b_{\text{map}}(\boldsymbol{\alpha}) \in \arg\min_b \frac{1}{n}\sum_{i=1}^{n} \xi_{\text{map}_i}(\boldsymbol{\alpha}, b) = \arg\min_b \frac{1}{n}\sum_{i=1}^{n} [y_i(g_i(\boldsymbol{\alpha}) - b)]_+ \quad (12.36)$$

Thus, b_{map} is defined by a one-dimensional optimization problem whose criterion takes the form

$$\begin{aligned} \rho(b) &= \frac{1}{n}\sum_{i=1}^{n} [y_i(g_i(\boldsymbol{\alpha}) - b)]_+ \\ &= \frac{1}{n}\sum_{i:y_i=1} [g_i(\boldsymbol{\alpha}) - b]_+ + \frac{1}{n}\sum_{i:y_i=-1} [-g_i(\boldsymbol{\alpha}) + b]_+ \end{aligned} \quad (12.37)$$

This criterion is a sum of hinge functions with slopes $-y_i/n$ and b-intercepts $g_i(\boldsymbol{\alpha})$ as shown in Figure 12.7. Thus ρ is a convex piecewise linear surface whose minima occur at corners, and also possibly along flat regions that have a corner at each end. Since the corners coincide with the points $g_i(\boldsymbol{\alpha})$, the set $\{g_i(\boldsymbol{\alpha}), i = 1, \ldots, n\}$ contains an optimal solution. Thus, it is sufficient to explore each member of this set and choose one that gives the minimum criterion value. A brute-force algorithm that computes the criterion from scratch for every member of this set requires $O(n^2)$ run time; but this can be reduced to $O(n \log n)$ using an algorithm that first sorts the values $g_i(\hat{\boldsymbol{\alpha}})$ and then visits them in order, using simple operations to update the criterion value at each step (see Exercise 12.6). Furthermore, when the minima occur along a flat region, it is easy for this more efficient algorithm to choose a midway point as the solution.

The last step in Figure 12.6 is to substitute $(\mathbf{w}_{\text{map}}, b_{\text{map}})$ into $f(\mathbf{x}) = \mathbf{w}^T \boldsymbol{\phi}(\mathbf{x}) + b$ to obtain a practical realization of $\hat{f}$. This gives

$$\hat{f}(\mathbf{x}) = \mathbf{w}_{\text{map}}(\hat{\boldsymbol{\alpha}})^T \boldsymbol{\phi}(\mathbf{x}) + b_{\text{map}}(\boldsymbol{\alpha}) = \sum_{j=1}^{n} \hat{\alpha}_j \boldsymbol{\phi}(\mathbf{x}_j)^T \boldsymbol{\phi}(\mathbf{x}) + \hat{b} = \sum_{j=1}^{n} \hat{\alpha}_j k(\mathbf{x}_j, \mathbf{x}) + \hat{b} \quad (12.38)$$

which can be implemented using computations entirely in the original data space through the use of kernel evaluations.

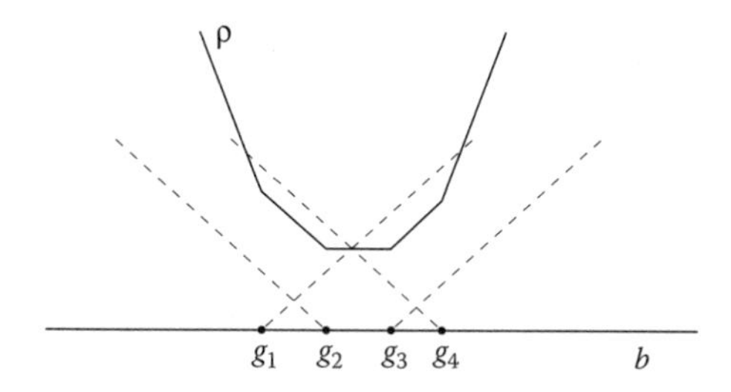

FIGURE 12.7
Example of the criterion function ρ for a problem with four data samples.

We have now completed steps 1–2, 4–5 in Figure 12.6. Thus, to compute an approximate minimizer of $\hat{s}_\lambda$ in practice, we must perform the following steps:

1. Use the training data to form a specific instance of the dual QP.
2. Compute an approximate solution $\hat{\boldsymbol{\alpha}}$ to the dual QP using the algorithm in Section 12.4.3.3.
3. Solve for $\hat{b} = b_{\text{map}}(\hat{\boldsymbol{\alpha}})$ using the aforementioned (and in Exercise 12.6) algorithm.
4. Then substitute $(\hat{\boldsymbol{\alpha}}, \hat{b})$ into the aforementioned expression for $\hat{f}$.

12.4.3.2 Margin, Support Vectors, and the Sparsity of Exact Solutions

Before we describe the dual QP algorithm, we review some important properties of (exact) SVM solutions. Recall that there is one coefficient α_i^* for each data sample $(\mathbf{x}_i, y_i)$. Simultaneous application of the Kuhn–Tucker condition in (12.24) and the box constraints in (12.28) yield the following necessary and sufficient conditions for an optimal solution (see Exercise 12.7):

$$\begin{aligned} y_i f^*(\mathbf{x}_i) &\le 1 \quad \text{when} \quad \alpha_i^* = \frac{y_i}{n} \\ y_i f^*(\mathbf{x}_i) &= 1 \quad \text{when} \quad 0 < y_i \alpha_i^* < \frac{1}{n} \\ y_i f^*(\mathbf{x}_i) &\ge 1 \quad \text{when} \quad \alpha_i^* = 0 \end{aligned} \tag{12.39}$$

The quantity $y_i f(\mathbf{x}_i)$ is often called the *margin*, and the aforementioned optimality conditions are characterized in terms of the so-called *margin cutoff* value of 1. These optimality conditions have the following implications for all kernels and are illustrated for a linear kernel in Figure 12.8. A sample $\mathbf{x}_i$ is assigned the correct label when the sign of $f(\mathbf{x}_i)$ agrees with the sign of y_i, that is, when its margin satisfies $y_i f(\mathbf{x}_i) > 0$. Similarly, $\mathbf{x}_i$ is assigned an incorrect label when its margin satisfies $y_i f(\mathbf{x}_i) < 0$. Thus, since all incorrectly classified samples satisfy $y_i f(\mathbf{x}_i) < 0 < 1$, the first condition in (12.39) implies that the coefficients for these samples satisfy $\alpha_i^* = y_i/n$. Furthermore, $\alpha_i^* = y_i/n$ also holds for correctly classified samples with margin less than 1, that is, $0 < y_i f^*(\mathbf{x}_i) < 1$. We say that these samples "fail to meet the margin cutoff." The only samples whose coefficients can take values between 0 and y_i/n (i.e., not at a bound) are samples whose margin falls exactly on the margin cutoff. Samples whose margin exceeds the margin cutoff have coefficient values of 0.

Samples with nonzero coefficients are called *support vectors*, and their corresponding coefficients α_i^* are called *support vector multipliers*. When the data are well separated by the SVM solution (i.e., most samples are classified correctly) then there may be a large number of samples whose margin exceeds the cutoff and, therefore, the SVM solution is very sparse, that is, it contains a small number of support vectors. Since $\mathbf{w}_{\text{map}}$ is an expansion over support vectors, the number of support vectors n_{SV} plays a dominant role in

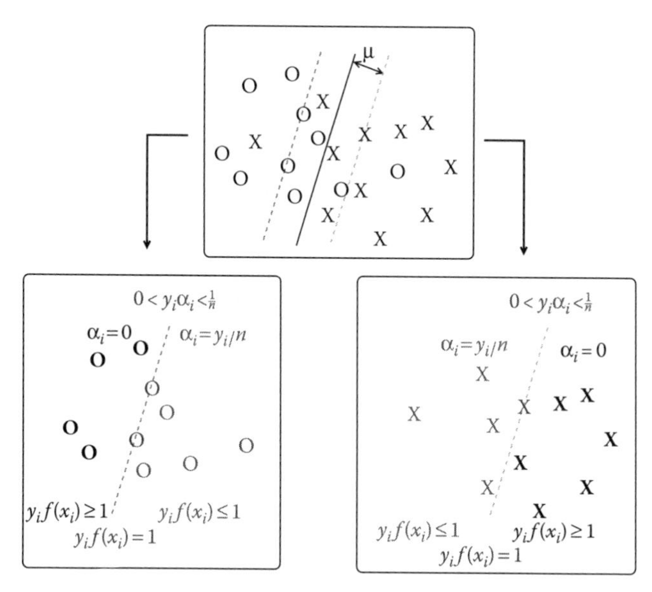

FIGURE 12.8
(See color insert.) Optimal solution characteristics illustrated for a linear kernel.

the implementation efficiency of the final solution. Thus, it is important to understand the conditions that lead to sparse solutions.

First, if λ is sufficiently large then nearly all the data points will be support vectors regardless of the data distribution or the kernel. However, these cases rarely correspond to the best classifiers (although they may be encountered in our search over λ). The following results are more typical of the sparse behavior for the best classifiers. They have been derived under mild restrictions on the distribution and the assumption that λ_n is chosen according to the guidelines for consistency described in Section 12.4.2.[44] Also, they were determined through an asymptotic analysis (i.e., a study of what happens as $n \to \infty$); but they provide a surprisingly accurate characterization of the finite sample results witnessed in practice. The main result, which holds for all kernels, states that

$$n_{SV} \geq 2e^* n$$

That is, the fraction of training samples that become support vectors is no less than twice the Bayes error. For universal kernels, when the Bayes error satisfies $e^* > 0$, this bound is actually achieved (asymptotically), that is, $n_{SV} = 2e^* n$. Thus, if the Bayes error is small, then the number of support vectors is small. On the other hand, if the Bayes error is large, then most of the training data will become support vectors. Finally, for universal kernels, if $e^* = 0$, then the number of support vectors tends to a *constant* for large n, that is, it becomes an infinitesimally small fraction of the number of training samples.

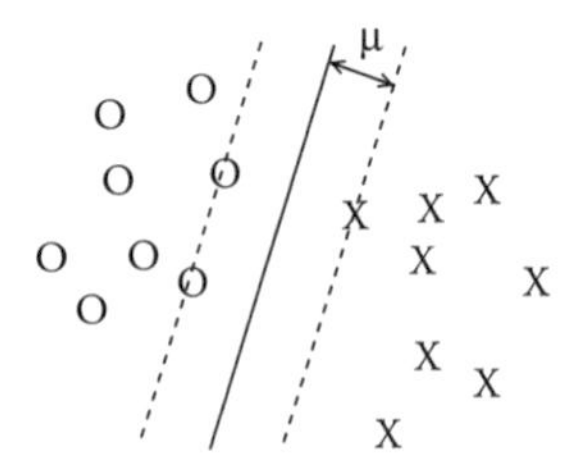

FIGURE 12.9
The margin μ of a linear support vector machine solution with linearly separable data.

In the special case when the training data are linearly separable and a linear kernel is used, we can obtain a nonasymptotic characterization of the number of support vectors. First consider modifying the primal QP problem (12.21) so that $\xi_i = 0, \forall i$, that is, all the slack variables are forced to take a value of zero. Because the data are linearly separable, the resulting QP problem is well-defined and has a unique solution $\mathbf{w}^*$. For this solution, the samples $(\mathbf{x}_i, y_i)$ that fall on the margin cutoff boundary, that is, $y_i f^*(\mathbf{x}_i) = 1$, are the only potential support vectors. An example is shown in Figure 12.9. It is easy to show that these support vectors lie at a distance $\mu = 1/|\mathbf{w}^*|$ from the decision boundary (this is true regardless of the kernel, see Exercise 12.8). This distance μ was Vapnik's original notion of margin, and he (and others) derived a number of properties based on it.[18,45,46] Although $\mathbf{w}^*$ is unique, the support vector expansion may not be, so the collection of optimal expansions may include different numbers of margin cutoff samples as support vectors. Let $\acute{n}_{SV}$ be the number of support vectors that appear in all possible expansions of $\mathbf{w}^*$. Then $\acute{n}_{SV}$ satisfies[18]

$$\acute{n}_{SV} \le d$$

In practice, solutions to this modified QP problem tend to produce a number of support vectors very close to $\acute{n}_{SV}$, which tends to be very close to the dimension of the space spanned by the data and is sometimes much smaller than d. Furthermore, solutions to the original QP problem (12.21), where we do not force the slack variable values to zero ahead of time, tend to exhibit this same behavior if λ is sufficiently small.

In the 1990s, there was great interest in using margin to characterize the performance of learning methods, inspired in part by Vapnik's introduction of SVMs. This work produced numerous results that supported the proposition that solution methods designed to optimize the margin could overcome the so-called "curse of dimensionality."[18,45,46] Roughly speaking, these results demonstrated that the estimation error can be controlled in a manner that is independent of dimension when the classifier has good margin properties. This served as the original motivation for SVMs, and this helps explain their robustness to dimension commonly witnessed in practice.

12.4.3.3 Decomposition Algorithms for the Dual Quadratic Programming Problem

In this section, we develop a solution algorithm for the dual QP. It will be convenient to express the box constraints $0 \le y_i \alpha_i \le \frac{1}{n}$ as follows:

$$l_i \le \alpha_i \le u_i, \quad i = 1, 2, \ldots, n \tag{12.40}$$

where

$$l_i = \begin{cases} 0, & y_i = 1 \\ -\frac{1}{n}, & y_i = -1 \end{cases}, \quad u_i = \begin{cases} \frac{1}{n}, & y_i = 1 \\ 0, & y_i = -1 \end{cases} \tag{12.41}$$

In principle, this problem can be solved using standard algorithms for convex QP problems[47] but these algorithms are often inadequate for the larger values of n encountered in practice. Here, we develop a simple and powerful decomposition algorithm that scales well with n. The basic idea is to solve the original size n problem by solving a sequence of smaller size q problems where $q \ll n$. In particular, each smaller problem optimizes a subset of $\boldsymbol{\alpha}$ components while keeping the remaining components fixed. Let $W = \{i_1, \ldots, i_q\}$ be the index set for a subset of $\boldsymbol{\alpha}$ components. The key is to select a so-called *working set* W^m at each iteration m so that the optimized components will guarantee sufficient progress toward the original problem solution. The smallest feasible working set size is $q = 2$ and algorithms that use this size at each iteration are traditionally called *sequential minimal optimization* (SMO) algorithms. The first SMO algorithm was introduced by Platt,[48] and numerous enhancements can be found in the literature.[49–54] The algorithm developed here is based largely on this body of work.

SMO algorithms have the advantage that there is a very simple closed-form solution for each two-variable QP problem. Indeed with $W^m = \{j, k\}$ and $g_j(\boldsymbol{\alpha}^m) > g_k(\boldsymbol{\alpha}^m)$, the updates for α_j and α_k take the form (see Exercise 12.11)

$$\alpha_j^{m+1} \leftarrow \alpha_j^m + \Delta^m, \quad \alpha_k^{m+1} \leftarrow \alpha_k^m - \Delta^m$$

where

$$\Delta^m \leftarrow \min\left(u_j - \alpha_j^m, \alpha_k^m - l_k, \frac{g_j(\boldsymbol{\alpha}^m) - g_k(\boldsymbol{\alpha}^m)}{Q_{jj} + Q_{kk} - 2Q_{jk}} \right)$$

To determine a suitable working set at each iteration, we consider the following optimality condition of the dual QP (see Exercise 12.10):

$$\boldsymbol{\alpha} \in \mathcal{A}^* \Leftrightarrow g_j(\boldsymbol{\alpha}) \le g_k(\boldsymbol{\alpha}) \quad \text{for all} \quad (j,k) \quad \text{where} \quad \alpha_j < u_j \quad \text{and} \quad \alpha_k > l_k \tag{12.42}$$

From this condition, we can deduce that the *nonoptimality* of $\boldsymbol{\alpha}$ is witnessed by index pairs (j, k) of the form

$$(j,k):\alpha_j < u_j, \alpha_k > l_k \quad \text{and} \quad g_j(\boldsymbol{\alpha}) > g_k(\boldsymbol{\alpha})$$

For nonoptimal $\boldsymbol{\alpha}$, the most extreme violator of the optimality conditions is a so-called *max-violating pair* given by

$$(j^*,k^*): j^* \in \arg\max_{i:\alpha_i<u_i} g_i(\boldsymbol{\alpha}), k^* \in \arg\min_{i:\alpha_i>l_i} g_i(\boldsymbol{\alpha}) \tag{12.43}$$

Choosing a max-violating pair for the working set at each iteration yields an SMO algorithm that is guaranteed to converge to a dual QP solution asymptotically.[55,56] Furthermore, stopping the algorithm at the first iteration $\acute{m}$ where

$$g_{j^*}(\boldsymbol{\alpha}^{\acute{m}}) - g_{k^*}(\boldsymbol{\alpha}^{\acute{m}}) \le \tau \tag{12.44}$$

is guaranteed to yield an approximate solution $\hat{\boldsymbol{\alpha}}$ whose dual accuracy satisfies $\epsilon_d = R^* - R(\hat{\boldsymbol{\alpha}}) \le \tau$.[57] In addition, if this approximate dual solution is mapped to an approximate primal solution $(\hat{\mathbf{w}}, \hat{b}, \hat{\xi})$ using the aforementioned functions $\mathbf{w}_{\text{map}}$, b_{map}, and ξ_{map} then this stopping condition also guarantees that primal accuracy satisfies $\epsilon_p = S(\hat{\mathbf{w}}, \hat{b}, \hat{\xi}) - S^* \le \tau$.[57] Thus, the user can guarantee an accuracy ε_p of the approximate primal solution through the choice of $\tau = \varepsilon_p$.

Implementation of this SMO algorithm requires the gradient vector at each iteration. A brute-force computation using (12.34) requires $O(n^2)$ operations; but it is much more efficient to *update* the gradient at each iteration using its previous value. Specifically, for a working set $W^m = \{j, k\}$, the sparsity of $(\boldsymbol{\alpha}^{m+1} - \boldsymbol{\alpha}^m)$ leads to the update

$$g(\boldsymbol{\alpha}^{m+1}) = g(\boldsymbol{\alpha}^m) - Q(\boldsymbol{\alpha}^{m+1} - \boldsymbol{\alpha}^m) = g(\boldsymbol{\alpha}^m) - \Delta^m(\boldsymbol{q}_j - \boldsymbol{q}_k) \tag{12.45}$$

where $\boldsymbol{q}_j$ and $\boldsymbol{q}_k$ are the jth and kth columns of Q.

A complete SMO algorithm that uses the max-violating pair to form the working sets and implements the stopping condition in (12.44) is shown in Figure 12.10. After computing an initial gradient vector, this algorithm iterates the process of computing a max-violating pair working set, solving the corresponding two-variable QP problem, updating the gradient vector, and testing the stopping condition. The computation performed in the main loop is dominated by steps 5 and 8. Both of these steps perform a simple computation for each of the n components of the gradient vector, and so the overall run time of the main loop is $O(n)$.

Procedure 2:

1. $\boldsymbol{\alpha}^0 \leftarrow 0$
2. $g^0 \leftarrow y$
3. $m \leftarrow 0$
4. **Repeat**
5. (j,k): $j \in \arg\max_{i:\alpha_i < u_i} g_i^m$, $k \in \arg\min_{i:\alpha_i > l_i} g_i^m$
6. $\Delta^m \leftarrow \min\left(u_j - \alpha_j^m, \alpha_k^m - l_k, \dfrac{g_j^m - g_k^m}{Q_{jj} + Q_{kk} - 2Q_{jk}}\right)$
7. $\alpha_j^{m+1} \leftarrow \alpha_j^m + \Delta^m$, $\alpha_k^{m+1} \leftarrow \alpha_k^m - \Delta^m$
8. $\mathbf{g}^{m+1} \leftarrow \mathbf{g}^m - \Delta^m (\boldsymbol{q}_j - \boldsymbol{q}_k)$
9. $m \leftarrow m + 1$
10. **Until** $(g_j^m - g_k^m \leq \tau)$

FIGURE 12.10
SMO algorithm based on the *max-violating pair.*

12.4.3.4 Rate Certifying Decomposition Algorithms

Although the SMO algorithm in Figure 12.10 is guaranteed to converge asymptotically, its convergence rate is unknown. With a little extra work, however, we can construct a decomposition algorithm that is guaranteed to converge at a fast rate and is often significantly faster in practice. This is accomplished by making the following two changes to the algorithm in Figure 12.10:

1. Instead of choosing a max-violating pair for the working set, we examine a small set of candidate pairs Ω, where $|\Omega| \leq n$, and choose a member of this set that yields the best stepwise improvement.[51,54] The set Ω can be formed in a way that guarantees the chosen pair is a *rate certifying pair,*[54,58] that is, a pair whose stepwise improvement is sufficient to establish a good convergence rate.
2. Replace the stopping rule $g_j^m - g_k^m \leq \tau$ with the new stopping rule $\varepsilon_{\text{bound}} \leq \tau$ where $\varepsilon_{\text{bound}}$ is a tighter bound on solution accuracy. This change is motivated by the fact that there are several candidates for $\varepsilon_{\text{bound}}$ that are easy to compute and typically provide a much tighter accuracy bound than $g_j^m - g_k^m$.[57] In practice, this new stopping rule usually stops the algorithm in far fewer iterations.

These two changes can be made without increasing the $O(n)$ run time of the main loop. The details are omitted here but can be found in other works.[54,57] The experiments in one work[54] suggest that rate certifying algorithms can reduce the number of iterations by approximately one order of magnitude and improve the overall run time by a factor of approximately two. Furthermore, the run time for these algorithms is *guaranteed* to satisfy the bound provided here, regardless of the input data, and therefore allows us to *guarantee* that the computation error is controlled with a computationally efficient algorithm. Specifically, the number of iterations $\acute{m}$ required for the rate certifying algorithm to achieve a dual accuracy ε_d is guaranteed to be no more than[54,59,60]

$$\acute{m} \leq \left\lceil 2n\left(\frac{2k_{\max}}{\lambda\varepsilon_d n} - 1 + \ln\left(\frac{\lambda n}{2k_{\max}}\right)\right)\right\rceil \tag{12.46}$$

where $k_{\max} = \max_i k(\mathbf{x}_i, x_i)$ and $\lceil . \rceil$ rounds up to the nearest integer. Note that this bound can be computed ahead of time and can therefore be used to stop the algorithm if the criterion $\epsilon^{m}_{\text{bound}} \leq \tau$ does not stop it sooner. When this result is combined with the $O(n)$ run time of the main loop and the $O(dn^2)$ time typically required to compute the kernel matrix Q ahead of time, the overall run time bound becomes

$$O\left(dn^2 + \frac{k_{\max}n}{\lambda\varepsilon_d} + n^2 \ln\left(\frac{\lambda n}{k_{\max}}\right)\right)$$

12.5 Multi-Class Classification

Pattern recognition problems with more than two class labels are called multiclass problems. For problems with M classes, the goal is to design an M-valued function g that assigns a label $y \in \{1, 2, \ldots, M\}$ to every point $\mathbf{x} \in \mathbb{R}^d$. There are numerous ways to design such a function, and most of them work by combining the outputs of multiple two-class classifiers.[2,5–11] One of the most common methods starts by designing M two-class classifiers $f_1, \ldots, f_M$, where each f_i is trained to discriminate between class i and all other $M - 1$ classes. Then the M-class classifier works by computing the function values $(f_1(\mathbf{x}), \ldots, f_M(\mathbf{x}))$ and assigning the label that corresponds to the largest function value, that is,

$$g(\mathbf{x}) = \arg \max_{i \in \{1,\ldots,M\}} f_i(\mathbf{x})$$

There are numerous alternatives for designing a multiclass SVM classifier, but this approach is perhaps the simplest and has been shown to work well in practice.[61]

12.6 MATLAB® Examples

In this section, we show how to use MATLAB to generate data samples; design and implement classifiers using the GML, LS-LINEAR, and SVM learning methods; and estimate the classifier error rates. To help with the examples in this section, we have created several MATLAB functions whose names begin with `DSPML_`. These functions are in the *functions* folder on the publisher's website for this book. Note that the minus symbol "–" cannot be used in a MATLAB variable name; so we use 0 instead of –1 to indicate class –1 variable names. For example, we use `MEAN0` for the class –1 mean vector.

We start with the data generation process. In general, samples of a random variable (X, Y) can be generated using the following two-step procedure: (1) Determine the class label y by drawing a sample from a Bernoulli distribution with success probability P_1, and (2) then determine the pattern vector $\mathbf{x}$ by drawing a sample from the class y probability density $p_{X|y}$. The MATLAB function

```
function [XDATA, YDATA] = DSPML_Generate2ClassGaussian(MEAN1,A1,
                              P1,MEAN0,A0,NSAMPLES)
```

uses this approach to generate labeled data samples for a two-class Gaussian problem, placing the $\mathbf{x}$ vectors in the rows of the `XDATA` array and the y labels in the rows of the `YDATA` array. The number of samples is specified by `NSAMPLES`, and the distribution is specified by the class probabilities (`P1, 1-P1`), the class means (`MEAN1, MEAN0`), and the transform matrices (`A1, A0`), which determine the class covariances by (Σ_1, Σ_{-1}) = (`A1'*A1, A0'*A0`).

Now we describe the learning methods: The GML, LS-LINEAR, and SVM learning methods are all implemented by MATLAB functions that accept the inputs (`XTRAIN, YTRAIN, LAMBDA`), where `XTRAIN` is an $n \times d$ array whose rows contain the $\mathbf{x}_i$ data vectors, `YTRAIN` is an $n \times 1$ array whose rows contain the $y_i \in \{-1,1\}$ labels, and `LAMBDA` is a scalar regularization parameter. The function

```
function [Q,W,B] = DSPML_GML_QUADRATIC(XTRAIN, YTRAIN, LAMBDA)
```

uses the GML learning method to compute the parameters (Q, w, b) of a quadratic classifier. It provides a straightforward implementation of the statistical estimates in (12.13) followed by a plug-in to (12.9), except that it allows the estimates $\hat{\Sigma}_y$ to be replaced with regularized estimates $\hat{\Sigma}_y + \lambda I$ so that the result is robust to cases where these matrices may be ill-conditioned or have reduced rank. Similarly, the MATLAB functions

```
function [W,B] = DSPML_GML_LINEAR(XTRAIN,YTRAIN,LAMBDA)
```

and

```
function [W,B] = DSPML_LS_LINEAR(XTRAIN,YTRAIN,LAMBDA)
```

use the GML and LS-LINEAR learning methods to compute the parameters $(\mathbf{w}, b)$ for a linear classifier.

The design and implementation of SVM classifiers is accomplished using the *libsvm* software package, which employs an SMO training algorithm[51] very much like the one described in Section 12.4.3.3. The instructions for downloading and installing *libsvm* for MATLAB can be found at the website listed in reference number 49. The *libsvm* software implements several types of SVMs, but defaults to the so-called `C-SVM`, which is equivalent to the SVM developed in this chapter. More specifically, the `C-SVM` algorithm is designed to optimize Vapnik's original criterion

$$\hat{s}_\lambda(\mathbf{w},b) = \frac{1}{2}\|\mathbf{w}\|^2 + C\sum_{i=1}^{n}[1 - y_i(\mathbf{w}^T\phi(\mathbf{x}_i) + b)]_+$$

which has a slightly different parameterization than the one developed here in (12.20). Except for scaling, however, these two criteria are identical, and with $C = 1/2\lambda n$ they have the same solutions. In *libsvm*, an SVM classifier is designed with the MATLAB function

```
function [SVM_MODEL] = svmtrain(YTRAIN', XTRAIN, OPTIONS),
```

where the `OPTIONS` argument is a text field used to specify the design parameters. This function implements an SMO algorithm that employs the max-violating pair stopping criterion shown in Figure 12.10 with a default tolerance value of $\tau = 0.001$. This default can be changed using the `-e` option. The kernel is selected with the `-t` option, and the value of C is specified with the `-c` option. Kernel parameters are specified with the `-g` option for the RBF parameter γ, and the (`-d`,`-g`,`-r`) options for the polynomial kernel parameters (q, c_1, c_0). The `svmtrain` function returns an approximate solution $(\hat{\boldsymbol{\alpha}}, \hat{b})$. More specifically, it returns $\hat{b}$, the nonzero coefficients of $\hat{\boldsymbol{\alpha}}$, and the corresponding support vectors $\{(x_i, y_i) : \hat{\alpha}_i \neq 0\}$. We have developed the MATLAB functions,

```
function [W,B] = DSPML_SVM_LINEAR(XTRAIN,YTRAIN,LAMBDA),
function [SVM_MODEL] = DSPML_SVM_POLY(XTRAIN,YTRAIN,ORDER,
                       LAMBDA,C1,C0),
```

and

```
function [SVM_MODEL] = DSPML_SVM_RBF(XTRAIN,YTRAIN,LAMBDA,
                       GAMMA),
```

which use the `svmtrain` function to implement the SVM learning method with regularization parameter λ = `LAMBDA` for the linear, polynomial, and RBF kernels, respectively. The additional polynomial and RBF kernel parameters are specified by (c_1, c_0, q) = (`C1,C0,ORDER`) and γ = `GAMMA`.

Once a linear, quadratic, or SVM classifier has been designed, it can be implemented using the MATLAB functions

```
function [YHAT,FX] = DSPML_LCLASS(W,B,XDATA),
function [YHAT,FX] = DSPML_QCLASS(Q,W,B,XDATA),
```

and

```
function [YHAT, ACCURACY, FX] = svmpredict(YDATA', XDATA,
                                SVM_MODEL),
```

where the YHAT array contains the predicted labels and the FX array contains the classifier function values $\hat{f}(x_i)$. The svmpredict function provided by *libsvm* accepts the input label array YDATA and returns the classification error estimate ACCURACY $= \hat{e}(\hat{f})$. In a true prediction scenario, where the labels for XDATA are not known, the svmpredict function can be called with dummy labels, for example,

```
[YHAT, ACCURACY, FX] = svmpredict(ones(size(XDATA,1),1),
                       XDATA, SVM_MODEL).
```

Obviously, the returned ACCURACY will have no bearing on the true accuracy in this case.

Finally, the MATLAB function

```
function [SVM_MODEL] = DSPML_SVM_RBF_ROBUST(XTRAIN,YTRAIN)
```

designs an RBF kernel classifier using the grid search algorithm in Figure 12.5, and the functions

```
function [Q,W,B] = DSPML_GML_QUADRATIC_ROBUST(XTRAIN,YTRAIN),
function [W,B] = DSPML_GML_LINEAR_ROBUST(XTRAIN,YTRAIN),
function [W,B] = DSPML_LS_LINEAR_ROBUST(XTRAIN,YTRAIN),
function [W,B] = DSPML_SVM_LINEAR_ROBUST(XTRAIN,YTRAIN), and
function [SVM_MODEL] = DSPML_SVM_POLY_ROBUST(XTRAIN,YTRAIN,
                       ORDER,C1,C0)
```

design classifiers using a similar grid search over the regularization parameter λ for each of the corresponding learning methods.

EXAMPLE 12.2

The MATLAB code here integrates the routines mentioned in Section 12.6 into a complete example that shows how to generate two-class Gaussian data, visualize the data with a scatterplot, design a quadratic classifier using the GML method, overlay the corresponding decision boundary on the scatterplot, and estimate the classifier error. This example also uses the formula $1.64\sqrt{\hat{e}(1-\hat{e})/n}$ from Section 12.3.1 to compute the accuracy of the error estimates:

```
%
% specify class probabilities, means, and coupling matrices
%
P1=.3; P0=1-P1; MEAN1=[0,0]; MEAN0=[2,1]; A1=[1,.5;-.5,2];
A0=[1,0;0,1];
%
% generate 50 training data samples and compute a quadratic
classifier using GML
%
NTRAIN=50;
[XTRAIN,YTRAIN] = Generate2ClassGaussian(MEAN1,A1,P1,MEAN0,A0,
NTRAIN);
[Q,W,B] = GML_QUADRATIC_ROBUST(XTRAIN,YTRAIN);
%
% show scatter plot and decision boundary
%
hold off
X1TRAIN=XTRAIN(find(YTRAIN==1),:);
X0TRAIN=XTRAIN(find(YTRAIN!=1),:);
plot(X1TRAIN(:,1),X1TRAIN(:,2),'bo',X0TRAIN(:,1),X0TRAIN(:,2),
'rx');
hold on
[x1,x2]=meshgrid(-6:.1:6,-8:.1:8);
FX=x1.*x1.*Q(1,1)+2*x1.*x2.*Q(1,2)+x2.*x2.*Q(2,2)+x1.*W(1)+x2.
*W(2)+B;
contour(x1,x2,FX,[0 0])
%
% generate 10000 test samples and use them to estimate
classification error
%
NTEST=10000;
[XTEST,YTEST] = Generate2ClassGaussian(MEAN1,A1,P1,MEAN0,A0,
NTEST);
[YHAT_TEST,FX_TEST] = QCLASS(Q,W,B,XTEST);
TEST_ERROR=size(find(YTEST!=YHAT_TEST),2)/NTEST;
ACCURACY=1.64*sqrt(TEST_ERROR*(1-TEST_ERROR)/NTEST);
printf("Generalization Error: %f (+- %f)\n",
TEST_ERROR,ACCURACY);
```

The results for one particular run of this code are shown in Figure 12.11.

EXAMPLE 12.3

This example applies the SVM learning strategy with RBF kernel to the two-class Gaussian problem described in Example 12.2. First, we use the `DSPML_SVM_RBF` function to design SVM-RBF classifiers with 3 different parameter settings, $(\lambda, \gamma) = (0.001,0.1)$, $(0.001,1)$, and $(0.001,10)$, on a training set with 50 samples.

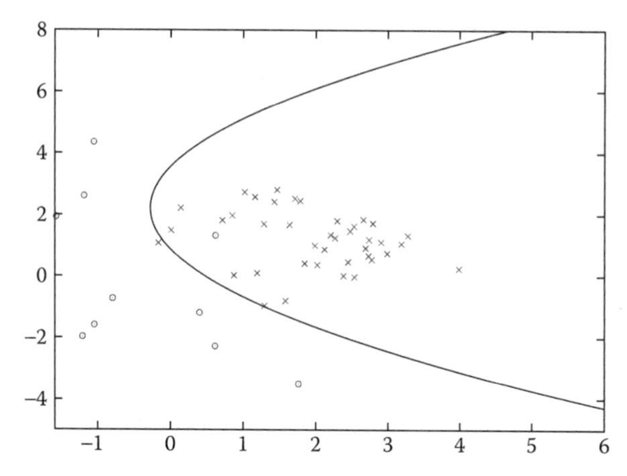

FIGURE 12.11
(See color insert.) Results corresponding to the design of a quadratic classifier using the Gaussian maximum likelihood method: The error estimate for this result is 0.114 with an accuracy of approximately 0.005.

The corresponding decision boundaries are shown in Figure 12.12a through 12.12c. It is clear that the complexity varies from simple to complex as γ is varied from 0.1 to 10. It is also apparent that the solution in Figure 12.12c has overfit the data. However, if we did not know ahead of time that the data were Gaussian then it would be far less obvious that the solutions in Figures 12.12a and b are also overfitting the data. On the other hand, if we employ the grid search algorithm in `DSPML_SVM_RBF_ROBUST` to determine the values of λ and γ, we obtain the result in Figure 12.12d, which has similar complexity to the (simpler) Bayes optimal solution. These results suggest that the SVM-RBF with grid search can obtain results that are comparable to the GML-QUADRATIC method without directly exploiting prior knowledge that the data are Gaussian. Since the SVM-RBF method can easily overfit the data, this illustrates the importance and power of the grid search procedure.

EXAMPLE 12.4

In this example, we provide a quantitative comparison of the six leaning methods: (1) LS-LINEAR, (2) GML-LINEAR, (3) SVM with linear kernel (SVM-LINEAR), (4) GML-QUAD, (5) SVM with polynomial kernel and $q = 2$, $c_0 = c_1 = 1$ (SVM-QUAD), and (6) SVM with RBF kernel (SVM-RBF). Training sets of size 50 and test sets of size 10,000 are generated according to the two-class Gaussian distribution described in Example 12.2. All six design methods use a grid search to robustly determine the regularization and/or kernel parameters. The decision boundaries and generalization errors obtained for these six methods are shown in Figure 12.13. As expected, the GML-QUAD method achieves the best performance since it most correctly exploits the prior knowledge that the data are Gaussian. The SVM-QUAD method achieves the second best performance because it also employs a learning strategy and function class that is well matched to the optimal solution form. The SVM-RBF method is capable of providing good approximations to the

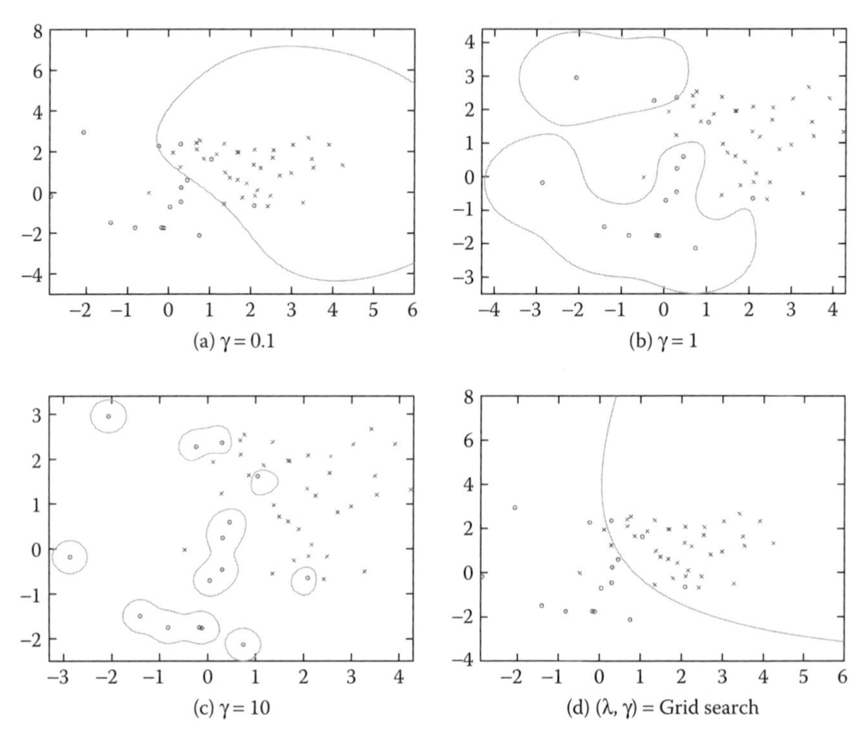

FIGURE 12.12
(See color insert.) Decision boundaries for support vector machine–radial basis function classifiers designed with four different values of (λ, γ): (a–c): $\lambda = 0.001$ and γ is given below the figure; (d): both λ and γ are chosen using the grid search procedure.

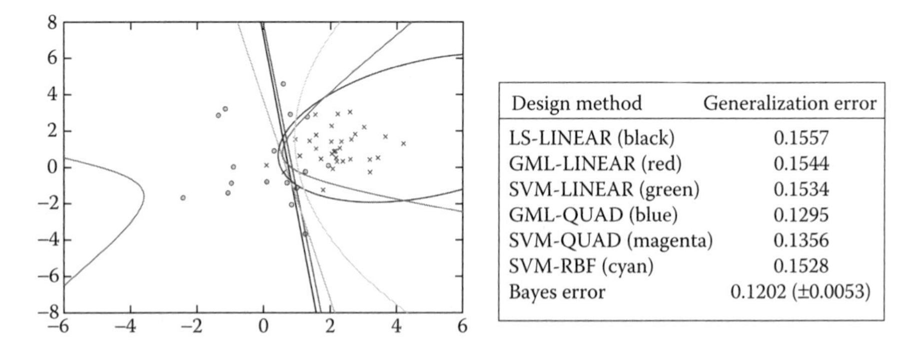

Design method	Generalization error
LS-LINEAR (black)	0.1557
GML-LINEAR (red)	0.1544
SVM-LINEAR (green)	0.1534
GML-QUAD (blue)	0.1295
SVM-QUAD (magenta)	0.1356
SVM-RBF (cyan)	0.1528
Bayes error	0.1202 (±0.0053)

FIGURE 12.13
(See color insert.) Results of the six classifier design methods for a two-class Gaussian problem.

optimal solution, but requires a larger training set to more correctly identify the optimal form. The results for the three linear classifier design methods are nearly identical, and although they are not capable of producing optimal solutions to this problem their performance is quite good considering the significant reduction in complexity that they provide.

EXAMPLE 12.5

This example repeats the comparison in Example 12.4, but uses data distributions that no longer follow the simple two-class Gaussian model. The new distributions are obtained by modifying the previous class −1 distribution and keeping the previous class 1 distribution the same. In particular, we add a second mode to the class −1 distribution so that it becomes a mixture of two Gaussians with means (2, 1) and (a, b) (the covariance matrix for both modes is I). We consider two different realizations by choosing two different values for (a, b). The first realization uses (a, b) = (10, 5), which places the new mode in a seemingly benign location far away from the overlap region as illustrated by the scatterplot in Figure 12.14. Since the new mode is far from the overlap region, and well within the region that would have been assigned the correct class label in Example 12.4, a successful design method would largely ignore the samples from the new mode and use the remaining samples to determine the decision boundary. The results in Figure 12.14 suggest that the LS-LINEAR and GML methods are not able to do this. Indeed, their decision boundaries are clearly biased toward the new mode and their classification performance suffers as a result. On the other hand, the SVM methods are much more robust to this new mode and therefore produce solutions with much smaller error. The second realization chooses (a, b) = (0, −6), which places the new mode in an adjacent region illustrated by the scatterplot in Figure 12.15. In this case, a successful design method will use the data from these two modes to carefully position the decision boundary in the two different overlap regions. The linear classifier design methods are unable to do this because their function class is too limited, and therefore their performance is poor. Quadratic classifiers are much better suited to this problem, although they are not optimal. The SVM quadratic classifier is much more successful than the GML quadratic classifier because the SVM learning strategy is better suited to this non-Gaussian problem. Finally, the SVM-RBF classifier is superior to all others in this example because its function class is capable of providing good approximations to the optimal classifier and its learning strategy is robust to distribution.

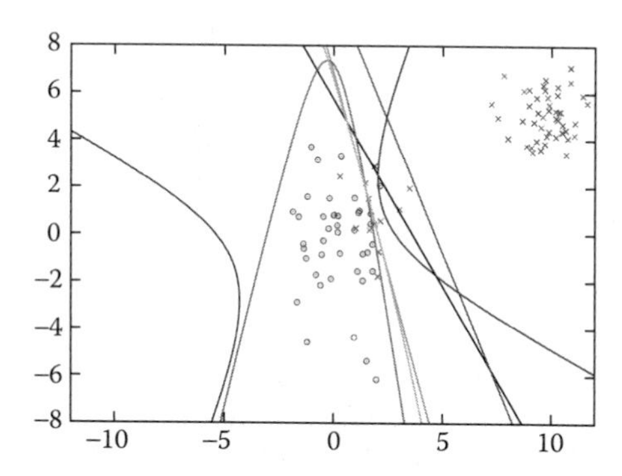

Design method	Generalization error
LS-LINEAR (black)	0.1561
GML-LINEAR (red)	0.1832
SVM-LINEAR (green)	0.0893
GML-QUAD (blue)	0.1262
SVM-QUAD (magenta)	0.0909
SVM-RBF (cyan)	0.0900

FIGURE 12.14
(See color insert.) Comparison of classifier design methods where blue samples are drawn from a Gaussian distribution and red samples are drawn from a mixture of two Gaussians, one that overlaps with the blue and one that is far away from the overlap region.

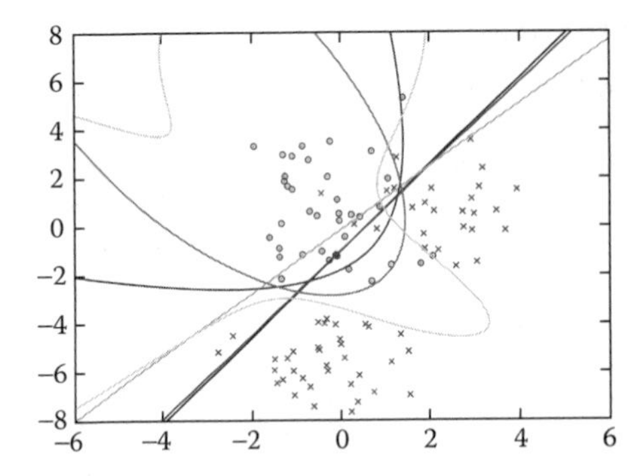

Design method	Generalization error
LS-LINEAR (black)	0.1897
GML-LINEAR (red)	0.1916
SVM-LINEAR (green)	0.2199
GML-QUAD (blue)	0.1641
SVM-QUAD (magenta)	0.1329
SVM-RBF (cyan)	0.0913

FIGURE 12.15
(See color insert.) Comparison of classifier design methods where blue samples are drawn from a Gaussian distribution, and red samples are drawn from a mixture of two Gaussians that overlap with different regions of the blue class.

EXAMPLE 12.6

This example compares the six classifier design methods listed in Example 12.4 on a network intrusion detection problem. Classifiers are designed using the KDD Cup 1999 data from the *UCI Machine Learning Repository*.[62] In particular, we use the data file named `kddcup.data_10_percent`. This file contains data vectors with 41 features, 3 of which are categorical variables represented by nonnumeric text fields. Since the GML and LS-LINEAR methods cannot directly accommodate nonnumerical data, and since we have not discussed kernels for nonnumerical data, we omit these three variables in our experiments here. This leaves a total of 38 numerical features that we extracted and placed in a file suitable for input to MATLAB. The label values in `kddcup.data_10_percent` are text fields that contain the word `normal` for data vectors associated with nonintrusive behavior and some other word for data vectors associated with intrusive behavior. In the experiment here, we convert the `normal` labels to $y = 1$ and all other labels to $y = -1$. This data set contains a total of 494,021 labeled data samples that we split randomly into training and test sets of size 13,400 and 480,621, respectively.

It is often useful to apply a normalization procedure to the data prior to training and testing, especially when there is a large variation in the dynamic ranges of the individual feature values. In this experiment, we transform the original data vectors $\mathbf{x}$ to normalized data vectors $\acute{\mathbf{x}}$ by shifting and scaling the individual components so that the new components have zero mean and unit variance. More specifically, we transform the data using

$$\acute{\mathbf{x}}_i = D(\mathbf{x}_i - \boldsymbol{\mu})$$

where $\boldsymbol{\mu}$ is the sample mean for the $\mathbf{x}$ training data and D is a diagonal matrix whose components take the form $D_{jj} = 1/s_j$ where s_j is the sample standard deviation for the jth component of the $\mathbf{x}$ training data. If the jth standard deviation is zero, we set $s_j = 1$. Note that the mean and standard deviations are computed over the aggregate training data, not the individual class data. Note also that only the training data are used to compute these estimates, but the normalization is applied to both the training and test data, and to all future data that are presented to the designed classifier.

TABLE 12.1

Generalization Error Estimates for the Network Intrusion Detection Problem

Design Method	Generalization Error
LS-LINEAR	0.0074
GML-LINEAR	0.0067
SVM-LINEAR	0.0039
GML-QUAD	0.0109
SVM-QUAD	0.0032
SVM-RBF	0.0026

Once the data has been normalized, the six methods listed in Example 12.4 are used to design classifiers. Each method uses a grid search to robustly determine the regularization and/or kernel parameters, and for this purpose the training data are split randomly into approximately $n_{\text{core}} = 10{,}000$ core training samples and $n_{\text{aux}} = 3{,}400$ auxiliary samples. Once the classifiers have been designed, they are evaluated against the 480,621 test samples and the results are shown in Table 12.1. These error estimates have an accuracy ≤ 0.00025 with probability .95 under the iid assumption. Because of the complex nature of this data set, the SVM classifiers significantly outperform the LS-LINEAR and GML classifiers and the SVM-RBF classifier provides the best performance overall.

The reader is encouraged to experiment with these design methods by applying them to the numerous benchmark data sets that can be found at the *UCI Machine Learning Repository*.[62] Some of these data sets have already been normalized and converted to the *libsvm* format and are available at the *libsvm* website.[49]

Exercises

12.1 Consider a two-class pattern recognition problem in which the class densities are Gaussian and the parameters of $p_{X,Y}$ are

$$\phi = (P_{-1}, \boldsymbol{\mu}_{-1}, \Sigma_{-1}, P_1, \boldsymbol{\mu}_1, \Sigma_1).$$

For iid data $D_n = ((\mathbf{x}_1, y_1), \ldots, (\mathbf{x}_n, y_n))$, the likelihood function is

$$\begin{aligned} L(\phi) = \textstyle\prod_{i=1}^{n} p_{X,Y}(\mathbf{x}_i, y_i \mid \phi) &= \textstyle\prod_{i=1}^{n} p_{X|Y}(\mathbf{x}_i \mid y_i, \phi) p_Y(y_i \mid \phi) \\ &= \textstyle\prod_{i=1}^{n} P_{y_i} p_{X|Y}(\mathbf{x}_i \mid y_i, \phi) \\ &= \left(\textstyle\prod_{i:y_i=1} P_1 p_{X|1}(\mathbf{x}_i \mid \boldsymbol{\mu}_1, \Sigma_1)\right)\left(\textstyle\prod_{i:y_i=-1} P_{-1} p_{X|-1}(\mathbf{x}_i \mid \boldsymbol{\mu}_{-1}, \Sigma_{-1})\right) \end{aligned}$$

Show that the log-likelihood is

$$\log[L(\phi)] = \sum_{i:y_i=1} \log[G_1(\mathbf{x}_i)] + \log(P_1) + \sum_{i:y_i=-1} \log[G_{-1}(\mathbf{x}_i)] + \log(P_{-1})$$

Show that choosing $(P_{-1}, \boldsymbol{\mu}_{-1}, \Sigma_{-1}, P_1, \boldsymbol{\mu}_1, \Sigma_1)$ to maximize $\log(L)$ subject to the constraint $P_1 + P_{-1} = 1$ gives the sample averages in (12.13).

(*Hint:* It may be helpful to employ maxtrix/vector identities from matrixcookbook.com or an equivalent textbook on the calculus of linear algebra.)

12.2 Show that choosing f to minimize $E_{X,Y}[(y - f(\mathbf{x}))^2]$ is equivalent to choosing f to minimize the mean-squared error $E_X[[f_0^*(\mathbf{x}) - f(\mathbf{x})]^2]$.

Hint: First show that

$$E_{X,Y}[[y - f(\mathbf{x})]^2] = \int [y - f(\mathbf{x})]^2 p_{X,Y}(\mathbf{x}, y)\, d\mathbf{x}\, dy$$
$$= 1 - 2\int f(\mathbf{x})[P_1 p_{X|-1}(\mathbf{x}) - P_{-1} p_{X|-1}(0)]\, d\mathbf{x} + \int f^2(\mathbf{x}) p_X(\mathbf{x})\, d\mathbf{x}.$$

Then use

$$f_0^*(\mathbf{x}) = P_{1|X}(\mathbf{x}) - P_{-1|X}(\mathbf{x}) = \frac{P_1 p_{X|1}(\mathbf{x}) - P_{-1} p_{X|-1}(\mathbf{x})}{p_X(\mathbf{x})}$$

to show that

$$E_{X,Y}[[y - f(\mathbf{x})]^2] = \int [f(\mathbf{x}) - f_0^*(\mathbf{x})]^2 p_X(\mathbf{x})\, d\mathbf{x} + \left[1 - \int f_0^{*2}(\mathbf{x}) p_X(\mathbf{x})\, d\mathbf{x}\right].$$

Then draw the conclusion by observing that the second term is independent of f.

12.3 Consider a supervised classification problem with $D_n = ((\mathbf{x}_1, y_1), \ldots, (\mathbf{x}_n, y_n))$. Show that if the data statistics satisfy $\hat{\Sigma}_1 = \hat{\Sigma}_{-1}$ and $\hat{P}_1 = \hat{P}_{-1} = 0.5$, then the LS solution for a linear classifier is identical to the GML solution. Furthermore, show that if $\hat{\Sigma}_1 = \hat{\Sigma}_{-1}$ but $\hat{P}_1 \neq \hat{P}_{-1}$ then the LS solution is a scaled version of the GML solution, except for the value of offset weight b.

Hint: Use the fact that regardless of the value of w there is always a scalar α such that

$$(\hat{\mu}_1 - \hat{\mu}_{-1})(\hat{\mu}_1 - \hat{\mu}_{-1})^T \mathrm{w} = \alpha(\hat{\mu}_1 - \hat{\mu}_{-1})$$

to simplify the derivation.

12.4 Consider a two-class pattern recognition problem in which the class conditional densities are Gaussian. The *Mahalanobis distance* from a point $\mathbf{x}$ to the class y mean $\boldsymbol{\mu}_y$ is defined as follows:

$$M_y(\mathbf{x}) := \sqrt{(\mathbf{x} - \boldsymbol{\mu}_y)^T \Sigma_y^{-1} (\mathbf{x} - \boldsymbol{\mu}_y)}$$

where Σ_y is the class y covariance matrix.

a. Show that the Bayes optimal classifier can be expressed as

$$f^*(\mathbf{x}) = M_{-1}^2(\mathbf{x}) - M_1^2(\mathbf{x}) + t$$

for some threshold t. Derive an expression for t.

b. Under what conditions does f^* assign the class label corresponding to the smallest Mahalanobis distance (i.e., under what conditions is $t = 0$)?

c. Under what conditions does f^* assign the class label corresponding to the smallest Euclidean distance (i.e., the class y for which $||\mathbf{x} - \boldsymbol{\mu}_y||$ is smallest)?

d. Under what conditions does f^* take the form of a *template-matching* classifier

$$f^*(\mathbf{x}) = \|\mathbf{x} - \boldsymbol{\mu}_1\| + t$$

for some threshold t?

e. Under what conditions does f^* take the form of a *matched filter* classifier

$$f^*(\mathbf{x}) = \mathbf{x}^T \boldsymbol{\mu}_1 + t$$

for some threshold t?

12.5 Consider the empirical SVM criterion $\hat{s}_\lambda$ in (12.20) and the primal QP problem in (12.21).

a. Show that for every solution $(\mathbf{w}^*, b^*, \boldsymbol{\xi}^*)$ of the primal QP problem, the pair $(\mathbf{w}^*, b^*)$ is a minimizer of $\hat{s}_\lambda$.

b. Show that for every minimizer $(\mathbf{w}^*, b^*)$ of $\hat{s}_\lambda$ there is a feasible $\boldsymbol{\xi}^*$ such that $(\mathbf{w}^*, b^*, \boldsymbol{\xi}^*)$ is a solution of the primal QP.

12.6 Consider the one-variable optimization problem $\min_b \rho(b)$ where ρ is defined in (12.37). Derive an $O(n \log n)$ algorithm for this problem.

Hint: Consider first sorting the values $g_i(\hat{\boldsymbol{\alpha}})$ and then visiting them in order, using constant time operations to update the criterion value at each step.

12.7 Apply the Kuhn–Tucker condition in (12.24) and the box constraints in (12.28) to obtain the optimality conditions in (12.39).

Hint: Apply the Kuhn–Tucker condition in (12.24) to the SVM problem to obtain

$$y_i\alpha_i^*[y_i[(\mathbf{w}^*)^T \boldsymbol{\phi}(\mathbf{x}_i) + b^*] + \xi_i^* - 1] = 0, \quad i = 1, 2, \ldots, n$$

$$\beta_i^*\xi_i^* = \left(\frac{1}{n} - y_i\alpha_i^*\right)\xi_i^* = 0, \quad i = 1, 2, \ldots, n$$

Then consider the implications of these results for the three cases: (1) $0 < y_i\alpha_i^* < 1/n$, (2) $y_i\alpha_i^* = 1/n$, and (3) $y_i\alpha_i^* = 0$.

12.8 For a function f^* that minimizes $\hat{s}_\lambda$ show that samples with $y_i f^*(\mathbf{x}_i) = 1$ lie at a distance $\mu = 1/|\mathbf{w}^*|$ from the decision boundary.

12.9 Prove that

$$\epsilon_{\text{gap}} = S(\hat{\mathbf{w}}, \hat{b}, \hat{\xi}) - R(\hat{\boldsymbol{\alpha}}) = \min_b \rho(b) - \hat{\boldsymbol{\alpha}}^T g(\hat{\boldsymbol{\alpha}})$$

where ρ is the criterion function defined in (12.37).

Hint: First show that

$$S(\hat{\mathbf{w}}, \hat{b}, \hat{\xi}) = \frac{1}{2}\hat{\boldsymbol{\alpha}}^T Q\,\hat{\boldsymbol{\alpha}} + \min_b \rho(b)$$

12.10 Consider the dual QP in (12.29) with the box constraints expressed as in (12.41). Apply Theorem 1 to verify the conditions in (12.42).

Hint: Re-write the dual QP as a minimization problem and then form the Lagrangian

$$L(\alpha,\beta,\gamma,\mu) = \frac{1}{2}\alpha^T Q\alpha - \alpha^T y - \sum_{i=1}^{n} \beta_i(\alpha_i - l_i) - \sum_{i=1}^{n} \gamma_i(u_i - \alpha_i) + \mu(\alpha^T 1)$$

where $\beta \geq 0$, $\gamma \geq 0$ and μ is an unconstrained scalar variable. Then combine the condition $\frac{\partial L}{\partial \alpha} = 0$ with the conditions in (12.23)–(12.24) to obtain the result.

12.11 Derive the two-variable update for the SMO algorithm. In particular for the working set $W^m = \{(j, k)\}$ with $g_j(\boldsymbol{\alpha}^m) > g_k(\boldsymbol{\alpha}^m)$, show that the updates for α_j and α_k take the form

$$\alpha_j^{m+1} \leftarrow \alpha_j^m + \Delta, \quad \alpha_k^{m+1} \leftarrow \alpha_k^m - \Delta$$

where

$$\Delta \leftarrow \min\left(u_j - \alpha_j^m, \quad \alpha_k^m - l_k, \quad \frac{g_j(\boldsymbol{\alpha}^m) - g_k(\boldsymbol{\alpha}^m)}{Q_{jj} + Q_{kk} - 2Q_{jk}} \right)$$

Hint: First show that since the equality constraint $\sum_{i=1}^{n} \alpha_i^m = 0$ holds for all m, the update can be written as follows:

$$\boldsymbol{\alpha}^{m+1} = \boldsymbol{\alpha}^m + \Delta\delta_{jk}$$

where Δ is a scalar that represents the amount of change and δ_{jk} is a vector with 1 in the jth position, –1 in the kth position, and zeros in all other positions. Then show that choosing Δ to maximize the stepwise improvement $R(\boldsymbol{\alpha}^m + \Delta\delta_{jk}) - R(\boldsymbol{\alpha}^m)$ subject to the box constraints $l_i \leq \alpha_i \leq u_i$ yields the aforementioned expression for Δ.

12.12 In this problem, you are asked to derive an SVM learning strategy for a two-class pattern recognition problem where the class 1 and class –1 errors have different costs. In particular, let c_y be the cost of a class y error and define the cost-sensitive classification error

$$e_c(f) = c_1 P_1 e_1(f) + c_{-1} P_{-1} e_{-1}(f)$$

and the Bayes optimal cost-sensitive error $e_c^* = \min_{\text{all } f} e_c(f)$. Now suppose you are given a collection $D_n = ((\mathbf{x}_1, y_1), \ldots, (\mathbf{x}_n, y_n))$ of iid samples from a density $p_{X,Y}$ and you want to determine a classifier $\hat{f}$ whose error $e_c(\hat{f})$ is as close to e_c^* as possible.

a. Derive an SVM learning strategy for this problem by following the steps in Section 12.4.2 to obtain the empirical SVM criterion:

$$\hat{s}_\lambda(\mathbf{w}, b) = \lambda \|\mathbf{w}\|^2 + \frac{1}{n}\left[\sum_{i:y_i=1} c_1[1 - y_i[\mathbf{w}^T\boldsymbol{\phi}(\mathbf{x}_i) + b]]_+ + \sum_{i:y_i=-1} c_{-1}[1 - y_i[\mathbf{w}^T\boldsymbol{\phi}(\mathbf{x}_i) + b]]_+\right]$$

b. Construct the primal QP for this problem.

c. Construct the dual QP for this problem.

d. Construct the dual-to-primal maps $\mathbf{w}_{\text{map}}$, b_{map}, and ξ_{map} for this problem.

e. Derive an SMO algorithm for the dual QP.

12.13 Repeat the experiment in Example 12.4 using Gaussians with different means, transform matrices, and class probabilities.

12.14 Repeat the experiment in Example 12.4, but vary the value of P_1 and observe the effect on classifier performance. Explore what happens when the classifiers are trained using data generated with one value of P_1 and evaluated using test data generated with a different value of P_1. What does this say about the application of these methods in practice?

12.15 Repeat the experiment in Example 12.5 using data distributions that are characterized by a mixture of two Gaussians for both classes. Experiment with the placement of these Gaussians and observe the effect on classifier performance.

12.16 Experiment with the six classifier design methods in Examples 12.4 through 12.6 by applying them to the benchmark data sets from the *UCI Machine Learning Repository*[62] and/or the *libsvm* website.[49]

References

1. Chen, C. H., Pau, L. F., and Wang, P. S. P. 1993. *Handbook of Pattern Recognition and Computer Vision*. Singapore: World Scientific Publishing Company.
2. Duda, R. O., and Hart, P. E. 1973. *Pattern Classification and Scene Analysis*. New York: John Wiley & Sons.
3. Theodoridis, S., and Koutroumbas, K. 2006. *Pattern Recognition*. Burlington, MA: Academic Press.
4. Young, T. Y., and Fu, K. S., eds. 1986. *Handbook of Pattern Recognition and Image Processing*. Orlando, FL: Academic Press.
5. Bishop, C. M. 2006. *Pattern Recognition and Machine Learning*. New York: Springer.
6. Duda, R. O., Hart, P. E., Stork, D. G. 2000. *Pattern Classification*. 2nd ed. John Wiley & Sons.
7. Fukunaga, K. 1990. *Introduction to Statistical Pattern Recognition*. 2nd ed. San Diego, CA: Academic Press.
8. Hastie, T., Tibshirani, R., and Friedman, J. 2009. *The Elements of Statistical Learning: Data Mining, Inference and Prediction*. 2nd ed. New York: Springer.
9. Haykin, S. 1998. *Neural Networks: A Comprehensive Foundation*. 2nd ed. Upper Saddle River, NJ: Prentice Hall.
10. Mitchell, T. 1997. *Machine Learning*. 1st ed. Boston: McGraw Hill.
11. Tou, J. T., and Gonzalez, R. C. 1974. *Pattern Recognition Principles*. Reading, MA: Addison–Wesley.
12. Devroye, L., Györfi, L., and Lugosi, G. 1996. *A Probabilistic Theory of Pattern Recognition*. New York: Springer.
13. Wolpert, D. H. 1996. The lack of a priori distinctions between learning algorithms. *Neural Comput* 8(7):1341–90.
14. Robert, C. P., and Casella, G. 1999. *Monte Carlo Statistical Methods*. New York: Springer-Verlag.
15. Boser, B., Guyon, I., and Vapnik, V. N. 1992. A training algorithm for optimal margin classifiers. In *Proceedings of the 5th Annual Conference on Learning Theory, COLT 1992*. 144–52. Burlington, MA: Morgan Kaufmann.
16. Cortes, C., and Vapnik, V. 1995. Support-vector networks. *Mach Learn* 20:273–97.
17. Vapnik, V. N. 1995. *The Nature of Statistical Learning Theory*. New York: Springer.
18. Vapnik, V. 1998. *Statistical Learning Theory*. New York: John Wiley & Sons.
19. Smola, A., Bartlett, P., Schlkopf, B., and Schuurmans, D., eds. 2000. *Advances in Large Margin Classifiers*. Cambridge, MA: MIT Press.
20. Burges, C. J. C. 1998. A tutorial on support vector machines for pattern recognition. *Data Min Knowl Discov* 2:121–67.
21. Cristianini, N., and Shawe-Taylor, J. 2000. *An Introduction to Support Vector Machines and Other Kernel-based Learning Methods*. 1st ed. Cambridge, UK: Cambridge University Press.

22. Herbrich, R. 2002. *Learning Kernel Classifiers.* Cambridge, MA: MIT Press.
23. Schlkopf, B., Burges, C., and Smola, A., ed. 1999. *Advances in Kernel Methods—Support Vector Learning.* Cambridge, MA: MIT Press.
24. Schlkopf, B., and Smola, A. 2002. *Learning with Kernels.* Cambridge, MA: MIT Press.
25. Shawe-Taylor, J., and Cristianini, N. 2004. *Kernel Methods for Pattern Analysis.* Cambridge, UK: Cambridge University Press.
26. Steinwart, I., and Christmann, A. 2008. *Support Vector Machines.* 1st ed. New York: Springer.
27. Vapnik, V. 2006. *Estimation of Dependences Based on Empirical Data.* 2nd ed. New York: Springer Verlag.
28. Berlinet, A. 2003. *Reproducing Kernel Hilbert Spaces in Probability and Statistics.* New York: Springer.
29. Aronszajn, N. 1950. Theory of reproducing kernels. *Trans Am Math Soc* 68(3):337–404.
30. Cortes, C., Haffner, P., and Mohri, M. 2004. Rational kernels: Theory and algorithms. *J Mach Learn Res* 5:1035–62.
31. Gartner, T. 2008. *Kernels for Structured Data.* Singapore: World Scientific Publishing Company.
32. Genton, M. G. 2001. Classes of kernels for machine learning: A statistics perspective. *J Mach Learn Res* 2:299–312.
33. Gneiting, T. 2001. Compactly supported correlation functions. *J Multivar Anal* 83(2):493–508.
34. Zhang, L., Zhou, W., and Jiao, L. 2004. Wavelet support vector machine. *IEEE Trans Syst Man Cybern* 34(1):34–9.
35. Courant, R., and Hilbert, D. 1993. *Methods of Mathematical Physics.* New York: Wiley-Interscience.
36. Bartlett, P. L., Jordan, M. I., and McAuliffe, J. D. 2006. Convexity, classification, and risk bounds. *J Am Stat Assoc* 101(473):138–56.
37. Steinwart, I. 2007. How to compare different loss functions and their risks. *Constructive Approximation* 26:225–87.
38. Zhang, T. 2004. Statistical behavior and consistency of classification methods based on convex risk minimization. *Ann Stat* 32(1):56–85.
39. Steinwart, I. 2005. Consistency of support vector machines and other regularized kernel classifiers. *IEEE Trans Inf Theory* 51:128–42.
40. Steinwart, I., and Scovel, C. 2007. Fast rates for support vector machines using Gaussian kernels. *Ann Stat* 35(2):575–607.
41. Steinwart, I., Hush, D., and Scovel, C. 2006. A new concentration result for regularized risk minimizers. *IMS Lect Notes Monogr Ser* 51:260–75.
42. Bertsekas, D. 2003. *Convex Analysis and Optimization.* Nashua, NH: Athena Scientific.
43. Fletcher, R. 1987. *Practical Methods of Optimization.* 2nd ed. New York: John Wiley & Sons.
44. Steinwart, I. 2003. Sparseness of support vector machines. *J Mach Learn Res* 4:1071–105.
45. Bartlett, P. L. 1998. The sample complexity of pattern classification with neural networks: The size of the weights is more important than the size of the network. *IEEE Trans Inf Theory* 44(2):525–36.

46. Schapire, R. E., Freund, Y., Bartlett, P., and Lee, W. S. 1998. Boosting the margin: A new explanation for the effectiveness of voting methods. *Ann Stat* 26(5):1651–86.
47. Gill, P. E., Murray, W., and Wright, M. H. 1981. *Practical optimization.* New York: Academic Press.
48. Platt, J. C. 1998. Fast training of support vector machines using sequential minimal optimization. In *Advances in Kernel Methods—Support Vector Learning,* ed. B. Schölkopf, C. J. C. Burges and A. J. Smola, 41–64. Cambridge, MA: MIT Press.
49. Chang, C. C., and Lin, C. J. 2001. LIBSVM: A library for support vector machines. Software available at http://www.csie.ntu.edu.tw/~cjlin/libsvm.
50. Chen, P. H., Fan, R. E., and Lin, C. J. 2006. A study on SMO-type decomposition methods for support vector machines. *IEEE Trans Neural Netw* 17:893–908.
51. Fan, R. E., Chen, P. H., and Lin, C. J. 2005. Working set selection using the second order information for training SVM. *J Mach Learn Res* 6:1889–918.
52. Joachims, T. 1998. Making large-scale SVM learning practical. In *Advances in Kernel Methods—Support Vector Learning,* ed. B. Schölkopf, C. J. C. Burges and A. J. Smola, 169–84. Cambridge, MA: MIT Press.
53. Keerthi, S. S., Shevade, S. K., Bhattacharyya, C., and Murthy, K. R. K. 2001. Improvements to Platt's SMO algorithm for SVM classifier design. *Neural Comput* 13:637–49.
54. Hush, D., Kelly, P., Scovel, C., and Steinwart, I. 2006. QP algorithms with guaranteed accuracy and run time for support vector machines. *J Mach Learn Res* 7:733–69.
55. Keerthi, S. S., and Gilbert, E. G. 2002. Convergence of a generalized SMO algorithm for SVM classifier design. *Mach Learn* 46:351–60.
56. Lin, C. J. 2002. Asymptotic convergence of an SMO algorithm without any assumptions. *IEEE Trans Neural Netw* 13:248–50.
57. List, N., Hush, D., Scovel, C., and Steinwart, I. 2007. Gaps in support vector optimization. In *Proceedings of the 20th Annual Conference on Learning Theory, COLT 2007, Lect Notes in Comput Science* 4539; 336–48.
58. Hush, D., and Scovel, C. 2003. Polynomial-time decomposition algorithms for support vector machines. *Mach Learn* 51:51–71.
59. List, N., and Simon, H. U. 2007. General polynomial time decomposition algorithms. *J Mach Learn Res* 8:303–21.
60. List, N., and Simon, H. U. 2005. General polynomial time decomposition algorithms. In *Proceedings of the 18th Annual Conference on Learning Theory, COLT 2005,* ed. P. Auer and R. Meir, 308–22.
61. Rifkin, R., and Klautau, A. 2004. In defense of one-vs-all classification. *J Mach Learn Res* 5:101–41.
62. Asuncion, A., and Newman, D. J. 2007. UCI Machine Learning Repository http://archive.ics.uci.edu/ml/ (accessed December 13, 2010).

Appendix: Table of Laplace and z Transforms

Introduction

The table below is like Table 3.1 in Chapter 3, except there are more transforms here than in Table 3.1 and the column with $h(t)$ is placed between the two transform columns, as if one were beginning with a system function, $H_a(s)$, then finding the impulse response, $h(t)$, and finally obtaining the z-transform, $H(z)$. The lettered and numbered lines have the same use as in Table 3.1.

Line	Laplace Transform	$h(t)$	z-Transform
A	$H_a(s) = \int_0^\infty h(t)e^{-st}dt$	$h(t)$	$H(z) = \sum_{m=0}^{\infty} h_m z^{-m}$
B	$AH_a(s)$	$Ah(t)$	$AH(z)$
C	$H_a(s) + G_a(s)$	$h(t) + g(t)$	$H(z) + G(z)$
D	$sH_a(s) - h(0^+)$	$\frac{d}{dt}h(t)$	–
E	$\frac{H_a(s)}{s}$	$\int_0^t h(\tau)\,d\tau$	–
F	$-\frac{d}{ds}H_a(s)$	$th(t)$	$-Tz\frac{d}{dz}H(z)$
G	$H_a(s+a)$	$e^{-at}h(t);\quad a>0$	$H(ze^{aT})$
H	$e^{-nsT}H_a(s)$	$h(t-nT);\quad n>0$	$z^{-n}H(z)$
I	$aH_a(as)$	$h\left(\frac{t}{a}\right);\quad a>0$	$H(z)$ with $\frac{T}{a} \to T$

Line	$H_a(s)$	$h(t)$ for $t \geq 0$	$H(z)$
100	1	$\delta(t);\quad h_m = 1 \quad \text{at} \quad m = 0$	1
101	$\frac{1}{s}$	$u(t);\quad h_m = 1 \quad \text{for} \quad m \geq 0$	$\frac{z}{z-1}$
102	$\frac{1}{s^2}$	T	$\frac{Tz}{(z-1)^2}$
103	$\frac{1}{s^3}$	$\frac{1}{2!}t^2$	$\frac{T^2 z(z+1)}{2(z-1)^3}$
104	$\frac{1}{s^4}$	$\frac{1}{3!}t^3$	$\frac{T^3 z(z^2+4z+1)}{6(z-1)^4}$
105	$\frac{1}{s^{k+1}}$	$\frac{1}{k!}t^k$	$\lim_{a\to 0}\frac{(-1)^k}{k!}\frac{\partial^k}{\partial a^k}\left(\frac{z}{z-e^{-aT}}\right)$
150	$\frac{1}{s-(1/T)\ln a}$	$a^{1/T}$	$\frac{z}{z-a}$
151	$\frac{1}{s+a}$	e^{-at}	$\frac{z}{z-e^{-aT}}$
152	$\frac{1}{(s+a)^2}$	te^{-at}	$\frac{Tze^{-aT}}{(z-e^{-aT})^2}$
153	$\frac{1}{(s+a)^3}$	$\frac{t^2}{2}e^{-at}$	$\frac{T^2 e^{-aT} z}{2(z-e^{-aT})^2}+\frac{T^2 e^{-2aT} z}{(z-e^{-aT})^3}$
154	$\frac{1}{(s+a)^{k+1}}$	$\frac{t^k}{k!}e^{-at}$	$\frac{(-1)^k}{k!}\frac{\partial^k}{\partial a^k}\left(\frac{z}{z-e^{-aT}}\right)$

155	$\frac{a}{s(s+a)}$	$1-e^{-at}$	$\frac{(1-e^{-aT})z}{(z-1)(z-e^{-aT})}$
156	$\frac{a}{s^2(s+a)}$	$t-\frac{1-e^{-at}}{a}$	$\frac{T_Z}{(z-1)^2}-\frac{(1-e^{-aT})z}{a(z-1)(z-e^{-aT})}$
157	$\frac{a}{s^3(s+a)}$	$\frac{1}{2!}\left(t^2-\frac{2}{a}t+\frac{2}{a^2}-\frac{2}{a^2}e^{-at}\right)$	$\frac{T^2z}{(z-1)^3}+\frac{(aT-2)Tz}{2a(z-1)^2}+\frac{z}{a^2(z-1)}-\frac{z}{a^2(z-e^{-aT})}$
158	$\frac{a}{s^{k+1}(s+a)}$	$\frac{1}{k!}\left[t^k-\frac{k}{a}t^{k-1}+\frac{k(k-1)}{a^2}t^{k-2}-\cdots+(-1)^{k-1}\frac{k!}{a^k}t+(-1)k\frac{k!}{a^k}\right]+(-1)^{k+1}\frac{e^{-at}}{a^k}$	$\frac{(-1)^{k+1}}{a^k}\frac{1}{1-e^{-aT}z^{-1}}+\frac{a}{k!}\lim_{x\to 0}\frac{\partial^k}{\partial x^k}\left[\frac{1}{(x+a)(1-e^{Tx}z^{-1})}\right]$
200	$\frac{b-a}{(s+a)(s+b)}$	$e^{-at}-e^{-bt}$	$\frac{z}{z-e^{-aT}}-\frac{z}{z-e^{-bT}}$
201	$\frac{(b-a)(s+c)}{(s+a)(s+b)}$	$(c-a)e^{-at}+(b-c)e^{-bt}$	$\frac{(c-a)z}{z-e^{-aT}}+\frac{(b-c)z}{z-e^{-bT}}$
202	$\frac{\beta}{s^2+\beta^2}$	$\sin\beta t$	$\frac{z\sin\beta T}{z^2-2z\cos\beta T+1}$
203	$\frac{s}{s^2+\beta^2}$	$\cos\beta t$	$\frac{z(z-\cos\beta T)}{z^2-2z\cos\beta T+1}$
204	$\frac{\beta}{s^2-\beta^2}$	$\sinh\beta t$	$\frac{z\sinh\beta T}{z^2-2z\cosh\beta T+1}$
205	$\frac{s}{s^2-\beta^2}$	$\cosh\beta t$	$\frac{z(z-\cosh\beta T)}{z^2-2z\cosh\beta T+1}$
206	$\frac{\beta}{(s+a)^2+\beta^2}$	$e^{-at}\sin\beta t$	$\frac{ze^{-aT}\sin\beta T}{z^2-2ze^{-aT}\cos\beta T+e^{-2aT}}$

(Continued)

Line	$H_a(s)$	$h(t)$ for $t \geq 0$	$H(z)$
207	$\frac{s+a}{(s+a)^2+\beta^2}$	$e^{-at}\cos\beta t$	$\frac{z^2 - ze^{-aT}\cos\beta T}{z^2 - 2ze^{-aT}\cos\beta T + e^{-2aT}}$
208	$\frac{\beta s}{(s+a)^2+\beta^2}$	$e^{-at}(\beta\cos\beta t - a\sin\beta t)$	$\frac{\beta z^2 - ze^{-aT}(\beta\cos\beta T + a\sin\beta T)}{z^2 - 2ze^{-aT}\cos\beta T + e^{-2aT}}$
300	$\frac{ab}{s(s+a)(s+b)}$	$1+\frac{b}{a-b}e^{-at} - \frac{a}{a-b}e^{-bt}$	$\frac{z}{z-1} + \frac{bz}{(a-b)(z-e^{-aT})} - \frac{az}{(a-b)(z-e^{-bT})}$
301	$\frac{ab(s+c)}{s(s+a)(s+b)}$	$c+\frac{b(c-a)}{a-b}e^{-at} + \frac{a(b-c)}{a-b}e^{-bt}$	$\frac{cz}{z-1} + \frac{b(c-a)z}{(a-b)(z-e^{-aT})} + \frac{a(b-c)z}{(a-b)(z-e^{-bT})}$
302	$\frac{1}{(s+a)(s+b)(s+c)}$	$\frac{e^{-at}}{(b-a)(c-a)} + \frac{e^{-bt}}{(a-b)(c-b)} + \frac{e^{-ct}}{(a-c)(b-c)}$	$\frac{z}{(b-c)(c-a)(z-e^{-aT})} + \frac{z}{(a-b)(c-b)(z-e^{-bT})}$ $+ \frac{z}{(a-c)(b-c)(z-e^{-cT})}$
303	$\frac{s+d}{(s+a)(s+b)(s+c)}$	$\frac{(d-a)}{(b-a)(c-a)}e^{-at} + \frac{(d-b)}{(a-b)(c-b)}e^{-bt} + \frac{(d-c)}{(a-c)(b-c)}e^{-ct}$	$\frac{(d-a)z}{(b-a)(c-a)(z-e^{-aT})} + \frac{(d-b)z}{(a-b)(c-b)(z-e^{-bT})}$ $+ \frac{(d-c)z}{(a-c)(b-c)(z-e^{-cT})}$
304	$\frac{a^2}{s(s+a)^2}$	$1-(1+at)e^{-at}$	$\frac{z}{z-1} - \frac{z}{z-e^{-aT}} - \frac{aTe^{-aT}z}{(z-e^{-aT})^2}$
305	$\frac{a^2(s+b)}{s(s+a)^2}$	$b-be^{-at} + a(a-b)te^{-at}$	$\frac{bz}{z-1} - \frac{bz}{z-e^{-aT}} + \frac{a(a-b)Te^{-aT}z}{(z-e^{-aT})^2}$
306	$\frac{(a-b)^2}{(s+b)(s+a)^2}$	$e^{-bt} - e^{-at} + (b-a)te^{-at}$	$\frac{z}{z-e^{-bT}} - \frac{z}{z-e^{-aT}} + \frac{(a-b)Te^{-aT}z}{(z-e^{-aT})^2}$

307	$\frac{(a-b)^2(s+c)}{(s+b)(s+a)^2}$	$(c-b)e^{-bt}+(b-c)e^{-at}-(a-b)(c-a)te^{-at}$	$\frac{(c-b)z}{z-e^{-bT}}+\frac{(b-c)z}{z-e^{-aT}}-\frac{(a-b)(c-a)Te^{-aT}z}{(z-e^{-aT})^2}$
308	$\frac{\beta^2}{s(s^2-\beta^2)}$	$\cosh\beta t-1$	$\frac{z(z-\cosh\beta T)}{z^2-2z\cosh\beta T+1}-\frac{z}{z-1}$
309	$\frac{\beta^2}{s(s^2+\beta^2)}$	$1-\cos\beta t$	$\frac{z}{z-1}-\frac{z(z-\cos\beta T)}{z^2-2z\cos\beta T+1}$
310	$\frac{\beta^2(s+a)}{s(s^2+\beta^2)}$	$a-a\sec\theta\cos(\beta t+\theta)$ where $\theta=\tan^{-1}\left(\frac{\beta}{a}\right)$	$\frac{az}{z-1}-\frac{az^2-az\sec\theta\cos(\beta T-\theta)}{z^2-2z\cos\beta T+1}$
311	$\frac{a^2+\beta^2}{s[(s+a)^2+\beta^2]}$	$1-e^{-at}\sec\theta\cos(\beta t+\theta)$ where $\theta=\tan^{-1}\left(-\frac{a}{\beta}\right)$	$\frac{z}{z-1}-\frac{z^2-ze^{-aT}\sec\theta\cos(\beta T-\theta)}{z^2-2ze^{-aT}\cos\beta T+e^{-2aT}}$
312	$\frac{(a^2+\beta^2)(s+b)}{s[(s+a)^2+\beta^2]}$	$b-be^{-at}\sec\theta\cos(\beta t+\theta)$ where $\theta=\tan^{-1}\left(\frac{a^2+\beta^2-ab}{b\beta}\right)$	$\frac{bz}{z-1}-\frac{b[z^2-ze^{-aT}\sec\theta\cos(\beta T-\theta)]}{z^2-2ze^{-aT}\cos\beta T+e^{-2aT}}$
313	$\frac{(a-b)^2+\beta^2}{(s+b)[(s+a)^2+\beta^2]}$	$e^{-bt}-e^{-at}\sec\theta\cos(\beta t+\theta)$ where $\theta=\tan^{-1}\left(\frac{b-a}{\beta}\right)$	$\frac{z}{z-e^{-bT}}-\frac{z^2-ze^{-aT}\sec\theta\cos(\beta T-\theta)}{z^2-2ze^{-aT}\cos\beta T+e^{-2aT}}$
314	$\frac{[(a-b)^2+\beta^2](s+\alpha)}{(s+b)[(s+a)^2+\beta^2]}$	$(\alpha-b)e^{-bt}-(\alpha-b)e^{-at}\sec\theta\cos(\beta t+\theta)$ where $\theta=\tan^{-1}\left(\frac{(\alpha-a)(b-a)+\beta^2}{(\alpha-b)\beta}\right)$	$\frac{(\alpha-b)z}{z-e^{-bT}}-\frac{(\alpha-b)[z^2-ze^{-aT}\sec\theta\cos(\beta T-\theta)]}{z^2-2ze^{-aT}\cos\beta T+e^{-2aT}}$
315	$\frac{[(a-b)^2+\beta^2](s^2+\alpha s+\gamma)}{(s+b)[(s+a)^2+\beta^2]}$	$(b^2-b\alpha+\gamma)e^{-bt}+k^2e^{-at}\sec\theta\cos(\beta t+\theta)$ where $k^2=a^2+\beta^2-2ab+b\alpha-\gamma$ $\theta=\tan^{-1}\left(\frac{ak^2-(a^2+\beta^2)(\alpha-b)+\gamma(2a-b)}{\beta k^2}\right)$	$\frac{(b^2-b\alpha+\gamma)z}{z-e^{-bT}}+\frac{k^2[z^2-ze^{-aT}\sec\theta\cos(\beta T-\theta)]}{z^2-2ze^{-aT}\cos\beta T+e^{-2aT}}$

(Continued)

Line	$H_a(s)$	$h(t)$ for $t \geq 0$	$H(z)$
400	$\frac{a^3}{s^2(s+a)^2}$	$at-2+(at+2)e^{-at}$	$\frac{(aT+2)z-2z^2}{(z-1)^2}+\frac{2z}{z-e^{-aT}}+\frac{aTe^{-aT}z}{(z-e^{-aT})^2}$
401	$\frac{a^2b^2}{s^2(s+a)(s+b)}$	$abt-(a+b)-\frac{b^2}{a-b}e^{-at}+\frac{a^2}{a-b}e^{-bt}$	$\frac{abTz}{(z-1)^2}-\frac{(a+b)z}{z-1}-\frac{b^2z}{(a-b)(z-e^{-aT})}+\frac{a^2z}{(a-b)(z-e^{-bT})}$
402	$\frac{a^2b^2(s+c)}{s^2(s+a)(s+b)}$	$abct+[ab-c(a+b)]-\frac{b^2(c-a)}{a-b}e^{-at}$ $-\frac{a^2(b-c)}{a-b}e^{-bt}$	$\frac{abcTz}{(z-1)^2}+\frac{ab-c(a+b)z}{z-1}-\frac{b^2(c-a)z}{(a-b)(z-e^{-aT})}-\frac{a^2(b-c)z}{(a-b)(z-e^{-bT})}$
403	$\frac{abc}{s(s+a)(s+b)(s+c)}$	$1-\frac{bc}{(b-a)(c-a)}e^{-at}-\frac{ca}{(c-b)(a-b)}e^{-bt}$ $-\frac{ab}{(a-c)(b-c)}e^{-ct}$	$\frac{z}{z-1}-\frac{bcz}{(b-a)(c-a)(z-e^{-aT})}-\frac{caz}{(c-b)(a-b)(z-e^{-bT})}$ $-\frac{abz}{(a-c)(b-c)(z-e^{-cT})}$
404	$\frac{abc(s+d)}{s(s+a)(s+b)(s+c)}$	$d-\frac{bc(d-a)}{(b-a)(c-a)}e^{-at}-\frac{ca(d-b)}{(c-b)(a-b)}e^{-bt}$ $-\frac{ab(d-c)}{(a-c)(b-c)}e^{-ct}$	$\frac{dz}{z-1}-\frac{bc(d-a)z}{(b-a)(c-a)(z-e^{-aT})}-\frac{ca(d-b)z}{(c-b)(a-b)(z-e^{-bT})}$ $-\frac{ab(d-c)z}{(a-c)(b-c)(z-e^{-cT})}$
405	$\frac{a^2b}{s(s+b)(s+a)^2}$	$1-\frac{a^2}{(a-b)^2}e^{-bt}+\frac{ab+b(a-b)}{(a-b)^2}e^{-at}+\frac{ab}{a-b}te^{-at}$	$\frac{z}{z-1}-\frac{a^2z}{(a-b)^2(z-e^{-bT})}+\frac{[ab+b(a-b)]z}{(a-b)^2(z-e^{-aT})}+\frac{abTe^{-aT}z}{(a-b)(z-e^{-aT})^2}$
406	$\frac{a^2b(s+c)}{s(s+b)(s+a)^2}$	$c+\frac{a^2(b-c)}{(a-b)^2}e^{-bt}+\frac{ab(c-a)+bc(a-b)}{(a-b)^2}e^{-at}$ $+\frac{ab(c-a)}{a-b}te^{-at}$	$\frac{cz}{z-1}+\frac{a^2(b-c)z}{(a-b)^2(z-e^{-bT})}+\frac{[ab(c-a)+bc(a-b)]z}{(a-b)^2(z-e^{-aT})}$ $+\frac{ab(c-a)Te^{-aT}z}{(a-b)(z-e^{-aT})^2}$

407	$\dfrac{(a^2+\beta^2)^2}{s^2[(s+a)^2+\beta^2]}$	$(a^2+\beta^2)t-2a+2ae^{-at}\sec\theta\cos(\beta t+\theta)$ where $\theta=\tan^{-1}\left(\dfrac{\beta^2-a^2}{2a\beta}\right)$	$\dfrac{[(a^2+\beta^2)T+2a]z-2az^2}{(z-1)^2}+\dfrac{2a[z^2-ze^{-aT}\sec\theta\cos(\beta T-\theta)]}{z^2-2ze^{-aT}\cos\beta T+e^{-2aT}}$
408	$\dfrac{(a^2+\beta^2)^2(s+b)}{s^2[(s+a)^2+\beta^2]}$	$b(a^2+\beta^2)t+k^2-k^2e^{-at}\sec\theta\cos(\beta t+\theta)$ where $k^2=a^2+\beta^2-2ab$ $\theta=\tan^{-1}\left(-\dfrac{ak^2+b(a^2+\beta^2)}{\beta k^2}\right)$	$\dfrac{[bT(a^2+\beta^2)-k^2]z+k^2z^2}{(z-1)^2}-\dfrac{k^2[z^2-ze^{-aT}\sec\theta\cos(\beta T-\theta)]}{z^2-2ze^{-aT}\cos\beta T+e^{-2aT}}$
500	$\dfrac{(abc)^2}{s^2(s+a)(s+b)(s+c)}$	$abct-(bc+ca+ab)+\dfrac{b^2c^2}{(b-a)(c-a)}e^{-at}+\dfrac{c^2a^2}{(c-b)(a-b)}e^{-bt}+\dfrac{a^2b^2}{(a-c)(b-c)}e^{-ct}$	$\dfrac{abcTz}{(z-1)^2}-\dfrac{(bc+ca+ab)z}{z-1}+\dfrac{b^2c^2z}{(b-a)(c-a)(z-e^{-aT})}+\dfrac{c^2a^2z}{(c-b)(a-b)(z-e^{-bT})}+\dfrac{a^2b^2z}{(a-c)(b-c)(z-e^{-cT})}$
501	$\dfrac{(abc)^2(s+d)}{s^2(s+a)(s+b)(s+c)}$	$abcdt+[abc-(bc+ca+ab)d]+\dfrac{b^2c^2(d-a)}{(b-a)(c-a)}e^{-at}+\dfrac{c^2a^2(d-b)}{(c-b)(a-b)}e^{-bt}+\dfrac{a^2b^2(d-c)}{(a-c)(b-c)}e^{-ct}$	$\dfrac{abcdTz}{(z-1)^2}+\dfrac{[abc-(bc+ca+ab)d]z}{z-1}+\dfrac{b^2c^2(d-a)z}{(b-a)(c-a)(z-e^{-aT})}+\dfrac{c^2a^2(d-b)z}{(c-b)(a-b)(z-e^{-bT})}+\dfrac{a^2b^2(d-c)z}{(a-c)(b-c)(z-e^{-cT})}$
502	$\dfrac{(a^2b)^2}{s^2(s+b)(s+a)^2}$	$a^2bt-[ab+a(a+b)]+\dfrac{a^4}{(a-b)^2}e^{-bt}-\dfrac{ab^2(3a-2b)}{(a-b)^2}e^{-at}-\dfrac{a^2b^2}{a-b}te^{-at}$	$\dfrac{a^2bTz}{(z-1)^2}-\dfrac{[ab+a(a+b)]z}{z-1}+\dfrac{a^4z}{(a-b)^2(z-e^{-bT})}-\dfrac{ab^2(3a-2b)z}{(a-b)^2(z-e^{-aT})}-\dfrac{a^2b^2Te^{-aT}z}{(a-b)(z-e^{-aT})^2}$

Index

A

Absolute gain error, for digital models, 387–389
Adaptation
  geometry of, 274
  as process, 274
  speed of, 296
Adaptive algorithm, 274, 301
Adaptive arithmetic coding, 327
Adaptive array, 300
Adaptive beamforming, 300
Adaptive filter, 274, 290, 300, 302, 304
Adaptive lattice, 301
Adaptive linear combiner, 300–303
Adaptive predictive coding, 314–319
Adaptive predictor, least-mean-square, 290
Adaptive signal compression, 290
Adaptive signal processing, 91, 239
  algorithms for
    direct descent, 291–296
    least-mean-square, 281–287, 291, 298, 304, 316
    one-step, 277
    other, 300–301
    random search, 275
    recursive least squares, 291–296
    steepest-descent, 281–287
  convergence in
    under ideal conditions, 278–279, 284
    learning curve for weights, 280–281
    parameter for, 277
    as performance measure, 279
    recursive least squares, 295, 298, 305
    steepest-descent algorithm, 283–286
    time constant for forgetting correlation estimate, 292
    time constant for LMS, 287
    time constant for MSE, 284, 287, 296
    time constant for weights, 278, 296
    time constant under ideal conditions, 284
  eigenvalues in, 282–287, 298
  exercises for, 301–306
  finite impulse response filters in, 275, 300–301
  introduction to, 273–274
  learning curve for, 279–281, 288, 294
  LMS examples of, 288–291
  mean-squared error in
    algorithm for, 276
    gradient of, 276
    least-mean-square convergence with, 287
    as performance measure, 279, 296–300
    performance surface of, 275–276
    search algorithms for, 275–281
    vector form of, 276
  performance measures of, 296–300
    misadjustment as, 296–299
    time constants as, 297–299
  performance surface of
    mean-squared error and, 275–276
    quadratic, 276, 279
    search algorithms for, 275–281
  prediction in, 280, 288
  predictive coding, 314–319
  in real time, 273, 276
  recursive least squares and, 291–296
  system performance measures, 296–300
  time constants in
    for forgetting correlation estimate, 292
    for mean-squared error, 284, 287, 296
    as system performance measure, 297–299
    for weight convergence, 278, 296
  weights in, 276–281

Additive noise, in least-squares system design, 234
a_decode, 327
a_encode, 326–327
Algebra
  in linear systems, 103
  vector and matrix, using MATLAB notation, 6–13
  in window formulas, for finite impulse response filters, 145
Algorithm(s), 226
  for adaptive signal processing
    direct descent, 291–296
    least-mean-square, 281–287, 291, 298, 304, 316
    one-step, 277
    other, 300–301
    random search, 275
    recursive least squares, 291–296
  for Fast Fourier transform, 45–47, 124–129
  Levinson's, 238
  for linear systems
    direct, 111–114
    fast Fourier transform, 124–129
    lattice, 116–124
    state-space, 114–116
  for mean-squared error minimization, 235
  steepest-descent, 281–287
Aliasing
  with multirate signal decomposition, 345–348
  of signal spectrum, 67–69
    cross-power, 223–226
    prevention of, 213, 344, 347–348
All-pass filter
  in finite impulse response, 148
  in infinite impulse response, 189
Amplitude distribution(s), 199–203
  continuous *vs.* discrete, 201–202
  of random signals, introduction to, 199–200
Amplitude response (gain)
  of averaging process, 192–193
  of finite impulse response differentiator, 153–154
  of infinite impulse response filters, 189
  in linear systems, 129–130
Amplitude spectrum, estimation of, 47–51
Analog-digital conversion, 363–364
  in finite impulse response filters, 141–142
  in infinite impulse response filters, bilinear transformation for, 177–180
  jitter in, 76–77
  of random signals, 201
Analog prefiltering, 213
Arithmetic coding, 324, 326
  adaptive, 327
Array
  adaptive, 300
  exponentiation, 12
  image, 5
  index, in mean-squared error minimization, 237
  in MATLAB language, 4–6
Array operations
  in algebra notation, 11–12
  summary of, 7
Array products
  in fading operation, 9–10
  in linear discrete system, 114
  types of, 9
Autocorrelation, 40
  in cross-power spectrum, 214–216
  in least-squares system design, 237, 245–246, 261
  of signal power spectrum, 213–216
Autocorrelation matrix, 237
  in adaptive signal processing
    inverse, 291, 294
    normal form of, 281
Autocorrelation vector, in mean-squared error minimization, 237
Autocovariance matrix, in least-squares system design, 243–244
  with broadband noise, 261–263
Average cross-power, 223
Average periodogram, in power spectral estimation, 217
Average periodogram method, of power spectral estimation, 218

Average power, of random stationary signals, 209–210
Average product, of two signals, 40
Averaging, spectral effects of, 189–193

B

Backslash, as algebraic operation, 12
Bandpass filter
  finite impulse response, 148–150
  infinite impulse response, 175, 179, 181
Bandstop filter
  finite impulse response, 148–150
  gains, 148
  infinite impulse response, 175, 179, 181
Bandwidth, in infinite impulse response frequency translation, 175
Bar graph, of power density, 211
Bayes error, 409, 431
Beamforming, adaptive, in signal processing, 300
Bessel function, in Kaiser window, for finite impulse response filters, 146
Bilinear approximation, 382, 384–386
Bilinear transformation, 175
  for analog-digital conversion, in infinite impulse response filters, 177–180
  in modeling of analog systems, 382, 384, 386
Binary code tree, in Huffman coding, 320
Binary digit, *see* Bits
Bits
  coding of, 319, *see also* Coding
  compression of, 309, *see also* Compression
  as information unit, 310
  two states of, 310
Blackman window, 145–146
  formula
    for finite impulse response filters, 145
    in spectral estimation, 221
Blind algorithms, in adaptive structures and algorithms, 301
Blind equalization, in adaptive signal processing, 301
Block adaptive process, 274
Block fast Fourier transform filtering, in linear systems, 128–129
Blocking effects, 343
Block processing, in least-squares system design, 242
"Boats," as image compression example, 333
Boxcar window
  for lowpass finite impulse response filters, 143–145
  in spectral estimation, 221–223
Broadband noise, effects on least-squares system design, 261–263
Butterworth filters, as infinite impulse response filters, 163–167
  Chebyshev *vs.*, 168–169, 172–173
Butterworth pole angles, 169
Butterworth transfer function, 165
Byte, as information unit, 310

C

Cascade filter structure
  in infinite impulse response filters, 166, 175, 181, 184–186, 189
  of linear systems, 109–111
Cascade system, 399–400
Causality
  of discrete linear systems, 92
  in finite impulse response filters, 140
Causal linear system, 236
Channel equalization, 247–250
Channel noise, in least-squares system design, 261
Characteristic equation, in steepest-descent algorithm, 282
Chebyshev filters
  as infinite impulse response filters, 167–173
    Butterworth *vs.*, 168–169, 172–173
  power gain curves, 172
  power gain poles, 169–170
  right-half σ-plane poles, 169
  transfer function, 171
Chirping signal, 78–80, 187

Chirp-z transform, 96–102
Closed-loop systems, simulating with nonlinearities, 391
Code lengths, in Huffman coding, 324, 328
Coding, of signals
  adaptive, 327
  amplitude distribution and, 207
  arithmetic, 324, 326
  entropy, 319–327
  equal-band, 353
  exercises for, 356–361
  fixed-symbol, 320
  Huffman, 320–324
  linear predictive, 257, 313, 335
  multirate processing for, 342–343, 349
  objective of, 328
  octave-band, 353–354
  predictive, 259–260, 314
  string, 314
  subband, 342–343
  terms used with, 310
  transform, 282, 328–335
  vector quantization as, 314
Coefficient(s)
  in discrete linear systems, 90
  in infinite impulse response filters, 181
  in least-squares system design
    identification example, 250
    for interference cancellation, 235
    for optimal mean-squared error minimization, 237–238
Coherence (*see magnitude-squared coherence*)
Column vector
  in algebra notation, 8
  in least squares, 22
Comb filter, 188
  infinite impulse response, 186–187
Command prompt, 4
Commands, in MATLAB language, 4
Complex plane, 13–14, 96–101
Complex products, in Fourier transforms, discrete *vs.* fast, 45
Complex signals
  finite impulse response filter processing of, 155
  instantaneous power of, 209
Complex Z-plane, 96–98
Components
  low-frequency *vs.* high-frequency, in image compression, 333
  in multirate signal decomposition, 343
Compression, of signals
  adaptive *vs.* nonadaptive, 290
  amplitude distribution and, 207
  entropy coding as, 319–327
  exercises for, 356–361
  multirate processing for, 342–343
  objective of, 309, 339
  power spectrum and, 213
  prediction error in, 314
  transform coding as, 314, 328–335
  two stages for, 314
Compression ratio, in least-squares system design, 260
Conjugate pairs
  of fractions, in linear systems, 109
  of poles, in infinite impulse response filters, 166, 180
Continuous amplitude distributions, 200–203
Continuous convolution, 368
Continuous form, of Fourier series, 32
Continuous Fourier transform
  in ideal lowpass filter, 138
  inverse, 139
  in linear continuous systems, 57–62
  of signal spectrum, 58
Continuous least-squares fit, 20
Continuous linear systems, 89
Continuous lowpass Butterworth filter, 166
Continuous periodic function, Fourier series for, 32
Continuous power gain spectra, 144
Continuous probability function, in uniform random signals, 204
Continuous systems
  digital models of, 372–373
  final value theorem for, 369
Continuous transforms, 57–60

Contour plot,
  clabel function in, 242
  in mean-squared error minimization, 240–241
Convergence, in adaptive signal processing
  under ideal conditions, 278–279, 284
  learning curve for weights, 280–281
  parameter for, 277
  as performance measure, 279
  recursive least squares and, 295, 298, 305
  steepest-descent algorithm for, 283–286
  time constant
    for forgetting correlation estimate, 292
    under ideal conditions, 284
    for LMS, 287
    for MSE, 284, 287, 296
    for weights, 278, 296
Convex programming problem, 425
Convex relaxation, SVM criterion, 420–421
Convolution
  analysis, 368
  continuous, 57–62
    integral, 61–62
  in least-squares system design, 245
  of linear equations discrete, 92–93
    fast Fourier transform algorithms and, 124–129
    periodic, 126, 128
    z-transform of, 94
  in lowpass finite impulse response filters, 142–143, 144
Correlation
  in adaptive signal processing, recursive estimate of, 292
  and covariance computation, 244–247
  in cross-power spectrum, 224
  exercises for, 84–87
  in least-squares equations, 239
    broadband noise effects on, 261–263
    computation of, 244–247
    system identification example, 250–253
  in signal processing, 40–42
Correlation detector, 41
Correlation estimate, recursive, in adaptive signal processing, 292
Correlation functions, in stationary conditions, 275
Cosh representation, 171
Cosine(s)
  basic applications of, 14
  in Fourier series, 26, 28–29, 39
  in infinite impulse response filters, 168–172
Covariance function, in least-squares equations, 243
  computation of, 244–247
  under noisy conditions, 262
  system identification example, 250–251
Cross-correlation function, 224
  least-squares computations with, 244–247
    broadband noise effects on, 261
Cross-covariance, in least-squares system design
  broadband noise effects on, 263
  computations with, 246–247
  equations for, 243
Cross-periodogram, 223–224
Cross-power density, 223
Cross-power spectrum, 223–226
C-Support vector machine (C-SVM), 438
Curse of dimensionality, 432
Cutoff frequency
  in finite impulse response filters, 138
  in infinite impulse response filters, 164, 166, 175

D

Data transformation, encryption as, 309–310
Data windows, *see* Window *entries*
  in spectral estimation, 221–223
d.c.-adjusted impulse-invariant model, 383
DCT, *see* Discrete cosine transform (DCT)
Decimation
  averaging, 142

in finite impulse response filters, 141–142
in multirate signal decomposition, 342, 346
in resampling, 75
Decoding, entropy, with multirate signal decomposition, 343
Decomposition, *see* Signal decomposition
Decomposition algorithms
for dual quadratic programming problem, 433–435
rate certifying, 435–436
Decorrelation, in least-squares system design, 233
Degrees of freedom
in power spectral estimation, 216
in sampled waveforms, 5
Dependent variable, in discrete linear systems, 90
Determinant, 12
DFT, *see* Discrete Fourier transform (DFT)
Diagonal matrix, in adaptive signal processing, 283, 285
Difference equations, 89
Differentiator, 152–154
Digital filters, 363
causal, 140
discrete linear systems and, 129–130
as finite impulse response filters, 137, 154
as infinite impulse response filters, 181–184
designing, 181
transformation to, 177–180
writing functions for, 180
transfer function of, 132, 364
Digital models, 372–373
absolute gain error for, 387–389
Digital resonator using infinite impulse response filters, 185–186
Digital signal processing (DSP), 89, 93
adaptive, *see* Adaptive signal processing
algebra review for, 6–13
geometry review for, 13–16
hardware and software, 2
introduction to, 1–3
language for, *see* MATLAB
MATLAB functions in, 16–17, *see also* Functions
text review techniques, 2
Direct algorithms, for linear systems, 111–114
conversion to lattice, 119–120
Direct descent algorithm, for adaptive signal processing, 291–296
Direct form, of linear systems, 111–114
Direction-finding system, 269
Direct-to-lattice conversion, in linear systems, 119–120
nonrecursive, 123
Discrete amplitude distributions, 200–201
Discrete convolution, of linear equations, 92–93
Discrete cosine transform (DCT)
short-time, in signal analysis, 353
in signal compression, 328
discrete Fourier transform function with, 329–330
fast Fourier transform computation of, 332, 337
inverse, 329, 332, 334
multirate processing *vs.*, 343
two-dimensional, 332–334
Discrete error model, 373–374
Discrete Fourier series
continuous form of, 32
exercises for, 33–37
fundamentals of, 26–27
as least-squares approximation, 27–31
MATLAB code for, 30–31
summary of, 30
Discrete Fourier transform (DFT), 209, 214, 221, 224, 364
complex products in, 45
continuous, 54, 67, 93
frequency increments in, 49–50
interpolation of time-domain in, 52, 54–56, 65
inverse, 51–52, 64, 125, 143, 209, 214, 337
in multirate signal decompression, 344–348

principles of, 42–43
properties of, 52–57
redundancy in, 43–44, 52, 65
resampling of, 52, 74
in spectral analysis, 47
z-transforms *vs.*, 93–96
Discrete impulse function, 148
Discrete least-squares
approximations, 20
process, 24
Discrete linear equations
continuous systems *vs.*, 89
convolution of, 92–93
properties of, 89–91
state-space, 114–116
Discrete linear systems, 124
causality of, 92
coefficients in, 90
digital filters and, 129–130
realizability of, 90
state-space equations for, 114–115
time domain in, 89–91, 94
variables in, 89–90
Discrete sine transform (DST), 335–341
definition, 335
inverse, 337–339
Discrete systems, final value theorem for, 369
Distortion, *see* Aliasing
Down-sampling
in finite impulse response filters, 141–142
with multirate signal decomposition, 342–346
DSP, *see* Digital signal processing (DSP)

E

EC, *see* Entropy coding (EC)
Eigenvalues, in adaptive signal processing, 282–287, 298
Eigenvector, in adaptive signal processing, 282–284
Eigenvector matrix, in adaptive signal processing, 283–284
Encoding, of signals, *see* Coding
Encryption
as data transformation, 309–310
with inverse transfer functions, 110–111
in least-squares system design, 260
Energy density, of signals, 211
Entropy, as information measure, 311
Entropy coding (EC), as signal compression, 313, 323
arithmetic scheme for, 324–328
Huffman scheme for, 320–324
in multirate decomposition, 343
Entropy decoding (ED), with multirate signal decomposition, 299–300, 343
Entropy reduction via quantization, 317
Envelope, exponential, imposed by array products, 10
Equal-bands
coding, 353
in multirate signal decomposition, 349–351
Equalization
blind, in adaptive signal processing, 301
in least-squares system design, 231, 233–234
methods for, 247–250
Equalizer, in least-squares system design, 247–250
Equal-ripple property, of infinite impulse response filters, 168
Equation error, for least-squares IIR system identification, 268
Equation error method, 268–269, 379
Error correction, in signals, 309
Error detection, in signals, 309
Error function, in Gaussian amplitude distributions, 206
Error(s)
quantizing, in sampling theorem, 63
total squared, in least squares, 19–21, 24
in waveform reconstruction, 66–67
Error surface, distributed minimum of, 242
Event, in information measurement, 311
Expansion, with multirate signal decomposition, 345–346, 349
Expected power, of stationary function, 209

Expected value, in correlation, 40
Exponential envelope, imposed by array products, 10
Exponentiation, as algebraic operation, 12
Expression simplification, geometric formulas for, 13–14
Extension
 periodic, in linear systems, 125
 zero, of discrete Fourier transform, 52–54

F

Fading operation, for array products, 10
Fast Fourier transform (FFT), 98, 101
 algorithms for, 45–47
  in linear systems, 124–129
 complex products in, 45–47
 discrete cosine transform computation with, 337
 discrete sine transform computation with, 337, 341
 filtering in linear systems
  block, 128–129
  nonrecursive, 127–128
Feedback control systems, 111
Feedback structure, of linear systems, 109–111
FFT, *see* Fast Fourier transform (FFT)
Filter gain, *see* Power gain, *see also* Amplitude response (gain)
Filtering
 in linear systems
  block fast Fourier transform, 128–129
  cascade, 109, 117
  digital, 129–130, 137, 154
  finite, *see* Finite impulse response (FIR) filters
  infinite, *see* Infinite impulse response (IIR) filters
  inverse, 257
  lattice, 121, 123–124
  nonrecursive fast Fourier transform, 127–128
  power gain of, 129–130
 multirate signal decomposition as, 342–351
Filter(s)
 in adaptive signal processing, 275, 300–301
 output component, 151
Final value theorems, 368–370
 continuous, 369
 discrete, 369
Finite impulse response (FIR) differentiator, 152–154
Finite impulse response (FIR) filters, 91
 in adaptive signal processing, 275, 300–301
 advantages of, 137–138
 all-pass, 148
 bandpass, 148–150
 bandstop, 148–150
 complete example of, 150–151
 digital, 137, 154
 disadvantages of, 137
 exercises for, 155–159
 Hibert transformer as, 154–155
 highpass, 148–150
 introduction to, 137–138
 least-squares design with, 252
 linear phase of, 137, 140
 lowpass
  conversion to other types, 148
  ideal characteristics of, 138–139
  window functions for, 142–147
 in multirate signal decomposition, 345, 348
 other types of, 152
 passband ripple in, 140, 147
 power gain, 140–141, 146, 147, 149, 150
  window spectrum as, 145
 realizable version of, 139–142
 stopband ripple in, 141–142
 truncation of weights in, 141, 147
 weights of, 137–140, 149–152
  formula for finding, 149
Finite signal vectors, in least-squares system design, 242–244
FIR filters, *see* Finite impulse response (FIR) filters
FIR (finite impulse response) differentiator, 152–154
First-order hold theorem, 377, 379
Fisher's linear discriminant, 415

Fixed-symbol coding, 320
Floating-point operations, in fast Fourier transforms, 47
Flops, in fast Fourier transforms, 47
Forward transform, in multirate signal decomposition, 346
Fourier coefficients, 30
  in correlation function, 42
Fourier series
  coefficients, equivalence of, 28
  discrete, *see* Discrete Fourier series
  harmonic analysis in, 13, 26
  introduction to, 19
  reconstruction, 64
  reconstruction of signals, 64–66
  theorems for, 35
Fourier spectra, 39
  analysis of, 47–51
  exercises for, 84–87
  frequency principles in, 43, 47–51
Fourier transforms (FT), of signal spectrum, 39, 152
  continuous, 54, 58–60, 138
  discrete, *see* Discrete Fourier transform (DFT)
  fast, *see* Fast Fourier transform (FFT)
  in ideal lowpass filter, 138–139
  inverse, 51–52, 59, 64
Fractions, in linear systems, 105, 109
Frequency and frequency domain
  of amplitude distributions, 200
  in discrete linear systems, 89, 93–94
  fundamental, in Fourier series, 27–28
  of lowpass finite impulse response filters, 143
  in multirate signal decomposition, 349
  signal power density and, 209–210
  in signal processing, 39
Frequency conversion factors, in spectral analysis, 49
Frequency-domain errors, 389
Frequency function, of digitized random signals, 201–202
Frequency increments, in spectral analysis, 50
Frequency resolution, in time-frequency analysis, 352–355
Frequency scales, in spectral analysis, 48–49
Frequency translations, analog-to-digital
  bilinear transformation for, 180, 384
  in infinite impulse response filters, 173–176
Frequency units, 48–49
  in finite impulse response filters, 138
  in signal power analysis, 212
FT, *see* Fourier transforms (FT)
Functions, MATLAB, 206, 437–438
  a_decode, 326–327
  a_encode, 326–327
  angle, 51
  autocorr, 246
  autocorr_mat, 246
  autocovar_mat, 246
  bar2, 226
  bessel_O, 147
  bilin, 181–182
  bw_analog_weights, 181–182
  bw_lowpass_size, 180–181
  bw_weights, 183
  ch_analog_weights, 181
  chirp_z, 131
  ch_lowpass_size, 181
  ch_weights, 183
  clabel, 242
  code_length, 324
  contour, 240–241
  conv, 128
  crosscorr, 246
  crosscovar, 247
  deconv, 128
  dir_to_lat, 120, 123
  display_data, 356
  in DSP, 4, 16–17
  erf, 206
  fft, 47, 339
  fft2, 332
  fftshift, 50, 55
  filter, 112, 128, 130–131, 184
  fir_weights, 147, 149
  flops, 47
  freq, 226, 312, 324, 327
  gain, 130
  gain_f, 131
  h codes, 324
  ifft, 51–52

image_scan, 335
imp_resp, 128
index_mat, 237
lat_filter, 121, 123
lat_to_dir, 118, 120
lms_filter, 289, 296
nr_dir_to_lat, 123
nr_lat_filter, 123
nr_lat_to_dir, 123
part_frac_exp, 106
pds, 226
power_gain, 131
pz_plot, 105
rand, 206
randn, 206
resamp, 65–66, 72–73, 75
residue, 105
rls_filter, 294
roots, 104–105
row_vec, 16
sp_dct, 332
sp_idct, 332
svmpredict, 439
svmtrain, 438
unwrap, 50–51
window, 104, 142–147
zigzag scan, 334
Fundamental frequency, in Fourier series, 27–28, 30
Fundamental period, in Fourier series, 26–27

G

Gain, 95, 130, *see also* Power gain
Gaussian distribution, of random signal amplitude, 204–209
white noise in, 215, 219, 224, 250
Gaussian maximum likelihood (GML) method, 414, 437
Gaussian probability density function, 205, 410
Gaussian RBF kernel, 419
Genetic algorithms, in adaptive structures and algorithms, 301
Geometric series, 13–16
Geometry
of adaptation, 274
for expression simplification, 13
Gradient, in adaptive signal processing, 276–277, 286, 291
Gradient vector, in LMS algorithm, 287
Gray-scale image, of amplitude distributions, 208–209
Grid search algorithm, 423, 439

H

Hamming window, 145–146, 151, 153–154
Hanning window
for finite impulse response filters, 145
in spectral estimation, 221
Harmonic analysis, in Fourier series, 13, 26, 33
Harmonic functions, in Fourier series, 26, 28–29
Help command, in MATLAB, 16
Hertz scale, in spectral analysis, 49
Highpass filters
with finite impulse response, 148–150
with infinite impulse response, 161, 345
in multirate signal decomposition, 342–343, 348
Hilbert transformer, as finite impulse response filter, 154–155
Huffman coding, 320, 324

I

Ideal conditions, in adaptive signal processing, convergence under, 278–279, 284
Ideal impulse response filters, 148, 152
Ideal lowpass filter gains, conversion of, 149
Ideal weight vector, 139
Identity matrix, definition of, 12
IIR filters, *see* Infinite impulse response (IIR) filters
Image array, in MATLAB language, 5
Image(s)
in MATLAB language, 5
mirror, in multirate signal decomposition, 348
transform coding of, 328–329, 333

Impulse-invariant approximation, 363–368
  adjusted, 382–384, 394
  examples, 365–367
  steps to obtain, 365
Impulse response, 61
  continuous, spectrum of, 61
  in linear systems, 89, 106
Independent broadband noise, effect of, 261–263
Independent noise, effects on least-squares system design, 234
Independent variable, in discrete linear systems, 90
Index array, in mean-squared error minimization, 237
Infinite impulse response (IIR) filters, 91
  in adaptive signal processing, 275
  all-pass, 189
  bandpass, 161, 175–176, 181
  bandstop, 161, 175–176, 181
  bilinear transformation of, 177–180
  Butterworth, 163–167
  Chebyshev, 167–173
  comb, 186–188
  digital, 181–184
    designing, 181
    transformation to, 175, 177–180
    writing functions for, 180
  exercises for, 193–196
  highpass, 161, 174–176
  introduction to, 161–162
  least-squares system identification and, 268–269
  linear phase of, 162–163
  lowpass, 161, 164–167, 173–174
    conversion to other types, 173–176
  other types of, 168, 175, 186
  passband ripple in, 168
  power gain of
    with analog-to-digital conversion, 177–178
    Chebyshev *vs.* Butterworth, 172–173, 178
  spectrogram and, 185, 187
  stopband ripple in, maximum gain in, 171
Infinite impulse response transfer function, 161
  linear phase of, 162–163
Information, *see* Signal information
Initial value theorem, 365
Input-invariant models, 374–382
Input signal, in linear systems, 61, 90, 129
Instantaneous frequency, in nonuniform and log-spaced sampling, 79
Instantaneous power, of random signals, 209
Integration
  digital, 189–193
  by parts, 16
  spectral effects of, 189–193
Interference canceling, 234–235, 253–256, *see also* Noise
  in least-squares system design
    broadband considerations for, 265–266
    example of, 234–235, 253, 256
    principles of, 234
Interpolation, *see* Resampling
Interpolation, of discrete Fourier transform, 52–56
Inverse, as algebraic operation, 11–12
Inverse autocorrelation, in adaptive signal processing, 291, 294
Inverse discrete cosine transform, 329–330, 332
Inverse discrete sine transform, 337
Inverse filtering, in least-squares system design, 257, 259
Inverse Fourier transforms, 139
  continuous, 59, 152
  discrete, 51–52, 64, 125, 244
Inverse modeling, least-squares system design for, 233–234
Inverse transfer functions, signal compression with, 111
Inverse transform, in multirate signal decomposition, 345, 347
Inverse *z*-transform, linear systems and, 107–109

J

Jitter, in analog-to-digital (A–D) conversion, 76–77

K

Kaiser window
  formula, for finite impulse response filters, 145
  lobe widths, 146
Kernel parameters, 438
Kuhn–Tucker theorem, 425, 430
κ weights, of lattice structure, 117, 124

L

Laplace transfer function, in infinite impulse response filters, 164
Laplace transform, 59–60, 375, 382, 453–459
Lattice algorithms, for linear systems, 116–124
Lattice filtering, in linear systems, 121, 124
Lattice stage, symmetric two-multiplier, 116–117
Lattice structure(s)
  adaptive, 301
  κ weights of, 117, 124
  symmetric
    nonrecursive, 121–122
    recursive, 117
Lattice-to-direct conversion, in linear systems, 118–119
  nonrecursive, 122
Learning curve
  for recursive least squares algorithm, 294–295
  for weight convergence, in adaptive signal processing, 279–281
Learning method, 413–416
Least-mean-square adaptive predictor, 290
Least-mean-square (LMS) algorithm, 281, 286
  for adaptive signal processing
    examples of, 288–291
    performance measure for, 297–298
    recursive least squares algorithm *vs.*, 294–295
Least-squares approximation, 19
  discrete, 20
    Fourier series as, 27, 31
Least-squares coefficient
  in Fourier series, 28–29
  vector, 22, 24, 26
Least-squares equalizer, 268
  for unknown channels, 247–248
Least-squares equations
  error identification in, 268–269
  system identification example, 250–253
  using correlation, 239
    broadband noise effects on, 261–263
    computation of, 244–247
  using covariance, 244
Least-squares interference canceling, 253, 256
Least squares (LS), 19–24
  applications of, 232–235, 245
  in digital signal processing systems, 231, 242
  exercises for, 33–37
  in pattern recognition, 414–415
  introduction to, 19
  polynomial, 20
  principle of, 19, 231
Least-squares predictor, for mean-squared error minimization, 239
Least-squares system design, 231
  applications of, 232–235
  channel equalization, 247–250
  correlation and covariance computation, 244–247
  correlation in, 242
    computation of, 244–247
  covariance in, 243
    computation of, 244–247
  equalization in, 231, 233–234
    of unknown channels, 247–250
  example of, 239–242
  exercises for, 263–270
  with finite signal vectors, 242–244
  illustration of, 232–235
  independent broadband noise, effect of, 261–263
  interference canceling in, 253–256
    example of, 253, 256
    principles of, 234
  introduction to, 231

least-squares predictor, 239
linear prediction and, 232, 257–260
example of, 257
for mean-squared error minimization, 232
linear prediction and recovery, 257–260
mean-squared error minimization in, 232, 235–239
modeling in, 231, 233
noise in, 233, 235
effects of independent broadband, 261–263
improvement methods for, 234–235, 254
one-step predictor, 239
system identification, 250–253
"Lena," as amplitude distribution image, 208–209
Levinson's algorithm, for mean-squared error minimization, 238
Libsvm software package, 438
Linear differential equations, 89
Linear digital filter, 137
Linear equations
discrete *vs.* continuous, 89
solving simultaneous, 12
transfer functions and, 93–96
Linear feedback, 110
Linearity, of discrete Fourier transform, 52
Linear kernel, optimal solution characteristics for, 431
Linear least-squares approximation, 20–21
Linear models, 382–386
comparison of, 386–389
configurations for, 399–400
Linear phase
of finite impulse response filters, 137, 140
of infinite impulse response filters, 162–163
Linear phase filter, 140, 162, 267
Linear phase shift
of discrete Fourier transform, 55–57
by finite impulse response filters, 137–138, 140, 153
by infinite impulse response filters, 137, 162–163, 183, 189
in multirate signal decomposition, 348–349
Linear predictive coding, 257, 313, 335
Linear predictor, 232–233
compression with, 233, 257
encoding with, 233, 257
in least-squares system design, 232–235
example of, 257–260
for mean-squared error minimization, 234
recovery and, 257–260
Linear support vector machine, margin of, 430, 432
Linear systems, 89–135
algorithms for
direct, 111–114
fast Fourier transform, 124–129
lattice, 116–124
convolution in
discrete, 92–93
fast Fourier transform algorithms and, 124–129
periodic, 126, 128
discrete *vs.* continuous, 89
exercises for, 131–135
poles and zeros in, 97–98, 101–105
processing speed of, 124
properties of, 89–91
state-space, 114–116
structural illustrations of, 109–111
transient *vs.* stable, 105–107
z-transform and, 93–96
inverse, 107–109
Linear transfer functions, 93–96
exercises for, 131–135
fractions in, 105, 109
structure options for, 109–111
LMS algorithm, *see* Least-mean-square (LMS) algorithm
Logarithmic sampling, 77–78
Log base 2, as information measure, 310–311
Log-spaced sampling, nonuniform sampling and, 76
logarithmic sampling in, 77–78
resamp_r function in, 81–84

Lossless signal compression, 314, 328
  with multirate decomposition, 343
Lossy signal compression, 314, 328
  using adaptive prediction, 318
Lowpass analog Chebyshev poles, 172
Lowpass filters
  design of, 266–267
  in finite impulse response
    conversion to other types, 148–150
    convolution of, 142–144
    effects on power gain, 145
    formulas for, 145
    ideal characteristics of, 138–139
    window functions for, 142–147
  in infinite impulse response
    conversion to other types, 175, 177–178
    ideal characteristics of, 161–162, 164, 174
  in multirate signal decomposition, 342–345
Lowpass transfer function, 292
LS-LINEAR learning method, 415

M

Magnitude, of discrete Fourier transform, 44
Magnitude-squared coherence (MSC), 224–225
Mathematical identities/operations, *see also specific type*
  summary of, 14–15
MATLAB
  in algebra notation, 6–13
  basic language of, 3–4
  element sequences processed by, 4–6
  examples, 439–445
  Fourier series implications for, 30–31
  functions in DSP, *see* Functions
  help command, 16
  introduction to, 3–4
  window formulas, for finite impulse response filters, 145
MATLAB code, 40
MATLAB functions, *see* Functions
MATLAB® language, 170
Matrix algebra, using MATLAB, 6–13
Matrix(ces)
  algebra review of, 6–13
  in least squares, 22–24
  in MATLAB language, 3–4
Max-violating pair, 434–435
Mean-squared error gradient, as adaptive signal processing vector, 276
Mean-squared error (MSE)
  in adaptive signal processing
    algorithm for, 276
    gradient of, 276
    least-mean-square convergence with, 287
    as performance measure, 279
    performance surface of, 275–276
    search algorithms for, 275–281
  autocorrelation matrix, 237
  causal linear system, 236
  contour plot of, 240–241
  of least-squares process
    applications of, 232, 235
    broadband noise effects on, 261–263
    equations for, 239
    minimization algorithms, 237–238
    minimization example, 239–242
    system design via, 235–239
  Levinson's algorithm, 238
  vector form of, 276
Mean value, in amplitude distributions, 202–203
  of Gaussian function, 205
  of uniform variate, 204
m-files, in MATLAB language, 4, 16
Mirror image, in multirate signal decomposition, 348
Misadjustment, in adaptive systems, 297
Modeling, in least-squares system design, 231, 233–234, 266, 270
Models of analog systems
  approaches to, 372–374
  classes of, 374
  comparison of, 386–389
  exercises for, 397–401
  input-invariant, 374–382
  introduction, 363–364
  linear models, 382–386
  multiple and nonlinear systems, 389–397

ramp-invariant, 376–377, 379–381
step-invariant, 375–377, 390, 393, 395, 398
Monte Carlo approximations, 413
MSC (magnitude-squared coherence), 224–225
MSE, *see* Mean-squared error (MSE)
Multilayer perceptron (MLP), 415
Multiple-input adaptive systems, 300
Multiple-input systems, modeling of, 389–397
Multirate processors, for signal coding and compression, 343, 348–349
Multirate signal decomposition, *see* Multirate signal processing
Multirate signal processing, for coding and compression
aliasing with, 345–348
decimation/down-sampling with, 342, 346
decomposition with, 342–351
discrete cosine transform *vs.*, 343
discrete Fourier transform in, 344–348
as filtering process, 342–349
forward, 346
Fourier transform in, 344
inverse, 345, 347
principles of, 342
recovery with, 343, 345–349
subband coding with, 342–351
time-frequency resolution of, 353–354
wavelets in, 342–343
Mutually orthogonal eigenvectors, 282

N

Neural networks, in adaptive signal processing, 301
Newton's method, in adaptive signal processing, 277
Node(s), in Huffman coding, 321–322
Noise
adaptive system performance and, 296–300
in cross-power spectrum, 223–226
effects of independent broadband, 261–263
exponents, 422
in least-squares system design, 231, 261, 263
Noise cancellation, with least-squares design
example of, 253–256
principles of, 234
Nonlinear filters, 130
Nonlinear modeling
of analog systems, 389–397
in least-squares system, 270
Nonrectangular gain curve, 141
Nonrectangular window, for lowpass finite impulse response filters, 142
Nonrecursive fast Fourier transform filtering, in linear systems, 128–129
Nonrecursive lattice, symmetric, 121–123
Nonrecursive linear equations, 90–91, *see also* Finite impulse response (FIR) filters
Nonsingular matrix, in algebra notation, 12
Nonstationary signals, 187, 199–200
in adaptive signal processing, 273, 290
Nonuniform sampling, and log-spaced sampling, 77
logarithmic sampling in, 77–78
resamp_r function in, 81–84
Normalized error function, in Gaussian amplitude distributions, 206
Normal probability function, in amplitude distributions, 204

O

Octave-band coding, 353–354
wavelet transforms as, 354–355
Octave bands, in multirate signal, decomposition, 311, 350–351
One, as bit state, 310
coding of, *see* Coding
One-step algorithm, for adaptive signal processing, 277
One-step predictor, for mean-squared error minimization, 239–240

Orthogonality, 19–37
exercises for, 33–37
in Fourier series, 28–30
introduction to, 19
principles of, 24–26
in steepest-descent algorithm, 281–287
Output signal, in linear systems, 90, 109
Overlapping segments, in power spectral estimation, 216–221

P

Parallel form, of transfer function, in linear systems, 109
Parallel structure, of linear systems, 109–111
Parameters
convergence, in adaptive signal processing, 277–278
for infinite impulse response filters, 142, 161, 181, 275
in linear systems
discrete, 91
for mean-squared error minimization, 235–239
Parseval's theorem, 210
Partial fractions, in linear systems, 105, 109
Partitioning, in power spectral estimation, 216–217
Parzen-kernel density, 414
Passband ripple
in finite impulse response filters, 142, 147, 164
in infinite impulse response filters, 161–162
Performance feedback, in adaptation process, 274
Performance surface
of mean-squared error
in adaptive signal processing, 275–276
search algorithms for, 275–281
quadratic, in adaptive signal processing, 276, 279
Periodic convolution, of linear equations, 125–126
Periodic extension
in Fourier series, 27
in linear systems, 125
Periodic function, in Fourier series, 27
Periodicity, of discrete Fourier transform, 52
Periodogram, 210–212, 216–217
in cross-power spectrum, 223
power density and, 210–211
in power spectral estimation, 216–221
Phase response (shift), in linear systems, 95, *see also* Linear phase shift
Phase spectrum, of signal vector, 47–48, 50
Phi, in least-squares system design, 237–238, 246
Pixel
coding and compression of, 310, 328, 332
in MATLAB language, 4
Plant noise, in least-squares system design, 261
Poles
analog to digital
mapping to *z*-plane, 370
in infinite impulse response filters, 137, 161, 175, 180
Butterworth, 163–168
Chebyshev, 167–173
conjugate pairs of, 166, 180
in linear systems, 101–105
mapping of, 370–372
s-plane, transformation to *z*-plane, 370–371
z-plane, transformation from *s*-plane, 370–371
Pole-zero comparisons, 370–372
Polynomial ratios/division
least-squares, 20
in linear systems, 128
Power density spectrum, 199–200
in terms of DFT, 209
energy spectra, 211
Parseval's theorem, 210
periodogram, 210–212
properties of, 213–216
of random signals
algorithms for, 226
cross-power, 223–226
data windows of, 221–223

estimation of, 216–221
introduction to, 199–200
properties of, 213–216
white noise in, 215
Power gain
curves for, 164–166
of digital filters, 129–130
of finite impulse response filters, 140, 143, 145
window spectrum effects on, 145
of infinite impulse response filters
Chebyshev *vs.* Butterworth, 163–164, 181
Power spectral estimation
principles of, 219–221
random signals and, 216–221
Power spectrum, *see* Power density spectrum
Prediction
in adaptive signal processing, 280, 288
in least-squares system design, 231–232
for mean-square error minimization, 235–239
recovery and, 257–260
linear, recovery in, 257–260
of seismic events, 257–258
Prediction error
in adaptive signal processing, 290
in least-squares system design, 257
in signal compression, 314
Predictive coding, 259–260, 314, 329
adaptive, 314–319
linear, 257, 313, 335
Prefiltered signals
for aliasing prevention, 67
in waveform reconstruction, 64–72
Prefix property, in Huffman coding, 322–323
Prewhitening, in least-squares system design, 260
Probability, in amplitude distributions, 201
Probability density function, 201
of amplitude distributions, 201–203
Gaussian form of, 204–209
Probability function
continuous, in uniform random signals, 204
normal, in amplitude distributions, 204
Processing speed, of linear systems, 124
Proper fraction, 363

Q

Qmfs, *see* Quadrature mirror filters (Qmfs)
Quadratic classifiers, 410
Quadratic error surface, in mean-squared error minimization, 296
Quadratic performance surface, in adaptive signal processing, 276, 279
Quadratic programming (QP), 423
primal, dual, and dual-to-primal map, constructing, 424–430
Quadrature filter, 269
Quadrature mirror filters (Qmfs), 348
Quantization
in adaptive predictive coding, 316–317
with multirate signal decomposition, 343
of prediction error, 316, 318–319
vector, as coding, 314
Quantizing errors, in sampling theorem, 63

R

Radial basis function (RBF), kernel, 419, 421–422
Radian-per-second scale, in spectral analysis, 49
Ramp functions
subtraction of, 73
sum of, 378
Ramp-invariant model, 376–377, 379–381
Random search algorithm, for adaptive signal processing, 275
Random signals
amplitude distribution of, 199–203
definition of, 200

exercises for, 226–229
Gaussian distribution of, 204–209
introduction to, 199–200
other distributions of, 204–209
power density spectra of
algorithms for, 226
cross-power, 223–226
data windows of, 221–223
definition of, 210
estimation of, 216–221
principles of, 219
properties of, 213–216
power of, 209–213
stationarity of, 199–200
statistical properties of, 199, 216
uniform distribution of, 204–209
RBF, *see* Radial basis function (RBF)
Realizability
of discrete linear systems, 90
of finite impulse response filters, 139–142
Realizable version
of FIR filters, 139–142
Real time
adaptive signal processing in, 273, 276
digital filters and phase shift in, 162
Reconstruction, *see* Resampling
Reconstruction error, in least-squares system design, 259
Recovery
with compression
in adaptive signal processing, 290
in multirate signal decomposition, 343, 345–349
linear prediction and, 257–260
Rectangular window
for lowpass filters
finite impulse response, 142–143
in spectral estimation, 223
Recursive adaptive filters, in signal processing, 301
Recursive estimate, in adaptive signal processing, 292
Recursive lattice, symmetric, 117
Recursive least squares (RLS) algorithm, for adaptive signal processing
convergence with, 294–296
correlation estimates with, 292
forms of, 291
learning curve for, 294
least-mean-square algorithm *vs.*, 294–295
Recursive linear equations, 91, *see also* Infinite impulse response (IIR) filters
Redundancy
of discrete Fourier transform, 43–44, 52, 65
removal of, 309–310, 329
Resampling
of discrete Fourier transform, 65, 74
of waveforms, 72–76
Residue theorem, 107
Resonator, digital, 185–188
cascade combination, 185–186
Reversal
of signal vectors, in digital filters, 163, 330, 335
time series, in infinite impulse response filters, 163
Reverse labels, 163
Right-half plane poles, 165, 169
RLS algorithm, *see* Recursive least squares (RLS) algorithm
Row vector, in algebra notation, 8, 16

S

Sampled waveform(s), in MATLAB language, 5
Sample plan, independent and identically distributed, 412–413
Sample vector, in least squares, 20
Sampling, log-spaced and nonuniform, 76–84
Sampling frequency, in spectral analysis, 49–50
Sampling interval, in MATLAB language, 5
Sampling theorem, 39, 62
exercises for, 84–87
Fourier transforms as proof of, 39, 54, 63
in waveform reconstruction, 62, 64–66, 68–69
Scanning, in image compression, 334–335

Second harmonic terms, in Fourier series, 28
Segments
in multirate signal decomposition, 351
in power spectral estimation
overlapping, 217–222
Seismic event(s), prediction and recovery of, 257–258
Sequential minimal optimization (SMO) algorithms, 433
implementation of, 434
Short-time discrete transform, of signal information, 352–353
Signal component, in discrete Fourier transform, 55–57
Signal decomposition, with multirate signal processing, 342–351
Signal energy, 210–211, 350
Signal information
coding, 309, 324, *see also* Coding
compressing, 312–314, *see also* Compression
exercises for, 356–361
introduction to, 309–310
measuring, 310–312
time-frequency analysis of, 352–356
wavelet transforms of, 354–356
Signal power, *see* Power *entries*
Signal processing
adaptive, *see* Adaptive signal processing
applications of, 201–202
digital, *see* Digital signal processing (DSP)
multirate, *see* Multirate signal processing
speed of
in adaptive systems, 296
in linear systems, 124
Signals
in digital filters, reversal of, 163
in linear systems, 90
in MATLAB language, 4–6
Signal space
in MATLAB language, 5
orthogonal vectors in, 25
Signal-to-noise ratio (SNR), 150
in cross-power spectrum, 225
in least-squares system design, 255–256
improvement methods for, 234–235, 254
Signal vector reversing, 163
Signal vectors, *see* Vector(s)
Sine(s)
basic applications of, 14
discrete sine transform (DST), 335–341
in Fourier series, 26, 28–29
in least squares, 23
SMO algorithms, *see* Sequential minimal optimization (SMO) algorithms
SNR, *see* Signal-to-noise ratio (SNR)
Spatial domain, spectrum mapping of, 39
Spectral analysis
discrete Fourier transform in, 47
exercises for, 84–87
transform coding and, 328–335
white noise in, 215
Spectral estimation, 199–229
algorithms for, 226
cross-power, 223–226
data windows in, 221–223
exercises for, 226–229
introduction to, 199–200
power, 221
average periodogram method, 217–218
periodogram, 216–217
white Gaussian noise, 219
Spectrogram
using infinite impulse response filters, 185–188
of signal information, 185–188, 352
Spectrum, in signal processing, 39
Speech signal
compression of, 316, 328
in interference canceling, 253–256
multirate signal decomposition of, 350–351
Speed, of processing
in adaptive systems, 296
in linear systems, 124
s-plane, in infinite impulse response filters, 165, 168–170
transformation to *z*-plane, 177

Stability, of linear systems, 106–107
discrete, 91
infinite impulse response filters and, 161, 165
Standard deviation, 202–203
in amplitude distributions
of Gaussian variate, 205
of uniform variate, 204
Startup sequence, for discrete convolution, 92
Startup time, of finite impulse response filters, 138
State-space equations, for discrete linear system, 114–115
State-space form, of linear systems, 114
State variable, of linear systems, 114
Stationarity, of signals, 199–200
Stationary function, instantaneous power of, 209
Stationary signal, 199–200
mean-squared error in, 237
total squared error in, 243
Statistical properties, of random signals, 199
Steepest-descent algorithm, for adaptive signal processing, 281–287
convergence criterion for, 284–286
eigenvalues in, 282–287, 298
goals of, 281
Step-function response
final value of, 369–370
using impulse-invariant model, 367
Step-invariant model, 375–377, 390, 393, 395, 398
Stopband ripple
in finite impulse response filters, 141–142, 147
in infinite impulse response filters, 171
maximum gain in, 164
String coding, 314
Subband coding, 342–351
with multirate signal decomposition, 342–343
Substitutional method, 382, 384–386
Superscripts, in multirate signal decomposition, 349–350
Support vector machine algorithms,
core, 423–424
grid search, 423
offset, 429
decomposition or SMO, 433–436
Support vector machines (SVM), 417
classifier, design and implementation, 436, 438–439
complexity regularization, 420–421
function class, 417–420
learning strategy, 420–423
margin, 430
multi-class classification, 436
Support vectors, 430–432
SVM, *see* Support vector machines (SVM)
Symbol code, 310, 314, 324, 327, *see also* Coding
Symbol entropy, 311, 313, 327
Symbol(s)
as coding term, 310
in entropy coding, 319–327
Symmetric nonrecursive lattice, 121
Symmetric recursive lattice, 117
Symmetric two-multiplier lattice stage, 116–117
System identification, least-squares system design for, 233, 250–253

T

Tapered rectangular window, in spectral estimation, 223
Time constant, in adaptive signal processing
for forgetting correlation estimate, 292
for learning curve, 279
for mean-squared error, 280, 284, 287, 296
as system performance measure, 297–299
for weight convergence, 278, 296
Time domain
of discrete Fourier transform
interpolation of, 52, 54–56, 65
zero extension in, 52–54
in discrete linear systems, 89–91, 94
errors, 389
signal power density and, 210

Time-frequency analysis, of signal information, 352–356
Time-frequency resolution, for multirate signal decomposition, 354–355
Time resolution, of signal information, 352–355
Time series reversal, in infinite impulse response filters, 163
Time step, in MATLAB language, 5
Toeplitz matrix, for mean-squared error minimization, 237
Total squared error (TSE), 243, 255
  in adaptive signal processing, 290
  in least squares
    broadband noise effects on, 262
    principle of, 19–21, 24
    system design applications, 243
    system identification example, 250, 252–253
Trace of a matrix, in steepest-descent algorithm, 285–286
Transfer function(s), 60, 89–131
  continuous, 60, 363, 372
  convolution of
    fast Fourier transform algorithms and, 124–129
    z-transform of, 94
  of digital filter, 364
  in least-squares system design, 233, 247, 250–251
  in linear systems, 94–95
    algorithms for, 111–129
    convolution of, 92–93
    discrete *vs.* continuous, 89
    exercises for, 131–135
    infinite impulse response, 91
    inversion of, 110–111
    poles and zeros in, 97, 101–105
    properties of, 89–91
    stability of, 105–107
    z-transform and, 93–96, 107–109
  principal, 373
Transformation, of data, 310
Transform coding, in signal compression, 314, 328–335
  discrete cosine transform in, 328–335
  discrete sine transform in, 335–341
  two-dimensional, 332–334
Transform components, in image compression, 328–329
Transient response, in linear systems, 106–107
Translated gradient, in adaptive signal processing, 284
Transpose, as algebraic operation, 11
Trigonometric identities, 14–15
Truncated ideal impulse response filters, 142–143, 148
Truncation, of finite impulse response filters, 141, 147
TSE, *see* Total squared error (TSE)
Tustin's approximation, 384
Two-dimensional transform
  in multirate signal decomposition, 349
  as signal compression, 332

U

Uncertainty principle, of time-frequency analysis, 353
Uncorrelated noise, in least-squares system design, 261
Uniform distribution, of random signal amplitude, 204, 206
Uniform probability density function, 204
Uniform variate, of random signal distribution, 204
Unit impulse, in linear systems, 106
Unit step function, 367
Unknown channels, in least-squares system design
  equalizer for, 247–248
  noise with, 261
Up-sampling, with multirate signal decomposition, 345

V

Variable(s)
  in discrete linear systems, 89–90
  in MATLAB language, 4–5
Variance, in amplitude distributions, 202–203
  of Gaussian function, 206–207
  of uniform variate, 204

Vector quantization, as coding, 314
Vector(s)
in adaptive signal processing, gradients as, 277, 287
algebra review of, 6–13
amplitude and phase spectra of, 47–51
information measurement in, 310–311
in least squares, 20–22
finite, 242–244
in MATLAB language, 4–6
orthogonal, 24–26
Video signals, 6, 316

W

Waveform reconstruction
aliased, 67, 344
Fourier series, 64–66, 73
in signal processing, 39, 67
methods for, 64–65
Waveform(s), sampled, in MATLAB language, 5
Waveform vectors
information measurement in, 310–311
in signal space, 5, 25
Wavelet coefficients, 55, 355
Wavelet(s), in multirate signal decomposition, 342–343, 353
Wavelet transforms, as octave-band coding, 353–354
Weight convergences
comparision of RLS and LMS, 295
ideal *vs.* LMS, 288, 294
Weights
in adaptive signal processing, 276–281
learning curve for, 279–281, 288, 294
LMS examples of, 288–291
recursive least squares and, 291–296
in system performance measures, 296–300
in discrete linear systems, 90, 130
in finite impulse response filters, 137–140, 149–152
formula for finding, 149
in infinite impulse response filters, 161, 173
in least-squares system design
as bias, 265
identification example, 250
for interference cancellation, 253–254, 256
for optimal mean-squared error minimization, 237–239
Whitening, in least-squares system design, 257–259
system identification example, 250
White noise
in adaptive signal processing, 298
in signal power spectrum, 215, 219, 224
least-squares system design and, 262
Whittaker's reconstruction, 70
Widrow-Hoff algorithm, *see* Least-mean-square (LMS) algorithm
Windowed weight vectors, 149
Window formulas, for finite impulse response filters, 145
Window function(s)
for lowpass finite impulse response filters, 142–146
effects on power gain, 145
formulas for, 145
in spectral estimation, 221–223
Window spectrum properties, effect of, 145
Word, as information unit, 310

Z

Zero extension (in time domain), of discrete Fourier transform, 53–54
Zero insertion (between samples), of discrete Fourier transform, 52
Zero-order hold theorem, 376
Zero(s)
as bit state, 310
coding of, *see* Coding
in finite impulse response filters, 137–138
in infinite impulse response filters, 161

in least-square system design, 242–243, 245–246, 261
with broadband noise, 261–263
in linear systems, 101–105
mapping of, 370–372
Zig-zag scanning, in image compression, 334–335
z-plane
complex, 96–101
in infinite impulse response filters, 161
transformation from *s*-plane, 177
in linear systems, 96–101
z-transform, 453–459
chirp, 96–102
in infinite impulse response filters, 162
bilinear transform, 177–180
linear transfer functions and, 93–96
inverse, 107–109
in multirate signal decomposition, 347
short table of, 108